Animal Microbiomes

Animal health and well-being is significantly affected by the associated microbiome. This volume reviews animal microbiomes in a variety of mostly domesticated animals such as dairy cattle, poultry, camels, buffalo, and pets. In addition, wild animal microbiomes in salmon, other marine fish, and shrimp are also reviewed.

Sunil Thomas is a research professor at the Lankenau Institute for Medical Research. He received his PhD from Cochin University of Science and Technology and has worked as a postdoctoral fellow at the Indian Institute of Science, Bangalore, and Mount Sinai School of Medicine, New York. His current research involves the development of immunotherapeutic and mRNA-based drugs for ulcerative colitis/colon cancer and Alzheimer's disease.

Animal Microbiomes

Edited by
Sunil Thomas

CRC Press
Taylor & Francis Group
Boca Raton London New York

CRC Press is an imprint of the
Taylor & Francis Group, an **informa** business

Designed cover image: Sunil Thomas

First edition published 2027
by CRC Press
2385 NW Executive Center Drive, Suite 320, Boca Raton FL 33431

and by CRC Press
4 Park Square, Milton Park, Abingdon, Oxon, OX14 4RN

CRC Press is an imprint of Taylor & Francis Group, LLC

ISBN: 978-1-041-07050-4 (hbk)
ISBN: 978-1-041-07051-1 (pbk)
ISBN: 978-1-003-63856-8 (ebk)

DOI: 10.1201/9781003638568

Typeset in Joanna
by Newgen Publishing UK

CONTENTS

Preface / vii
List of Contributors / ix

1 • **METAGENOMICS IN LIVESTOCK: UNRAVELING MICROBIAL COMMUNITIES TO ENHANCE PRODUCTIVITY AND HEALTH / 1**
Mottakunja Deepthi, Preety Sweta Hembrom, Bhagya Mappurath, Narchikundil Reeja, Shemmy Sadanandan, Gayatri Krishna S., Ginny Antony, and Tony Grace

2 • **MICROBIOMES IN DOMESTICATED ANIMALS: ECOLOGY, HEALTH, PRODUCTIVITY, AND ONE HEALTH / 26**
Veysel Bay and Berrin Özdil

3 • **MICROBIAL ECOLOGY OF BOVINE MILK AND ITS HEALTH IMPLICATIONS ON DAIRY PROCESSING / 53**
Anuradha Chakroborty, Soumitra Nath, and Indu Sharma

4 • **ROLE OF THE GUT MICROBIOME IN ANIMAL HEALTH AND DISEASE: MECHANISTIC INSIGHTS AND APPLICATIONS / 75**
Kun Li, Sijia Lu, Qing He, Chang Xu, and Zubair Azhar Nomi

5 • **LIVESTOCK MICROBIOME / 91**
Alex Muppasani and Sunil Thomas

6 • **BIRDS, MICROBES, AND DISEASE: A REVIEW OF THE AVIAN MICROBIOME AND ITS IMPLICATIONS / 127**
Zi Yan, Huiyan Xu, Jinfeng Zhao, and Longxian Zhang

7 • **AN OVERVIEW OF POULTRY LITTER MICROBIAL ECOLOGY AND APPLICATION OF MICROBIOME ANALYSES / 149**
Steven C. Ricke, Elena G. Olson, Carson T. Priddle, Maxim A. Peckenschneider, and Brett M. Hale

8 • **THE MICROBIOME OF COMPANION ANIMALS / 157**
Alex Muppasani and Sunil Thomas

9 • **THE IGUANID MICROBIOME AND ITS PHYSIOLOGICAL SIGNIFICANCE / 182**
Kwanho C. Ki and Erin L. Lewis

10 • **ASSESSING THE INFLUENCE OF MICROORGANISMS ON RODENT WELL-BEING AND HEALTH / 192**
Fu Yin, Jinfeng Zhao, Huiyan Xu, and Longxian Zhang

11 • **THE MARINE MICROBIOME / 209**
Olivia Lutz and Sunil Thomas

Index / 241

PREFACE

Microorganisms represent the earliest and most enduring forms of life on Earth. Long before the emergence of complex multicellular organisms, microbial life shaped the planet's atmosphere, soils, waters, and biogeochemical cycles. Even today, all eukaryotic life—including plants, animals, and humans—remains fundamentally rooted in microbial ancestry. Modern eukaryotic cells themselves are conserved amalgamations of ancient microbial partnerships, a testament to the central role microorganisms have played in the evolution of life.

Higher organisms cannot exist in isolation from microorganisms. From plants drawing nutrients through symbiotic soil microbes to animals relying on gut bacteria for digestion, microorganisms are indispensable partners in survival. The collective community of microorganisms that inhabit a host or environment is known as the microbiome. These microbial communities are not passive inhabitants; rather, they form dynamic, interactive systems that profoundly influence the health, development, and resilience of their hosts.

The microbiome—comprising trillions of bacteria, archaea, fungi, viruses, and protozoa—acts much like a vital organ. In humans and animals, it supports digestion, synthesizes essential vitamins, regulates metabolism, trains the immune system to distinguish friend from foe, and influences inflammation, behavior, mood, and disease susceptibility. In plants, the microbiome enhances nutrient uptake, improves stress tolerance, and protects against pathogens. Microbial diversity is a defining feature of a healthy microbiome, conferring stability and adaptability to changing conditions.

Beneficial microorganisms produce a wide array of secondary metabolites that are critical for maintaining physiological balance and overall well-being. When this balance is disrupted—a condition known as dysbiosis—the consequences can be severe. Dysbiosis has been linked to infections, allergies, autoimmune diseases, metabolic disorders, reduced crop productivity, and even neurological and behavioral conditions. Understanding the mechanisms that govern microbial balance is therefore essential for advancing health, agriculture, and environmental sustainability.

Recent breakthroughs in high-throughput sequencing, metagenomics, and bioinformatics have revolutionized microbiome research. These technologies allow rapid identification and functional analysis of microbial communities associated with different hosts, diets, environments, and lifestyles. What once took decades to uncover can now be achieved in a fraction of the second, opening new frontiers in medicine, nutrition, agriculture, ecology, and biotechnology.

The six interconnected volumes listed below present a comprehensive exploration of microbial life across biological systems and ecosystems:

Volume 1: The Human Microbiome

Volume 2: Diet and the Microbiome

Volume 3: Animal Microbiomes

Volume 4: Plant Microbiomes

Volume 5: Insect Microbiomes

Volume 6: The Environmental Microbiome

Together, these volumes provide an integrated perspective on how microbial communities shape life on Earth.

I extend my sincere and heartfelt gratitude to all the authors who contributed chapters to this book series. Their expertise, dedication, and scholarly insight form the foundation of these volumes. Each contribution reflects a commitment to scientific rigor and a shared passion for advancing knowledge in the rapidly evolving field of microbiome research.

The collective efforts of these authors have enriched the depth, breadth, and quality of this series, transforming it into a comprehensive and authoritative resource. Their willingness to share ideas, collaborate across disciplines, and invest their time and intellect has been invaluable. This work stands as a testament to their contributions, and I am deeply thankful for their role in bringing this series to fruition.

This year marks the 100th anniversary of the Lankenau Institute for Medical Research. The anniversary represents decades of unwavering commitment to scientific inquiry, education, and the advancement of medical knowledge. The Institute has played a pivotal role in shaping and supporting my scientific endeavors. Its culture of intellectual rigor, encouragement, and dedication to research excellence has provided an environment in which curiosity could thrive, and ideas could evolve into meaningful scientific contributions.

This six-volume book series is presented as a tribute to the Lankenau Institute for Medical Research. It stands as a testament to the Institute's enduring legacy, its profound impact on generations of researchers, and its mission to advance science for the betterment of human health and society. Through this work, I honor its past achievements, acknowledge its present influence, and celebrate its continued inspiration for future scientific discovery.

CONTRIBUTORS

GINNY ANTONY
Department of Plant Science
School of Biological Sciences
Central University of Kerala,
Kasaragod, Kerala, India

VEYSEL BAY
Faculty of Agriculture
Department of Animal Science
Ege University
Bornova-Izmir, Turkey

ANURADHA CHAKROBORTY
Department of Microbiology
Assam University
Silchar, Assam, India

MOTTAKUNJA DEEPTHI
Department of Genomic Science
School of Biological Sciences
Central University of Kerala
Kasaragod, Kerala, India

TONY GRACE
Department of Genomic Science
School of Biological Sciences
Central University of Kerala
Kasaragod, Kerala, India

BRETT M. HALE
AgriGro
Doniphan, MO, USA

QING HE
College of Veterinary Medicine
Nanjing Agricultural University
Nanjing, China

PREETY SWETA HEMBROM
Department of Genomic Science
School of Biological Sciences
Central University of Kerala
Kasaragod, Kerala, India

KWANHO C. KI
Westminster University
Salt Lake City, UT, USA

ERIN L. LEWIS
Utah State University
Logan, UT, USA

KUN LI
College of Veterinary Medicine
Nanjing Agricultural University
Nanjing, China

SIJIA LU
College of Veterinary Medicine
Nanjing Agricultural University
Nanjing, China

OLIVIA LUTZ
Lankenau Institute for Medical Research
Wynnewood, PA, USA

BHAGYA MAPPURATH
Department of Genomic Science
School of Biological Sciences
Central University of Kerala
Kasaragod, Kerala, India

ALEX MUPPASANI
Lankenau Institute for Medical Research
Wynnewood, PA, USA

SOUMITRA NATH
Department of Biotechnology
Gurucharan University
Silchar, Assam, India

ZUBAIR AZHAR NOMI
Faculty of Veterinary and Animal Sciences
Muhammad Nawaz Shareef University of Agriculture
Multan, Pakistan

ELENA G. OLSON
Department of Animal and Dairy Sciences
University of Wisconsin–Madison
Madison, WI, USA

BERRIN ÖZDIL
Faculty of Medicine
Department of Biophysics
Bornova-Izmir, Turkey

MAXIM A. PECKENSCHNEIDER
Department of Animal and Dairy Sciences Meat Science and Animal Biologics
University of Wisconsin–Madison
Madison, WI, USA

CARSON T. PRIDDLE
AgriGro
Doniphan, MO, USA

NARCHIKUNDIL REEJA
Department of Genomic Science
School of Biological Sciences
Central University of Kerala
Kasaragod, Kerala, India

STEVEN C. RICKE
Department of Animal and Dairy Sciences
University of Wisconsin-Madison
Madison, WI, USA

GAYATRI KRISHNA S.
Department of Genomic Science
School of Biological Sciences
Central University of Kerala
Kasaragod, Kerala, India

SHEMMY SADANANDAN
Department of Genomic Science
School of Biological Sciences
Central University of Kerala
Kasaragod, Kerala, India

INDU SHARMA
Department of Microbiology
Assam University
Silchar, Assam, India

SUNIL THOMAS
Lankenau Institute for Medical Research
Wynnewood, PA, USA

CHANG XU
College of Veterinary Medicine
Nanjing Agricultural University
Nanjing, China

HUIYAN XU
College of Veterinary Medicine
Henan Agricultural University
Zhengzhou, China

ZI YAN
College of Veterinary Medicine
Henan Agricultural University
Zhengzhou, China

FU YIN
College of Veterinary Medicine
Henan Agricultural University
Zhengzhou, China

LONGXIAN ZHANG
College of Veterinary Medicine
Henan Agricultural University
Zhengzhou, China

JINFENG ZHAO
College of Veterinary Medicine
Henan Agricultural University
Zhengzhou, China

CHAPTER ONE

Metagenomics in Livestock

UNRAVELING MICROBIAL COMMUNITIES TO ENHANCE PRODUCTIVITY AND HEALTH

Mottakunja Deepthi, Preety Sweta Hembrom, Bhagya Mappurath, Narchikundil Reeja, Shemmy Sadanandan, Gayatri Krishna S., Ginny Antony, and Tony Grace

1.1 INTRODUCTION

The livestock microbiome has emerged as a foundational lens through which animal health, productivity, welfare, and environmental sustainability can be understood and improved. Enabled by advances in DNA and RNA sequencing, metagenomics has transformed our view from culture-dependent snapshots of a few organisms to comprehensive, high-resolution portraits of entire microbial communities and their functional capacities.

In production animals, these communities occupy diverse anatomical niches—including skin, oral and respiratory tracts, reproductive tissues, and, most intensively, the gastrointestinal system and rumen—where they regulate nutrient acquisition, immune development, barrier integrity, pathogen resistance, and host physiology. Across species and niches, microbiome states are associated with disease resistance, udder and respiratory health, reproductive outcomes, and environmental resilience.

Contemporary metagenomic approaches have proven essential in connecting microbial structure and function to economically important traits such as feed efficiency, milk yield and composition, growth performance, meat quality, and methane emissions, laying the groundwork for precision management and microbiome-informed breeding. Integrative multiomics extends sequence-based inference to molecular activity.

These insights motivate a growing toolbox for microbiome engineering—spanning probiotics, prebiotics and synbiotics, postbiotics, fecal microbiota transplantation, phage and bacteriocins, dietary and management interventions, and synthetic biology—that aims to steer communities toward desirable functions under commercial conditions.

1.2 METAGENOMICS APPROACHES IN LIVESTOCK MICROBIOME STUDIES

Animal microbiome studies have emerged as a new discipline with the advent of novel sequencing approaches (Wani et al., 2023). Advancements in sequencing techniques have enabled the transition from conventional culture-based methods to high-throughput metagenomic approaches. Metagenomics involves the study of complete microbial communities, providing insights into disease resistance, gut health, and productivity enhancement (Pallen, 2014). Conventional culture-dependent microbial methods have proven insufficient for understanding livestock–microbiome interactions; therefore, metagenomic studies have been a significant advancement in deciphering these interactions. The major metagenomic

DOI: 10.1201/9781003638568-1

approaches applied in livestock microbiome research include 16S rRNA amplicon sequencing and shotgun metagenomics, long-read sequencing, and integrative multiomics platforms that enable in-depth functional analysis (Table 1.1). By leveraging metagenomic approaches, integrating microbiome data with host genetic and phenotypic information can elucidate the role of gut microbes in promoting more sustainable livestock management.

1.2.1 Amplicon Sequencing

Amplicon sequencing is the most widely used approach for characterizing host-associated microbiomes, in which the aim is targeted amplification of hypervariable regions, the V2–V3 and V3–V4 16S rRNA for bacteria and archaea (Abellan-Schneyder et al., 2021; García-Martínez et al., 1999), the V4 and V9 18S rRNA hypervariable regions for protists (Choi & Park, 2020), and the internal transcribed spacer (ITS) rRNA regions (ITS1 and ITS2) for fungi (Schoch et al., 2012; Nilsson et al., 2019). This PCR-based method relies on a dual-indexing mechanism to simultaneously process large numbers of individual libraries and then taxonomically assign them using curated reference databases, including Greengenes v13_8 (McDonald et al., 2012), SILVA 138 (Quast et al., 2013), and RDP (Cole et al., 2014). This cost-effective methodology provides information on microbial abundance and diversity, characterized by operational taxonomic units (OTUs), in which sequences with ≥97% similarity were clustered into species-like groups. Recently, amplicon sequence variants (ASVs) have replaced OTUs, providing single-nucleotide resolution and greater reproducibility across studies. QIIME2 (Bolyen et al., 2019) and DADA2 (Callahan et al., 2016) have become standards for processing and analyzing microbial diversity; QIIME2 is a versatile tool that supports command-line workflows for phylogenetic analyses and alpha–beta diversity metrics, while DADA2 excels at error correction and accurately identifies microbial populations. The development of other software tools, such as Tax4Fun (Aßhauer et al., 2015) and PiCRUSt2 (Douglas et al., 2020), has enabled preliminary functional prediction from 16S rRNA sequence data by providing rough estimates of KEGG pathways and the metabolic diversity present in the overall microbial community.

The majority of livestock microbiome studies have relied on the 16S rRNA approach; for example, high-forage diets increased populations of Ruminococcus and Fibrobacter, enhancing fiber digestion, whereas high-grain diets favored Prevotella and lactate-producing bacteria, linked to ruminal acidosis (Henderson et al., 2015). In broiler chickens, 16S profiling revealed that antibiotic-free birds harbored higher levels of Lactobacillus. In contrast, antibiotic-treated birds had increased Enterococcus and Escherichia-Shigella, suggesting microbiome disruption and increased antimicrobial resistance (AMR) risk (Oakley et al., 2014). In piglets exposed to early weaning stress, reduced microbial diversity and a rise in *Clostridium perfringens* were observed, correlating with diarrhea incidence (Slifierz et al., 2015). However, several limitations remain, including PCR biases and restricted species-level resolution. To overcome these challenges, whole-genome approaches such as shotgun metagenomics are increasingly applied, offering more comprehensive resolution and deeper functional insights into microbial communities.

1.2.2 Shotgun Metagenomics Sequencing

Shotgun metagenomics sequencing, whereby entire genomic DNA is sequenced, can provide detailed insights into the taxonomic composition, microbial functional genes, metabolic pathways, and host–microbe interactions (Handelsman, 2004; Denman et al., 2018). It prevents amplification errors and, in turn, has become the standard approach for gut microbiome studies aimed at understanding host–microbe interactions. High-coverage metagenomic sequencing can also facilitate the construction of metagenome-assembled genomes (MAGs). It provides detailed insights into the types of enzymes and metabolic pathways the microbiome may express.

In livestock research, feces and manure samples are usually collected from pen floors, fresh feces, or manure pits, transported on ice, and stored at −80°C before processing to reduce DNA degradation (Wegl et al., 2021; Chen et al., 2020). Total DNA is extracted using standard protocol and optimized methods to account for complex environmental matrices, and extracts are confirmed for quality and quantity before sequencing (Chen et al., 2020). Metagenomic sequencing libraries are constructed by fragmenting DNA, ligating the sequencing adapters, and then either following a PCR-free method or performing limited-cycle amplification to mitigate bias, depending on the sequencing protocol (Wegl et al., 2021).

Sequencing is done predominantly on Illumina platforms, with paired-end 150 bp reads having the depth needed to capture diversity in communities and low-abundance functional genes (Chen et al., 2020; Wegl et al., 2021).

In shotgun metagenomics, raw reads are first quality-filtered using tools such as Trimmomatic to remove adapters and low-quality bases. Taxonomic profiling is then performed using fast k-mer–based classifiers such as Kraken2 and Sourmash, or marker-gene–based methods such as MetaPhlAn4, while newer hashing-based methods, such as Skani, are being developed for large-scale datasets. Short-read assembly into contigs is typically performed with MEGAHIT or metaSPAdes, followed by binning with MetaBAT2, MaxBin2, or CONCOCT to reconstruct MAGs. Quality assessment using CheckM and refinement with MetaWRAP ensures the reliability of bins, which is critically important for uncovering novel livestock-associated microbes from complex habitats like the rumen. Functional annotation integrates similarity-based searches (BLAST, DIAMOND), profile HMM approaches (HMMER), and increasingly structure-guided predictions (AlphaFold) to assign putative functions. Genome-scale pipelines such as DRAM and METABOLIC facilitate in-depth reconstruction of metabolic pathways, linking MAGs to ecological functions like fiber degradation, methane production, and nitrogen cycling. Furthermore, livestock research often focuses on resistome and virulome profiling using databases such as CARD, VFDB, and SARG, and annotation pipelines such as ARG-OAP, enabling the detection of AMR genes, mobile genetic elements, and virulence factors.

Together, these bioinformatics workflows convert raw sequence data into ecological and functional insights and have played a pivotal role in livestock microbiome studies. For example, in cattle, whole metagenomic sequencing identified more than 27,000 carbohydrate-active enzymes (CAZymes), many of which were novel, to explain how the rumen efficiently breaks down cellulose and hemicellulose (Stewart et al., 2018). AMR surveillance in pigs revealed the presence of diverse AMR genes in the gut, including tetracycline and macrolide resistance, most of which were carried on plasmids, underscoring the risk of horizontal gene transfer in intensive farming (Munk et al., 2018). Shotgun metagenomics associated Lactobacillus and Bacteroides species with the production of short-chain fatty acids (SCFAs), directly connecting microbiome composition to host energy metabolism in poultry (Shang et al., 2018).

Despite significant advancements in short-read metagenomics, many challenges remain, including high sequencing costs, fragmented assemblies, limited strain-level resolution, and high computational requirements that can lead to discrepancies in microbial community identification. These challenges can be addressed by integrating long-read sequencing platforms (e.g., Oxford Nanopore, PacBio HiFi), which improve assembly contiguity and enable recovery of full-length transcripts and operons (Liu et al., 2024). Additionally, the vast amount of metadata generated during sequencing complicates data analysis, requiring more precise diagnostic tools and expertise to interpret results and integrate all functional multiomics effectively.

1.2.3 Long-Read Sequencing

Long-read sequencing technologies, such as Single-Molecule Real-Time (SMRT) sequencing, Pacific Biosciences (PacBio), and Oxford Nanopore Technologies (ONT), generate significantly longer reads (10–50 kbp for PacBio; >10 kbp for ONT), enabling the resolution of full-length RNA transcripts, repetitive regions, and complex genomic structures (Steijger et al., 2013; Gao et al., 2016; Lang et al., 2020). Long-read sequencing is now more reliable and applicable to metagenomics and other omics research due to recent improvements in base-calling techniques and sequencing chemistry, such as ONT Q20+ and PacBio HiFi circular consensus sequencing, which have significantly increased read accuracy (>99%). It enhances taxonomic resolution from genus to species and even strain level, reducing classification biases and improving phylogenetic analyses (Buetas et al., 2024). To improve read accuracy and assembly quality, tools such as metaFlye, Guppy, Canu, DeepConsensus, and Unicycler are commonly employed, while SVIM and Sniffles2 help to identify large genomic rearrangements. For instance, a comparison of long-read sequencing platforms Nanopore and PacBio reported N50 read lengths of approximately 14,056 bp for PacBio and 6,873 bp for Nanopore in the feces of a cow treated with ceftiofur. About 74% of BLASTN alignments had coverage of at least 60%, while PacBio produced more gene alignments with coverage of at least 80% (Watson et al., 2021). Although it has its own benefits, long-read sequencing also comes with challenges, such as high cost and computational requirements.

However, as long-read sequencing technologies advance, the demand for high-contiguity genome assembly and structural variant annotation databases is ever-increasing. As these are further developed, their application in microbial genomics, taxonomy, and functional analysis will continue to grow stronger and become solidified as indispensable tools in microbiome research.

1.2.4 Metatranscriptomics, Metaproteomics and Metabolomics

Multiomics analysis methods have emerged as an effective tool for integrating data to identify microbiomes from diverse habitats and environments. Metatranscriptomics can help address this by analyzing the complete set of RNA transcripts to indicate which genes are expressed by microbes (Shakya et al., 2019). It is important to recognize that gene abundance does not directly reflect protein levels or functional enzyme activities (Edfors et al., 2016). It has revealed vital metabolic pathways within the rumen microbiome; for example, Mann et al. (2018) identified 1,607 functional potential activities, comprising central metabolism, galactose metabolism, and starch and sucrose metabolism pathways in rumen wall microbial communities.

Whereas metaproteomics and metabolomics can be used to evaluate and quantify the proteins and metabolites produced by microbial communities, respectively. Xie et al. (2022) demonstrated the upregulation of pathways involved in fiber breakdown, glycolysis, and the pentose phosphate pathway, underscoring the method's ability to characterize active microbial processes and nutrient cycling within the gut ecosystem. These results illustrate how diet-induced alterations, including the effectiveness of fiber degradation and microbial metabolic interactions, can be examined using metatranscriptomics.

Metabolomics focuses on the direct measurement of metabolic outcomes and linking microbial activity to functional potential. For instance, the production of SCFAs, such as acetate, propionate, and butyrate, plays a critical role in host energy metabolism and productivity (Lee & Hase, 2014). These SCFAs are important indicators of microbial interactions with the host, reflecting the efficiency of fiber degradation and fermentation processes. Comprehensive profiling is complicated by the dynamic, diverse range of metabolites produced by rumen microbes. For precise metabolite identification and quantification, advanced analytical techniques like mass spectrometry and nuclear magnetic resonance (NMR) spectroscopy are crucial. Integrating metabolic data with other omics platforms provides a detailed understanding of microbial metabolites and their interactions with the host.

Metaproteomics identifies and quantifies microbial proteins, revealing insights into their functional potential and their roles in microbial activity. For example, identifying enzymes involved in carbohydrate metabolism can help elucidate how the microbiome processes dietary components (Hagen et al., 2021; Andersen et al., 2023). However, there are analytical difficulties due to the complexity of proteins and their posttranslational modifications. Advanced analytical pipelines and further expansion of various protein databases tailored to the gut microbiome will continue to increase the accuracy and complexity of metaproteomic research (Table 1.1).

1.3 MICROBIAL COMMUNITIES IN DIFFERENT LIVESTOCK SYSTEMS

Body surfaces that are constantly in contact with the external environment, particularly the oral cavity, skin, respiratory, gastrointestinal, and reproductive systems, are colonized by diverse microorganisms. Among these niches, the gastrointestinal system offers a particularly conducive environment for microbial colonization, due to its rich, continuous nutrient supply. The abundance and composition of microbial communities differ across body systems and are influenced by various factors, such as diet, age, host genetics, environment, and physiological conditions.

1.3.1 Skin

The skin is the largest and the outermost barrier forming the first line of defense against infections in animal and human bodies. It performs multiple functions, including the ability to protect the internal body environment from external stress and pressure, regulating body temperature, reducing microorganisms' invasions, limiting radiation exposure impact, and preventing the entry of different chemicals into the body (Montagna, 1962; Grice & Segre, 2011). However, due to its large surface area, the skin is continuously exposed to physical and chemical agents. It is susceptible to various injuries that can facilitate the invasion

TABLE 1.1

Comparison between different sequencing approaches and multiomics analysis essential for exploring microbial communities in livestock

Approach	Target molecule	Key outputs	Common tools/platforms	Applications	Advantages	Limitations
16S/18S rRNA & ITS amplicon sequencing	Marker genes (16S, 18S, ITS)	Taxonomic composition, metabolic pathways (KEGG)	QIIME2, DADA2, Greengenes, SILVA, RDP,PICRUSt2,Tax4Fun	Diet-associated microbial loads in cattle, poultry, pigs; and its functional profiling; dysbiosis detection	Cost-effective, high throughput	PCR bias, limited species/strain resolution, indirect functional inference
Shotgun metagenomics	Total genomic DNA	Taxonomy, functional genes, metabolic pathways; ARGs, MGEs, virulence factors	Kraken2, MetaPhlAn4, MEGAHIT, metaSPAdes; CARD, VFDB, SARG, ARG-OAP	SCFA production in poultry, fiber degradation in ruminants, AMR surveillance	High-resolution taxonomy and function, unbiased	High cost, computationally intensive
Metagenome-assembled genomes (MAGs)	Assembled contigs	Complete microbial genomes	MetaBAT2, MaxBin2, CONCOCT, CheckM	Discovery of novel rumen microbes and CAZymes	Genome-level functional studies	Fragmentation, strain level uncertainty
Long-read metagenomics	High-molecular-weight DNA	Full-length genomes, operons, SVs	PacBio HiFi, ONT, metaFlye, Canu	Improved rumen genome assemblies, plasmid resolution	High contiguity, strain-level resolution	Expensive, high computational demand
Metatranscriptomics	Total RNA	Actively expressed genes	HISAT2, Salmon, featureCounts	Diet-responsive gene expression in rumen microbiota	Captures functional activity	RNA instability, high cost
Metaproteomics	Proteins	Enzyme abundance, metabolic activity	MaxQuant, MetaProteome Analyzer	Carbohydrate metabolism and fiber degradation	Direct functional evidence	Protein complexity, PTMs
Metabolomics	Metabolites	SCFAs, fermentation products	LC–MS, GC–MS, NMR	Host–microbiome metabolic interactions	Links microbiome to phenotype	Metabolite diversity, interpretation challenges

of infectious agents. Despite constant exposure, skin health is well maintained by the presence of a complex, balanced community of resident microorganisms.

Additionally, commensal organisms, along with the skin microbiota, are crucial for maintaining cutaneous homeostasis and immunological competence. Specific opportunistic pathogens, such as *Staphylococcus aureus*, are commensal bacteria on humans' and animals' skin and mucous membranes. However, they can cause disease when they invade the body, either due to injury or a compromised immune system. Colonization of S. aureus on the skin is also responsible for many bacterial infections in livestock, ranging from local lesions such as boils to life-threatening sepsis. The most well-characterized of these infections is bovine mastitis in cattle (Mala et al., 2021).

The skin functions as a complex ecosystem of diverse microenvironments, including numerous folds, invaginations, and specialized niches that support a broad array of microorganisms, including bacteria (e.g., Proteobacteria, Corynebacterium, and Staphylococcus spp.), fungi (e.g., Malassezia), and viruses (e.g., Capripox) (Grice et al., 2011). These microorganisms play important roles in the host defense against pathogens and in the development of the host immune system (Mala et al., 2021). A healthy skin microbiota contributes to host fitness by occupying pathogen adhesion sites and producing pathogen inhibitors (Grice et al., 2011). For example, *Staphylococcus aureus* is a common pathogen that can cause both localized and systemic infections. However, some bacteria, including *Staphylococcus epidermis* and *Corynebacterium* spp., can inhibit or reverse the growth of *S. aureus* (Cogen et al., 2010). Both beneficial and pathogenic skin microorganisms may play a role in protecting livestock from disease. The microbial composition varies between species because, unlike humans, the skin of livestock and other animals is usually covered with dense hair, fur, or feathers. In addition to these species-specific traits, other factors such as body location, sex, age, geographic location, diet, habitat, proximity to other animals, maternal transfer, and disease conditions strongly shape the microbial community structure in animals. Previous studies have shown that different skin regions are predominantly occupied by *Staphylococcus* spp. and *Corynebacterium* spp., while *Propionibacterium* spp. predominated in sebaceous areas and Gram-negative (G−) organisms (such as Betaproteobacteria) colonized dry skin areas, such as the forearm and leg (Mala et al., 2021).

1.3.2 Gut

The gastrointestinal tract of animals, particularly in cattle, harbors a vibrant and diverse microbial community, comprising more than 1,800 genera and approximately 40,000 bacterial species. The total number of microbial cells in the gut is estimated to be 10–100 times greater than that of host cells, highlighting their integral role in bovine physiology. These microorganisms contribute significantly to nutrient digestion and absorption, help maintain gut stability, and support both production efficiency and the animal's overall health (Liu et al., 2023).

In cattle, ruminal fermentation is a central metabolic process, driven primarily by microbial activity, during which rumen microorganisms produce a wide array of enzymes that degrade complex feed components such as cellulose, proteins, lipids, and other plant polymers. The end products of this fermentation process supply nearly 70% of the host's energy requirements, mainly through volatile fatty acids (VFAs). In contrast, the remaining energy is derived from nutrients absorbed along the digestive tract (Bergman, 1990). Beyond their nutritional role, gut microorganisms also profoundly influence immune regulation and metabolic functions, thereby contributing to the host's resilience and productivity. Understanding the composition and functional dynamics of the bovine gut microbiota can provide valuable insights to enhance dairy cattle breeding strategies and improve herd health management.

Interestingly, variations in the Firmicutes/Bacteroidetes ratio within the cow's gastrointestinal tract have been shown to correlate positively with milk fat yield. Shifts in the abundance of these microbial groups are also linked to genes involved in milk fat synthesis, indicating a close relationship between gut microbial structure and dairy performance traits (Barathan et al., 2024, O'Hara et al., 2020).

1.3.3 Rumen

The rumen is the largest and most important of the four compartments comprising a ruminant's stomach: reticulum, rumen, omasum, and abomasum. The rumen is sterile at birth, but is rapidly sequentially colonized by environmental bacteria, beginning with facultative organisms. Anaerobic bacteria colonize within a few hours after birth, followed by cellulolytic bacteria and methanogenic organisms by 2–4 days of age, anaerobic fungi by the second week, and ciliate protozoa during the third week (Fonty et al., 1987; Jami et al., 2013). In adult cattle, the rumen can reach a volume of up to 180 L and contains approximately 10^{10}~10^{11} bacteria/mL, 10^{6}~10^{8} archaea/mL, 10^{4}~10^{6} protozoa/mL, and 10^{3}~10^{6} fungi/mL (Liu et al., 2023). This compartment hosts a dense and diverse microbial ecosystem that plays a crucial role in feed digestion, maintaining animal health, and supporting overall productivity. The microbial population is significantly influenced by the animal's diet, which in turn affects key production traits such as feed efficiency, methane emissions, and immune function (Perez et al., 2024). The relationship between ruminal microbes and the host is mutualistic and symbiotic, essential for the survival and optimal performance of both. Predominant bacterial genera in the rumen include Ruminococcus, Fibrobacter, Prevotella, Selenomonas, Streptococcus, and Ruminobacter (Russell, 2002). Ruminants effectively degrade fibrous feed through the metabolic activities of colonized anaerobic microbiota, including bacteria, fungi, and protozoa (Chaucheyras-Durand & Ossa, 2014). About half of the total biomass is made up of bacteria, which are a significant source of protein, contributing 75–80% of the animal's metabolizable protein. Likewise, they produce the enzymes that break down sugar, starch, and fiber (cellulose and hemicellulose). Ciliate protozoa, which can make up as much as 50% of the total microbial biomass, contribute to the digestion of plant particles and help mitigate acidosis by utilizing lactic acid. Anaerobic fungi, which make up 8–10% of the microbial biomass, are pivotal in fiber degradation (McSweeney & Mackie, 2012). Acidity levels and oxygen content decrease as the digestive contents move from the rumen through the small intestine to the hindgut, while microbial abundance increases. The most prevalent microorganisms in the rumen are bacteria, which can be categorized into four subgroups: liquid-phase bacteria, solid-phase bacteria, rumen epithelial-adherent bacteria, and eukaryotic-associated bacteria. The dominant bacterial phyla in the rumen include Firmicutes and Bacteroidetes, with major genera such as Prevotella, Fibrobacter, and Succinivibrio. Studies have shown that the rumen harbors the highest diversity of bacterial species in the ruminant digestive tract, with 47 genera identified, including 16 ruminal-specific genera. (Kim et al., 2022; Wang et al., 2022). In the small intestine, bacteria predominate at approximately 10^{9}–10^{10}/mL, with Firmicutes and Proteobacteria dominant. In the hindgut, bacterial populations can reach 10^{12}–10^{14}/mL, dominated by Firmicutes, Bacteroidetes, Actinobacteria, and Verrucomicrobia (Abeyta et al., 2023; Bach et al., 2017). In cows with lower feed conversion rates, the rumen harbors higher abundances of microorganisms, including *Clostridium aerogenes*, Actinomyces, and Methanobrevibacter, which ferment cellulose to produce gases that help maintain normal rumen pressure, promoting digestion and passage of food. However, Methanobrevibacter, while playing a functional role in digestion, also produces significant quantities of methane and other greenhouse gases, thereby lowering feed conversion efficiency and contributing negatively to global climate change (Liu et al., 2023).

1.3.4 Oral

The oral cavity of ruminants represents a highly dynamic and multifaceted microbial environment composed of numerous microhabitats that support distinct microbial communities. In particular, the nonshedding dental surfaces facilitate the formation of complex polymicrobial biofilms, whose organization and functionality are determined by the collective interactions and metabolic activities of the resident microbes. The composition of the oral microbiome is shaped throughout an individual's lifetime by a combination of host genetic factors, maternal microbial transmission, dietary practices, oral hygiene habits, medication use, and various systemic influences (Sedghi et al., 2021). Genera such as Streptococcus, members of the Pasteurellaceae family, and Veillonella are differentially abundant in ruminant oral fluid samples, reflecting the intricate ecological balance within this habitat.

In cattle, a process called rumination, characterized by the regurgitation and rechewing of rumen contents, enables the transfer of ruminal microorganisms and their genetic material into the oral cavity (Marcos et al., 2024). This close microbial interchange has positioned saliva sampling as a valuable, noninvasive method for exploring the structure and functional dynamics of the rumen microbiome. Despite this, traditional interpretations often treat oral bacteria identified in rumen samples as mere transient passengers with little metabolic activity. However, recent studies suggest that metabolically active oral microbes may contribute to the degradation of plant material within the rumen. In vitro cultivation-based analyses have further confirmed that the ruminant oral cavity harbors a substantial population of anaerobic, cellulolytic bacteria, predominantly of the genus Streptococcus, that remain active under ruminal conditions (Edwards et al., 2023). This overlap between oral and ruminal microbial communities not only supports functional interconnectivity between the two ecosystems but also highlights a potential pathway for the transmission of antibiotic resistance genes between cattle and humans.

1.3.5 Respiratory Tract

The bovine respiratory tract harbors a complex and dynamic microbial ecosystem that plays a critical role in maintaining respiratory health. It is predominantly composed of five bacterial phyla, Proteobacteria, Firmicutes, Tenericutes, Actinobacteria, and Bacteroidetes, whose relative abundances vary with the animal's age and production system. Distinct microbial communities inhabit the upper and lower respiratory tracts, with the nasopharyngeal microbiota contributing most significantly to the microbial composition of the lower respiratory tract. Consequently, the nasopharynx represents a key target for sampling and modulation strategies aimed at improving respiratory health (Timsit et al., 2018).

Microbiotas enriched in beneficial bacteria, such as Lactobacillus, have been associated with good respiratory health. In contrast, enrichment with pathogenic taxa, including *Mycoplasma bovis*, *Mannheimia haemolytica*, and *Pasteurella multocida*, is commonly linked to bovine respiratory disease (BRD). Studies have shown that intranasal administration of selected Lactobacillus strains alters the composition of the nasopharyngeal microbiota, enhancing colonization resistance against opportunistic pathogens such as M. haemolytica. Indeed, cattle that remained healthy exhibited a higher relative abundance of Lactobacillaceae and Bacillaceae, both of which are capable of inhibiting the in vitro growth of *M. haemolytica*, *P. multocida*, and Histophilussomni (Holman et al., 2015; Amat et al., 2017; Corbeil et al., 1985).

Although *M. haemolytica*, *P. multocida*, *H. somni*, and *M. bovis* have been associated with BRD-related mortalities, these bacteria are also commonly found in the upper respiratory tract of healthy animals. The ability of cattle to resist disease despite harboring potential pathogens underscores the importance of the respiratory tract's innate defense mechanisms. These include the mucosal layer, epithelial barrier, and complex immune signaling networks that maintain microbial homeostasis, clear pathogens, and coordinate immune responses (Centeno-Martinez et al., 2022).

Moreover, environmental factors can influence respiratory health and microbiome composition. Emissions from livestock farms comprising gaseous pollutants and particulate matter may introduce microbial elements into the surrounding environment, potentially influencing microbial transmission and respiratory disease dynamics (de Rooij et al., 2019; Gao et al., 2023).

1.3.6 Reproductive Tract

The microbial communities of the bovine reproductive tract play a crucial role in maintaining reproductive health and fertility. Recent studies have demonstrated that bacterial populations differ between the uterus and vagina and fluctuate throughout the estrous cycle in response to hormonal changes. Under healthy conditions, these bacterial communities interact harmoniously with the uterine environment, helping to maintain an optimal pH, supplying and metabolizing essential nutrients, and modulating the immune responses of the reproductive tract. However, disturbances in this balance, particularly the overgrowth of potentially pathogenic bacteria, can adversely impact uterine health and fertility outcomes.

In humans, the reproductive tract microbiome accounts for approximately 9% of the total body microbial community (NIH HMP Working Group, 2009). Compared with other microbial ecosystems, such as the gastrointestinal or respiratory tracts, the human reproductive tract displays relatively low microbial diversity. This reduced diversity is mainly due to the dominance of Lactobacillus species, which typically account for over 90% of the vaginal bacterial population in healthy individuals (The Human Microbiome Project, 2012).

Similarly, in cattle, the reproductive tract is highly susceptible to microbial invasion, particularly during parturition, when exposure to the external environment increases. This exposure can allow opportunistic and pathogenic bacteria to enter and colonize the uterus. Common pathogens associated with uterine infections include *Escherichia coli*, *Trueperella pyogenes* (formerly

Arcanobacterium pyogenes), and *Fusobacterium necrophorum*, all of which are frequently linked to postpartum uterine diseases and impaired fertility (Ault-Seay et al., 2023).

1.3.7 Comparative Microbiome Diversity across Livestock Systems

1.3.7.1 Ruminants (Cattle, Sheep, and Goats)

Ruminants such as cattle, sheep, and goats possess a highly specialized digestive system that enables them to utilize fibrous plant material efficiently. Their stomach is divided into four compartments: the rumen, reticulum, omasum, and abomasum, each contributing to the fermentation and digestion process (Denman et al., 2018). Unlike monogastric animals, which depend mainly on enzymatic digestion, ruminants rely heavily on microbial fermentation to break down complex carbohydrates. The rumen provides an anaerobic environment that supports a vast and diverse microbial community responsible for degrading structural carbohydrates, such as cellulose and hemicellulose (Matthews et al., 2019).

Among these microbial inhabitants, bacteria form the dominant group, with Firmicutes and Bacteroidetes representing the most abundant phyla. These include key fiber-degrading species such as *Fibrobacter succinogenes*, *Ruminococcus flavefaciens*, and *Ruminococcus albus*, which specialize in cellulose degradation, and Prevotella, Butyrivibrio, and Pseudobutyrivibrio, which primarily degrade hemicellulose (Chen et al., 2022). These bacteria collectively produce VFAs as end products of fermentation, which are absorbed through the rumen wall and serve as primary energy sources for gluconeogenesis and milk fat synthesis in ruminants. When these microbes pass into the abomasum and intestines, they are digested by host enzymes, providing a rich source of microbial protein for the animal (Chen et al., 2022).

Although bacteria are the predominant microorganisms, fungi also play a vital role in the rumen ecosystem. Anaerobic fungi belonging to the phylum Neocallimastigomycota, such as Piromyces and Neocallimastix, are capable of degrading recalcitrant plant materials using a wide array of hydrolytic enzymes (Tardiolo et al., 2025). These fungi form synergistic relationships with fibrolytic bacteria and methanogenic archaea, collectively enhancing the breakdown of lignocellulose in the rumen (Hess et al., 2020).

The rumen environment maintains a pH between 5.5 and 7.0, which is optimal for fiber-digesting microbes such as Fibrobacter and Ruminococcus (Russell & Rychlik, 2001). The extended residence time of digesta, typically 30–50 hours, allows for efficient fermentation and nutrient extraction (Malmuthuge et al., 2013), whereas monogastric animals exhibit shorter digestion times, limiting microbial fermentation. Although the gut is anaerobic, specific regions, such as the intestinal epithelium, support microaerophilic niches where facultative anaerobes, such as Escherichia coli, can thrive, especially during host metabolic changes (Shin et al., 2015). These physiological and environmental factors together shape the unique microbial composition observed across species.

1.3.7.2 Monogastrics (Pigs, Poultry, Horses, and Rabbits)

Monogastric animals, including pigs (*Sus scrofa domesticus*), poultry, horses (*Equus ferus caballus*), and rabbits (*Oryctolagus cuniculus*), are primarily composed of a simpler, single-chambered stomach that relies on enzymatic digestion (Dal Pont et al., 2022). This anatomical difference significantly influences their feeding behavior, nutrient absorption, and the structure of their gut microbiota. In monogastrics, microbial fermentation occurs mainly in the large intestine or cecum rather than the stomach, resulting in a microbial community distinct from that of ruminants.

In these animals, fungal species such as Candida and Saccharomyces are typically commensal members of the gut microbiota, though their overgrowth can lead to dysbiosis and digestive disturbances (Iliev et al., 2013). While most studies have focused on bacterial populations, understanding the interactions between bacteria, fungi, and bacteriophages is essential for a complete picture of gut ecosystem function.

In equines, the gut microbiota is dominated by Firmicutes (45–68%) and Bacteroidetes (14–37%) in healthy individuals (Tardiolo et al., 2025). However, other studies have reported higher proportions of Bacteroidetes in both young and aged horses, including those affected by colitis. While considering equine gut, Proteobacteria, a phylum of Gram-negative bacteria, ranks as the second most prevalent group, particularly in the upper intestinal regions, with peak abundance in the ileum (approximately 33%) and elevated levels have been associated with intestinal inflammation and disorders such as equine colic (McKinney et al., 2020, Chaucheyras-Durand et al., 2022).

The acidic conditions of the monogastric stomach (pH 1.5–3.5) favor acid-tolerant bacteria such as Lactobacillus and Streptococcus, which play a role in initial carbohydrate fermentation and in inhibiting pathogens (Tugnoli et al., 2020). The shorter transit time in the monogastric digestive system limits microbial fermentation compared to ruminants, but the microbial community remains vital for nutrient metabolism, immune regulation, and gut health.

1.3.8 Functional Aspects of Livestock Microbiomes

Advances in high-throughput sequencing, coupled with robust computational analysis, have enabled the exploration of entire microbial communities, including those that cannot be cultured in laboratory conditions. These approaches have uncovered uncharacterized microbes and identified key microbial players associated with productivity, disease resistance, and health disorders in livestock. The strategic use of beneficial microorganisms offers a sustainable approach to overcome nutritional limitations, enhancing disease resistance, improving feed utilization and digestion, and ultimately promoting overall animal health and welfare.

Numerous studies have examined the impact of probiotics and prebiotics on fecal and seminal microbiome profiles, demonstrating their potential to enhance gut health by increasing microbial diversity. Targeted dietary strategies, particularly the inclusion of probiotics and prebiotics, have been shown to influence the composition of the equine fecal microbiota, and across all treatments and sampling periods in miniature pony stallions, Firmicutes, Bacteroidetes, and Verrucomicrobia consistently remained the dominant bacterial phyla in fecal samples (Cooke et al., 2024).

1.4 FUNCTIONAL INSIGHTS: LINKING MICROBIOMES TO DESIRABLE TRAITS

Advances in metagenomic studies have revealed the microbiome's role in shaping key traits in livestock, including feed efficiency, disease resistance, product quality, methane emissions, and reproduction (Kaur et al., 2023). A better understanding of these microbial communities opens new opportunities to improve desirable traits in livestock by modulating microbiome composition, advancing outcomes that benefit both the industry and the environment (Andrade et al., 2022; Forcan et al., 2022).

1.4.1 Optimizing Feed Efficiency and Nutrient Utilization

Ruminants are vital to human communities because of their unique ability to convert low-quality feed into valuable animal products, such as milk and meat (Lee & Hase, 2014). This process is mediated by the gastrointestinal tract (GIT) microbiome, which plays an essential role in transforming fibrous, low-quality diets into energy-rich substrates. By exerting a profound influence on nutrient metabolism, the microbiome also determines the quality and quantity of animal products (Martinez Boggio et al., 2024; Monteiro et al., 2024).

Studies have consistently demonstrated that variation in feed conversion efficiency is associated with changes in the abundance and function of specific microbial groups. Lin et al. (2024) reported a higher abundance of cellulolytic bacteria, such as *Ruminococcus* spp. and *Fibrobacter* spp., in feed-efficient cattle compared with less efficient counterparts. Similarly, hemicellulose-degrading taxa such as *Prevotella* spp. and Cryptobacteroides were enriched in efficient cattle (Lin et al., 2024). In native Indian cattle breeds, Anaerovibrio, Succinivibrio, and Roseburia were more abundant, taxa associated with high feed efficiency despite lower feed intake. In contrast, in crossbred cattle, these taxa were less abundant (Deepthi et al., 2023).

Research on Nellore steers revealed a greater relative abundance of Bacteroidetes alongside reduced levels of Firmicutes and methanogenic archaea, correlating with improved feed efficiency (Lopes et al., 2021). Beyond cattle, a significant prevalence of carbohydrate-degrading Actinobacillus was observed in the ileum and caecum of feed-efficient pigs (McCormack et al., 2019). Akkermansia, a next-generation beneficial microbe known for its role in maintaining gut barrier integrity and influencing glucose and lipid metabolism, was enriched in pigs with enhanced feed efficiency (Yang et al., 2017; Cani et al., 2022).

These microbial groups are essential targets in gut microbiome research, as they directly contribute to feed digestion and energy extraction, thereby influencing animal productivity and health. Manipulating microbial populations to favor fiber-degrading and SCFA-producing bacteria holds significant promise in reducing feed costs and enhancing animal performance. Martínez Álvaro et al. (2024) demonstrated the genetic impact of microbial metabolism on beef cattle growth, suggesting that microbiome-informed breeding could enhance growth rates. Likewise, improving feed efficiency is an important goal in animal breeding programs, with microbial markers emerging as selection criteria. Rabee et al. (2025) reported that rumen microbial traits could be used to identify feed-efficient Egyptian goats, offering opportunities to integrate microbiome knowledge into breeding strategies.

Rather than viewing the microbiome as a collection of individual taxa, it may be more appropriate to consider it as a functional unit. Selecting for production traits based on microbial functionality—or modulating overall microbiome function rather than specific taxa—could yield more sustainable improvements (Gardiner et al., 2020). If successful, such approaches could reshape breeding and nutritional interventions, ultimately improving feed utilization while reducing resource inputs.

1.4.2 Disease Resistance and Immune Modulation

The gut microbiome plays a pivotal role in shaping host immunity, influencing both innate and adaptive responses (Okolie et al., 2025). Symbiotic microorganisms produce metabolites such as SCFAs, indoles, and secondary bile acids, which regulate epithelial barrier integrity and modulate immune cell differentiation (Ornelas et al., 2022). In ruminants, SCFAs like butyrate are particularly important, as they enhance epithelial health, reduce pathogen colonization, and regulate the production of anti-inflammatory cytokines, thereby contributing to overall disease resilience (Rawal et al., 2024).

Taxonomic shifts in the gut microbiome have been linked with altered immune competence and susceptibility to infections (El-Khoury et al., 2025). For instance, enrichment of Lactobacillus and Bifidobacterium species is associated with improved mucosal immunity and protection against enteric pathogens (Iqbal et al., 2021). In cattle, *Faecalibacterium prausnitzii*, a well-known anti-inflammatory bacterium, has been negatively correlated with the incidence of diarrhea in calves (Du et al., 2023). Similarly, higher abundance of Prevotella species has been linked with increased production of immunomodulatory metabolites, supporting disease resistance (Iljazovic et al., 2021).

Functional metagenomics has further revealed that immune-related traits are not solely dependent on microbial composition, but also on the presence of genes involved in pathways such as butyrate production, vitamin synthesis, and antimicrobial peptide biosynthesis (Jin et al., 2025). These functional attributes enhance host immune resilience, reduce pathogen load, and may serve as biomarkers for selecting disease-resistant animals (Bai et al., 2022). Future strategies such as microbiome-informed probiotics, prebiotics, and microbiota transplants could therefore be integrated into livestock management to enhance immunity and reduce reliance on antimicrobials.

1.4.3 Better Milk Production and Quality

The microbiome significantly contributes to milk yield and quality across diverse dairy species (Kaur et al., 2023; Wang et al., 2023). The rumen microbiota provides fermentation products, primarily VFAs, which serve as precursors for milk fat and lactose synthesis. In contrast, amino acid metabolism supports milk protein production (Stergiadis et al., 2021). Recent multiomics studies in dairy cattle have shown that distinct rumen microbial enterotypes correlate with differences in milk protein yield, with pathways related to amino acid metabolism as the key drivers (Zhang et al., 2023; Xue et al., 2020). In goats, modulation of the rumen microbiome has been linked with the milk fat content and fat globule size (Zhang et al., 2024). A study reported that dietary supplementation with herbal additives improved milk fat percentage and altered the abundance of rumen bacteria (Hassan et al., 2021). Similarly, ewe milk yield and quality traits have been tied to changes in microbial composition under different feeding systems (Zhao et al., 2023). A recent investigation demonstrated that differences in buffalo rumen microbial communities correlated with variation in milk fat yield (Zhao et al., 2024).

Beyond rumen function, the milk microbiome itself has emerged as a determinant of milk quality. In dairy cows, supplementation with multistrain probiotics improved yield and milk components, including protein and fat, while also reducing somatic cell counts, thereby enhancing hygienic quality (Olchowy et al., 2019). Studies in dairy cows have reported that probiotic supplementation improved udder health and reduced mastitis incidence, thereby contributing indirectly to better milk composition (Wang et al., 2022).

Collectively, these studies emphasize that across livestock species, the microbiome affects milk not only by modulating microbial communities through breeding, diet, and microbial supplementation quantity (yield), but also by influencing quality (composition, hygienic parameters, and nutraceutical properties). Targeted, therefore, offers a promising avenue for optimizing milk production and quality in a sustainable and species-specific manner.

1.4.4 Reduced Methane Emissions

Methane production in ruminants is primarily regulated by the rumen microbiome, in which methanogenic archaea use hydrogen and carbon dioxide to produce methane (Kelly et al., 2019). Manipulating microbial communities to redirect hydrogen flow toward alternative fermentation pathways, or to reduce methanogen abundance and activity, has emerged as a sustainable strategy to mitigate methane emissions without compromising productivity (Liu et al., 2023). Studies have shown that rumen microbiomes with higher abundance of Succinivibrio, Megasphaera, and Propionibacterium species exhibit lower methane yields due to increased propionate formation (Chen et al., 2020; Ogunade et al., 2020).Targeting the methanogenic archaea directly through microbiome modulation is another promising strategy to mitigate methane emission (Patra et al., 2017). Selective inhibition of dominant genera, such as Methanobrevibacter, has been achieved through dietary interventions that foster bacterial consortia that suppress methanogen colonization (Cristobal-Carballo et al., 2021). In ruminants, the introduction of feed additives has been linked to both reduced methane and improved rumen fermentation efficiency (Sun et al., 2021). Functional metagenomic studies further highlight that pathways associated with methanogenesis can be downregulated by dietary modulation of the microbiome (Zhao et al., 2024). For instance, supplementation with plant secondary metabolites selectively alters microbial composition, reducing methanogenesis-related transcripts while improving protein and fiber degradation and reducing feed energy loss (Ku-Vera et al., 2020).

Overall, methane mitigation through microbiome manipulation offers a dual benefit: lowering greenhouse gas emissions while improving feed energy efficiency. Future work combining functional microbiome profiling, selective breeding, and microbiome engineering could establish low-emission, high-performance livestock populations, contributing significantly to climate-smart animal agriculture.

1.4.5 Enhanced Growth Performance

Growth performance in livestock is a multifactorial trait influenced by genetics, nutrition, and the gut microbiome (Negash, 2022). Microbial communities in the gastrointestinal tract influence body weight gain and muscle development (Han et al., 2022). Recent evidence indicates that alterations in the gut microbiome correlate with changes in the growth rate of aquatic animals, providing a conceptual framework for enhancing growth rate through intensified gut microbiome research (Zhang et al., 2024).

In cattle, growth efficiency has been linked with increased abundance of Treponema, Roseburia, and Lactobacillus, microbes that produce propionate and butyrate, energy substrates for muscle growth (Gardiner et al., 2020). Functional metagenomics revealed enrichment of genes involved in amino acid metabolism, SCFA production, and the growth performance in livestock, underlining the importance of microbial functional capacity rather than taxonomic diversity alone (Kou et al., 2024). Similarly, in pigs, enrichment of carbohydrate-degrading genera such as Cryptobacteroides, Gemmiger, Vescimonas, and Faecousia has been correlated with superior feed efficiency and higher body weight gain (Yang et al., 2024).These findings indicate that microbiome composition and functionality directly influence growth trajectories across species.

Microbiome modulation strategies are increasingly being tested to enhance growth performance (Zhang et al., 2025). Probiotics and prebiotics serve as feed additives and have been shown to selectively enrich beneficial microbes, improve nutrient utilization, and stimulate growth performance (Direkvandi et al., 2020). Moreover, fecal microbiota transplantation experiments demonstrate that transferring "fast-growing" microbial communities can promote better growth performance in recipient animals (Cheng et al., 2022). Such findings open new avenues for incorporating microbiome-based markers in breeding programs and for designing dietary strategies tailored to enhance livestock growth.

1.4.6 Improved Meat Quality

Microbiome composition plays a growingly appreciated role in shaping meat quality traits in livestock, including its color and nutritional value (Zhang et al., 2023). A recent study in beef cattle found significant associations between rumen bacterial genera and carcass and meat quality attributes: Selenomonas spp. were positively correlated with hot carcass weight (HCW) and lower purge loss, whereas Christensenellaceae R-7, Moryella, and Prevotella were positively correlated with dressing percentage and intramuscular fat content. Conversely, Acidaminococcus and Succinivibrionaceae UCG-001 were negatively correlated with marbling, moisture content, and IMF, suggesting that microbial community structure affects both the amount and distribution of fat in muscle (Holman et al., 2024).

In pigs, manipulation of the microbiome via dietary or transplantation interventions has been shown to modify lipid profiles and IMF. For example, in Laiwu pigs (known for high IMF) compared with commercial DLY pigs, fecal microbiota transplantation (FMT) from Laiwu to DLY piglets increased IMF in the longissimus dorsi muscle. Multiomics analysis highlighted key microbial species—such as *Bacteroides uniformis*, *Sphaerochaeta globosa*, *Hydrogenoanaerobacterium saccharovorans*, and Pyramidobacterpiscolens—as potentially important in lipid deposition. These species upregulated genes such as FADS1, FABP4, and FASN and downregulated AMPK, thereby enhancing fat deposition.

Beyond the IMF, the gut–muscle axis (interactions among the gut microbiome, metabolites, and muscle physiology) is increasingly being studied. A review summarizing recent work (Wen et al., 2024) indicates that interventions such as probiotics, prebiotics, and FMT can impact muscle fiber type composition (fast vs. slow twitch), intramuscular fat content, and oxidative stability, thereby improving sensory properties such as tenderness and flavor. The microbial metabolites (SCFAs and amino acid derivatives) act as both substrates and signaling molecules in muscle tissue, impacting fiber metabolism and overall meat quality.

Dietary interventions also shape meat quality by altering the microbiome in finishing pigs and by altering the diet with silage, which in turn alters gut microbiota composition, improving traits such as water-holding capacity, color, and oxidative stability of pork. These changes are proposed to result from shifts in the abundance of beneficial microbial taxa and from enhanced production of antioxidative metabolites.

Another recent study compared two pig breeds (Songliao Black and Large White × Landrace) and found breed-specific differences in gut microbiota that corresponded with differences in fat deposition and meat quality traits (Kumar et al., 2025). Metabolomic profiling revealed that specific gut microbial metabolites might underlie variation in meat quality between breeds, suggesting that both breed and microbiota jointly modulate meat quality traits.

1.5 STRATEGIES FOR MICROBIOME ENGINEERING IN LIVESTOCK

Microbiome engineering has emerged as a promising approach for enhancing livestock health and performance by manipulating microbial communities in a targeted manner. A variety of methods contribute to this field, including probiotics, prebiotics, FMT, synthetic biology, and genome editing, each aimed at modifying microbial composition or function to improve outcomes (Berhanu et al., 2021). The rapid progress in microbial ecology, genomics, and biotechnology has further expanded opportunities to design interventions that not only support animal health but also improve overall production efficiency (Mueller & Sachs, 2015). Building on these advances, microbiome engineering strategies can be broadly classified into six categories: direct microbial interventions, genetic and synthetic biology approaches, nutritional modulation, phage therapy and bacteriocins, host-targeted methods, and environmental or management-based practices (Berhanu et al., 2021; Peixoto et al., 2022). Collectively, these strategies represent a versatile toolkit for driving innovation in livestock systems, offering solutions that balance productivity, animal welfare, and sustainability.

1.5.1 Direct Microbial Interventions

Direct microbial interventions are among the most widely explored strategies for microbiome engineering in livestock. Probiotics, which are live microorganisms, have been extensively applied in farming systems due to their strain-specific ability to enhance gut health, improve feed conversion efficiency, strengthen immunity, and exclude pathogenic microbes through competitive interactions and the production of antibacterial substances (Liu et al., 2023). Building on this, next-generation probiotics have been developed to deliver more precise and targeted effects. Prebiotics, nondigestible feed additives, selectively stimulate the growth of beneficial gut microbes, support nutrient absorption, and suppress pathogens (Kouhounde et al., 2022). Their combination with probiotics forms synbiotics, which produce synergistic effects that counter stressors and pathogens, thereby improving animal health and productivity (Vinderola et al., 2022). Postbiotics, consisting of microbial metabolites and bioactive molecules, provide similar health-promoting benefits without the need for live organisms, offering additional safety and stability advantages (Reuben et al., 2021). Another innovative approach, FMT, involves transferring microbiomes from healthy animals to restore balance in recipients. Although FMT shows promise for correcting dysbiosis and supporting growth, its large-scale adoption in livestock production is constrained by practical and regulatory challenges (Wang et al., 2023). Collectively, these microbial-based interventions represent sustainable alternatives to antibiotics, with the potential to improve animal welfare, disease resistance, and production efficiency while reducing the environmental footprint of livestock systems.

1.5.2 Genetic and Synthetic Biology Approaches

Genetic and synthetic biology approaches are increasingly being recognized as powerful tools for advancing livestock health and productivity. Animal diseases remain a significant constraint on the sector, accounting for up to 50% of turnover losses in developing countries and around 17% in developed nations (G. Hu et al., 2020). The traditional reliance on antibiotics is becoming less viable due to rising AMR and growing consumer demand for alternatives, underscoring the need for innovative strategies. Biotechnology offers solutions through genetic selection for disease-resistant animals and genetic engineering to introduce protective traits from novel sources, thereby reducing disease incidence and mitigating the risks of zoonotic transmission, such as avian influenza (Knap & Doeschl-Wilson, 2020). Building on this foundation, synthetic biology provides an interdisciplinary platform that integrates biotechnology, molecular biology, and systems biology to design new genomes, optimize metabolic pathways, and improve traits related to nutrition, reproduction, and immunity (Bishop & Woolliams, 2014). Within the context of microbiome engineering, these advances have enabled the development of engineered probiotics, which are genetically modified microbes that enhance nutrient metabolism or inhibit pathogens, as well as the application of CRISPR-based genome editing to selectively modulate microbial communities (Menchaca et al., 2020). Furthermore, synthetic microbial consortia are being designed to deliver stable, predictable benefits under farm conditions. Collectively, these innovations offer transformative potential for disease resistance, improved productivity, and the development of precise diagnostic and therapeutic tools in livestock systems (Levit et al., 2022).

1.5.3 Nutritional Modulation

Nutritional modulation is a powerful strategy for enhancing livestock health, as it influences both host immunity and the composition of microbial communities. Specific nutrients, such as vitamins D and E, omega-3 polyunsaturated fatty acids, and minerals like zinc, have been shown to modulate immune responses, although their optimal levels vary (Bobeck, 2020). In ruminants, nutritional status not only affects resistance to gastrointestinal parasites but also determines the ability to tolerate pathogen burden, with targeted supplementation improving host resilience (Walkden-Brown & Kahn, 2002). Similarly, in poultry, nutrition impacts disease resistance through multiple mechanisms, including immune system development, provision of substrates for immune activity, limitation of pathogen growth, and regulation of immune cell function (Hu et al., 2023). Beyond these direct effects on immunity, diet also plays a crucial role in shaping the livestock microbiome. Feed formulation strategies, such as incorporating dietary fiber, phytochemicals, fermented feeds, and bioactive compounds, can significantly influence microbial diversity and activity (Jin et al., 2025). The inclusion of feed enzymes further enhances fermentation efficiency and nutrient utilization, indirectly steering microbial structure. Phytobiotics and other plant-derived additives are increasingly recognized as sustainable alternatives to antibiotics, as they combine antimicrobial and immune-modulating properties with the ability to maintain gastrointestinal integrity and microbiome balance (Abdelli et al., 2021). Together, these nutritional approaches highlight the central role of diet in the diet–disease agent–host triad, offering integrated benefits for animal health, productivity, and resilience.

1.5.4 Phage Therapy and Bacteriocins

The rise of multidrug-resistant bacteria driven by antibiotic overuse in livestock has intensified the search for alternative antimicrobial strategies, with phage therapy and bacteriocins emerging as promising solutions (Ferriol-González & Domingo-Calap, 2021). Phage therapy utilizes viruses that specifically infect and lyse bacterial pathogens, providing a precision tool for microbiome modulation while sparing beneficial microbes. Their ability to evolve alongside bacteria, replicate at the site of infection, and act with high specificity makes them attractive candidates for livestock applications (Dec et al., 2020). Experimental trials have demonstrated encouraging results, with phages showing high success rates against infections caused by pathogens such as Staphylococcus, Pseudomonas, Enterobacter, and Enterococcus (Dec et al., 2020). Complementing this, bacteriocins, proteinaceous antimicrobial compounds naturally produced by bacteria, can effectively suppress pathogenic populations and support a healthier microbial balance (Seal et al., 2018). Nisin, for instance, remains the only bacteriocin currently approved for food use, but many others are under investigation for livestock production (Joerger, 2003). Despite their potential, both phage- and bacteriocin-based approaches face practical limitations, including narrow host ranges, the risk of resistance development, regulatory barriers, and high production costs, all of which must be addressed before their widespread adoption in farming systems (Ferriol-González & Domingo-Calap, 2021).

1.5.5 Host-Targeted Approaches

Host-targeted approaches in livestock health are emerging as innovative strategies to reduce reliance on antibiotics while enhancing animal resilience against diseases. These methods encompass diverse research areas, including improved infection detection, regulation of systemic and mucosal inflammation, strengthening of intestinal barrier function, and strategic modulation of immune defenses (Cook et al., 2016). Selective breeding programs are increasingly used to propagate animals with naturally favorable microbiome profiles. At the same time, immunomodulation through vaccines or immune-enhancing feed additives can promote beneficial microbial colonization and limit pathogen burden (Doeschl-Wilson, 2011). Advanced tools, such as genomic analysis, provide insights into complex host–pathogen–vector interactions, with applications in controlling challenges such as livestock tick pathogen dynamics (Jensen et al., 2007). Likewise, RNA interference (RNAi) offers a sustainable and targeted biocontrol option against pests and ectoparasites, with reduced environmental impact compared to chemical interventions (Menezes et al., 2022). Mathematical modeling further aids in deciphering host–pathogen interactions, though livestock models still lag behind human systems in integrating both health outcomes and production impacts (Doeschl-Wilson, 2011). Importantly, these approaches are guided by considerations of societal value, commercial viability, consumer acceptance, and safety, ensuring their responsible application in modern animal production systems (Cook et al., 2016).

1.5.6 Environmental and Management Practices

Sustainable livestock management is central to balancing productivity with environmental conservation in the face of rising global demand for animal products (Rivero & Lee, 2022). Practices such as rotational grazing, integrated crop–livestock systems, precision feeding, and improved waste management through composting or anaerobic digestion enhance resource efficiency, lower greenhouse gas emissions, and reduce ecological footprints while sustaining production (Tully & Ryals, 2017). Over the past decade, advanced methods for assessing environmental impacts and technical innovations have further supported these goals. At the same time, management practices play a critical role in shaping the livestock microbiome, which directly influences animal health and productivity (Tully & Ryals, 2017). Early-life microbial interventions, including colostrum feeding, maternal microbial transfer, and inoculation with beneficial microbes, help establish a stable gut microbiome during the neonatal period (Herrero et al., 2015). Similarly, housing conditions, hygiene, and stress management contribute to microbial resilience. The integration of scientific advances with precision livestock farming tools now enables real-time monitoring of microbiome shifts, allowing timely interventions that align productivity targets with environmental sustainability (Rivero & Lee, 2022). Achieving these outcomes, however, requires addressing economic and sociocultural barriers, supported by collaborative stakeholder engagement, farmer education, and enabling policies (Wathes et al., 2013).

1.6 METAGENOMICS IN DISEASE SURVEILLANCE AND ONE HEALTH PERSPECTIVE

Infectious disease places a significant burden on human health. There are hundreds of pathogens, which vary in their pathogenesis, epidemiology, and therapeutic vulnerabilities. Clinical diagnosis is challenging because patients can present with multiple sets of symptoms that are similar across multiple pathogens. In low- and middle-income countries, overlapping symptoms significantly increase disease burden. These countries also have limited molecular diagnostic tests. It does not negate the fact that other pathogens are absent, even if the test can confirm the presence of a specific pathogen (Oguzie et al., 2023).

Disease surveillance is a continuous effort by the healthcare sector to collect, organize, and evaluate data on disease occurrence in specific regions to inform public health policies. Such surveillance can occur in hospitals and research institutes and helps lay the foundation for infection control programs (Osterholm & Hedberg, 2014). Efficient and routine disease surveillance involves mandatory reporting of specific illnesses by licensed practitioners, clinics, and hospitals in the region. Public health departments are responsible for properly evaluating, reporting, and distributing resources to reduce disease burdens on the healthcare sector. Disease surveillance faces its fair share of challenges, as the tests available at the local level do not process and analyze data quickly enough to detect suspicious trends.

In most cases, they also lack the sensitivity to detect minute changes in the local population's health, leading to missed signs that would otherwise have helped prevent a disease epidemic. The emergence of zoonotic viruses has originated in livestock, and the expansion of livestock farming requires greater surveillance to effectively predict disease trends, thereby improving our preparedness against epidemics and pandemics. Metagenomics can efficiently analyze samples from the environment (such as soil and water) and from animals (such as feces and saliva) to carefully monitor microbial diversity and the transmission of pathogens (Osterholm & Hedberg, 2014).

The most considerable burden of emerging, neglected, and endemic pathogenic diseases falls on the public healthcare system of middle- and low-income countries. Metagenomic approaches are becoming more economical, with high-quality research aimed at making them more affordable for low-income communities, who suffer the most in a multi-pathogen health scare scenario. They can be the future of effective disease surveillance and monitoring of molecular epidemiology. Metagenomic agnostic sequencing enables the identification of recognized and unrecognized pathogens in a single sample. It is much more cost-effective than standard PCR, which detects pathogens only with specific primers. Using these approaches, one can easily identify and detect known and unknown emerging and zoonotic pathogens, and promptly provide this data to lawmakers, researchers, and authorities to establish appropriate containment zones and prevent an epidemic-like situation, thereby minimizing the burden on the public healthcare sector (von Fricken et al., 2023). One can reasonably identify all potential causal agents of an illness by employing metagenomic analysis (Oguzie et al., 2023).

A growing trend in recent years has been that most novel infectious diseases have emerged from wildlife. It is not easy to handle all these emerging diseases while also juggling the burden of preexisting noninfectious diseases, especially amid significant changes in ecosystems, agricultural land use, and urbanization. All these changes have also been seen as responsible for the rapid increase in novel infectious diseases. It led to the beginning of the One Health approach, a collaborative effort to combine animal, human, and environmental health so that existing boundaries are blurred and aid understanding and the fight against emerging diseases (Mackenzie & Jeggo, 2019). At present, disease surveillance prioritizes diseases with adequate historical data and clinical syndromes that the healthcare system is adept at managing. However, the issue lies with the infectious zoonotic pathogens, which can spread exponentially (Aarestrup et al., 2021). The human–livestock mode of transmission is highly conducive to pathogen mutation, facilitating the passage of mobile elements, such as plasmids, via clonal or horizontal gene transfer. All this increases the likelihood of pathogen spread and the severity of the infection, along with resistance to antibiotic medications (von Fricken et al., 2023). It is important to tackle them effectively; adequate early detection strategies are in place to make it easier to catch pathogens before they cause human disease. This situation calls for a quick warning surveillance system at the human, animal, and environmental interface. Metagenomics is one such tool that is very handy in these cases, as it is a comprehensive strategy rather than a detection and sequencing tool that focuses on information from prior diseases. It can serve as the first pillar of preparedness against pandemics. Before the COVID-19 pandemic, genomic pathogen surveillance was slow but was seeping into public health and diagnostics. However, after the pandemic, genomic surveillance growth increased by a factor of 10. Genomic surveillance helped identify and sequence a large number of genomes, resulting in more SARS-CoV-2 genomes available than those of HIV, influenza, and all foodborne bacterial species (Aarestrup et al., 2021). By utilizing metagenomics, researchers could also ascertain the human-to-human pathogen spread and the rate at which the virus was mutating, which helped drive this spread (von Fricken et al., 2023). Once the power of genomic surveillance became clear, it emerged as an essential component of the public health toolbox—driving major investments to build the infrastructure and tools needed to strengthen disease preparedness and to better detect, control, and prevent future outbreaks and pandemics (Aarestrup et al., 2021).

In the past three decades with the emergence of severe acute respiratory disease (SARS) in beginning of 2003 and later the epidemic of highly infectious and pathogenic avian influenza (H5N1 virus), researchers were able to identify a link between human and animal health which was documented in the form of the "Manhattan principles" derived at a meeting of the wildlife conservation society in 2004. It was a turning point in the healthcare sector, highlighting the need for cross-disciplinary approaches to respond to and control emerging and resurgent diseases. Including animal health in essential disease prevention, surveillance, control, and mitigation is important (Mackenzie & Jeggo, 2019). The One Health approach uses metagenomics to combat AMR, as most people live alongside small-holding animals, facilitating the transmission of pathogens from animals to humans. Metagenomics helps identify pathogen hotspots in wildlife that can transmit to humans (Napit et al., 2025).

Metagenomics has become essential for assessing the potential for AMR within gut microbiomes and for uncovering environmental reservoirs that may harbor AMR bacteria and resistance genes (Napit et al., 2025). Metagenomic techniques offer a powerful tool for detecting rare or previously unidentified pathogens that conventional methods may miss. During the COVID-19 pandemic, viral infections in kidney transplant recipients (KTRs) were investigated. A next-generation sequencing (NGS) approach was used to identify uncommon viruses across sample types, including bronchoalveolar lavage fluid (BALF), urine, and blood. This strategy helped identify infections and coinfections in immunocompromised KTRs, providing critical insights that guided appropriate treatment decisions. During the ebola pandemic of 2014–2016, portable sequencing technology was introduced as a pilot project to accelerate detection of the virus and support public health efforts to control its transmission (von Fricken et al., 2023). Similarly, during the COVID-19 pandemic, a modified whole-genome sequencing assay was developed, producing high-quality metagenomic data quickly in a low-budget setting (Arévalo et al., 2022). Metagenomics is also used to track fecal contamination in water sources. This contamination can occur due to the indiscriminate release of wastewater and animal waste into the water bodies. Human and animal fecal contamination poses a considerable risk to public health. All this can result in foodborne diseases, livestock health, and deteriorating environmental health. Viral metagenomics is an emerging tool that can characterize a sample's virome, helping researchers identify which pathogens are associated with specific hosts and have entered the system through contamination (Hollister, 2023). It has been observed that novel pathogens were present in human and animal populations for quite some time before they could be successfully detected. Hence, in such situations, metagenomics will give researchers the upper hand in analyzing samples rapidly and providing an overview of potential transmission and reservoirs of such pathogens (Aarestrup et al., 2021).

The interdependent interactions between animals, humans, and the environment have led to the emergence of infectious diseases whose likelihood is ever-increasing. Epidemics and Pandemics are the new type of war this world is waging and will continue waging if a proper understanding of such complex situations is not achieved (Aarestrup et al., 2021). The increasing occurrence of pandemics and epidemics shows that, when it comes to global security, the system is broken, highlighting a need for advanced skills, refined and efficient tools, and techniques (von Fricken et al., 2023). Metagenomics is critical in such scenarios, as it can effectively initiate collaboration beyond traditional infrastructure and build new networks and disciplines. Understanding this field's great potential and utilizing it to build novel essential partnerships together is important (Aarestrup et al., 2021).

1.7 CONCLUSION

Metagenomics has redefined how livestock microbiomes are characterized and leveraged. Across anatomical niches—from skin and respiratory tract to the rumen and hindgut—amplicon, shotgun, and long-read sequencing reveal taxonomic and strain-level structure. In contrast, multiomics approaches reveal the activities and metabolic outputs that shape feed efficiency, disease resilience, methane reduction, milk production and quality, growth rates, and meat traits. Together, these advances establish microbiome science as a practical lever for improving productivity and animal welfare while reducing environmental impacts.

However, important challenges remain. PCR and database biases, fragmented short-read assemblies, and limited strain resolution can hinder accurate interpretation of key ecological and epidemiological dynamics. Addressing these gaps will require standardized sampling, metadata, and analysis workflows; continued adoption of high-accuracy long-read sequencing; curated, livestock-specific reference resources for genes, genomes, transcripts, proteins, and metabolites; and robust benchmarking across pipelines. By coupling methodological rigor with translational intent, livestock metagenomics can help deliver climate-smart, resource-efficient, and health-promoting production systems. The convergence of accurate sequencing, interoperable data standards, and integrative analytics positions the field to move from associ ation to intervention—turning microbial insight into measurable gains for agriculture and public health.

REFERENCES

Aarestrup, F. M., Bonten, M., & Koopmans, M. (2021). Pandemics–One Health preparedness for the next. *The Lancet Regional Health–Europe*, 9, 100210.

Abdelli, N., Solà-Oriol, D., & Pérez, J. F. (2021). Phytogenic feed additives in poultry: Achievements, prospective and challenges. *Animals (Basel)*, 11(12), 3471.

Abellan-Schneyder, I., Matchado, M. S., Reitmeier, S., Sommer, A., Sewald, Z., Baumbach, J., List, M., & Neuhaus, K. (2021). Primer, pipelines, parameters: Issues in 16S rRNA gene sequencing. *mSphere*, 6(1), e01202-20.

Abeyta, M. A., Horst, E. A., Mayorga, E. J., Goetz, B. M., al-Qaisi, M., McCarthy, C. S., ... & Baumgard, L. H. (2023). Effects of hindgut acidosis on metabolism, inflammation, and production in dairy cows consuming a standard lactation diet. *Journal of Dairy Science*, 106, 1429–1440. doi: 10.3168/jds.2022-22303

Aßhauer, K. P., Wemheuer, B., Daniel, R., & Meinicke, P. (2015). Tax4Fun: predicting functional profiles from metagenomic 16S rRNA data. *Bioinformatics*, 31(17), 2882–2884. https://doi.org/10.1093/bioinformatics/btv287

Amat, S., Subramanian, S., Timsit, E., & Alexander, T. W. (2017). Probiotic bacteria inhibit the bovine respiratory pathogen *Mannheimia haemolytica* serotype 1 in vitro. *Letters in Applied Microbiology*, 64(5), 343–349.

Andersen, H., Kirkegaard, R. H., Karst, S. M., Albertsen, M., & Nielsen, P. H. (2023). Metaproteomics: uncovering microbial community activity in complex environments. *Trends in Microbiology*, 31(2), 134–147. https://doi.org/10.1016/j.tim.2022.08.005

Andrade, B. G. N., de Souza, M. M., Amat, S., Regitano, L. C. A., & de Oliveira, P. S. N. (2022). Editorial: Microbiome genomics for livestock production. *Frontiers in Genetics*, 13, 1000749.

Arévalo, M. T., Karavis, M. A., Katoski, S. E., Harris, J. V., Hill, J. M., Deshpande, S. V., ... & Bernhards, R. C. (2022). A rapid, whole genome sequencing assay for detection and characterization of novel coronavirus (SARS-CoV-2) clinical specimens using nanopore sequencing. *Frontiers in Microbiology*, 13, 910955. https://doi.org/10.3389/fmicb.2022.910955

Ault-Seay, T. B., Moorey, S. E., Mathew, D. J., Schrick, F. N., Pohler, K. G., McLean, K. J., & Myer, P. R. (2023). Importance of the female reproductive tract microbiome and its relationship with the uterine environment for health and productivity in cattle: a review. *Frontiers in Animal Science*, 4, 1111636.

Bach, A., Aris, A., Vidal, M., Fabregas, F., & Terre, M. (2017). Influence of milk processing temperature on growth performance, nitrogen retention, and hindgut's inflammatory status and bacterial populations in a calf model. *Journal of Dairy Research*, 84:355–359. doi: 10.1017/S0022029917000401

Bai, X., & Plastow, G. S. (2022). Breeding for disease resilience: opportunities to manage polymicrobial challenge and improve commercial performance in the pig industry. *CABI Agriculture and Bioscience*, 3(1), 6.

Barathan, M., Ng, S. L., Lokanathan, Y., Ng, M. H., & Law, J. X. (2024). The profound influence of gut microbiome and extracellular vesicles on animal health and disease. *International Journal of Molecular Sciences*, 25(7), 4024.

Bergman, E. N. (1990). Energy contributions of volatile fatty acids from the gastrointestinal tract in various species. *Physiological Reviews*, 70(2), 567–590.

Bishop, S. C., & Woolliams, J. A. (2014). Genomics and disease resistance studies in livestock. *Livestock Science*, 166, 190–198.

Bobeck, E. A. (2020). Nutrition and health: companion animal applications: Functional nutrition in livestock and companion animals to modulate the immune response. *Journal of Animal Science*, 98(3), skaa035.

Bolyen, E., Rideout, J. R., Dillon, M. R., Bokulich, N. A., Abnet, C. C., Al-Ghalith, G. A., ... & Caporaso, J. G. (2019). Reproducible, interactive, scalable, and extensible microbiome data science using QIIME 2. *Nature Biotechnology*, 37(8), 852–857. https://doi.org/10.1038/s41587-019-0209-9

Buetas, E., Jordán-López, M., López-Roldán, A., D'Auria, G., Martínez-Priego, L., De Marco, G., ... & Mira, A. (2024). Full-length 16S rRNA gene sequencing by PacBio improves taxonomic resolution in human microbiome samples. *BMC Genomics*, 25(1), 310.

Callahan, B. J., McMurdie, P. J., Rosen, M. J., Han, A. W., Johnson, A. J., & Holmes, S. P. (2016). DADA2: High-resolution sample inference from Illumina amplicon data. *Nature Methods*, 13(7), 581–583. https://doi.org/10.1038/nmeth.3869

Cani, P. D., Depommier, C., Derrien, M., Everard, A., & de Vos, W. M. (2022). Akkermansia muciniphila: paradigm for next-generation beneficial microorganisms. *Nature Reviews Gastroenterology & Hepatology*, 19(10), 625–637.

Centeno-Martinez, R. E., Glidden, N., Mohan, S., Davidson, J. L., Fernández-Juricic, E., Boerman, J. P., ... & Johnson, T. A. (2022). Identification of bovine respiratory disease through the nasal microbiome. *Animal Microbiome*, 4(1), 15.

Chaucheyras-Durand, F., & Ossa, F. (2014). The rumen microbiome: composition, abundance, diversity, and new investigative tools. *The Professional Animal Scientist*, 30(1), 1–12.

Chaucheyras-Durand, F., Sacy, A., Karges, K., & Apper, E. (2022). Gastro-intestinal microbiota in equines and its role in health and disease: The black box opens. *Microorganisms*, 10(12), 2517.

Chen, J., Harstad, O. M., McAllister, T., Dörsch, P., & Holo, H. (2020). Propionic acid bacteria enhance ruminal feed degradation and reduce methane production in vitro. *Acta Agriculturae Scandinavica, Section A—Animal Science*, 69(3), 169–175.

Chen, X., Yan, F., Liu, T., Zhang, Y., Li, X., Wang, M., ... & Wu, S. (2022). Ruminal microbiota determines the high-fiber utilization of ruminants: evidence from the ruminal microbiota transplant. *Microbiology Spectrum*, 10(4), e00446-22.

Cheng, J., Zhang, X., Zhang, D., Zhang, Y., Li, X., Zhao, Y., ... & Wang, W. (2022). Sheep fecal transplantation affects growth performance in mouse models by altering gut microbiota. *Journal of Animal Science*, 100(11), skac303.

Choi, J., & Park, S. C. (2020). 18S rRNA gene-based metagenomic analysis of protistan communities. *Microorganisms*, 8(12), 1903. https://doi.org/10.3390/microorganisms8121903

Cogen, A. L., Yamasaki, K., Sanchez, K. M., Dorschner, R. A., Lai, Y., MacLeod, D. T., ... & Gallo, R. L. (2010). Selective antimicrobial action is provided by phenol-soluble modulins derived from *Staphylococcus epidermidis*, a normal resident of the skin. *Journal of Investigative Dermatology*, 130(1), 192–200.

Cole, J. R., Wang, Q., Fish, J. A., Chai, B., McGarrell, D. M., Sun, Y., ... & Tiedje, J. M. (2014). Ribosomal Database Project: data and tools for high-throughput rRNA analysis. *Nucleic Acids Research*, 42(D1), D633–D642. https://doi.org/10.1093/nar/gkt1244

Cook, M. E., Bütz, D. E., Yang, M., & Sand, J. M. (2016). Host-targeted approaches to managing animal health: Old problems and new tools. *Domestic Animal Endocrinology*, 56 Suppl, S11–22.

Cooke, C. G., Gibb, Z., Grupen, C. G., Schemann, K., Deshpande, N., & Harnett, J. E. (2024). Effect of probiotics and prebiotics on the composition of the equine fecal and seminal microbiomes and sperm quality: a pilot study. *Journal of Equine Veterinary Science*, 135, 105032. doi: 10.1016/j.jevs.2024.105032

Corbeil, L. B., Blau, K. E. N. T., Prieur, D. J., & Ward, A. C. (1985). Serum susceptibility of *Haemophilus somnus* from bovine clinical cases and carriers. *Journal of Clinical Microbiology*, 22(2), 192–198.

Cristobal-Carballo, O., McCoard, S. A., Cookson, A. L., Ganesh, S., Lowe, K., Laven, R. A., & Muetzel, S. (2021). Effect of methane inhibitors on ruminal microbiota during early life and its relationship with ruminal metabolism and growth in calves. *Frontiers in Microbiology*, 12, 710914.

Dal Pont, G. C., Eyng, C., Bortoluzzi, C., & Kogut, M. H. (2022). Enzymes and gut health in monogastric animals: effects beyond digestibility. In *Gut microbiota, immunity, and health in production animals* (pp. 33–55). Springer International Publishing.

de Rooij, M. M., Smit, L. A., Erbrink, H. J., Hagenaars, T. J., Hoek, G., Ogink, N. W., ... & Wouters, I. M. (2019). Endotoxin and particulate matter emitted by livestock farms and respiratory health effects in neighboring residents. *Environment International*, 132, 105009.

Dec, M., Wernicki, A., & Urban-Chmiel, R. (2020). Efficacy of experimental phage therapies in livestock. *Animal Health Research Reviews*, 21(1), 69–83.

Deepthi, M., Arvind, K., Saxena, R., Pulikkan, J., Sharma, V. K., & Grace, T. (2023). Exploring variation in the fecal microbial communities of Kasaragod Dwarf and Holstein crossbred cattle. *Antonie Van Leeuwenhoek*, 116(1), 53–65.

Denman, S. E., Martinez Fernandez, G., Shinkai, T., Mitsumori, M., & McSweeney, C. S. (2015). Metagenomics analysis of the rumen microbial community following inhibition of methane formation by a halogenated methane analog. *Frontiers in Microbiology*, 9, 222. https://doi.org/10.3389/fmicb.2018.00222

Direkvandi, E., Mohammadabadi, T., & Salem, A. Z. (2020). Effect of microbial feed additives on growth performance, microbial protein synthesis, and rumen microbial population in growing lambs. *Translational Animal Science*, 4(4), txaa203.

Doeschl-Wilson, A. B. (2011). The role of mathematical models of host-pathogen interactions for livestock health and production – a review. *Animal*, 5(6), 895–910.

Douglas, G. M., Maffei, V. J., Zaneveld, J., Yunteng, H., White, J. R., Richter, C., … & Langille, M. G. I. (2020). PICRUSt2 for prediction of metagenome functions. *Nature Biotechnology*, 38(6), 685–688. https://doi.org/10.1038/s41587-020-0548-6

Du, W., Wang, X., Hu, M., Hou, J., Du, Y., Si, W., ... & Xu, Q. (2023). Modulating gastrointestinal microbiota to alleviate diarrhea in calves. *Frontiers in Microbiology*, 14, 1181545.

Edfors, F., Danielsson, F., Hallström, B. M., Uhlén, M., Pontén, F., & Fagerberg, L. (2016). Gene-specific correlation of RNA and protein levels in human cells and tissues. *Molecular Systems Biology*, 12(10), 883. https://doi.org/10.15252/msb.20167144

Edwards, J. E., Kim, E. J., Davies, D. R., Hanafy, R., & Kingston-Smith, A. H. (2023). Ruminant salivary microbes: passenger or player in the rumen? *Microorganisms*, 11(10), 2390. https://doi.org/10.3390/microorganisms11102390

El-Khoury, G., Hajjar, C., Geitani, R., Karam Sarkis, D., Butel, M. J., Barbut, F., ... & Kapel, N. (2025). Gut microbiota and viral respiratory infections: microbial alterations, immune modulation, and impact on disease severity. A narrative review. *Frontiers in Microbiology*, 16, 1605143.

Ferriol-González, C., & Domingo-Calap, P. (2021). Phage therapy in livestock and companion animals. *Antibiotics (Basel)*, 10(5), 559.

Fonty, G., Gouet, P., Jouany, J. P., & Senaud, J. (1987). Establishment of the microflora and anaerobic fungi in the rumen of lambs. *Journal of General Microbiology*, 133(7), 1835–1843.

Forcina, G., Pérez-Pardal, L., Carvalheira, J., & Beja-Pereira, A. (2022). Gut microbiome studies in livestock: Achievements, challenges, and perspectives. *Animals (Basel)*, 12(23), 3375.

Gao, F. Z., He, L. Y., Bai, H., He, L. X., Zhang, M., Chen, Z. Y., ... & Ying, G. G. (2023). Airborne bacterial community and antibiotic resistome in the swine farming environment: metagenomic insights into livestock relevance, pathogen hosts and public risks. *Environment International*, 172, 107751.

Gao, R., Li, Z., Ruan, J., Zhang, G., Chen, K., & Xuan, Z. (2016). Long-read sequencing for genomic structural variation detection. *Genomics, Proteomics & Bioinformatics*, 14(1), 4–16. https://doi.org/10.1016/j.gpb.2016.01.002

García-Martínez, J., Acinas, S. G., Massana, R., & Rodríguez-Valera, F. (1999). Comparison of molecular methods for the study of natural communities of marine bacterioplankton. *FEMS Microbiology Ecology*, 28(2), 131–139. https://doi.org/10.1111/j.1574-6941.1999.tb00568.x

Gardiner, G. E., Metzler-Zebeli, B. U., & Lawlor, P. G. (2020). Impact of intestinal microbiota on growth and feed efficiency in pigs: a review. *Microorganisms*, 8(12), 1886.

Grice, E. A., & Segre, J. A. (2011). The skin microbiome. *Nature Reviews Microbiology*, 9(4), 244–253. doi: 10.1038/nrmicro2537

Hagen, L. H., Frank, J. A., Zamanzadeh, M., Eijsink, V. G. H., Pope, P. B., & Arntzen, M. Ø. (2021). Quantitative metaproteomics highlights the metabolic contributions of uncultured phylotypes in mixed microbial communities. *mSystems*, 6(2), e01300-20. https://doi.org/10.1128/mSystems.01300-20

Han, Q., Huang, X., Yan, F., Yin, J., & Xiao, Y. (2022). The role of gut microbiota in the skeletal muscle development and fat deposition in pigs. *Antibiotics*, 11(6), 793.

Handelsman, J. (2004). Metagenomics: application of genomics to uncultured microorganisms. *Microbiology and Molecular Biology Reviews*, 68(4), 669–685. https://doi.org/10.1128/MMBR.68.4.669-685.2004

Hassan, F., Tang, Z., Ebeid, H. M., Li, M., Peng, K., Liang, X., & Yang, C. (2021). Consequences of herbal mixture supplementation on milk performance, ruminal fermentation, and bacterial diversity in water buffaloes. *PeerJ*, 9, e11241.

Henderson, G., Cox, F., Ganesh, S., Jonker, A., Young, W., Abecia, L., … & Janssen, P. H. (2015). Rumen microbial community composition varies with diet and host, but a core microbiome is found across a wide geographical range. *Scientific Reports*, 5, 14567. https://doi.org/10.1038/srep14567

Herrero, M., Wirsenius, S., Henderson, B., Rigolot, C., Thornton, P., Havlík, P., De Boer, I., & Gerber, P. (2015). Livestock and the environment: What have we learned in the past decade? *Annual Review of Environment and Resources*, 40, 177–202.

Hess, M., Paul, S. S., Puniya, A. K., Van der Giezen, M., Shaw, C., Edwards, J. E., & Fliegerová, K. (2020). Anaerobic fungi: past, present, and future. *Frontiers in Microbiology*, 11, 584893.

Holman, D. B., Gzyl, K. E., Scott, H., Cara Service, Prieto, N., & López-Campos, Ó. (2024). Associations between the rumen microbiota and carcass merit and meat quality in beef cattle. *Applied Microbiology and Biotechnology*, 108(1), 287.

Holman, D. B., McAllister, T. A., Topp, E., Wright, A. D. G., & Alexander, T. W. (2015). The nasopharyngeal microbiota of feedlot cattle that develop bovine respiratory disease. *Veterinary Microbiology*, 180(1–2), 90–95.

Hu, G., Do, D. N., Gray, J., & Miar, Y. (2020). Selection for favorable health traits: A potential approach to cope with diseases in farm animals. *Animals*, 10(9), 1717.

The Human Microbiome Project Consortium. (2012). Structure, function and diversity of the healthy human microbiome. *Nature*, 486, 207–214.

Iliev, I. D., & Underhill, D. M. (2013). Striking a balance: Fungal commensalism versus pathogenesis. *Current Opinion in Microbiology*, 16, 366–373. doi: 10.1016/j.mib.2013.05.004

Iljazovic, A., Roy, U., Gálvez, E. J., Lesker, T. R., Zhao, B., Gronow, A., … & Strowig, T. (2021). Perturbation of the gut microbiome by *Prevotella* spp. enhances host susceptibility to mucosal inflammation. *Mucosal Immunology*, 14(1), 113–124.

Iqbal, Z., Ahmed, S., Tabassum, N., Bhattacharya, R., & Bose, D. (2021). Role of probiotics in prevention and treatment of enteric infections: A comprehensive review. *3 Biotech*, 11(5), 242.

Jami, E., Israel, A., Kotser, A., & Mizrahi, I. (2013). Exploring the bovine rumen bacterial community from birth to adulthood. *The ISME Journal*, 7(6), 1069–1079.

Jin, Y., Gong, T., Lu, X., Wang, Y., & Cheng, Y. (2025). Effects of gut microbiota and metabolites on the host defense peptide expression. *Applied Microbiology and Biotechnology*, 109(1), 10.

Joerger, R. D. (2003). Alternatives to antibiotics: Bacteriocins, antimicrobial peptides and bacteriophages. *Poultry Science*, 82(4), 640–647.

Kaur, H., Kaur, G., Gupta, T., Mittal, D., & Ali, S. A. (2023). Integrating omics technologies for a comprehensive understanding of the microbiome and its impact on cattle production. *Biology*, 12(9), 1200.

Kelly, W. J., Leahy, S. C., Kamke, J., Soni, P., Koike, S., Mackie, R., … & Attwood, G.T. (2019). Occurrence and expression of genes encoding methyl-compound production in rumen bacteria. *Animal Microbiome*, 1(1), 15.

Kim, S. H., Ramos, S. C., Valencia, R. A., Cho, Y. I., & Lee, S. S. (2022). Heat stress: effects on rumen microbes and host physiology, and strategies to alleviate the negative impacts on lactating dairy cows. *Frontiers in Microbiology*, 13, 804562.

Knap, P. W., & Doeschl-Wilson, A. (2020). Why breed disease-resilient livestock, and how? *Genetics Selection Evolution*, 52(1), 60.

Kou, X., Ma, Q., Liu, Y., Khan, M. Z., Wu, B., Chen, W., Liu, X., Wang, C., & Li, Y. (2024). Exploring the effect of gastrointestinal Prevotella on growth performance traits in livestock animals. *Animals*, 14(13), 1965.

Kouhounde, S., Adéoti, K., Mounir, M., Giusti, A., Refinetti, P., Otu, A., Effa, E., Ebenso, B., Adetimirin, V. O., Barceló, J. M., Thiare, O., Rabetafika, H. N., & Razafindralambo, H. L. (2022). Applications of probiotic-based multi-components to human, animal and ecosystem health: Concepts, methodologies, and action mechanisms. *Microorganisms*, 10(9), 1700.

Kumar, S. T., Zhang, Y., Zhang, Q., Azeem, R. M., Jing, Z., Pan, L., … & Zhang, S. M. (2025). Breed-specific differences of gut microbiota and metabolomic insights into fat deposition and meat quality in Chinese Songliao Black Pig and Large White× Landrace pig breeds. *BMC Microbiology*, 25(1), 334.

Ku-Vera, J. C., Jiménez-Ocampo, R., Valencia-Salazar, S. S., Montoya-Flores, M. D., Molina-Botero, I. C., Arango, J., … & Solorio-Sánchez, F. J. (2020). Role of secondary plant metabolites on enteric methane mitigation in ruminants. *Frontiers in Veterinary Science*, 7, 584.

Lang D, Zhang S, Ren P, Liang F, Sun Z, Meng G, Tan Y, Li X, Lai Q, Han L, Wang D, Hu F, Wang W, & Liu S. Comparison of the two up-to-date sequencing technologies for genome assembly: HiFi reads of Pacific Biosciences Sequel II system and ultralong reads of Oxford Nanopore. *Gigascience*. 2020 Dec 15;9(12):giaa123.

Lee, W. J., & Hase, K. (2014). Gut microbiota–generated metabolites in animal health and disease. *Nature Chemical Biology*, 10(6), 416–424. https://doi.org/10.1038/nchembio.1535.

Levit, R., Cortes-Perez, N. G., de Moreno de Leblanc, A., Loiseau, J., Aucouturier, A., Langella, P., LeBlanc, J. G., & Bermúdez-Humarán, L. G. (2022). Use of genetically modified lactic acid bacteria and bifidobacteria as live delivery vectors for human and animal health. *Gut Microbes*, 14(1), 2110821.

Lin, L., Ma, H., Zhang, J., Yang, H., Zhang, J., et al. (2024). Lignocellulolytic microbiomes orchestrating degradation cascades in the rumen of dairy cattle and their diet-influenced key degradation phases. *Animal Advances*, 1, e002.

Liu, R., Shen, Y., Ma, H., Li, Y., Lambo, M.T., Dai, B., ... & Zhang, Y. (2023). Silibinin reduces in vitro methane production by regulating the rumen microbiome and metabolites. *Frontiers in Microbiology*, 14, 1225643.

Liu, X., Zheng, J., Ding, J., Wu, J., Zuo, F., & Zhang, G. (2024). When livestock genomes meet third-generation sequencing technology: from opportunities to applications. *Genes*, 15(2), 245. https://doi.org/10.3390/genes15020245

Lopes, S. A., de Lana Ferreira, M. F., Costa e Silva, L. F., et al. (2022). Evaluation of nonlinear models to predict milk yield and composition of beef cows: A meta-analysis. *Animal Feed Science and Technology*, 294, 115455.

Mackenzie, J. S., & Jeggo, M. (2019). The One Health approach—Why is it so important? *Tropical Medicine and Infectious Disease*, 4(2), 88. https://doi.org/10.3390/tropicalmed4020088

Mala, L., Lalouckova, K., & Skrivanova, E. (2021). Bacterial skin infections in livestock and plant-based alternatives to their antibiotic treatment. *Animals: An Open Access Journal from MDPI*, 11(8), 2473. https://doi.org/10.3390/ani11082473

Malmuthuge, N., Li, M., Goonewardene, L. A., Oba, M., & Guan, L. L. (2013). Effect of calf starter feeding on gut microbial diversity and expression of genes involved in host immune responses and tight junctions in dairy calves during weaning transition. *Journal of Dairy Science*, 96(5), 3189–3200.

Mann, E., Wetzels, S. U., Wagner, M., Zebeli, Q., & Schmitz-Esser, S. (2018). Metatranscriptome sequencing reveals insights into active rumen microbial community functions. *Frontiers in Microbiology*, 9, 92. https://doi.org/10.3389/fmicb.2018.00092

Marcos, C. N., Bach, A., Gutiérrez-Rivas, M., & González-Recio, O. (2024). The oral microbiome as a proxy for feed intake in dairy cattle. *Journal of Dairy Science*, 107(8), 5881–5896.

Martinez Boggio, G., Monteiro, H. F., Lima, F. S., Figueiredo, C. C., Bisinotto, R. S., Santos, J. E. P., Mion, B., Schenkel, F. S., Ribeiro, E. S., Weigel, K. A., & Peñagaricano, F. (2024). Host and rumen microbiome contributions to feed efficiency traits in Holstein cows. *Journal of Dairy Science*, 107(5), 3090–3103.

Martínez-Álvaro, M., Mattock, J., González-Recio, Ó., Saborío-Montero, A., Weng, Z., Lima, J., Duthie, C. A., Dewhurst, R., Cleveland, M. A., Watson, M., & Roehe, R. (2024). Including microbiome information in a multi-trait genomic evaluation: A case study on longitudinal growth performance in beef cattle. *Genetics Selection Evolution*, 56(1), 19.

Matthews, C., Crispie, F., Lewis, E., Reid, M., O'Toole, P. W., & Cotter, P. D. (2019). The rumen microbiome: a crucial consideration when optimising milk and meat production and nitrogen utilisation efficiency. *Gut Microbes*, 10(2), 115–132. https://doi.org/10.1080/19490976.2018.1505176

McCormack, U. M., Curião, T., Metzler-Zebeli, B. U., Wilkinson, T., Magowan, E., Berry, D. P. ... & O'Shea, C. J. (2019). Porcine feed e ciency-associated intestinal microbiota and physiological traits: finding consistent cross-locational biomarkers for residual feed intake. *mSystems*, 4, e00324-18.

McDonald, D., Price, M. N., Goodrich, J., Kottman, R., Amatur, C., Baertsch, R., ... & Knight, R. (2012). An improved Greengenes taxonomy with explicit ranks for ecological and evolutionary analyses of bacteria and archaea. *The ISME Journal*, 6(3), 610–618. https://doi.org/10.1038/ismej.2011.139

McKinney, C. A., Oliveira, B. C., Bedenice, D., Paradis, M. R., Mazan, M., Sage, S., ... & Widmer, G. (2020). The fecal microbiota of healthy donor horses and geriatric recipients undergoing fecal microbial transplantation for the treatment of diarrhea. *PloS One*, 15(3), e0230148.

McSweeney, C., & Mackie, R. (2012). Micro-organisms and ruminant digestion: state of knowledge, trends and future prospects. Paper No. 61. Commission of Genetic Resources for Food and Agriculture.

Menchaca, A., dos Santos-Neto, P. C., Mulet, A. P., & Crispo, M. (2020). CRISPR in livestock: From editing to printing. *Theriogenology*, 150, 247–254.

Menezes, P. S., Yan, Y., Yang, Y., Mitter, N., Mahony, T. J., & Mody, K. T. (2022). RNAi-based biocontrol of pests to improve the productivity and welfare of livestock production. *Applied Biosciences*, 1(3), 229–243.

Montagna, W. (1962). The structure and function of skin. NIH HMP Working Group, 2009.

Monteiro, H. F., Figueiredo, C. C., Mion, B., Santos, J. E. P., Bisinotto, R. S., Peñagaricano, F., ... & Lima, F. S. (2024). An artificial intelligence approach of feature engineering and ensemble methods depicts the rumen microbiome contribution to feed efficiency in dairy cows. *Animal Microbiome*, 6(1), 5.

Mueller, U. G., & Sachs, J. L. (2015). Engineering microbiomes to improve plant and animal health. *Trends in Microbiology*, 23(10), 606–617.

Munk, P., Knudsen, B. E., Lukjancenko, O., Duarte, A. S. R., Van Gompel, L., Luiken, R. E. C., ... & Aarestrup, F. M. (2018). Abundance and diversity of the faecal resistome in slaughter pigs and broilers in nine European countries. *Nature Microbiology*, 3(8), 898–908. https://doi.org/10.1038/s41564-018-0192-9

Napit, R., Gurung, A., Poudel, A., Chaudhary, A., Manandhar, P., Sharma, A. N., ... & Karmacharya, D. B. (2025). Metagenomic analysis of human, animal, and environmental samples identifies potential emerging pathogens, profiles antibiotic resistance genes, and reveals horizontal gene transfer dynamics. *Scientific Reports*, 15(1), 1–15. https://doi.org/10.1038/s41598-025-90777-8

Negash, A. (2022). Gut microbiota ecology role in animal nutrition and health performance. *Journal of Clinical Microbiology and Antimicrobials*, 6, 1–14.

NIH HMP Working Group, Peterson, J., Garges, S., et al. (2009). The NIH Human Microbiome Project. *Genome Research*, 19(12), 2317–2323.

Nilsson, R. H., Larsson, K. H., Taylor, A. F. S., Bengtsson-Palme, J., Jeppesen, T. S., Schigel, D., ... & Abarenkov, K. (2019). The UNITE database for molecular identification of fungi: handling dark taxa and parallel taxonomic classifications. *Nucleic Acids Research*, 47(D1), D259–D264. https://doi.org/10.1093/nar/gky1022

Oakley, B. B., Lillehoj, H. S., Kogut, M. H., Kim, W. K., Maurer, J. J., Pedroso, A., ... & Lee, M. D. (2014). The chicken gastrointestinal microbiome. *FEMS Microbiology Letters*, 360(2), 100–112. https://doi.org/10.1111/1574-6968.12608

Ogunade, I. M., McCoun, M., Idowu, M. D., & Peters, S. O. (2020). Comparative effects of two multispecies direct-fed microbial products on energy status, nutrient digestibility, and ruminal fermentation, bacterial community, and metabolome of beef steers. *Journal of Animal Science*, 98(9), skaa201.

Oguzie, J. U., Petros, B. A., Oluniyi, P. E., Mehta, S. B., Eromon, P. E., Nair, P., ... & Happi, C. T. (2023). Metagenomic surveillance uncovers diverse and novel viral taxa in febrile patients from Nigeria. *Nature Communications*, 14(1), 1–14. https://doi.org/10.1038/s41467-023-40247-4

O'Hara, E., Neves, A. L., Song, Y., & Guan, L. L. (2020). The role of the gut microbiome in cattle production and health: driver or passenger? *Annual Review of Animal Biosciences*, 8(1), 199–220.

Okolie, M. C., Edo, G. I., Ainyanbhor, I. E., Jikah, A. N., Akpoghelie, P. O., Yousif, E., ... & Umar, H. (2025). Gut microbiota and immunity in health and diseases: a review. *Proceedings of the Indian National Science Academy*, 91(2), 397–414.

Olchowy, T. W. J., Soust, M., & Alawneh, J. (2019). The effect of a commercial probiotic product on the milk quality of dairy cows. *Journal of Dairy Science*, 102(3), 2188–2195.

Ornelas, A., Dowdell, A. S., Lee, J. S., & Colgan, S. P. (2022). Microbial metabolite regulation of epithelial cell-cell interactions and barrier function. *Cells*, 11(6), 944.

Osterholm, M. T., & Hedberg, C. W. (2014). Epidemiologic principles. In *Mandell, Douglas, and Bennett's principles and practice of infectious diseases* (pp. 146–157). Elsevier.

Pallen, M. J. (2014). Microbial bioinformatics and metagenomics. In P. J. Turnbaugh (Ed.), *Methods in microbiology* (Vol. 41, pp. 1–22). Academic Press. https://doi.org/10.1016/bs.mim.2014.06.002

Patra, A., Park, T., Kim, M., & Yu, Z. (2017). Rumen methanogens and mitigation of methane emission by anti-methanogenic compounds and substances. *Journal of Animal Science and Biotechnology*, 8(1), 13.

Peixoto, R. S., Voolstra, C. R., Sweet, M., Duarte, C. M., Carvalho, S., Villela, H., Lunshof, J. E., Gram, L., Woodhams, D. C., Walter, J., Roik, A., Hentschel, U., Thurber, R. V., Daisley, B., Ushijima, B., Daffonchio, D., Costa, R., Keller-Costa, T., Bowman, J. S., Rosado, A. S., Reid, G., Mason, C. E., Walke, J. B., Thomas, T., & Berg, G. (2022). Harnessing the microbiome to prevent global biodiversity loss. *Nature Microbiology*, 7(11), 1726–1735.

Perez, H. G., Stevenson, C. K., Lourenco, J. M., & Callaway, T. R. (2024). Understanding rumen microbiology: an overview. *Encyclopedia*, 4(1), 148–157. https://doi.org/10.3390/encyclopedia4010013

Quast, C., Pruesse, E., Yilmaz, P., Gerken, J., Schweer, T., Yarza, P., ... & Glöckner, F. O. (2013). The SILVA ribosomal RNA gene database project: improved data processing and web-based tools. *Nucleic Acids Research*, 41(D1), D590–D596. https://doi.org/10.1093/nar/gks1219

Rabee, A. E., Abou-souliman, I., Yousif, A. I., Lamara, M., El-Sherbieny, M.A., Elwakeel, E. A. ... & Sallam, A. M. (2025). Variations in rumen microbiota and host genome impacted feed efficiency in goat breeds. Frontiers in Microbiology, 16, 1492742. doi: 10.3389/fmicb.2025.1492742

Rawal, S., Kaur, H., Bhathan, S., Mittal, D., Kaur, G., & Ali, S. A. (2024). Ruminant gut microbiota: interplay, implications, and innovations for sustainable livestock production. In *Sustainable Agriculture Reviews: Animal Biotechnology for Livestock Production 4* (pp. 205–228). Springer Nature Switzerland.

Reuben, R. C., Sarkar, S. L., Roy, P. C., Anwar, A., Hossain, M. A., & Jahid, I. K. (2021). Prebiotics, probiotics and postbiotics for sustainable poultry production. *World's Poultry Science Journal*, 77(4), 825–882.

Rivero, M. J., & Lee, M. R. F. (2022). A perspective on animal welfare of grazing ruminants and its relationship with sustainability. *Animal Production Science*, 62, 1739.

Russell, J. B. (2002). *Rumen microbiology and its role in ruminant nutrition*. Department of Microbiology, Cornell University.

Russell, J. B., & Rychlik, J. L., (2001). Factors that alter rumen microbial ecology. *Science*, 292 (5519), 1119–1122.

Santiago-Rodriguez, T. M., & Hollister, E. B. (2023). Viral metagenomics as a tool to track sources of fecal contamination: A One Health approach. *Viruses*, 15(1), 236. https://doi.org/10.3390/v15010236

Schoch, C. L., Seifert, K. A., Huhndorf, S., Robert, V., Spouge, J. L., Levesque, C. A., ... & Schindel, D. (2012). Nuclear ribosomal internal transcribed spacer (ITS) region as a universal DNA barcode marker for Fungi. *PNAS*, 109(16), 6241–6246. https://doi.org/10.1073/pnas.1117018109

Seal, B. S., Drider, D., Oakley, B. B., Brüssow, H., Bikard, D., Rich, J. O., Miller, S., Devillard, E., Kwan, J., Bertin, G., Reeves, S., Swift, S. M., Raicek, M., & Gay, C. G. (2018). Microbial-derived products as potential new antimicrobials. *Veterinary Research*, 49(1), 66.

Sedghi, L., DiMassa, V., Harrington, A., Lynch, S. V., & Kapila, Y. L. (2021). The oral microbiome: Role of key organisms and complex networks in oral health and disease. *Periodontology 2000*, 87(1), 107–131. https://doi.org/10.1111/prd.12393

Shakya, M., Lo, C. C., & Chain, P. S. (2019). Advances and challenges in metatranscriptomic analysis. *Frontiers in Genetics*, 10, 904. https://doi.org/10.3389/fgene.2019.00904

Shang, Y., Kumar, S., Oakley, B., & Kim, W. K. (2018). Chicken gut microbiota: importance and detection technology. *Frontiers in Veterinary Science*, 5, 254. https://doi.org/10.3389/fvets.2018.00254

Shin, N. R., Whon, T. W., & Bae, J. W. (2015). Proteobacteria: microbial signature of dysbiosis in gut microbiota. *Trends in Biotechnology*, 33(9), 496–503.

Slifierz, M. J., Friendship, R. M., & Weese, J. S. (2015). Longitudinal study of the early-life fecal and nasal microbiotas of pigs. *Applied and Environmental Microbiology*, 81(9), 3005–3016.

Steijger, T., Abril, J. F., Engström, P. G., Kokocinski, F., Akerman, M., Alioto, T., ... & Guigó, R. (2013). Assessment of transcript reconstruction methods for RNA-seq. *Nature Methods*, 10(12), 1177–1184. https://doi.org/10.1038/nmeth.2714

Stergiadis, S., Cabeza-Luna, I., Mora-Ortiz, M., Stewart, R. D., Dewhurst, R. J., Humphries, D. J., ... & Auffret, M. D. (2021). Unravelling the role of rumen microbial communities, genes, and activities on milk fatty acid profile using a combination of omics approaches. *Frontiers in Microbiology*, 11, 590441.

Stewart, R. D., Auffret, M. D., Warr, A., Walker, A. W., Roehe, R., Watson, M., ... & Snelling, J. M. (2018). Assembly of 913 microbial genomes from metagenomic sequencing of the cow rumen. *Nature Communications*, 9(1), 870. https://doi.org/10.1038/s41467-018-03317-6

Sun, K., Liu, H., Fan, H., Liu, T., & Zheng, C. (2021). Research progress on the application of feed additives in ruminal methane emission reduction: a review. *PeerJ*, 9, e11151.

Tardiolo, G., La Fauci, D., Riggio, V., Daghio, M., Di Salvo, E., Zumbo, A., & Sutera, A. M. (2025). Gut microbiota of ruminants and monogastric livestock: an overview. *Animals*, 15(5), 758. https://doi.org/10.3390/ani15050758

Timsit, E., Workentine, M., van der Meer, F., & Alexander, T. (2018). Distinct bacterial metacommunities inhabit the upper and lower respiratory tracts of healthy feedlot cattle and those diagnosed with bronchopneumonia. *Veterinary Microbiology*, 221, 105–113.

Tugnoli, B., Giovagnoni, G., Piva, A., & Grilli, E. (2020). From acidifiers to intestinal health enhancers: how organic acids can improve growth efficiency of pigs. *Animals*, 10, 134. doi: 10.3390/ani10010134

Tully, K., & Ryals, R. (2017). Agroecology and sustainable food systems nutrient cycling in agroecosystems: Balancing food and environmental objectives. *Agroecology and Sustainable Food Systems*, 41(7), 761–798.

Vinderola, G., Sanders, M. E., Salminen, S., & Szajewska, H. (2022). Postbiotics: The concept and their use in healthy populations. *Frontiers in Nutrition*, 9, 1002213.

von Fricken ME, Melendrez MC, Linton Y-M and Takhampunya R (2023). Editorial: Metagenomics for epidemiological surveillance in One Health. *Front. Microbiol.* 14:1191946. doi: 10.3389/fmicb.2023.1191946

Walkden-Brown, S.W., & Kahn, L. P. (2002). Nutritional modulation of resistance and resilience to gastrointestinal nematode infection – a review. *Asian-Australasian Journal of Animal Sciences*, 15(6), 912–924.

Wang, D., Chen, L., Tang, G., Yu, J., Chen, J., Li, Z., ... & Yao, J. (2023). Multi-omics revealed the long-term effect of ruminal keystone bacteria and the microbial metabolome on lactation performance in adult dairy goats. *Microbiome*, 11(1), 215.

Wang, Y. L., Wang, W. K., Wu, Q. C., & Yang, H. J. (2022). The release and catabolism of ferulic acid in plant cell wall by rumen microbes: a review. *Animal Nutrition*, 9, 335–344. doi: 10.1016/j.aninu.2022.02.003

Wani, A. R., Rather, M. A., & Jan, N. (2023). Advances in microbiome research: implications for animal health and production. *Frontiers in Veterinary Science*, 10, 1123456.

Watson, M., Laing, C., Johnston, C., Turner, A., Walker, A. W., Roehe, R., ... & Snelling, J. M. (2021). A comparison of Nanopore and PacBio sequencing for microbial metagenomics. *BMC Genomics*, 22, 780.

Wegl, G., Grabner, N., Köstelbauer, A., Klose, V., & Ghanbari, M. (2021). Toward best practice in livestock microbiota research: a comprehensive comparison of sample storage and DNA extraction strategies. *Frontiers in Microbiology*, 12, 627539. https://doi.org/10.3389/fmicb.2021.627539

Wen, C., Wang, Q., Gu, S., Jin, J., & Yang, N. (2024). Emerging perspectives in the gut–muscle axis: The gut microbiota and its metabolites as important modulators of meat quality. *Microbial Biotechnology*, 17(1), e14361.

Xie, R., Sun, Y., Wu, J., Bi, Y., Mao, S., Cheng, Y., ... & Zhu, W. (2022). Metaproteomics reveals the functional roles of rumen microbiota in dairy cows fed different diets. *Frontiers in Microbiology*, 13, 862795. https://doi.org/10.1186/s42523-022-00170-3

Xue, M. Y., Sun, H. Z., Wu, X. H., Liu, J. X., & Guan, L. L. (2020). Multi-omics reveals that the rumen microbiome and its metabolome together with the host metabolome contribute to individualized dairy cow performance. *Microbiome*, 8(1), 64.

Yang, H., Huang, X., Fang, S., He, M., Zhao, Y., Wu, Z., ... & Huang, L. (2017). Unravelling the fecal microbiota and metagenomic functional capacity associated with feed e ciency in pigs. *Frontiers in Microbiology*, 8, 1555.

Yang, J., Fan, Y., Jin, R., Peng, Y., Chai, J., Wei, X., ... & Li, Y. (2024). Exploring the intestinal microbial community of lantang pigs through metagenome-assembled genomes and carbohydrate degradation Genes. *Fermentation*, 10(4), 207.

Zhang, M., Liu, Z., Wu, K., Zhang, C., Fu, T., Sun, Y., ... & Han, L. (2024). The ruminal microbiome alterations associated with diet-induced milk fat depression and milk Fat globule size reduction in dairy goats. *Animals*, 14(17), 2614.

Zhang, X., Han, L., Gui, L., Raza, S. H. A., Hou, S., Yang, B., ... & Ibrahim, S. F. (2023). Metabolome and microbiome analysis revealed the effect mechanism of different feeding modes on the meat quality of Black Tibetan sheep. *Frontiers in Microbiology*, 13, 1076675.

Zhang, Z., Yang, Q., Liu, H., Jin, J., Yang, Y., Zhu, X., ... & Xie, S. (2025). Potential functions of the gut microbiome and modulation strategies for improving aquatic animal growth. *Reviews in Aquaculture*, 17(1), e12959.

Zhao, H., Bai, S., Tan, J., Liu, M., Zhao, Y., & Jiang, L. (2024). Can meta-omics revolutionize our understanding of rumen methane emissions? *Animal Nutriomics*, 1, e14.

Zhao, X., Ali, S., Hassan, M. F., Bashir, M. A., Ni, X., Lv, C., ... & Quan, G. (2023). Effects of graded levels of dietary protein supplementation on milk yield, body weight gain, blood biochemical parameters, and gut microbiota in lactating ewes. *Frontiers in Veterinary Science*, 10, 1223450.

CHAPTER TWO

Microbiomes in Domesticated Animals

ECOLOGY, HEALTH, PRODUCTIVITY, AND ONE HEALTH

Veysel Bay and Berrin Özdil

2.1 INTRODUCTION: THE HOLOBIONT AND THE SCOPE OF MICROBIAL PARTNERSHIP

The scientific understanding of domesticated animals has increasingly shifted, recognizing that the microbial communities residing within and upon these hosts are not passive inhabitants but dynamic partners that profoundly influence host biology. These microbial communities, collectively known as microbiomes, form an invisible yet vital layer of life that links animals to their surrounding environments. The scope of this partnership is vast, covering essential physiological functions such as digestion, immunity, metabolism, and even behavior. The health, welfare, and overall productivity of major livestock species, including cattle, sheep, pigs, poultry, and fish, depend crucially on the successful maintenance of these microbial systems. Consequently, understanding animal microbiomes is central to achieving the sustainability of global food systems (Aruwa et al., 2021; S. Chen et al., 2021; O'Hara et al., 2020; Tardiolo et al., 2025).

The composition and functional activity of these microbial communities, which include bacteria, archaea, fungi, protozoa, and viruses, are highly sensitive to various external and internal factors. Diet composition, genetics, housing conditions, the age of the animal, and exposure to antimicrobial agents are all critical factors that shape the structure and output of these microbiomes. In turn, the microbiome acts as a key mediator between external environmental conditions and the internal physiological responses of the host (Bay et al., 2023; Flandroy et al., 2018; Gupta & Gaur, 2025; Hills et al., 2019; Woodhams et al., 2020). Specific measurable outcomes influenced by the microbiome include feed efficiency, growth rate, resilience to disease, reproductive success, and, notably, methane emissions in ruminants. For instance, the rumen microbiome is known to modulate both milk yield and methane production in cattle, while the cecal microbiome in chickens plays a central role in both nutrient utilization and gut health (Badhan et al., 2025; Wallace et al., 2019).

From an ecological viewpoint, the host animal and its associated microbiome are best viewed as a holobiont, which is a functional unit where the host and its microbes coevolve (Rosenberg & Zilber-Rosenberg, 2016). The balance maintained within this holobiont ultimately determines the overall health and performance metrics of the animal. The holobiont concept moves the scientific perspective beyond viewing microbes as isolated entities and instead frames them as integral

DOI: 10.1201/9781003638568-2

components of animal physiology (Esser et al., 2018; H. Zhou et al., 2022). The scope of this field centers particularly on domesticated species vital for food and fiber production, such as ruminants (cattle, sheep, goats), monogastric species (pigs, poultry), and aquaculture species (fish, shrimp) (Kaur et al., 2023; Lorgen-Ritchie et al., 2023; Tardiolo et al., 2025). Companion animals (dogs, cats, horses) are also considered, especially when the context involves shared human–animal microbiomes (Branck et al., 2024). In all these production systems, the microbiome functions dually as a driver of productivity and a sentinel of welfare.

Crucially, the impact of animal microbiomes extends significantly beyond the individual animal, directly engaging in the One Health framework. This framework recognizes the fundamental linkage between the health of humans, the health of animals, and the health of the environment they share. Microbial systems connect these sectors through the shared circulation of microorganisms and genes. Changes or shifts in livestock microbiota can alter the spread of antimicrobial resistance (AMR) in farm environments and affect the microbial ecology of soil and water through processes like manure application. This interconnectedness highlights that the study of animal microbiomes is not merely an agricultural pursuit but also a public health necessity. AMR genes, for example, can circulate across animal and human microbiomes via the food chain and the environment. Furthermore, zoonotic pathogens like Salmonella or Campylobacter utilize the ecological overlaps that exist between human and animal microbial reservoirs (Koutsoumanis et al., 2021; Lingxin et al., 2023; Sessitsch et al., 2023; Trinh et al., 2018).

Despite the rapid expansion of this research area, the field often remains descriptive. Much research still emphasizes identifying "who is there" rather than investigating "what they are doing." Functional studies, which aim to link microbial metabolism and gene expression directly to host physiology and productivity, remain relatively rare. Furthermore, the translation of microbiome science into scalable interventions for commercial production systems is still limited. This lack of translational application persists despite compelling evidence suggesting that microbial modulation can successfully improve animal health and performance while simultaneously reducing environmental footprints (Kaur et al., 2023; Warne & Kirschman, 2025; Ye et al., 2025).

Microbial communities themselves are dynamic assemblages rather than static collections, shaped by complex ecological interactions including competition, cooperation, and metabolic exchange. Beneficial symbiotic bacteria can enhance vital processes such as fiber digestion and energy yield. Conversely, opportunistic species can exploit stress or dietary imbalances, potentially leading to disease. Thus, studying these microbiomes requires an ecological mindset that views the community as an evolving, self-regulating system in continuous dialogue with its host. Achieving a deeper understanding of these microbial ecosystems is crucial for the development of both sustainable and health-oriented farming practices (Culp & Goodman, 2023; Lorgen-Ritchie et al., 2023; Q. Xu et al., 2021). The core aim of this scientific endeavor is to bridge the gap between microbiome ecology and applied animal science, encouraging professionals to view the animal as a host–microbe partnership whose balance dictates the success of modern agriculture.

2.2 COMPARATIVE ANATOMY AND PHYSIOLOGY RELEVANT TO MICROBIOMES

The characteristics of the host animal's anatomy and physiology exert a profound influence on the establishment and functionality of its microbiome. Each species provides a unique ecological landscape for microorganisms, defined by its specific digestive architecture, patterns of nutrient flow, pH gradients, and local oxygen availability. These anatomical and physiological features determine where microbes can successfully colonize, which specific metabolic pathways will predominate, and how the resulting microbial metabolites interact with the host's tissues. Therefore, interpreting microbiome data, designing effective feeding strategies, and developing targeted microbiome-based interventions all require a foundational understanding of these anatomical differences (Culp & Goodman, 2023; Lorgen-Ritchie et al., 2023).

2.2.1 Ruminants: Complex Foregut Fermentation Systems

Ruminant species, including cattle, sheep, and goats, are characterized by a highly specialized digestive system comprising four distinct compartments: rumen, reticulum, omasum, and abomasum. Rumen is the centerpiece, functioning as a vast, continuous anaerobic fermentation chamber. It hosts dense microbial consortia that specializes in the degradation of structural plant polysaccharides. The primary beneficial output of this fermentation process is volatile fatty acids (VFAs), which can supply up to 70% of the host's energy requirements (Gharechahi et al., 2020; Guo et al., 2021).

The complex compartmentalization of the ruminant digestive system supports a highly diverse microbiota, encompassing bacteria, archaea, protozoa, and anaerobic fungi. Among bacteria, prominent genera include Prevotella, Ruminococcus, and Fibrobacter, which are the essential drivers of cellulose and hemicellulose degradation. Furthermore, the rumen community includes methanogenic archaea, particularly species belonging to Methanobrevibacter. These archaea are responsible for converting fermentation end products into methane, thereby establishing a direct link between the rumen microbiome and global greenhouse gas emission (Henderson et al., 2015; Kumar et al., 2015).

Rumen physiology actively shapes its microbial ecology through specific conditions: it maintains a warm, strictly anaerobic environment with a near-neutral pH, and features a slow passage rate. This slow passage rate is crucial as it allows for extensive microbial growth and the formation of biofilms on feed particles. VFAs produced by fermentation are absorbed through the epithelial papillae of the rumen, a process vital for maintaining the host's acid–base balance and energy flux. Microbial composition shifts markedly along the remainder of the gastrointestinal tract, with the abomasum (the true stomach) and the intestines harboring communities adapted to either more acidic or more oxygenated conditions. Because the rumen microbiome fundamentally mediates nutrient extraction and methane production, it is central to strategic efforts aimed at improving feed efficiency and reducing environmental emissions. Diet-induced modifications, such as shifting from forage-based feeding to concentrate feeding, can drastically remodel these microbial networks, generally favoring amylolytic bacteria and thus altering the fermentation profiles (Na & Guan, 2022; Neves et al., 2025; Newbold & Ramos-Morales, 2020).

2.2.2 Monogastric Mammals: The Hindgut Fermentation Strategy

Monogastric animals, a group that includes pigs, horses, and companion species, possess a single-chamber stomach. Unlike ruminants, these animals rely primarily on microbial activity occurring within the large intestine and colon for essential fermentation processes. In pigs, the hindgut supports dense microbial populations, typically dominated by genera such as Lactobacillus, Bacteroides, Ruminococcus, and Faecalibacterium. These microbes specialize in metabolizing dietary fibers and resistant starches into short-chain fatty acids (SCFAs), including acetate, propionate, and butyrate (Branck et al., 2024; Kumar et al., 2015; Wallace et al., 2019).

Monogastric animals feature faster intestinal transit time and exhibit less structural complexity in their digestive systems compared to ruminants. This results in smaller, though still highly metabolically active, microbial niches. The highly acidic environment of the stomach (pH 2–4) functions as a strong antimicrobial barrier. Subsequent segments of the small intestine, containing bile and digestive enzymes, restrict microbial density. Consequently, the vast majority of vital microbial fermentation occurs in the cecum and colon, where the anaerobic environment supports complex carbohydrate breakdown and essential vitamin synthesis (Jha et al., 2019; Tardiolo et al., 2025). The pig microbiome is known to play crucial roles in nutrient absorption, mucosal immunity, and resistance to enteric pathogens. Studies have established links between microbial diversity and the abundance of beneficial species like Lactobacillus and Bifidobacterium with measurable improvements in feed conversion and gut barrier integrity (Liao et al., 2024; Quan et al., 2018). Therefore, feeding strategies aimed at modulating the hindgut microbial composition, such as the inclusion of prebiotics or the adjustment of dietary fiber levels, offer practical avenues to improve both productivity and animal welfare.

2.2.3 Poultry: Paired Cecal Fermentation Chambers

The poultry digestive system includes several specialized organs that contribute to unique microbial habitats: the crop, proventriculus, gizzard, small intestine, paired ceca, and large intestine. The paired ceca stand out as the principal fermentation chambers. They house dense microbial communities responsible for degrading nondigestible carbohydrates and synthesizing important SCFAs, amino acids, and vitamins (Binek et al., 2017; Sergeant et al., 2014; Sun et al., 2021).

Microbial colonization in chickens is a rapid process that begins immediately after hatching, stabilizing generally within the first few weeks of life. Early colonizers are influential, shaping long-term immune development and pathogen resistance. The cecal microbiome is typically dominated by groups such as Bacteroides, Ruminococcus, and Clostridium clusters, which are adapted to a semi-anaerobic, nutrient-rich environment. However, factors like feed composition, housing, and general management practices can significantly reshape these communities. For example, studies have shown that diets incorporating alternative ingredients, such as black soldier fly (BSF) meal, can enrich beneficial bacteria populations while simultaneously reducing the presence of potential pathogens. The relatively short digestive transit time characteristic of poultry necessitates rapid microbial turnover. Since poultry do not ruminate, their microbial processes must act swiftly to recover energy from feed residues before the material is excreted (Biasato et al., 2020; Özkan et al., 2024; Shehata et al., 2021; Takeshita et al., 2021; S.-S. Xiao et al., 2021).

2.2.4 Aquaculture Species: Microbiomes in Water-Linked Digestive Systems

In aquaculture species like fish and shrimp, the microbiome is shaped not only by the host's anatomy but also significantly by the surrounding aquatic environment. The gastrointestinal tract of teleost fish usually includes a stomach (though not in all species) and an intestine of variable length, often complemented by pyloric caeca to increase the absorptive surface area. Microbial fermentation in these species is typically limited compared to terrestrial animals. Nevertheless, these microbial communities are critical for lipid metabolism, immune modulation, and defense against pathogens (Huang et al., 2020; Tarnecki et al., 2017).

A defining feature of the aquaculture environment is the constant exchange of microbes between the fish and the surrounding water. Consequently, the water microbiome influences the composition of the gut community more strongly than the environment influences land animals. Parameters such as temperature, salinity, stocking density, and feed type all modulate microbial diversity in the host. For instance, the use of probiotic supplementation in aquaculture systems can enhance nutrient uptake and disease resistance by actively stabilizing gut microbial populations. Similarly, in shrimp, the intestinal microbiome serves a vital role in immune defense, acting as a primary protection against pathogens such as Vibrio species (Kanika et al., 2025; Khanjani et al., 2024; Rahayu et al., 2024). Therefore, maintaining microbial balance through the use of prebiotics, feed additives, or specific water management techniques is essential for the sustainability of aquaculture.

2.2.5 Comparative Insights across Species

Despite the considerable anatomical diversity observed among domesticated animals (foregut fermentation in ruminants vs. hindgut in monogastrics), a set of unifying ecological principles governs the host–microbe relationship.

1. *Fermentation Site and Retention Time:* The key anatomical variable determining microbial density and diversity is the size and retention time of the fermentation chamber. Ruminants utilize a slow, large-scale fermentation process in the rumen, while poultry and monogastrics rely on smaller, faster-processing systems like the ceca and colon (Tardiolo et al., 2025).
2. *Substrate Availability:* Diet dictates microbial selection. Herbivorous diets rich in cellulose select for specialized fibrolytic microbes, whereas omnivorous or grain-based diets promote amylolytic and proteolytic taxa. The biochemical nature of the diet is what ultimately defines which microbial metabolic pathways are active (Deusch et al., 2017).
3. *Host Control:* Host physiological mechanisms act as selective filters that shape microbial colonization and persistence. These mechanisms include variations in pH, bile salt concentrations, immune surveillance, and the rate of epithelial turnover. These differences in physiological control explain why microbiomes can differ so markedly, even among animals receiving similar diets (Weimer, 2015).

4. *Microbial Zonation:* The gastrointestinal tract is not monolithic; each segment constitutes a distinct ecological niche. For example, microbial diversity characteristically peaks in the rumen of cattle and in the ceca of poultry. These zones share common characteristics of low oxygen tension and sufficient retention time to permit extensive fermentation and nutrient recycling (W. Li et al., 2025).

Understanding these complex host–microbe interactions require appreciating anatomy as both a physical scaffold and a selective environment that drives microbial evolution. This foundation sets the stage for examining how microbial communities assemble developmentally across different life stages, a process that profoundly shapes long-term health and productivity (Table 2.1).

2.3 MICROBIOME ASSEMBLY ACROSS LIFE STAGES

The establishment of the microbiome is a crucial developmental process that unfolds across the entire lifespan of domesticated animals (Figure 2.1). This process commences either before or at birth, progresses through several critical early-life transitions, and ultimately stabilizes during adulthood. Throughout this assembly, the microbiome remains highly sensitive to the animal's age, diet, environment, and management decisions. Understanding this maturation sequence is essential for successfully designing interventions that support long-term health, maximize productivity, and build resilience in the animals (Schoultz et al., 2025).

Microbiome assembly is characterized by sequential waves of colonizers and involves highly dynamic ecological interactions. It is not

TABLE 2.1

A comparative overview of digestive system anatomies across domesticated animals, illustrating how distinct physiological niches shape unique microbial communities and their functional roles

Species group	Primary fermentation site	Key physiological condition	Essential microbial function	Representative microbes/traits
Ruminants (cattle, sheep, goats)	Rumen (foregut)	Large, anaerobic, warm, near-neutral pH, slow passage rate	Degradation of structural plant polysaccharides into VFAs (up to 70% of host energy)	Fibrobacter, Ruminococcus (cellulose degraders), Methanobrevibacter (methane producers)
Monogastric mammals (pigs, horses)	Large intestine/colon (hindgut)	Single stomach with low pH (antimicrobial barrier), most activity is anaerobic in hindgut	Metabolize dietary fibers and resistant starches into SCFAs (acetate, propionate, butyrate)	Lactobacillus, Bacteroides, Faecalibacterium
Poultry (chickens, turkeys)	Paired ceca	Short digestive transit time, semi-anaerobic environment, rapid microbial turnover	Degradation of nondigestible carbohydrates, synthesis of SCFAs, amino acids, and vitamins	Bacteroides, Ruminococcus, Clostridium clusters
Aquaculture (fish, shrimp)	Intestine	Constant microbial exchange with water environment, limited fermentation compared to terrestrial species	Lipid metabolism, immune modulation, pathogen defense	Strongly influenced by water quality and feed type, including Lactobacillus, Vibrio, Cetobacterium

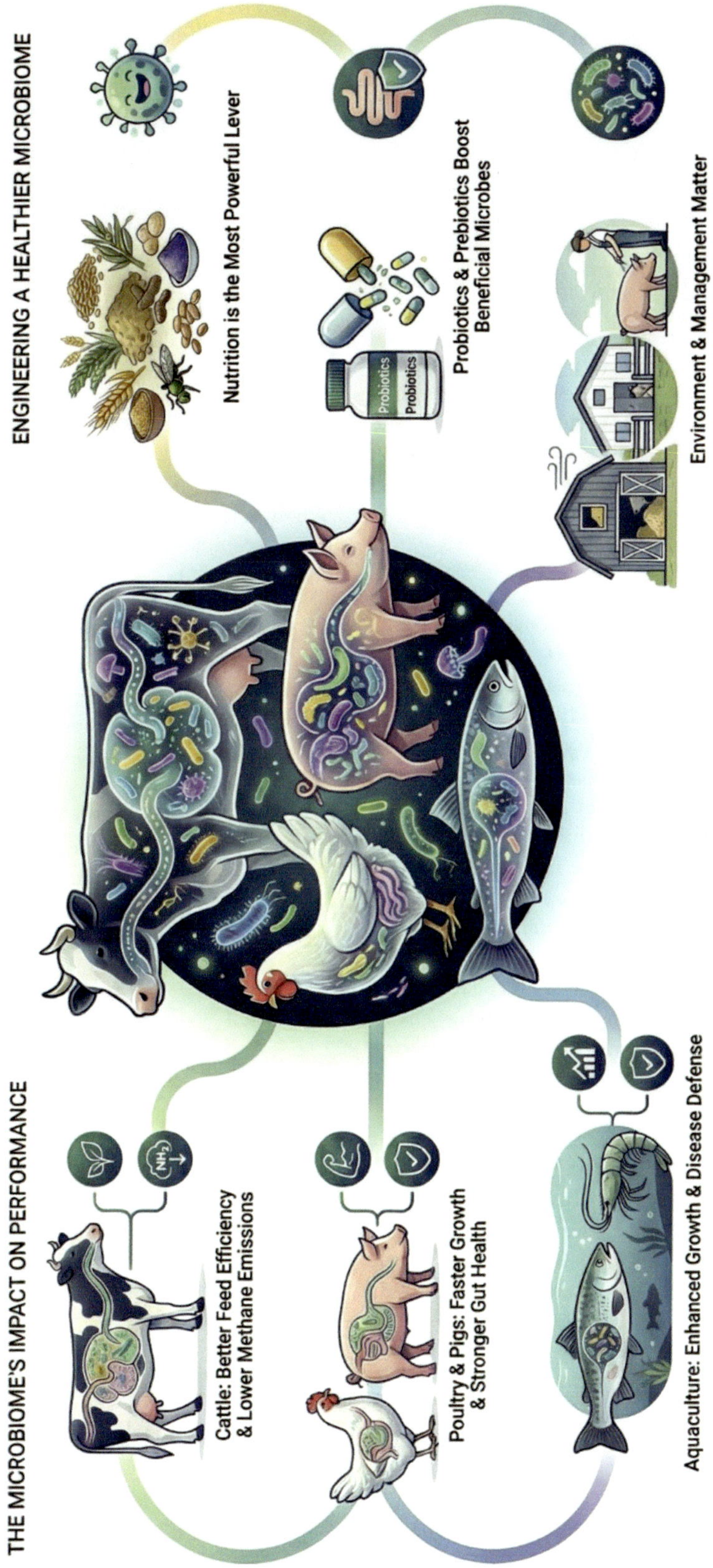

Figure 2.1 The dynamic life course of the gut microbiome.

a simple linear process but a constant negotiation between the microbial communities and the host's developing physiology. Each distinct life stage provides a unique ecological context that determines which microbes are able to colonize, how they interact with the host, and what specific functions they perform (Choudhury et al., 2021; Mercer et al., 2024).

2.3.1 Prenatal and Perinatal Influences

While the fetus was historically considered to be in a sterile environment, emerging research challenges this view. There is evidence suggesting that microbial components, including DNA and metabolites, can potentially reach the fetal environment, traversing the placenta, amniotic fluid, or maternal bloodstream. However, the exact extent of viable microbial colonization before birth remains a subject of ongoing debate (Perez-Muñoz et al., 2017; L. Xiao & Zhao, 2023).

Irrespective of whether colonization occurs prenatally, it is unequivocally clear that maternal factors deeply influence the initial microbiome encountered by the newborn animal. The maternal gut, skin, mammary gland, and reproductive tract function as major reservoirs for early microbial transmission. For instance, in both cattle and sheep, the maternal vaginal microbes are significant in shaping the initial microbiome of newborn calves and lambs. In pigs, the early colonization process is predominantly dictated by maternal fecal and skin microbes encountered during farrowing. In contrast, poultry lack direct maternal contact, making the hatching environment and the litter microbiome the primary sources of initial colonizers immediately after hatching. The first burst of colonization immediately after birth is crucial because it acts as the trigger for the development of the innate and adaptive immune systems. This initial exposure also shapes mucosal tolerance and activates the metabolic pathways necessary for neonate growth (Ferretti et al., 2018; Suárez-Martínez et al., 2023; Zhuang et al., 2024).

2.3.2 Early-Life Colonization and Succession

The early life period represents a critical window where environmental inputs have the capacity to produce long-lasting consequences. During the first days and weeks, the developing microbiome is highly dynamic and extremely sensitive to changes in diet, hygiene levels, antimicrobial exposure, and housing conditions.

Milk Feeding and Colostrum: Colostrum is highly rich in immunoglobulins, antimicrobial peptides, and oligosaccharides. These components serve as selective substrates for the growth of beneficial bacteria. In calves, colostrum intake actively promotes the growth of beneficial species like Bifidobacterium and lactic acid bacteria, thus providing critical support for immune development. Similar patterns are seen in piglets, where sow milk influences the early microbial structure and associated metabolism (Arshad et al., 2021; Guevarra et al., 2019; Zhuang et al., 2025).

Environmental Exposure and Antibiotic Impacts: The early environment, including farm dust, bedding material, and litter, introduces a wide diversity of microbial species. In poultry, the litter microbiota can influence cecal colonization within just hours of hatching. Conversely, early antimicrobial exposure can leave persistent, detrimental signatures. Perinatal antibiotic administration in pigs, for example, has been shown to alter gut community composition for months and is linked to an increased abundance of AMR genes. As microbial succession proceeds during early life, the community structure gradually shifts from initial facultative anaerobes toward more strictly anaerobic taxa. This fundamental trend reflects the stabilization of gut oxygen levels, the increased availability of complex carbohydrates, and the successful development of mucosal immunity (Guevarra et al., 2019; Huda et al., 2025).

2.3.3 Weaning Transition: A Major Ecological Turning Point

The process of weaning is recognized as one of the most abrupt and significant ecological transitions in the life of domesticated animals. It involves simultaneous and drastic changes in dietary intake, stress levels, microbial exposure, and gastrointestinal physiology.

In young ruminants (calves, lambs) and pigs, the change from milk to solid feed introduces completely new microbial substrates. This shift strongly favors bacteria that specialize in fiber-degrading and starch-degrading processes. The result of this abrupt change is frequently temporary dysbiosis, accompanied by gut inflammation and often a growth check, which is collectively termed postweaning stress. Successful dietary adaptation and gut maturation are often reflected in the balance between lactic acid bacteria and Prevotella species (Guevarra et al., 2018; S. Meale et al., 2017; Wei et al., 2021).

In poultry, while true weaning does not occur, the transition from starter diets to grower or finisher diets still significantly alters cecal fermentation patterns. This dietary shift affects the abundance of critical species like Ruminococcus and Bacteroides (Kouzounis et al., 2022). Similarly, in aquaculture species, developmental stages such as metamorphosis or the transition from live-feed to formulated feed represent comparable ecological shifts that reshape gut communities and influence both immunity and survival rates (Ghosh et al., 2025).

2.3.4 Maturation and Stability in Adulthood

Upon reaching adulthood, the microbiome achieves a greater level of stability, although it continues to respond dynamically to diet, environment, and management factors. Adult microbial communities typically exhibit greater alpha diversity, display stronger host specificity, feature the enrichment of specialized fermentative groups, and maintain stable microbial networks and metabolic functions.

In cattle, the adult rumen community maintains a characteristic structure, primarily dominated by fibrolytic and amylolytic taxa, depending predictably on the diet. This stability is vital as it permits efficient fermentation and ensures predictable VFA production. In pigs, the adult hindgut microbiome stabilizes into a community dominated by Bacteroides, Prevotella, and Faecalibacterium, which are essential for supporting immune tolerance and optimal energy harvest. However, even stable adult microbiomes remain sensitive to stressors. Environmental stressors such as heat, transportation, crowding, and acute dietary disruptions can still cause shifts. For instance, heat stress in poultry has been documented to reduce microbial diversity and shift fermentation profiles toward less favorable compositions (Feehan et al., 2022; Newbold & Ramos-Morales, 2020; Upadhaya & Kim, 2022).

2.3.5 Aging and Late-Life Microbiome Changes

In animals that reach older ages, the microbiome undergoes gradual changes, typically influenced by factors such as reduced feed intake, slower intestinal transit time, immune senescence, and increased oxidative stress.

Studies conducted in older cattle and pigs reveal consistent trends, including reduced microbial diversity, an increased abundance of proinflammatory taxa, and a decrease in vital butyrate-producing bacteria. These changes are associated with heightened susceptibility to pathogens. Aging also exacerbates vulnerability to dysbiosis, especially when combined with stressful conditions or poor nutrition. Understanding these aging-associated microbial shifts is gaining relevance as breeding animals are often maintained for longer productive periods (Estrada et al., 2024; Lim et al., 2019; G. Zhang et al., 2019).

2.3.6 Ecological Summary across Life Stages

Microbiome assembly is governed by a defined sequence of ecological filters:

1. Initial exposure, derived from maternal sources and the immediate external environment.
2. Selective pressures, imposed by diet, physiological conditions (pH, oxygen gradients), and the maturation of the immune system.
3. Transitions, such as the abrupt stress of weaning, deliberate dietary shifts, or significant environmental changes.
4. Long-term stability, achieved through successful host–microbe coadaptation.
5. Late-life decline, resulting from factors related to immunity and physiological aging.

Each stage in this assembly process presents distinct opportunities for targeted interventions. Early life offers high plasticity, adulthood reflects microbial resilience, and the aging phase highlights inherent vulnerabilities.

2.4 DETERMINANTS OF MICROBIOME STRUCTURE

The intricate structure of the microbiome in domesticated animals is determined by a complex interplay between host, environmental, and management factors. Although anatomical and physiological constraints are species-specific, many determinants operate broadly across livestock, poultry, and aquaculture systems. These determinants influence microbial colonization, community composition, metabolic activity, and overall resilience. By understanding how these factors work, opportunities arise to design proactive strategies that promote health, enhance productivity, and support sustainable animal agriculture (Figure 2.2).

2.4.1 Host Genetics and Heritable Microbial Traits

Host genetics contributes to a nonexclusive but important influence on microbiome structure. Gene variations related to immunity, mucosal barrier function, epithelial glycosylation, and digestive enzyme activity can create specific selective pressures that favor the colonization of particular microbial taxa (Ryu & Davenport, 2022).

Research in cattle, for instance, has demonstrated that certain rumen microbial taxa exhibit moderate heritability. This observation strongly suggests that host genetic factors contribute meaningfully to the structure of the microbial community. Similar heritable patterns have been documented in pigs, where the abundance of species like Campylobacter, Lactobacillus, and Prevotella has been successfully linked to specific host genomic regions (P. Fan et al., 2021; F. Li et al., 2019; W. Wang et al., 2024).

These findings lend support to the concept of host–microbe coadaptation and open the door to the potential integration of microbiome-associated traits into breeding programs. Genetic selection could be used to favor animals that naturally support more robust or favorable microbial communities, potentially leading to improved feed efficiency, reduced methane emissions, or enhanced disease resistance. However, it is essential to note that genetic influences are typically considered modest when compared to the stronger effects exerted by environmental and dietary factors. This necessitates that genetic selection strategies must be carefully integrated with targeted management and nutritional interventions (Abbas et al., 2020; Ryu & Davenport, 2022; W. Wang et al., 2024).

2.4.2 Nutrition and Feed Composition: The Most Powerful Determinant

Diet is recognized as one of the most powerful determinants of microbiome structure across all domesticated species. Feed ingredients supply the primary substrates that microbial communities utilize for growth and metabolic activity. Even minor changes in the composition of the diet can elicit measurable shifts in the microbial population.

Ruminants: Dietary factors drastically alter the rumen. High-forage diets lead to the enrichment of fibrolytic bacteria (e.g., Fibrobacter and Ruminococcus). Conversely, high-grain or starch-rich diets significantly increase the populations of amylolytic bacteria (e.g., Prevotella). A rapid shift from forage to concentrate feeding can critically reduce rumen pH, predisposing the animal to dysbiosis and acidosis (H. Chen et al., 2021; Ramos et al., 2021).

Pigs and Monogastric Mammals: In monogastrics, the type and digestibility of dietary fiber are highly influential on hindgut communities. Specific fermentable compounds, such as resistant starch, arabinoxylans, and inulin, are known to enrich beneficial butyrate-producing

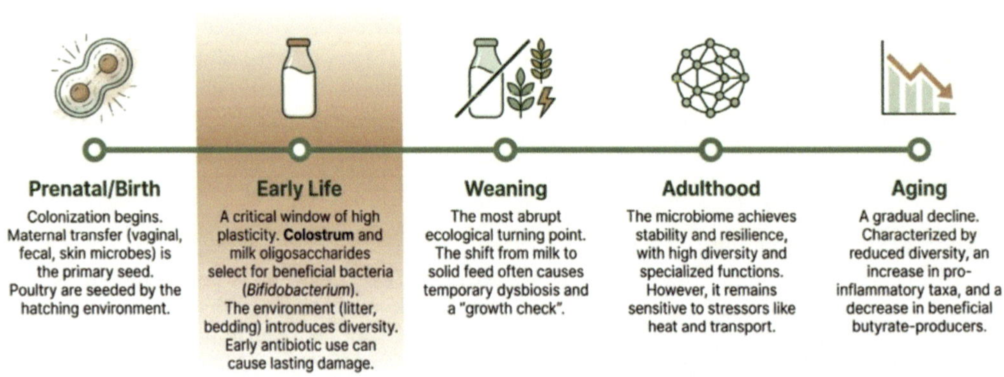

Figure 2.2 The role of the gut microbiome in animal performance and sustainable production systems.

bacteria (e.g., Faecalibacterium and Roseburia). In contrast, diets that are overly rich in protein can increase putrefactive fermentation and lead to metabolites associated with intestinal inflammation (Jha et al., 2019; Pieper et al., 2016).

Poultry: The cecal microbiome of poultry responds rapidly to adjustments in feed formulation. Incorporating alternative protein sources, such as BSF larvae, sunflower meal, or wheat middlings, can shift microbial composition toward metabolic pathways that improve nutrient utilization or strengthen gut integrity (Jha et al., 2019; Özkan et al., 2024).

Aquaculture: In fish and shrimp, the form of the feed and the lipid composition are strong determinants of gut microbiota. For example, marine-derived oils favor different microbial taxa compared to plant-derived oils, while specific feed additives are used to modulate immune function and pathogen resistance. Across all species, diet not only shapes the specific microbial taxa present but also dictates the functional pathways that are active, thereby influencing the production of SCFAs, lactate, ammonia, and other secondary metabolites that directly affect host physiology (Marcharla et al., 2025; Nadal et al., 2020).

2.4.3 Environment and Housing Conditions

The physical environment exerts significant control over the microbiome, acting as a determinant through factors such as temperature, humidity, bedding type, stocking density, and sanitation. These conditions influence both microbial colonization and the host's immune responses that regulate those communities.

For instance, heat stress in poultry and pigs is documented to decrease microbiome diversity and promote the growth of potentially pathogenic taxa. This pattern is linked to systemic immune suppression and oxidative stress. In cattle, the type of housing affects exposure to environmental microbes, influencing the microbiota of the udder and teat skin, which has direct implications for the risk of mastitis (Kikusato & Toyomizu, 2023; Patra & Kar, 2021).

In aquaculture, water quality is perhaps the most central environmental determinant. Parameters such as dissolved oxygen levels, pH, salinity, and the microbial load present in the water environment all affect the colonization of the gut, skin, and gills. Animals raised in environments with richer microbial diversity often develop more diverse microbiomes themselves, which is typically associated with stronger immune responses compared to animals raised in overly sterile or highly sanitized settings (Sehnal et al., 2021).

2.4.4 Antimicrobial Use and Pharmaceutical Interventions

The administration of antibiotics can rapidly and profoundly alter the microbiome structure. The negative effects observed include a reduction in overall diversity, the elimination of beneficial taxa, the enrichment of opportunistic pathogens, and a significant expansion of AMR genes.

In pigs and poultry, the use of in-feed antimicrobials, historically used for growth promotion, has been proven to reshape gut communities and reduce overall microbial resilience. In cattle, antimicrobial treatment impacts both the rumen and the hindgut microbiota, leading to consequences for nutrient metabolism and pathogen susceptibility. Because microbial communities can take weeks or even months to recover after antibiotic exposure, this determinant is crucial for discussions regarding animal welfare and the development of sustainable production strategies. Consequently, alternatives such as bacteriophages, probiotics, organic acids, and phytochemicals are increasingly being investigated as tools to support microbial balance while simultaneously reducing the need for routine antimicrobial use (H.-Y. Liu et al., 2024; Rochegüe et al., 2021; Silvestro et al., 2025).

2.4.5 Management Practices and Production Systems

Management decisions are levers that shape the microbiome indirectly through their influence on the determinants already discussed, including diet, environment, stress levels, and microbial exposure. Key management factors that exert influence include:

- Weaning strategies employed in pigs, calves, and lambs (C. Li et al., 2025; Y. Li et al., 2018).
- The type of litter used and its turnover frequency in poultry houses (Ekunseitan et al., 2025; L. Wang et al., 2016).
- The use of grazing systems versus confinement housing in cattle (Jiang et al., 2025; Yuan et al., 2023).

- Stocking density and biosecurity protocols, and the levels of stress induced by handling and transportation (C. Li et al., 2025; Yuan et al., 2023).

For example, early weaning in pigs can result in abrupt microbial shifts and intestinal inflammation, whereas gradual weaning or creep feeding strategies can facilitate more stable transitions (Guevarra et al., 2019; W. Tang et al., 2021). In poultry, the reuse of litter introduces a consistent microbial inoculum that can stabilize cecal communities, but this practice also carries the risk of increasing the pathogen load if hygiene standards are not maintained (L. Wang et al., 2016). Production systems themselves create microbial differences; grass-fed cattle show distinct rumen microbial compositions compared to grain-fed cattle, which reflects the variances in fiber availability, pH, and metabolic pathways (Jiang et al., 2025; J. Liu et al., 2022).

2.4.6 Social Interactions and Group Microbiomes

Animals housed in groups readily share microbes through direct contact, grooming behavior, coprophagy, and airborne exchange. These social behaviors accelerate the microbiome convergence within a herd or flock. For example, cohoused poultry birds rapidly share cecal microbial strains, leading to the formation of flock-level microbial patterns. Similar phenomena are observed in cattle and pigs, where pen-mates often develop more similar microbiomes than animals housed separately. While these social microbial networks can support community stability, they also inherently facilitate pathogen transmission, underscoring the dual nature of social interactions as determinants of microbial structure (Petrullo et al., 2025; Sarkar et al., 2020, 2024).

2.5 MICROBIOME–PHENOTYPE LINKS IN PRIORITY DOMESTICATED SPECIES

Microbial communities actively influence a comprehensive spectrum of phenotypes in domesticated animals. These links encompass growth, feed efficiency, immune competence, metabolic balance, behavior, product quality, and disease susceptibility. The relationship extends beyond simple correlation; increasing evidence derived from multiomics studies (metagenomics, metabolomics), gnotobiotic models, and longitudinal studies demonstrate that the microbiome is an active participant in shaping these measurable host traits (I. Khan et al., 2024; Xue et al., 2020; Ye et al., 2025).

2.5.1 Cattle: Feed Efficiency, Methane Emissions, and Metabolic Health

The cattle microbiome is central to both productivity and environmental impact.

Feed Efficiency and Growth: Differences in feed efficiency among cattle populations are strongly correlated with the composition of the rumen microbiome. Animals that display higher feed efficiency typically exhibit a lower abundance of methanogenic archaea and a greater prevalence of propionate-producing bacteria. They also tend to possess more stable microbial networks. Rumen communities enriched in genera such as Prevotella, Succinivibrio, and Megasphaera are repeatedly associated with improved energy harvest and a reduced need for feed intake to achieve the same growth level (Xie et al., 2022; Xue et al., 2022).

Methane Emissions: Methane production is fundamentally driven by a complex ecological interaction involving fibrolytic bacteria, hydrogen-producing fermenters, and methanogenic archaea. Cattle categorized as high methane emitters show elevated levels of Methanobrevibacter species and a rumen microbiome functionally geared toward high acetate and hydrogen production. Reducing methane emissions requires strategically shifting microbial pathways, either toward propionate formation or by utilizing alternative hydrogen sinks (Shinkai et al., 2023; Xie et al., 2022).

Metabolic Disorders (Ketosis): Both subclinical and clinical ketosis, a major metabolic disorder, are associated with distinct microbial fingerprints. Cows with high levels of beta hydroxybutyrate (BHB) often show reduced diversity of rumen bacteria, an enrichment of lactic acid producers like Streptococcus and Lactobacillus, and a lower abundance of fibrolytic taxa. These microbial changes lead to reduced fiber fermentation, alter the crucial proportions of VFAs, and can consequently exacerbate the negative energy balance of the host (Badhan et al., 2025; Xie et al., 2022).

Hoof and Skin Health: The infectious lameness condition known as Digital Dermatitis is linked to specific microbial communities on the bovine foot skin, particularly dominated by Treponema species. Changes in the skin microbiome can serve as indicators of early disease states, inflammatory responses, and the outcomes of treatment (Bay et al., 2018, 2023).

2.5.2 Poultry: Growth, Feed Conversion Ratio (FCR), and Enteric Health

Poultry cecum is a key site linking microbial functions to host performance.

Growth and FCR: Birds demonstrating superior FCR typically exhibit specific cecal traits: higher microbial evenness and a significant increase in the abundance of butyrate producers, including Ruminococcus, Faecalibacterium, and Butyricicoccus species. This microbial profile leads to elevated short-chain fatty acid synthesis. These beneficial fermenters are crucial because they support efficient nutrient extraction and work to reduce gut inflammation (Bortoluzzi et al., 2025; Deryabin et al., 2024).

Immune Maturation: The early development of the cecal microbiome fundamentally shapes the bird's immune competence. Successful colonization by lactic acid bacteria and Bacteroides species has been shown to enhance mucosal tolerance, promote the crucial maturation of goblet cells, and significantly reduce pathogen susceptibility (Y.-H. Fan et al., 2023; N. Liu et al., 2025).

Enteric Diseases: Conditions like necrotic enteritis and general dysbiosis are characterized by a pronounced increase in the abundance of *Clostridium perfringens*, a parallel reduction in butyrate-producing bacteria, and a general disruption of beneficial anaerobic fermentation. Intervention studies, such as those using BSF meal, have successfully demonstrated that this alternative protein source can improve gut microbial stability and enhance resistance against enteric pathogens (Bortoluzzi et al., 2025).

2.5.3 Pigs: Growth, Immune Regulation, and Postweaning Performance

The pig hindgut microbiome is central to the fermentation of resistant starches and fiber.

Growth and Feed Efficiency: Efficient growth in pigs is strongly associated with a microbial profile featuring a greater abundance of Prevotella, Lactobacillus, and Faecalibacterium. This community structure results in increased butyrate production and reduced markers indicative of intestinal inflammation (Amat et al., 2020; Yang et al., 2016; M. Zhou et al., 2025).

Postweaning Diarrhea (PWD) and Stress: Weaning is characterized by dramatic alterations in microbial communities. The typical pathological hallmarks of PWD include rapid shifts toward *Escherichia coli* and a damaging loss of beneficial anaerobes. Management interventions, such as gradual weaning or the inclusion of fiber-enriched starter diets, effectively stabilize early microbiome development and subsequently reduce disease incidence (Gresse et al., 2017; Han et al., 2024).

Metabolic Traits: Microbial communities have been successfully linked to commercially important traits in swine genetics, including backfat thickness, average daily gain, and lipid metabolism. The microbiome explains a measurable portion of the phenotypic variance observed in these growth traits, thus complementing host genetics (Aliakbari et al., 2022; S. Tang et al., 2020).

2.5.4 Small Ruminants: Parasite Resistance and Fiber Digestion

The microbiomes of sheep and goats mediate crucial aspects of health, particularly in extensive systems.

Parasite Resistance: In sheep and goats, resistance against gastrointestinal nematodes is correlated with increased microbial diversity and the enrichment of butyrate-producing bacteria. These beneficial microbes are thought to function by strengthening mucosal immunity, thereby reducing the parasite burden on the host (Cortés et al., 2019; F. Liu et al., 2025; Paz et al., 2021).

Fiber Utilization: Small ruminants displaying high fiber digestibility possess an increased abundance of cellulolytic taxa, such as *Ruminococcus albus* and *Fibrobacter succinogenes*. This microbial capacity enables these animals to extract substantially more energy from low-quality forages, a trait vital for the success of extensive grazing systems (Langda et al., 2020; K. Zhang et al., 2024).

2.5.5 Aquaculture Species: Growth, Immune Defense, and Survival

The gut microbiota in farmed fish and shrimp is critical for performance and survival.

Growth Performance: Growth rate is linked to the microbial capacity for lipid metabolism and protein digestion. Faster-growing fish often harbor higher amounts of Lactobacillus, Vibrio, and Cetobacterium species, which facilitate increased nutrient absorption (Ringø et al., 2022; Yukgehnaish et al., 2020; Z. Zhang et al., 2024).

Immune Defense: Gut microbiota profoundly influences key immune markers, including lysozyme activity, complement proteins, and cytokine expression. Communities modulated by probiotics are known to improve resistance to specific pathogens such as Aeromonas and Vibrio species (Ringø et al., 2022; Yukgehnaish et al., 2020).

Larval Survival: Microbiome stability is absolutely essential during the highly sensitive early aquaculture stages. The microbial composition of live-feed and water shapes the larval gut communities, and successful, stable colonization is linked to improved survival rates and enhanced stress tolerance (Borges et al., 2020; Shan et al., 2025; Vadstein et al., 2018).

2.5.6 Integrated Phenotype Understanding through Multiomics

The ability to link microbiomes reliably to host phenotypes is significantly strengthened by combining data from multiple -omics technologies: metagenomics, metatranscriptomics, metabolomics, host transcriptomics, and cytokine/immune assays. For example, studies integrating metabolomics and metagenomics in dairy cows have successfully identified microbial pathways associated with important traits like milk fat yield, somatic cell count, and overall energy balance. In pigs, the integration of host genetic data with microbiome data has been shown to improve prediction accuracy for growth traits compared to using host genetics alone. These comprehensive multiomics approaches are necessary to distinguish true causation from mere correlation and to precisely identify microbial functions that exert a direct phenotypic impact (Xue et al., 2020; Ye et al., 2025).

2.6 MICROBIOME INTERVENTIONS AND ENGINEERING STRATEGIES

Targeted interventions offer a powerful and practical opportunity to intentionally reshape the microbial communities of domesticated animals, promoting improvements in health, productivity, and sustainability. Given that microbial communities are highly responsive to diet, environment, and management, strategies can be designed to shift microbiome structure and function in predictable and desirable ways. Interventions span a wide range, from traditional nutritional approaches to advanced biotechnological tools.

2.6.1 Nutritional Modulation: The Primary Intervention Lever

Diet is universally recognized as the most powerful and practical tool for engineering the microbiome across all domesticated species. Feed ingredients define the substrates available for microbial metabolism, thereby shaping community composition and metabolic end products.

Fiber and Carbohydrate Manipulation: Adjusting the type of dietary fiber selectively enriches beneficial taxa. In cattle, diets rich in structural carbohydrates enhance populations of Ruminococcus and Fibrobacter, which are essential for cellulose degradation (Lin et al., 2024). In pigs, the provision of resistant starch specifically increases butyrate-producing bacteria, which strengthens gut integrity and barrier function (Miller et al., 2025). In poultry, fermentable carbohydrates like arabinoxylans and nonstarch polysaccharides stimulate fermentation pathways that support crucial immune development (Singh & Kim, 2021).

Alternative Protein Sources (e.g., BSF Meal): Novel and alternative protein sources can fundamentally reshape microbial metabolism. For instance, inclusion of BSF larvae meal in poultry diets has been shown to increase lactic acid bacteria while simultaneously reducing the abundance of potential pathogens (Ndotono et al., 2022; Özkan et al., 2024). In pigs, novel proteins alter nitrogen and amino acid fermentation, affecting harmful metabolites (Kar et al., 2021).

Fats and Oils: The composition of dietary fats influences microbial profiles. Omega-3 rich feeds in aquaculture, for example, shift microbial communities toward anti-inflammatory profiles. SCFAs, whether produced endogenously or supplemented externally, modulate gut immunity and pathogen resistance (Jadhav et al., 2025; Waliaula et al., 2025).

Feed Additives (Phytochemicals): Various additives, including phytochemicals, organic acids, and enzymes, can enhance fermentation efficiency or suppress pathogens. Tannins, for example, are effective at reducing methanogenic archaea and hydrogen-producing bacteria in ruminants, thereby shifting rumen fermentation profiles (Ismail et al., 2025; Kholif & Olafadehan,

2021). Organic acids, by lowering gut pH, can inhibit pathogens like Campylobacter in poultry (Dittoe et al., 2018). Nutritional strategies thus offer a high degree of precision engineering based on specific microbial substrates and metabolic pathways.

2.6.2 Probiotics, Prebiotics, and Synbiotics

These approaches involve the direct introduction of microbes or selective substrates to modulate populations.

Probiotics: These are live microorganisms administered to confer a health benefit. Commonly used species include Lactobacillus, Bifidobacterium, Bacillus, and Enterococcus. Evidence supports their use for improved growth and FCR in poultry, reduced postweaning diarrhea in pigs, enhanced rumen fermentation and milk yield in cattle, and improved pathogen resistance in fish (Koo et al., 2025; Yue et al., 2025).

Prebiotics: These are nondigestible compounds (e.g., inulin, fructooligosaccharides, mannan oligosaccharides) that selectively promote the growth of beneficial resident microbes. Mannan oligosaccharides, for instance, improve immune function in poultry and pigs. Inulin increases vital butyrate production in pigs and cattle (Dunisławska et al., 2017; Yue et al., 2025).

Synbiotics: Synbiotics combine both probiotics and prebiotics, enhancing the colonization success and persistence of the beneficial strains. They show particular promise for early-life interventions intended to stabilize gut development (Dunisławska et al., 2017; Koo et al., 2025; X. Xu et al., 2025).

2.6.3 Microbial Transplantation and Community Replacement

Introducing entire microbial communities can be used to correct severe dysbiosis.

Rumen Transfaunation: The transfer of rumen fluid from a healthy donor is a long-standing practice used to stabilize fermentation during digestive disorders in cattle. Modern studies confirm that this technique leads to improved rumen pH, the restoration of fibrolytic bacterial populations, and recovery from conditions like acidosis or poor feed efficiency (Mu et al., 2021; M. Zhou et al., 2018).

Fecal Microbiota Transplantation (FMT): FMT in piglets has been demonstrated to increase microbial diversity, improve growth, and reduce pathogen susceptibility. However, community replacement strategies necessitate strict pathogen screening and the use of standardized protocols to mitigate risks of pathogen transmission and ensure consistent colonization success (Rahman et al., 2025).

2.6.4 Selective Breeding for Microbiome-Associated Traits

Given that certain microbiome traits exhibit measurable heritability, selective breeding offers a long-term approach to microbiome management.

- Cattle with lower methane emissions possess heritable differences in methanogen abundance (Arshad et al., 2021).
- Pigs show heritable variation in Prevotella abundance that is correlated with growth traits (Amat et al., 2020; Bergamaschi et al., 2020).
- Chickens exhibit breed-specific cecal microbial signatures that influence digestion and immunity (Lan et al., 2025).

Breeding programs can therefore incorporate these traits to enhance efficiency and improve sustainability.

2.6.5 Bacteriophages and Competitive Exclusion

These tools provide targeted control or prevention of specific pathogens.

Bacteriophages (Phages): These are viruses that specifically target bacteria. They can be used to suppress pathogens with high specificity, ideally without disrupting the beneficial, nontarget microbial taxa. Applications include the successful reduction of Salmonella in poultry (M. A. S. Khan & Rahman, 2022), *Escherichia coli* infections in pigs (Jamalludeen et al., 2009), and Vibrio species in aquaculture (Silva et al., 2014).

Competitive Exclusion (CE): CE involves administering complex microbial consortia that effectively outcompete pathogens by occupying ecological niches. This strategy is highly effective and scalable, particularly for controlling Salmonella, *E. coli* O157, and *C. jejuni* colonization in poultry (Mead, 2000).

2.6.6 Environmental and Husbandry-Based Microbiome Engineering

Intervention extends to managing the environment, which is a major microbial reservoir.

- Improved ventilation and heat mitigation are essential to preserve gut microbial stability during periods of heat stress in poultry and cattle (Vitt et al., 2017).
- Bedding management significantly alters skin and hoof microbiota in cattle, impacting conditions like mastitis risk (van Gastelen et al., 2011).
- In aquaculture, water filtration enhances microbial quality for the host's gut, skin, and gill communities (Tayyab et al., 2025).
- Gradual weaning strategies support favorable microbial succession and reduce dysbiosis in pigs and calves (S. J. Meale et al., 2016; St-Pierre et al., 2023).

2.6.7 Emerging Biotechnologies for Precision Engineering

Cutting-edge technologies allow for high-precision modulation of microbial communities (Table 2.2).

Engineered Probiotics: Synthetic biology enables the design of probiotic strains that are capable of producing specific metabolites, delivering vaccines, secreting antimicrobial peptides, or regulating host gene expression (Neil et al., 2021).

CRISPR-Based Modulation: CRISPR bacteriophages are being investigated for their ability to selectively remove antibiotic resistance genes or highly deleterious bacterial strains without damaging the broader beneficial microbiome (Mkilima, 2025).

Defined Microbial Consortia: Instead of using single-species probiotics, defined multispecies communities are developed to mimic natural microbial networks. These consortia offer greater metabolic diversity, improved colonization

TABLE 2.2

Overview of microbiome-directed interventions. Summary of tools and approaches available to reshape microbial communities for improved animal performance and sustainability

Intervention category	Specific example	Target effect and benefit	Target species
Nutritional modulation (additives)	Tannins	Reduce methanogenic archaea and shift fermentation toward propionate	Ruminants (cattle, sheep)
Nutritional modulation (protein)	BSF meal	Increases lactic acid bacteria and reduces potential pathogens, improving performance.	Poultry, pigs, fish
Microbial supplementation	Probiotics (Lactobacillus, Bacillus)	Introduce live beneficial microbes to improve growth, FCR, enhance immune function, and prevent enteric disease	Poultry, pigs, cattle, aquaculture
Microbial transplantation	Rumen transfaunation (RT)	Corrects severe dysbiosis, restores fibrolytic bacterial populations, improves recovery from acidosis	Ruminants (cattle, sheep)
Microbial targeting	Bacteriophages (phages)	Targeted suppression of specific pathogens (e.g., Salmonella, *E. coli*, Vibrio) without disrupting beneficial taxa	Poultry, pigs, aquaculture
Breeding strategies	Selecting for heritable traits	Favors animals that naturally support favorable communities (e.g., low methanogen abundance or high Prevotella abundance)	Cattle, pigs
Emerging technology	Engineered probiotics	Synthetic biology strains produce specific metabolites or deliver vaccines, offering high-precision modulation	All species (emerging)

stability, and more predictable functional outputs (Mead, 2000).

2.7 TRANSLATIONAL, REGULATORY, AND ETHICAL CONSIDERATIONS

Translating scientific insights into real-world farm practice is a multifaceted challenge. It requires not just technical innovation but also robust regulatory pathways, alignment with global AMR policies, and careful attention to animal welfare and broader One Health responsibilities.

2.7.1 Translating Microbiome Knowledge into Farm Practice

Microbiome research is most impactful when it generates actionable information for veterinarians, nutritionists, and farmers.

Microbiome-Informed Nutrition: Feed formulation is already guided by energy and protein requirements, but microbiome data adds critical information on how ingredients shape fermentation profiles, short-chain fatty acid production, and pathogen suppression. For dairy cows, rumen profiles associated with better energy balance or reduced ketosis risk can be used to refine specific fresh cow diets (Kong et al., 2025). In poultry, cecal microbial signatures of efficient birds help identify dietary components that support butyrate producers and mitigate dysbiosis (He et al., 2023).

Microbiome-Guided Breeding: Since some microbial traits show moderate heritability (e.g., rumen communities linked to methane output or pig gut profiles associated with growth), there is potential to include microbiome-associated phenotypes in genetic selection indices. However, practical, widespread adoption requires validated, affordable microbiome assays and statistically robust models (Bay et al., 2023; Martínez-Álvaro et al., 2022).

Diagnostics and Risk Prediction: Microbiome-based biomarkers are being explored to predict disease risk. Specific rumen configurations may appear before clinical ketosis in dairy cows (Kong et al., 2025), while certain microbial patterns precede postweaning diarrhea or necrotic enteritis in pigs and poultry (Antonissen et al., 2016; Canibe et al., 2022). Integrating these signatures into herd health programs allows for the early identification of high-risk groups and enables targeted, timely interventions.

Decision Support Tools: For practical translation, microbiome metrics must be integrated into actionable tools, such as feed formulation software that predicts microbial responses, or breeding value evaluations that incorporate microbial data. The primary translational challenges remain the standardization of sampling and analysis, the establishment of meaningful microbial indicators, and the robust demonstration that microbiome-informed decisions consistently improve outcomes beyond existing best practices (Akintan et al., 2025; Ferrocino et al., 2023; Monteiro et al., 2024).

2.7.2 Regulatory Pathways for Microbiome Interventions

Microbiome products often fall under existing regulatory categories (feed additives, veterinary medicines). Regulatory authorities are adapting guidance to address safety, efficacy, and environmental impact.

Safety and Characterization: Agencies like the EFSA (European Food Safety Authority) FEEDAP Panel provide detailed guidance on characterizing microorganisms used as feed additives. This includes requirements for full taxonomic identification, genome-based assessment of virulence factors, and screening for AMR genes. Testing bacterial susceptibility to antimicrobials of veterinary and human importance is central to preventing the spread of resistance determinants via probiotic strains (More et al., 2022).

Efficacy Standards: Regulators demand robust evidence that microbial products deliver claimed benefits under realistic farm conditions. Updated guidance requires rigorous study designs, clear endpoints, and statistical proof for demonstrating efficacy, particularly for products targeting gut health, performance, or reduced emissions (Rodriguez et al., 2025).

Global Alignment and Probiotics: Joint FAO and WHO expert consultations established definitions and evaluation criteria for probiotics, emphasizing strain identity, safety, and evidence from controlled studies. These principles influence the evaluation of animal probiotics, particularly regarding minimum safety and efficacy evidence (Hoffmann et al., 2014).

Engineered Microbes: The regulation of engineered probiotics and gene-edited microbial communities is less harmonized. Some jurisdictions apply existing GMO (Genetically

Modified Organism) frameworks, while others are developing new categories. Safety concerns focus on gene flow, ecological persistence, and unintended effects on nontarget organisms. Regulators are still formulating guidance for these emerging tools, making early engagement essential for developers (Ahmad et al., 2024).

2.7.3 AMR, One Health, and Microbiome Policies

Microbiome interventions are deeply intertwined with the global threat of AMR. A substantial portion of antimicrobials worldwide is used in animals.

One Health Framing: International organizations promote the One Health perspective, integrating human, animal, and environmental sectors to design AMR strategies. Resistant bacteria and genes move across food chains, water, and shared environments, making the reduction of antimicrobial use in livestock a critical part of global action plans (Khairullah et al., 2025; Milijasevic et al., 2024).

Alternatives to Antibiotics: Probiotics, prebiotics, competitive exclusion cultures, and vaccines are increasingly promoted as crucial tools to reduce reliance on routine antimicrobials in food animals. When properly validated, these strategies maintain growth and health while supporting AMR stewardship goals. However, policymakers emphasize that microbiome products must not be used to mask poor underlying husbandry or welfare issues (Ibeagha-Awemu et al., 2025; Khairullah et al., 2025; Milijasevic et al., 2024).

Risk of Unintended Selection: Microbiome strategies are not automatically neutral regarding AMR. Introducing strains that carry mobile resistance genes or using bacteriophages that select for resistant bacterial variants could inadvertently generate new evolutionary pressures. This risk necessitates genomic screening of all candidate strains and the careful integration of microbiome monitoring into comprehensive AMR surveillance programs (Peixoto et al., 2022; Rodriguez et al., 2025; Simon et al., 2025).

2.7.4 Ethical and Animal Welfare Considerations

Microbiome interventions must balance production goals with ethical responsibility.

Welfare in Experimental Work: Animal experiments require strict ethical review, applying core principles like the three R's (Replacement, Reduction, Refinement). Ethical committees increasingly demand justification for the welfare implications of intensive sampling and challenge infection studies (Clark, 2017; Sneddon et al., 2017).

Intensification Concerns: A key ethical concern is whether microbiome engineering primarily serves to further intensify production, for instance, by making animals more tolerant of high density or crowded housing, without genuinely addressing underlying welfare deficits. Ethical analyses question using host-associated microbiota as a manipulation target when the goal is supporting high-throughput production rather than fundamental welfare improvement (Foo et al., 2017; Smith, 2020).

Ecological Responsibilities: Alterations to the animal microbiome can have downstream effects on the environmental microbial communities. Therefore, researchers have a responsibility to consider ecological consequences. Precision engineering may alter transmission dynamics to wildlife, soil, and water, requiring consideration of both individual and collective goods, and anticipation of long-term ecological changes (Peixoto et al., 2022; Trinh et al., 2018).

2.8 CONCLUSION

Domesticated animals and their microbiomes form a tightly integrated biological system. Microbial communities are not peripheral details but central partners that influence digestion, immunity, metabolism, behavior, product quality, and the overall environmental impact of the host. Rumen microbes determine a cow's feed conversion efficiency and methane output, while cecal communities in poultry dictate whether a flock achieves efficient growth or succumbs to enteric disease. Hindgut microbes in pigs and small ruminants shape postweaning resilience, resistance to parasites, and long-term performance.

Across all species, diet, environment, and management are the strongest factors shaping microbiome structure, with feed composition and quality being the most powerful and practical levers for engineering these communities. While host genetics play a role, their effect is smaller than nutritional and environmental determinants, emphasizing the need for an integrated approach involving both breeding and management. Crucially, microbial diversity and functional richness generally correlate with resilience, metabolic stability, and lower disease risk, while dysbiosis often precedes or accompanies major clinical problems.

Interventions are rapidly transitioning from concept to widespread practice, including nutritional adjustments, the use of probiotics and prebiotics, competitive exclusion, and improved husbandry. Advanced approaches, such as microbial transplantation and selection for heritable microbiome-associated traits, point toward a future where microbiome management becomes a standard component of veterinary and nutritional decision-making. These tools, however, must be developed responsibly, addressing critical regulatory, ethical, and One Health concerns regarding AMR, ecological spillover, and technological equity.

The future success of the field depends on moving from correlation to causation, from mere taxonomic description to functional understanding, and achieving standardization of methods. Ultimately, recognizing the animal as a consortium (holobiont) whose collective host and microbe functions determine its success offers a powerful route toward more sustainable, welfare-oriented, and scientifically informed animal agriculture, aligning directly with the broad goals of One Health.

ACKNOWLEDGMENTS

The authors acknowledge the use of NotebookLM (Google LLC) as an AI-assisted tool for figure design support. All scientific content and final interpretations remain the responsibility of the authors.

REFERENCES

Abbas, W., Howard, J., Paz, H., Hales, K., Wells, J., Kuehn, L., Erickson, G., Spangler, M., & Fernando, S. (2020). Influence of host genetics in shaping the rumen bacterial community in beef cattle. *Scientific Reports*, 10, 15101. https://doi.org/10.1038/s41598-020-72011-9

Ahmad, J., Grunden, A., & Kuzma, J. (2024). Biotechnology executive order opens door for regulatory reform and social acceptance of genetically engineered microbes in agriculture. *GM Crops & Food*, 15, 248–261. https://doi.org/10.1080/21645698.2024.2381294

Akintan, O. A., Gebremedhin, K. G., & Uyeh, D. D. (2025). Linking animal feed formulation to milk quantity, quality, and animal health through data-driven decision-making. *Animals*, 15(2), 162. https://doi.org/10.3390/ani15020162

Aliakbari, A., Zemb, O., Cauquil, L., Barilly, C., Billon, Y., & Gilbert, H. (2022). Microbiability and microbiome-wide association analyses of feed efficiency and performance traits in pigs. *Genetics, Selection, Evolution: GSE*, 54, 29. https://doi.org/10.1186/s12711-022-00717-7

Amat, S., Lantz, H., Munyaka, P., & Willing, B. (2020). Prevotella in pigs: the positive and negative associations with production and health. *Microorganisms*, 8, 1584. https://doi.org/10.3390/microorganisms8101584

Antonissen, G., Eeckhaut, V., Van Driessche, K., Onrust, L., Haesebrouck, F., Ducatelle, R., Moore, R. J., & Van Immerseel, F. (2016). Microbial shifts associated with necrotic enteritis. *Avian Pathology*, 45(3), 308–312. https://doi.org/10.1080/03079457.2016.1152625

Arshad, M., Hassan, F., Rehman, M., Huws, S., Cheng, Y., & Din, A. (2021). Gut microbiome colonization and development in neonatal ruminants: strategies, prospects, and opportunities. *Animal Nutrition*, 7, 883–895. https://doi.org/10.1016/j.aninu.2021.03.004

Aruwa, C., Pillay, C., Nyaga, M., & Sabiu, S. (2021). Poultry gut health – microbiome functions, environmental impacts, microbiome engineering and advancements in characterization technologies. *Journal of Animal Science and Biotechnology*, 12, 119. https://doi.org/10.1186/s40104-021-00640-9

Badhan, A., Wang, Y., Terry, S., Gruninger, R., Guan, L., & McAllister, T. (2025). Interplay of rumen microbiome and the cattle host in modulating feed efficiency and methane emissions. *Journal of Dairy Science*, 108(6), 5489–5501. https://doi.org/10.3168/jds.2024-26063

Bay, V., Gillespie, A., Ganda, E., Evans, N. J., Carter, S. D., Lenzi, L., Lucaci, A., Haldenby, S., Barden, M., Griffiths, B. E., Sánchez-Molano, E., Bicalho, R., Banos, G., Darby, A., & Oikonomou, G. (2023). The bovine foot skin microbiota is associated with host genotype and the development of infectious digital dermatitis lesions. *Microbiome*, 11(1), 4. https://doi.org/10.1186/s40168-022-01440-7

Bay, V., Griffiths, B., Carter, S., Evans, N. J., Lenzi, L., Bicalho, R. C., & Oikonomou, G. (2018). 16S rRNA amplicon sequencing reveals a polymicrobial nature of complicated claw horn disruption lesions and interdigital phlegmon in dairy cattle. *Scientific Reports*, 8(1), 15529. https://doi.org/10.1038/s41598-018-33993-9

Bergamaschi, M., Maltecca, C., Schillebeeckx, C., McNulty, N. P., Schwab, C., Shull, C., Fix, J., & Tiezzi, F. (2020). Heritability and genome-wide association of swine gut microbiome features with growth and fatness parameters. *Scientific Reports*, 10(1), 10134. https://doi.org/10.1038/s41598-020-66791-3

Biasato, I., Ferrocino, I., Dabbou, S., Evangelista, R., Gai, F., Gasco, L., Cocolin, L., Capucchio, M. T., & Schiavone, A. (2020). Black soldier fly and gut health in broiler chickens: insights into the relationship between cecal microbiota and intestinal mucin composition. *Journal of Animal Science and Biotechnology*, 11(1), 1–12. https://doi.org/10.1186/s40104-019-0413-y

Binek, M., Cisek, A., Rzewuska, M., Chrobak-Chmiel, D., Stefańska, I., & Kizerwatter-Swida, M. (2017). Chicken intestinal microbiome: development and function. *Medycyna Weterynaryjna*, 73, 618–625. https://doi.org/10.21521/mw.5790

Borges, N., Keller-Costa, T., Sanches-Fernandes, G., Louvado, A., Gomes, N., & Costa, R. (2020). Bacteriome structure, function, and probiotics in fish larviculture: the good, the bad, and the gaps. *Annual Review of Animal Biosciences*, 9(1), 423–452. https://doi.org/10.1146/annurev-animal-062920-113114

Bortoluzzi, C., Watson, M., Iuspa, M., Lumpkins, B., Mathis, G., Jones, M., & Hofacre, C. (2025). Precision biotics enhance growth performance in broiler chickens by selectively modifying their intestinal microbiome to better respond to enteric challenges. *Poultry Science*, 104, 105454. https://doi.org/10.1016/j.psj.2025.105454

Branck, T., Hu, Z., Nickols, W., Walsh, A., Bhosle, A., Short, M., Nearing, J., Asnicar, F., McIver, L., Maharjan, S., Rahnavard, A., Louyakis, A., Badri, D., Brockel, C., Thompson, K., & Huttenhower, C. (2024). Comprehensive profile of the companion animal gut microbiome integrating reference-based and reference-free methods. *The ISME Journal*, 18, wrae201. https://doi.org/10.1093/ismejo/wrae201

Canibe, N., Højberg, O., Kongsted, H., Vodolazska, D., Lauridsen, C., Nielsen, T. S., & Schönherz, A. A. (2022). Review on preventive measures to reduce post-weaning diarrhoea in piglets. *Animals*, 12(19), 2585. https://doi.org/10.3390/ani12192585

Chen, H., Wang, C.-J., Huasai, S., & Chen, A. (2021). Effects of dietary forage to concentrate ratio on nutrient digestibility, ruminal fermentation and rumen bacterial composition in Angus cows. *Scientific Reports*, 11, 17023. https://doi.org/10.1038/s41598-021-96580-5

Chen, S., Luo, S., & Yan, C. (2021). Gut microbiota implications for health and welfare in farm animals: a review. *Animals: An Open Access Journal from MDPI*, 12, 93. https://doi.org/10.3390/ani12010093

Choudhury, R., Middelkoop, A., Boekhorst, J., Gerrits, W., Kemp, B., Bolhuis, J., & Kleerebezem, M. (2021). Early life feeding accelerates gut microbiome maturation and suppresses acute post-weaning stress in piglets. *Environmental Microbiology*, 23, 7201–7213. https://doi.org/10.1111/1462-2920.15791

Clark, J. M. (2017). The 3Rs in research: a contemporary approach to replacement, reduction and refinement. *British Journal of Nutrition*, 120, S1–S7. https://doi.org/10.1017/s0007114517002227

Cortés, A., Wills, J., Su, X., Hewitt, R., Robertson, J., Scotti, R., Price, D., Bartley, Y., McNeilly, T., Krause, L., Powell, J., Nisbet, A., & Cantacessi, C. (2019). Infection with the sheep gastrointestinal nematode *Teladorsagia circumcincta* increases luminal pathobionts. *Microbiome*, 8, 60. https://doi.org/10.1186/s40168-020-00818-9

Culp, E., & Goodman, A. (2023). Cross-feeding in the gut microbiome: ecology and mechanisms. *Cell Host & Microbe*, 31(4), 485–499. https://doi.org/10.1016/j.chom.2023.03.016

Deryabin, D., Lazebnik, C., Vlasenko, L., Karimov, I., Kosyan, D., Zatevalov, A., & Duskaev, G. (2024). Broiler chicken cecal microbiome and poultry farming productivity: a meta-analysis. *Microorganisms*, 12, 747. https://doi.org/10.3390/microorganisms12040747

Deusch, S., Camarinha-Silva, A., Conrad, J., Beifuss, U., Rodehutscord, M., & Seifert, J. (2017). A structural and functional elucidation of the rumen microbiome influenced by various diets and microenvironments. *Frontiers in Microbiology*, 8, 1605. https://doi.org/10.3389/fmicb.2017.01605

Dittoe, D. K., Ricke, S. C., & Kiess, A. S. (2018). Organic acids and potential for modifying the avian gastrointestinal tract and reducing pathogens and disease. *Frontiers in Veterinary Science*, 5, 216. https://doi.org/https://doi.org/10.3389/fvets.2018.00216

Dunisławska, A., Sławinska, A., Stadnicka, K., Bednarczyk, M., Gulewicz, P., Józefiak, D., & Siwek, M. (2017). Synbiotics for broiler chickens—in vitro design and evaluation of the influence on host and selected microbiota populations following in ovo delivery. *PLoS One*, 12, e0168587. https://doi.org/10.1371/journal.pone.0168587

Ekunseitan, D., Harrison, S., Ogunade, I., & Fasina, Y. (2025). Influence of different litter regimens on ceca microbiota profiles in *Salmonella*-challenged broiler chicks. *Animals: An Open Access Journal from MDPI*, 15, 2039. https://doi.org/10.3390/ani15142039

Esser, D., Lange, J., Marinos, G., Sieber, M., Best, L., Prasse, D., Bathia, J., Rühlemann, M., Boersch, K., Jaspers, C., & Sommer, F. (2018). Functions of the microbiota for the physiology of animal metaorganisms. *Journal of Innate Immunity*, 11, 393–404. https://doi.org/10.1159/000495115

Estrada, R., Romero, Y., Figueroa, D., Coila, P., Hañari-Quispe, R. D., Aliaga, M., Galindo, W., Alvarado, W., Casanova, D., & Quilcate, C. (2024). Effects of age in fecal microbiota and correlations with blood parameters in genetic nucleus of cattle. *Microorganisms*, 12, 1331. https://doi.org/10.3390/microorganisms12071331

Fan, P., Nelson, C., Driver, J., Elzo, M., Peñagaricano, F., & Jeong, K. (2021). Host genetics exerts lifelong effects upon hindgut microbiota and its association with bovine growth and immunity. *The ISME Journal*, 15, 2306–2321. https://doi.org/10.1038/s41396-021-00925-x

Fan, Y.-H., Ju, T., Bhardwaj, T., Korver, D., & Willing, B. (2023). Week-old chicks with high Bacteroides abundance have increased short-chain fatty acids and reduced markers of gut inflammation. *Microbiology Spectrum*, 11, e03616-22. https://doi.org/10.1128/spectrum.03616-22

Feehan, B., Ran, Q., Dorman, V., Rumback, K., Pogranichniy, S., Ward, K., Goodband, R., Niederwerder, M., Summers, K., & Lee, S. (2022). Stability and volatility shape the gut bacteriome and *Kazachstania slooffiae* dynamics in preweaning, nursery and adult pigs. *Scientific Reports*, 12, 15080. https://doi.org/10.1038/s41598-022-19093-9

Ferretti, P., Pasolli, E., Tett, A., Asnicar, F., Gorfer, V., Fedi, S., Armanini, F., Truong, D. T., Manara, S., Zolfo, M., Beghini, F., Bertorelli, R., De Sanctis, V., Bariletti, I., Canto, R., Clementi, R., Cologna, M., Crifò, T., Cusumano, G., ... Segata, N. (2018). Mother-to-infant microbial transmission from different body sites shapes the developing infant gut microbiome. *Cell Host & Microbe*, 24, 133–145. https://doi.org/10.1016/j.chom.2018.06.005

Ferrocino, I., Rantsiou, K., McClure, R., Kostic, T., de Souza, R. S. C., Lange, L., FitzGerald, J., Kriaa, A., Cotter, P., Maguin, E., Schelkle, B., Schloter, M., Berg, G., Sessitsch, A., Cocolin, L., & Consortium, T. M. (2023). The need for an integrated multi-OMICs approach in microbiome science in the food system. *Comprehensive Reviews in Food Science and Food Safety*, 22(2), 1082–1103. https://doi.org/https://doi.org/10.1111/1541-4337.13103

Flandroy, L., Poutahidis, T., Berg, G., Clarke, G., Dao, M., Decaestecker, E., Furman, E., Haahtela, T., Massart, S., Plovier, H., Sanz, Y., & Rook, G. (2018). The impact of human activities and lifestyles on the interlinked microbiota and health of humans and of ecosystems. *The Science of the Total Environment*, 627, 1018–1038. https://doi.org/10.1016/j.scitotenv.2018.01.288

Foo, J., Ling, H., Lee, Y. S., & Chang, M. (2017). Microbiome engineering: current applications and its future. *Biotechnology Journal*, 12(3), 1600099. https://doi.org/10.1002/biot.201600099

Gharechahi, J., Vahidi, M. F., Bahram, M., Han, J.-L., Ding, X., & Salekdeh, G. (2020). Metagenomic analysis reveals a dynamic microbiome with diversified adaptive functions to utilize high lignocellulosic forages in the cattle rumen. *The ISME Journal*, 15, 1108–1120. https://doi.org/10.1038/s41396-020-00837-2

Ghosh, S. K., Wong, M. K., Hyodo, S., & Hamasaki, K. (2025). Shifting seas and first feeds: gut microbiome dynamics in juvenile chum salmon (*Oncorhynchus keta*) and their climate vulnerability. *Current Research in Microbial Sciences*, 9, 100452. https://doi.org/10.1016/j.crmicr.2025.100452

Gresse, R., Chaucheyras-Durand, F., Fleury, M., Van De Wiele, T., Forano, E., & Blanquet-Diot, S. (2017). Gut microbiota dysbiosis in postweaning piglets: understanding the keys to health. *Trends in Microbiology*, 25(10), 851–873. https://doi.org/10.1016/j.tim.2017.05.004

Guevarra, R., Hong, S., Cho, J., Kim, B.-R., Shin, J., Lee, J. H., Kang, B., Kim, Y. H., Wattanaphansak, S., Isaacson, R., Song, M., & Kim, H. (2018). The dynamics of the piglet gut microbiome during the weaning transition in association with health and nutrition. *Journal of Animal Science and Biotechnology*, 9, 54. https://doi.org/10.1186/s40104-018-0269-6

Guevarra, R., Lee, J. H., Lee, S., Seok, M.-J., Kim, D., Kang, B., Johnson, T., Isaacson, R., & Kim, H. (2019). Piglet gut microbial shifts early in life: causes and effects. *Journal of Animal Science and Biotechnology*, 10, 1. https://doi.org/10.1186/s40104-018-0308-3

Guo, L., Yao, J., & Cao, Y. (2021). Regulation of pancreatic exocrine in ruminants and the related mechanism: the signal transduction and more. *Animal Nutrition*, 7, 1145–1151. https://doi.org/10.1016/j.aninu.2021.09.004

Gupta, R., & Gaur, S. (2025). Effect of diet and lifestyle on microbiome composition. *International Review of Cell and Molecular Biology*, 395, 157–174. https://doi.org/10.1016/bs.ircmb.2024.12.008

Han, X., Hu, X., Jin, W., & Liu, G. (2024). Dietary nutrition, intestinal microbiota dysbiosis and post-weaning diarrhea in piglets. *Animal Nutrition*, 17, 188–207. https://doi.org/10.1016/j.aninu.2023.12.010

He, Z., Liu, R., Wang, M., Wang, Q., Zheng, J., Ding, J., Wen, J., Fahey, A. G., & Zhao, G. (2023). Combined effect of microbially derived cecal SCFA and host genetics on feed efficiency in broiler chickens. *Microbiome*, 11(1), 198. https://doi.org/10.1186/s40168-023-01627-6

Henderson, G., Cox, F., Ganesh, S., Jonker, A., Young, W., Abecia, L., Angarita, E., Aravena, P., Arenas, G. N., Ariza, C., Attwood, G., Avila, J. M., Ávila-Stagno, J., Bannink, A., Barahona, R., Batistotti, M., Bertelsen, M., Brown-Kav, A., Carvajal, A., ... Janssen, P. (2015). Rumen microbial community composition varies with diet and host, but a core microbiome is found across a wide geographical range. *Scientific Reports*, 5, 14567. https://doi.org/10.1038/srep14567

Hills, R., Pontefract, B., Mishcon, H., Black, C., Sutton, S., & Theberge, C. (2019). Gut microbiome: profound implications for diet and disease. *Nutrients*, 11, 1613. https://doi.org/10.1159/000523712

Hoffmann, D. E., Fraser, C. M., Palumbo, F., Ravel, J., Rowthorn, V., & Schwartz, J. (2014). Probiotics: achieving a better regulatory fit. *Food and Drug Law Journal*, 69(2), 237–272, ii.

Huang, Q., Sham, R., Deng, Y., Mao, Y., Wang, C., Zhang, T., & Leung, K. (2020). Diversity of gut microbiomes in marine fishes is shaped by host-related factors. *Molecular Ecology*, 29, 5019–5034. https://doi.org/10.1111/mec.15699

Huda, M., Kelly, E., Barron, K., Xue, J., Valdar, W., Tarantino, L., Schoenrock, S., Ideraabdullah, F., & Bennett, B. (2025). The impact of early-life exposures on growth and adult gut microbiome composition is dependent on genetic strain and parent- of- origin. *Microbiome*, 13, 143. https://doi.org/10.1186/s40168-025-02130-w

Ibeagha-Awemu, E., Omonijo, F., Piché, L., & Vincent, A. (2025). Alternatives to antibiotics for sustainable livestock production in the context of the One Health approach: tackling a common foe. *Frontiers in Veterinary Science*, 12, 1605215. https://doi.org/10.3389/fvets.2025.1605215

Ismail, A., Ibrahim, M., Abdl-Rahman, M., & Sawiress, F. (2025). Comparative physiological assessment of three natural feed additives on in-vitro rumen fermentation efficiency. *Veterinary Medical Journal (Giza)*, 71(1), 69–83.

Jadhav, V., Fasina, Y., & Harrison, S. (2025). Dietary polyunsaturated fatty acids effect on cecal microbiome profile of maturing broiler chicken. *Poultry Science*, 104, 105167. https://doi.org/10.1016/j.psj.2025.105167

Jamalludeen, N., Johnson, R. P., Shewen, P. E., & Gyles, C. L. (2009). Evaluation of bacteriophages for prevention and treatment of diarrhea due to experimental enterotoxigenic *Escherichia coli* O149 infection of pigs. *Veterinary Microbiology*, 136(1), 135–141. https://doi.org/https://doi.org/10.1016/j.vetmic.2008.10.021

Jha, R., Fouhse, J., Tiwari, U., Li, L., & Willing, B. (2019). Dietary fiber and intestinal health of monogastric animals. *Frontiers in Veterinary Science*, 6, 48. https://doi.org/10.3389/fvets.2019.00048

Jiang, K., Jianfei, Xu, J., Zhang, Y., & Niu, H. (2025). Multi-omics revealed the effects of different feeding systems on rumen microorganisms, cellulose degradation, and metabolites in mongolian cattle. *Animals: An Open Access Journal from MDPI*, 15, 1774. https://doi.org/10.3390/ani15121774

Kanika, N., Liaqat, N., Chen, H., Ke, J.-P., Lu, G., Wang, J., & Wang, C. (2025). Fish gut microbiome and its application in aquaculture and biological conservation. *Frontiers in Microbiology*, 15, 1521048. https://doi.org/10.3389/fmicb.2024.1521048

Kar, S. K., Schokker, D., Harms, A. C., Kruijt, L., Smits, M. A., & Jansman, A. J. M. (2021). Local intestinal microbiota response and systemic effects of feeding black soldier fly larvae to replace soybean meal in growing pigs. *Scientific Reports*, 11(1), 15088.

Kaur, H., Kaur, G., Gupta, T., Mittal, D., & Ali, S. A. (2023). Integrating omics technologies for a comprehensive understanding of the microbiome and its impact on cattle production. *Biology*, 12, 1200. https://doi.org/10.3390/biology12091200

Khairullah, A., Moses, I., Yanestria, S., Dameanti, F. N. A. E. P., Effendi, M., Tang, J. Y. H., Tyasningsih, W., Budiastuti, B., Kusala, M. K. J., Kurniasih, D. A. A., Wardhani, B. W. K., Wibowo, S., Ma'ruf, I., Fauziah, I., Ahmad, R., & Latifah, L. (2025). Potential of the livestock industry environment as a reservoir for spreading antimicrobial resistance. *Open Veterinary Journal*, 15, 504–518. https://doi.org/10.5455/ovj.2025.v15.i2.2

Khan, I., Nassar, N., Chang, H., Khan, S., Cheng, M., Wang, Z., & Xiang, X. (2024). The microbiota: a key regulator of health, productivity, and reproductive success in mammals. *Frontiers in Microbiology*, 15, 1480811. https://doi.org/10.3389/fmicb.2024.1480811

Khan, M. A. S., & Rahman, S. (2022). Use of phages to treat antimicrobial-resistant *Salmonella* infections in poultry. *Veterinary Sciences*, 9, 438. https://doi.org/10.3390/vetsci9080438

Khanjani, M., Sharifinia, M., Akhavan-Bahabadi, M., & Emerenciano, M. (2024). Probiotics and phytobiotics as dietary and water supplements in biofloc aquaculture systems. *Aquaculture Nutrition*, 2024, 3089887. https://doi.org/10.1155/anu/3089887

Kholif, A., & Olafadehan, O. (2021). Essential oils and phytogenic feed additives in ruminant diet: chemistry, ruminal microbiota and fermentation, feed utilization and productive performance. *Phytochemistry Reviews*, 20, 1087–1108. https://doi.org/10.1007/s11101-021-09739-3

Kikusato, M., & Toyomizu, M. (2023). Mechanisms underlying the effects of heat stress on intestinal integrity, inflammation, and microbiota in chickens. *The Journal of Poultry Science*, 60, 2023021. https://doi.org/10.2141/jpsa.2023021

Kong, F., Wang, S., Zhang, Y., Li, C., Dai, D., Guo, C., Wang, Y., Cao, Z., Yang, H., Bi, Y., Wang, W., & Li, S. (2025). Rumen microbiome associates with postpartum ketosis development in dairy cows: a prospective nested case–control study. *Microbiome*, 13(1), 69. https://doi.org/10.1186/s40168-025-02072-3

Koo, S.-M., Lee, J.-H., Oh, S.-H., & Jang, J.-C. (2025). Impacts of Bacillus-based biotics and an enzyme cocktail on growth performance, immunity, and gut pathogenic microorganisms of nursery pigs under commercial conditions. *Frontiers in Veterinary Science*, 12, 1627739. https://doi.org/10.3389/fvets.2025.1627739

Koutsoumanis, K., Allende, A., Álvarez-Ordoñez, A., Bolton, D., Bover-Cid, S., Chemaly, M., Davies, R., De Cesare, A., Herman, L., Hilbert, F., Lindqvist, R., Nauta, M., Ru, G., Simmons, M., Skandamis, P., Suffredini, E., Argüello, H., Berendonk, T., Cavaco, L., ... Peixe, L. (2021). Role played by the environment in the emergence and spread of antimicrobial resistance (AMR) through the food chain. *EFSA Journal*, 19, e06651. https://doi.org/10.2903/j.efsa.2021.6651

Kouzounis, D., Kers, J., Soares, N., Smidt, H., Kabel, M., & Schols, H. (2022). Cereal type and combined xylanase/glucanase supplementation influence the cecal microbiota composition in broilers. *Journal of Animal Science and Biotechnology*, 13, 51. https://doi.org/10.1186/s40104-022-00702-6

Kumar, S., Indugu, N., Vecchiarelli, B., & Pitta, D. (2015). Associative patterns among anaerobic fungi, methanogenic archaea, and bacterial communities in response to changes in diet and age in the rumen of dairy cows. *Frontiers in Microbiology*, 6, 781. https://doi.org/10.3389/fmicb.2015.00781

Lan, F., Wang, X., Zhou, Q., Li, X., Jin, J., Zhang, W., Wen, C., Wu, G., Li, G., Yan, Y., Yang, N., & Sun, C. (2025). Deciphering the coordinated roles of the host genome, duodenal mucosal genes, and microbiota in regulating complex traits in chickens. *Microbiome*, 13, 62. https://doi.org/10.1186/s40168-025-02054-5

Langda, S., Zhang, C., Zhang, K., Gui, B., Ji, D., Deji, C., Cuoji, A., Wang, X., & Wu, Y. (2020). Diversity and composition of rumen bacteria, fungi, and protozoa in goats and sheep living in the same high-altitude pasture. *Animals: An Open Access Journal from MDPI*, 10, 186. https://doi.org/10.3390/ani10020186

Li, C., Xu, Y., Jia, J., Weng, X., Zhang, Y., Peng, J., An, X., & Wang, G. (2025). Impacts of early weaning on lamb gut health and immune function: short-term and long-term effects. *Animals: An Open Access Journal from MDPI*, 15, 2135. https://doi.org/10.3390/ani15142135

Li, F., Li, C., Chen, Y., Liu, J., Zhang, C.-Y., Irving, B., Fitzsimmons, C., Plastow, G., & Guan, L. (2019). Host genetics influence the rumen microbiota and heritable rumen microbial features associate with feed efficiency in cattle. *Microbiome*, 7, 92. https://doi.org/10.1186/s40168-019-0699-1

Li, Y., Guo, Y., Wen, Z., Jiang, X., Xin, M., & Han, X. (2018). Weaning stress perturbs gut microbiome and its metabolic profile in piglets. *Scientific Reports*, 8, 18068. https://doi.org/10.1038/s41598-018-33649-8

Liao, S., Ji, F., Fan, P., & Denryter, K. (2024). Swine gastrointestinal microbiota and the effects of dietary amino acids on its composition and metabolism. *International Journal of Molecular Sciences*, 25, 1237. https://doi.org/10.3390/ijms25021237

Lim, M., Song, E.-J., Kang, K., & Nam, Y. (2019). Age-related compositional and functional changes in micro-pig gut microbiome. *GeroScience*, 41, 935–944. https://doi.org/10.1007/s11357-019-00121-y

Lin, L., Ma, H., Zhang, J., Yang, H., Zhang, J., Lai, Z., Qi, W., Xie, F., Zhu, W., & Mao, S. (2024). Lignocellulolytic microbiomes orchestrating degradation cascades in the rumen of dairy cattle and their diet-influenced key degradation phases. *Animal Advances*, 1(1), e002. https://doi.org/https://doi.org/10.48130/animadv-0024-0002

Liu, F., McNally, J., Fleming, D., Ingham, A., Hunt, P., & Li, R. (2025). *Escherichia coli* is implicated in the development and manifestation of host susceptibility to the roundworm *Trichostrongylus colubriformis* infections in sheep. *Veterinary Research*, 56, 133. https://doi.org/10.1186/s13567-025-01565-1

Liu, H.-Y., Zhu, C., Zhu, M., Yuan, L., Li, S., Gu, F., Hu, P., Chen, S., & Cai, D. (2024). Alternatives to antibiotics in pig production: looking through the lens of immunophysiology. *Stress Biology*, 4, 1. https://doi.org/10.1007/s44154-023-00134-w

Liu, J., Bai, Y., Liu, F., Kohn, R., Tadesse, D., Sarria, S., Li, R., & Song, J. (2022). Rumen microbial predictors for short-chain fatty acid levels and the grass-fed regimen in Angus cattle. *Animals: An Open Access Journal from MDPI*, 12, 2995. https://doi.org/10.3390/ani12212995

Liu, N., Zhang, Z., Zhang, J., Ma, W., & Wang, C. (2025). Effects of N-Carbamylglutamate supplementation on cecal morphology, microbiota composition, and short-chain fatty acids contents of broiler breeder roosters. *Scientific Reports*, 15, 7489. https://doi.org/10.1038/s41598-025-91577-w

Lorgen-Ritchie, M., Webster, T. U., McMurtrie, J., Bass, D., Tyler, C., Rowley, A., & Martin, S. (2023). Microbiomes in the context of developing sustainable intensified aquaculture. *Frontiers in Microbiology*, 14, 1200997. https://doi.org/10.3389/fmicb.2023.1200997

Ma LC, Zhao HQ, Wu LB, Cheng ZL, & Liu C (2023). Impact of the microbiome on human, animal, and environmental health from a One Health perspective. *Science in One Health*, 2, 100037. https://doi.org/10.1016/j.soh.2023.100037

Marcharla, E., Vishnuprasadh, A., Gnanasekaran, L., Vinayagam, S., Sundaram, T., & Ganesan, S. (2025). The role of functional feed in modulating fish gut microbiome to enhance resistance against aquaculture pathogens. *Probiotics and Antimicrobial Proteins*. https://doi.org/10.1007/s12602-025-10660-w

Martínez-Álvaro, M., Mattock, J., Auffret, M., Weng, Z., Duthie, C.-A., Dewhurst, R. J., Cleveland, M. A., Watson, M., & Roehe, R. (2022). Microbiome-driven breeding strategy potentially improves beef fatty acid profile benefiting human health and reduces methane emissions. *Microbiome*, 10(1), 166. https://doi.org/10.1186/s40168-022-01352-6

Mead, G. (2000). Prospects for "competitive exclusion" treatment to control salmonellas and other foodborne pathogens in poultry. *Veterinary Journal*, 159(2), 111–123. https://doi.org/10.1053/tvjl.1999.0423

Meale, S., Li, S., Azevedo, P., Derakhshani, H., DeVries, T., Plaizier, J., Steele, M., & Khafipour, E. (2017). Weaning age influences the severity of gastrointestinal microbiome shifts in dairy calves. *Scientific Reports*, 7, 198. https://doi.org/10.1038/s41598-017-00223-7

Meale, S. J., Li, S., Azevedo, P., Derakhshani, H., Plaizier, J. C., Khafipour, E., & Steele, M. A. (2016). Development of ruminal and fecal microbiomes are affected by weaning but not weaning strategy in dairy calves. *Frontiers in Microbiology*, 7, 582. www.frontiersin.org/journals/microbiology/articles/10.3389/fmicb.2016.00582

Mercer, E., Ramay, H., Moossavi, S., Laforest-Lapointe, I., Reyna, M., Becker, A., Simons, E., Mandhane, P., Turvey, S., Moraes, T., Sears, M., Subbarao, P., Azad, M., & Arrieta, M. C. (2024). Divergent maturational patterns of the infant bacterial and fungal gut microbiome in the first year of life are associated with inter-kingdom community dynamics and infant nutrition. *Microbiome*, 12, 22. https://doi.org/10.1186/s40168-023-01735-3

Milijasevic, M., Vesković-Moračanin, S., Milijasevic, B., Petrović, J., & Nastasijević, I. (2024). Antimicrobial resistance in aquaculture: risk mitigation within the One Health context. *Foods*, 13, 2448. https://doi.org/10.3390/foods13152448

Miller, H., Schroer, L., Petzel, E., Anderson, C., Schmitz-Esser, S., Ericsson, A., & Petry, A. (2025). Impact of insoluble corn-based fiber on fecal microbiome and short chain fatty acid composition in growing pigs. *Journal of Animal Science*, 103, 96–97. https://doi.org/10.1093/jas/skaf102.103

Mkilima, T. (2025). Synthetic biology approaches for restoring gut microbial balance and engineering disease-specific microbiome therapeutics. *Microbial Pathogenesis*, 207, 107931. https://doi.org/https://doi.org/10.1016/j.micpath.2025.107931

Monteiro, H. F., Figueiredo, C. C., Mion, B., Santos, J. E. P., Bisinotto, R. S., Peñagaricano, F., Ribeiro, E. S., Marinho, M. N., Zimpel, R., da Silva, A. C., Oyebade, A., Lobo, R. R., Coelho, W. M., Jr, Peixoto, P. M. G., Ugarte Marin, M. B., Umaña-Sedó, S. G., Rojas, T. D. G., Elvir-Hernandez, M., Schenkel, F. S., … Lima, F. S. (2024). An artificial intelligence approach of feature engineering and ensemble methods depicts the rumen microbiome contribution to feed efficiency in dairy cows. *Animal Microbiome*, 6(1), 5. https://doi.org/10.1186/s42523-024-00289-5

More, S., Bampidis, V., Benford, D., Bragard, C., Halldorsson, T., Hernandez-Jerez, A., Bennekou, S., Koutsoumanis, K., Lambré, C., Machera, K., Mullins, E., Nielsen, S., Schlatter, J., Schrenk, D., Turck, D., Younes, M., Herman, L., Peláez, C., Van Loveren, H., … Cocconcelli, P. (2022). Evaluation of existing guidelines for their adequacy for the food and feed risk assessment of microorganisms obtained through synthetic biology. *EFSA Journal*, 20, e07479. https://doi.org/10.2903/j.efsa.2022.7479

Mu, Y., Qi, W., Zhang, T., Zhang, J., Mei, S., & Mao, S. (2021). Changes in rumen fermentation and bacterial community in lactating dairy cows with subacute rumen acidosis following rumen content transplantation. *Journal of Dairy Science*, 104(10), 10780–10795. https://doi.org/10.3168/jds.2021-20490

Na, S., & Guan, L. (2022). Understanding the role of rumen epithelial host-microbe interactions in cattle feed efficiency. *Animal Nutrition*, 10, 41–53. https://doi.org/10.1016/j.aninu.2022.04.002

Nadal, A. L., Ikeda-Ohtsubo, W., Sipkema, D., Peggs, D., McGurk, C., Forlenza, M., Wiegertjes, G., & Brugman, S. (2020). Feed, microbiota, and gut immunity: using the zebrafish model to understand fish health. *Frontiers in Immunology*, 11, 114. https://doi.org/10.3389/fimmu.2020.00114

Ndotono, E. W., Khamis, F. M., Bargul, J. L., & Tanga, C. M. (2022). Gut microbiota shift in layer pullets fed on black soldier fly larvae-based feeds towards enhancing healthy gut microbial community. *Scientific Reports*, 12(1), 16714. https://doi.org/10.1038/s41598-022-20736-0

Neil, K., Allard, N., Roy, P., Grenier, F., Menendez, A., Burrus, V., & Rodrigue, S. (2021). High-efficiency delivery of CRISPR-Cas9 by engineered probiotics enables precise microbiome editing. *Molecular Systems Biology*, 17, MSB202110335. https://doi.org/10.15252/msb.202110335

Neves, A., Vieira, R. A. M., Vargas-Bello-Pérez, E., Chen, Y., McAllister, T., Ominski, K., Lin, L., & Guan, L. (2025). Impact of feed composition on rumen microbial dynamics and phenotypic traits in beef cattle. *Microorganisms*, 13, 310. https://doi.org/10.3390/microorganisms13020310

Newbold, C., & Ramos-Morales, E. (2020). Review: ruminal microbiome and microbial metabolome: effects of diet and ruminant host. *Animal: An International Journal of Animal Bioscience*, 14(S1), s78–s86. https://doi.org/10.1017/s1751731119003252

O'Hara, E., Neves, A., Song, Y., & Guan, L. (2020). The role of the gut microbiome in cattle production and health: driver or passenger? *Annual Review of Animal Biosciences*, 8, 199–220. https://doi.org/10.1146/annurev-animal-021419-083952

Özkan, S., Bay, V., Cömert Acar, M., & Yalcın, S. (2024). Partial replacement of soybean with local alternative sources: effects on behavior, cecal microbiota, and intestinal histomorphometry of local chickens. *Frontiers in Veterinary Science*, 11, 1463301. https://doi.org/10.3389/fvets.2024.1463301

Patra, A., & Kar, I. (2021). Heat stress on microbiota composition, barrier integrity, and nutrient transport in gut, production performance, and its amelioration in farm animals. *Journal of Animal Science and Technology*, 63, 211–247. https://doi.org/10.5187/jast.2021.e48

Paz, E., Chua, E., Hassan, S. U., Greeff, J., Palmer, D., Liu, S., Lamichhane, B., Sepúlveda, N., Liu, J., Tay, C., & Martin, G. (2021). Bacterial communities in the gastrointestinal tract segments of helminth-resistant and helminth-susceptible sheep. *Animal Microbiome*, 4, 23. https://doi.org/10.1186/s42523-022-00172-1

Peixoto, R., Voolstra, C., Sweet, M., Duarte, C., Carvalho, S., Villela, H., Lunshof, J., Gram, L., Woodhams, D., Walter, J., Roik, A., Hentschel, U., Thurber, R., Daisley, B., Ushijima, B., Daffonchio, D., Costa, R., Keller-Costa, T., Bowman, J., ... Berg, G. (2022). Harnessing the microbiome to prevent global biodiversity loss. *Nature Microbiology*, 7, 1726–1735. https://doi.org/10.1038/s41564-022-01173-1

Perez-Muñoz, M., Arrieta, M., Ramer-Tait, A., & Walter, J. (2017). A critical assessment of the "sterile womb" and "in utero colonization" hypotheses: implications for research on the pioneer infant microbiome. *Microbiome*, 5, 48. https://doi.org/10.1186/s40168-017-0268-4

Petrullo, L., Webber, Q., Raulo, A., Boutin, S., Lane, J., McAdam, A., & Dantzer, B. (2025). Social microbial transmission in a solitary mammal. *Ecology Letters*, 28, e70186. https://doi.org/10.1111/ele.70186

Pieper, R., Tudela, C., Taciak, M., Bindelle, J., Pérez, J., & Zentek, J. (2016). Health relevance of intestinal protein fermentation in young pigs. *Animal Health Research Reviews*, 17, 137–147. https://doi.org/10.1017/s1466252316000141

Quan, J., Cai, G., Ye, J., Yang, M., Ding, R., Wang, X., Zheng, E., Fu, D., Li, S., Zhou, S., Liu, D., Yang, J., & Wu, Z. (2018). A global comparison of the microbiome compositions of three gut locations in commercial pigs with extreme feed conversion ratios. *Scientific Reports*, 8, 4536. https://doi.org/10.1038/s41598-018-22692-0

Rahayu, S., Amoah, K., Huang, Y., Cai, J., Wang, B., Shija, V. M., Jin, X., Anokyewaa, M., & Jiang, M. (2024). Probiotics application in aquaculture: its potential effects, current status in China and future prospects. *Frontiers in Marine Science*, 11, 1455905. https://doi.org/10.3389/fmars.2024.1455905

Rahman, R., Marcolla, C., & Willing, B. (2025). Fecal microbiota transplantation in pigs: current status and future perspective. *Animal Microbiome*, 7, 76. https://doi.org/10.1186/s42523-025-00440-w

Ramos, S., Jeong, C., Mamuad, L., Kim, S. H., Kang, S., Kim, E., Cho, Y.-I., Lee, S.-S., & Lee, S. (2021). Diet transition from high-forage to high-concentrate alters rumen bacterial community composition, epithelial transcriptomes and ruminal fermentation parameters in dairy cows. *Animals: An Open Access Journal from MDPI*, 11, 838. https://doi.org/10.3390/ani11030838

Ringø, E., Harikrishnan, R., Soltani, M., & Ghosh, K. (2022). The effect of gut microbiota and probiotics on metabolism in fish and shrimp. *Animals: An Open Access Journal from MDPI*, 12, 3016. https://doi.org/10.3390/ani12213016

Rochegüe, T., Haenni, M., Mondot, S., Astruc, C., Cazeau, G., Ferry, T., Madec, J., & Lupo, A. (2021). Impact of antibiotic therapies on resistance genes dynamic and composition of the animal gut microbiota. *Animals: An Open Access Journal from MDPI*, 11, 3280. https://doi.org/10.3390/ani11113280

Rodriguez, J., Cordaillat-Simmons, M., Pot, B., & Druart, C. (2025). The regulatory framework for microbiome-based therapies: insights into European regulatory developments. *NPJ Biofilms and Microbiomes*, 11, 53. https://doi.org/10.1038/s41522-025-00683-0

Rosenberg, E., & Zilber-Rosenberg, I. (2016). Microbes drive evolution of animals and plants: the hologenome concept. *MBio*, 7, 10-1128. https://doi.org/10.1128/mbio.01395-15

Ryu, E., & Davenport, E. (2022). Host genetic determinants of the microbiome across animals: from *Caenorhabditis elegans* to cattle. *Annual Review of Animal Biosciences*, 10, 203–226. https://doi.org/10.1146/annurev-animal-020420-032054

Sarkar, A., Harty, S., Johnson, K., Moeller, A., Archie, E., Schell, L., Carmody, R., Clutton-Brock, T., Dunbar, R., & Burnet, P. (2020). Microbial transmission in animal social networks and the social microbiome. *Nature Ecology & Evolution*, 4, 1020–1035. https://doi.org/10.1038/s41559-020-1220-8

Sarkar, A., McInroy, C. J. A., Harty, S., Raulo, A., Ibata, N. G. O., Valles-Colomer, M., Johnson, K. V. A., Brito, I., Henrich, J., Archie, E., Barreiro, L., Gazzaniga, F., Finlay, B., Koonin, E., Carmody, R., & Moeller, A. (2024). Microbial transmission in the social microbiome and host health and disease. *Cell*, 187, 17–43. https://doi.org/10.1016/j.cell.2023.12.014

Schoultz, I., Claesson, M., Dominguez-Bello, M., Hållenius, F. F., Konturek, P., Korpela, K., Laursen, M., Penders, J., Roager, H., Vatanen, T., Öhman, L., & Jenmalm, M. (2025). Gut microbiota development across the lifespan: disease links and health-promoting interventions. *Journal of Internal Medicine*, 297, 560–583. https://doi.org/10.1111/joim.20089

Sehnal, L., Brammer-Robbins, E., Wormington, A., Bláha, L., Bisesi, J., Larkin, I., Martyniuk, C., Simonin, M., & Adamovsky, O. (2021). Microbiome composition and function in aquatic vertebrates: small organisms making big impacts on aquatic animal health. *Frontiers in Microbiology*, 12, 567408. https://doi.org/10.3389/fmicb.2021.567408

Sergeant, M., Constantinidou, C., Cogan, T., Bedford, M., Penn, C., & Pallen, M. (2014). Extensive microbial and functional diversity within the chicken cecal microbiome. *PLoS One*, 9, e91941. https://doi.org/10.1371/journal.pone.0091941

Sessitsch, A., Wakelin, S., Schloter, M., Maguin, E., Cernava, T., Champomier-Vergès, M., Charles, T., Cotter, P., Ferrocino, I., Kriaa, A., Lebre, P., Cowan, D., Lange, L., Kiran, S., Markiewicz, L., Meisner, A., Olivares, M., Sarand, I., Schelkle, B., … Kostić, T. (2023). Microbiome interconnectedness throughout environments with major consequences for healthy people and a healthy planet. *Microbiology and Molecular Biology Reviews*, 87, e00212-22. https://doi.org/10.1128/mmbr.00212-22

Shan, X., Li, K., Stadler, P., Borbor, M., Reyes, G., Solórzano, R., Chamorro, E., Bayot, B., & Cordero, O. (2025). Microbiome determinants of productivity in aquaculture of whiteleg shrimp. *Applied and Environmental Microbiology*, 91. https://doi.org/10.1128/aem.02420-24

Shehata, A., Paswan, V., Attia, Y., Abdel-Moneim, A., Abougabal, M., Sharaf, M., Elmazoudy, R., Alghafari, W., Osman, M., Farag, M., & Alagawany, M. (2021). Managing gut microbiota through in ovo nutrition influences early-life programming in broiler chickens. *Animals: An Open Access Journal from MDPI*, 11, 3491. https://doi.org/10.3390/ani11123491

Shinkai, T., Takizawa, S., Fujimori, M., & Mitsumori, M. (2023). The role of rumen microbiota in enteric methane mitigation for sustainable ruminant production. *Animal Bioscience*, 37, 360–369. https://doi.org/10.5713/ab.23.0301

Silva, Y. J., Costa, L., Pereira, C., Mateus, C., Cunha, Â., Calado, R., Gomes, N. C. M., Pardo, M. A., Hernandez, I., & Almeida, A. (2014). Phage therapy as an approach to prevent Vibrio anguillarum infections in fish larvae production. *PLOS One*, 9(12), e114197. https://doi.org/10.1371/journal.pone.0114197

Silvestro, S., Biondo, C., Midiri, A., Lucia, B., & Mancuso, G. (2025). The role of livestock antibiotic use in microbiota dysbiosis and neuroinflammation. *Antibiotics*, 14, 608. https://doi.org/10.3390/antibiotics14060608

Simon, B. O., Nnaji, N., Anumudu, C. K., Aleke, J., Ekwueme, C., Uhegwu, C. C., Ihenetu, F., Obioha, P., Ifedinezi, O. V., Ezechukwu, P. S., & Onyeaka, H. (2025). Microbiome-based interventions for food safety and environmental health. *Applied Sciences*, 15, 5219. https://doi.org/10.3390/app15095219

Singh, A., & Kim, W. (2021). Effects of dietary fiber on nutrients utilization and gut health of poultry: a review of challenges and opportunities. *Animals: An Open Access Journal from MDPI*, 11, 181. https://doi.org/10.3390/ani11010181

Smith, H. (2020). An ethical investigation into the microbiome: the intersection of agriculture, genetics, and the obesity epidemic. *Gut Microbes*, 12, 1760712. https://doi.org/10.1080/19490976.2020.1760712

Sneddon, L., Halsey, L., & Bury, N. (2017). Considering aspects of the 3Rs principles within experimental animal biology. *Journal of Experimental Biology*, 220, 3007–3016. https://doi.org/10.1242/jeb.147058

St-Pierre, B., Perez Palencia, J. Y., & Samuel, R. S. (2023). Impact of early weaning on development of the swine gut microbiome. *Microorganisms*, 11(7), 1753. https://doi.org/10.3390/microorganisms11071753

Suárez-Martínez, C., Santaella-Pascual, M., Yagüe-Guirao, G., & Martínez-Graciá, C. (2023). Infant gut microbiota colonization: influence of prenatal and postnatal factors, focusing on diet. *Frontiers in Microbiology*, 14, 1236245. https://doi.org/10.3389/fmicb.2023.1236254

Sun, B., Hou, L., & Yang, Y. (2021). The development of the gut microbiota and short-chain fatty acids of layer chickens in different growth periods. *Frontiers in Veterinary Science*, 8, 666535. https://doi.org/10.3389/fvets.2021.666535

Takeshita, N., Watanabe, T., Ishida-Kuroki, K., & Sekizaki, T. (2021). Transition of microbiota in chicken cecal droppings from commercial broiler farms. *BMC Veterinary Research*, 17, 10. https://doi.org/10.1186/s12917-020-02688-7

Tang, S., Xin, Y., Yunlong, Xu, X., Zhao, S., & Cao, J. (2020). Screening of microbes associated with swine growth and fat deposition traits across the intestinal tract. *Frontiers in Microbiology*, 11, 586776. https://doi.org/10.3389/fmicb.2020.586776

Tang, W., Liu, J., Yanfei, Wei, Y., Liu, J., & Wang, H. (2021). Impairment of intestinal barrier function induced by early weaning via autophagy and apoptosis associated with gut microbiome and metabolites. *Frontiers in Immunology*, 12, 804870. https://doi.org/10.3389/fimmu.2021.804870

Tardiolo, G., La Fauci, D., Riggio, V., Daghio, M., Di Salvo, E., Zumbo, A., & Sutera, A. (2025). Gut microbiota of ruminants and monogastric livestock: an overview. *Animals: An Open Access Journal from MDPI*, 15, 758. https://doi.org/10.3390/ani15050758

Tarnecki, A., Burgos, F., Ray, C., Arias, C., & Dupree, H. (2017). Fish intestinal microbiome: diversity and symbiosis unravelled by metagenomics. *Journal of Applied Microbiology*, 123, 2–17. https://doi.org/10.1111/jam.13415

Tayyab, M., Zhao, Y., & Zhang, Y. (2025). Microbiome engineering to enhance disease resistance in aquaculture: current strategies and future directions. *Frontiers in Microbiology*, 16, 1625265. https://doi.org/10.3389/fmicb.2025.1625265

Trinh, P., Zaneveld, J., Safranek, S., & Rabinowitz, P. (2018). One Health relationships between human, animal, and environmental microbiomes: a mini-review. *Frontiers in Public Health*, 6, 235. https://doi.org/10.3389/fpubh.2018.00235

Upadhaya, S., & Kim, I. (2022). Maintenance of gut microbiome stability for optimum intestinal health in pigs – a review. *Journal of Animal Science and Biotechnology*, 13, 140. https://doi.org/10.1186/s40104-022-00790-4

Vadstein, O., Attramadal, K., Bakke, I., Forberg, T., Olsen, Y., Verdegem, M., Giatsis, C., Skjermo, J., Aasen, I., Gatesoupe, F., Dierckens, K., Sorgeloos, P., & Bossier, P. (2018). Managing the microbial community of marine fish larvae: a holistic perspective for larviculture. *Frontiers in Microbiology*, 9, 140. https://doi.org/10.3389/fmicb.2018.01820

van Gastelen, S., Westerlaan, B., Houwers, D. J., & van Eerdenburg, F. J. C. M. (2011). A study on cow comfort and risk for lameness and mastitis in relation to different types of bedding materials. *Journal of Dairy Science*, 94(10), 4878–4888. https://doi.org/https://doi.org/10.3168/jds.2010-4019

Vitt, R., Weber, L., Zollitsch, W., Hörtenhuber, S. J., Baumgartner, J., Niebuhr, K., Piringer, M., Anders, I., Andre, K., Hennig-Pauka, I., Schönhart, M., & Schauberger, G. (2017). Modelled performance of energy saving air treatment devices to mitigate heat stress for confined livestock buildings in Central Europe. *Biosystems Engineering*, 164, 85–97. https://doi.org/https://doi.org/10.1016/j.biosystemseng.2017.09.013

Waliaula, P., Yin, X., Lepp, D., Clairmont, L., Kiarie, E., & Diarra, M. (2025). Cecal bacterial communities and lung immunity in layers fed omega-3 fatty acids or yeast bioactives under different space allowance. *Poultry Science*, 104, 105866. https://doi.org/10.1016/j.psj.2025.105866

Wallace, R., Sasson, G., Garnsworthy, P., Tapio, I., Gregson, E., Bani, P., Huhtanen, P., Bayat, A., Strozzi, F., Biscarini, F., Snelling, T., Saunders, N., Potterton, S., Craigon, J., Minuti, A., Trevisi, E., Callegari, M., Cappelli, F., Cabezas-Garcia, E., … Mizrahi, I. (2019). A heritable subset of the core rumen microbiome dictates dairy cow productivity and emissions. *Science Advances*, 5, eaav8391. https://doi.org/10.1126/sciadv.aav8391

Wang, L., Lilburn, M., & Yu, Z. (2016). Intestinal microbiota of broiler chickens as affected by litter management regimens. *Frontiers in Microbiology*, 7, 593. https://doi.org/10.3389/fmicb.2016.00593

Wang, W., Wei, Z., Li, Z., Ren, J., Song, Y., Xu, J., Liu, A., Li, X., Li, M., Fan, H., Jin, L., Niyazbekova, Z., Wang, W., Gao, Y., Jiang, Y., Yao, J., Li, F., Wu, S., & Wang, Y. (2024). Integrating genome- and transcriptome-wide association studies to uncover the host–microbiome interactions in bovine rumen methanogenesis. *IMeta*, 3, e234. https://doi.org/10.1002/imt2.234

Warne, R., & Kirschman, L. (2025). Microbiomes in animal ecology: advances, assumptions, and opportunities. *Integrative and Comparative Biology*, 65(3), 772–782. https://doi.org/10.1093/icb/icaf112

Wei, X., Tsai, T., Howe, S., & Zhao, J. (2021). Weaning induced gut dysfunction and nutritional interventions in nursery pigs: a partial review. *Animals: An Open Access Journal from MDPI*, 11, 1279. https://doi.org/10.3390/ani11051279

Weimer, P. (2015). Redundancy, resilience, and host specificity of the ruminal microbiota: implications for engineering improved ruminal fermentations. *Frontiers in Microbiology*, 6, 296. https://doi.org/10.3389/fmicb.2015.00296

Woodhams, D., Woodhams, D., Bletz, M., Becker, C., Bender, H., Buitrago-Rosas, D., Buitrago-Rosas, D., Diebboll, H., Huynh, R., Kearns, P., Kueneman, J., Kurosawa, E., LaBumbard, B., Lyons, C., McNally, K., McNally, K., Schliep, K., Shankar, N., Tokash-Peters, A., … Whetstone, R. (2020). Host-associated microbiomes are predicted by immune system complexity and climate. *Genome Biology*, 21, 23. https://doi.org/10.1186/s13059-019-1908-8

Xiao, L., & Zhao, F. (2023). Microbial transmission, colonisation and succession: from pregnancy to infancy. *Gut*, 72, 772–786. https://doi.org/10.1136/gutjnl-2022-328970

Xiao, S.-S., Mi, J., Mei, L., Liang, J., Feng, K., Wu, Y., Liao, X.-D., & Wang, Y. (2021). Microbial diversity and community variation in the intestines of layer chickens. *Animals: An Open Access Journal from MDPI*, 11, 840. https://doi.org/10.3390/ani11030840

Xie, Y., Sun, H., Xue, M., & Liu, J. (2022). Metagenomics reveals differences in microbial composition and metabolic functions in the rumen of dairy cows with different residual feed intake. *Animal Microbiome*, 4, 19. https://doi.org/10.1186/s42523-022-00170-3

Xu, Q., Qiao, Q., Gao, Y., Hou, J., Hu, M., Du, Y., Zhao, K., & Li, X. (2021). Gut microbiota and their role in health and metabolic disease of dairy cow. *Frontiers in Nutrition*, 8, 701511. https://doi.org/10.3389/fnut.2021.701511

Xu, X., Zhou, Z., Zhou, Z., Ma, Y., Luo, D., Zhang, S., Yang, P., An, T., & Sun, Q. (2025). Effect of *Lactiplantibacillus plantarum* N-1 and isomaltose-oligosaccharide on promoting growth performance and modulating the gastrointestinal microbiota in newborn Hu sheep. *Animal Microbiome*, 7, 25. https://doi.org/10.1186/s42523-025-00392-1

Xue, M., Sun, H., Wu, X., Liu, J.-X., & Guan, L. (2020). Multi-omics reveals that the rumen microbiome and its metabolome together with the host metabolome contribute to individualized dairy cow performance. *Microbiome*, 8, 64. https://doi.org/10.1186/s40168-020-00819-8

Xue, M., Xie, Y., Zhong, Y., Ma, X.-J., Sun, H., & Liu, J. (2022). Integrated meta-omics reveals new ruminal microbial features associated with feed efficiency in dairy cattle. *Microbiome*, 10, 32. https://doi.org/10.1186/s40168-022-01228-9

Yang, H., Huang, X., Fang, S., Xin, W., Huang, L., & Chen, C. (2016). Uncovering the composition of microbial community structure and metagenomics among three gut locations in pigs with distinct fatness. *Scientific Reports*, 6, 27427. https://doi.org/10.1038/srep27427

Ye, X., Sahana, G., Lund, M., & Cai, Z. (2025). The crosstalk between host and rumen microbiome in cattle: insights from multi-omics approaches and genome-wide association studies. *World Journal of Microbiology & Biotechnology*, 41, 267. https://doi.org/10.1007/s11274-025-04504-6

Yuan, N., Wang, Y., Pan, Q., Zhao, L., Qi, X., Sun, S., Suolang, Q., Ciren, L., Danzeng, L., Liu, Y., Zhang, L., Gao, T., Basang, Z., Lian, H., & Sun, Y. (2023). From the perspective of rumen microbiome and host metabolome, revealing the effects of feeding strategies on Jersey Cows on the Tibetan Plateau. *PeerJ*, 11, e16010. https://doi.org/10.7717/peerj.16010

Yue, T., Lu, Y.-N., Ding, W., Xu, B., Zhang, C., Li, L., Jian, F., & Huang, S. (2025). The role of probiotics, prebiotics, synbiotics, and postbiotics in livestock and poultry gut health: a review. *Metabolites*, 15, 478. https://doi.org/10.3390/metabo15070478

Yukgehnaish, K., Kumar, P., Sivachandran, P., Marimuthu, K., Arshad, A., Paray, B., & Arockiaraj, J. (2020). Gut microbiota metagenomics in aquaculture: factors influencing gut microbiome and its physiological role in fish. *Reviews in Aquaculture*, 12, 1903–1927. https://doi.org/10.1111/raq.12416

Zhang, G., Wang, Y., Luo, H., Qiu, W., Zhang, H., Hu, L., Wang, Y., Dong, G., & Guo, G. (2019). The association between inflammaging and age-related changes in the ruminal and fecal microbiota among lactating holstein cows. *Frontiers in Microbiology*, 10, 1803. https://doi.org/10.3389/fmicb.2019.01803

Zhang, K., He, C., Wang, L., Suo, L., Guo, M., Guo, J., Zhang, T., Xu, Y., Lei, Y., Liu, G., Qian, Q., Mao, Y., Kalds, P., Wu, Y., Cuoji, A., Yang, Y., Brugger, D., Gan, S., Wang, M., ... Chen, Y. (2024). Compendium of 5810 genomes of sheep and goat gut microbiomes provides new insights into the glycan and mucin utilization. *Microbiome*, 12, 104. https://doi.org/10.1186/s40168-024-01806-z

Zhang, Z., Yang, Q., Liu, H., Jin, J., Yang, Y., Zhu, X., Han, D., Zhou, Z., & Xie, S. (2024). Potential functions of the gut microbiome and modulation strategies for improving aquatic animal growth. *Reviews in Aquaculture*, 17(1), e12959. https://doi.org/10.1111/raq.12959

Zhou, H., Yang, L., Ding, J., Dai, R., He, C., Xu, K., Luo, L., Xiao, L., Zheng, Y., Han, C., Akinyemi, F., Honaker, C., Zhang, Y., Siegel, P., & Meng, H. (2022). Intestinal microbiota and host cooperate for adaptation as a hologenome. *MSystems*, 7, e01261-21. https://doi.org/10.1128/msystems.01261-21

Zhou, M., Peng, Y., Chen, Y., Klinger, C., Oba, M., Liu, J.-X., & Guan, L. (2018). Assessment of microbiome changes after rumen transfaunation: implications on improving feed efficiency in beef cattle. *Microbiome*, 6. https://doi.org/10.1186/s40168-018-0447-y

Zhou, M., Wu, L., Sun, X., Liu, M., Wang, Y., Yang, B., Ai, H., Chen, C., & Huang, L. (2025). Assessing the relationship between the gut microbiota and growth traits in Chinese indigenous pig breeds. *BMC Veterinary Research*, 21. https://doi.org/10.1186/s12917-025-04739-3

Zhuang, Y., Dong, X., Chen, T., Liu, S., Wang, J., Xiao, J., Ma, M., Wang, W., Li, M., Li, S., Cao, Z., Wang, Y., & Ma, J. (2025). Pasteurized colostrum improves blood immunity and gastrointestinal microbiota in dairy calves from birth to 180 days of age. *Microorganisms*, 13, 2089. https://doi.org/10.3390/microorganisms13092089

Zhuang, Y., Liu, S., Gao, D., Xu, Y., Jiang, W., Hou, G., Li, S., Zhao, X., Chen, T., Li, S., Zhang, S., Huang, Y., Wang, J., Xiao, J., Li, M., Wang, W., Li, S., & Cao, Z. (2024). Maternal gastrointestinal microbiome shapes gut microbial function and resistome of newborns in a cow-to-calf model. *Microbiome*, 12, 216. https://doi.org/10.1186/s40168-024-01943-5

CHAPTER THREE

Microbial Ecology of Bovine Milk and Its Health Implications on Dairy Processing

Anuradha Chakroborty, Soumitra Nath, and Indu Sharma

3.1 INTRODUCTION

Bovine milk is a nutritionally rich biological fluid that supports the growth and development of mammals and serves as a staple component of the human diet (Quigley et al., 2013). Traditionally, milk was considered sterile, but advances in culture-independent molecular methods have revealed that it contains a diverse and dynamic microbiome (Guo et al., 2024; Steinberg et al., 2022; Yan et al., 2022). This microbial community includes bacteria, yeasts, and fungi originating from the mammary gland, the environment, and milking equipment (Wouters et al., 2022; Yuan et al., 2022). The balance among these microorganisms influences milk composition, fermentation potential, flavour development, and safety (Coelho et al., 2022).

Beneficial microbes, particularly lactic acid bacteria (LAB), contribute to the nutritional and functional properties of milk through fermentation and the production of anti-microbial compounds that inhibit spoilage and pathogenic organisms (Quigley et al., 2013; Galli et al., 2022). However, the milk ecosystem can also harbour opportunistic and pathogenic species such as *Staphylococcus aureus*, *Escherichia coli*, and *Pseudomonas fluorescens*, which may cause mastitis and food-borne illness if not properly managed (Rešetar et al., 2023; Büthe et al., 2025; Chakroborty et al., 2025; Tusa et al., 2024). Next-generation sequencing has further enhanced the understanding of microbial interactions and their implications for dairy processing and product quality (Parente et al., 2023).

Previous studies emphasise that environmental and management practices, such as hygiene, feeding regime, antibiotic use and storage conditions, play critical roles in shaping the milk microbiome (De Jonghe et al., 2011; Vissers et al., 2007; Yuan et al., 2018; Lettat et al., 2013; Li et al., 2018). A detailed characterisation of both beneficial and harmful microbial communities helps enhance fermentation and eliminate pathogens, promoting more functional, safer dairy products. The insights from microbial profiling also support precision management practices that maintain milk hygiene and stability throughout production and processing. The objective of this review is to understand the diversity, ecological roles, and functional significance of the milk microbiota, and to explore the factors that influence its composition and impact on dairy processing.

3.2 NUTRITIONAL VALUE AND HEALTH BENEFITS OF RAW BOVINE MILK

Milk and dairy products are rich sources of essential nutrients, especially for children and pregnant women. They provide high-quality protein and digestible calcium and are a good source of key vitamins such as vitamin B_{12}, thiamine (B_1), and riboflavin (B_2), all of which support growth, energy metabolism, and overall health (Figure 3.1). Along with other animal-source foods, milk consumption has been found to improve anthropometric indices and cognitive function in children and to reduce nutritional deficiencies (Grace

DOI: 10.1201/9781003638568-3

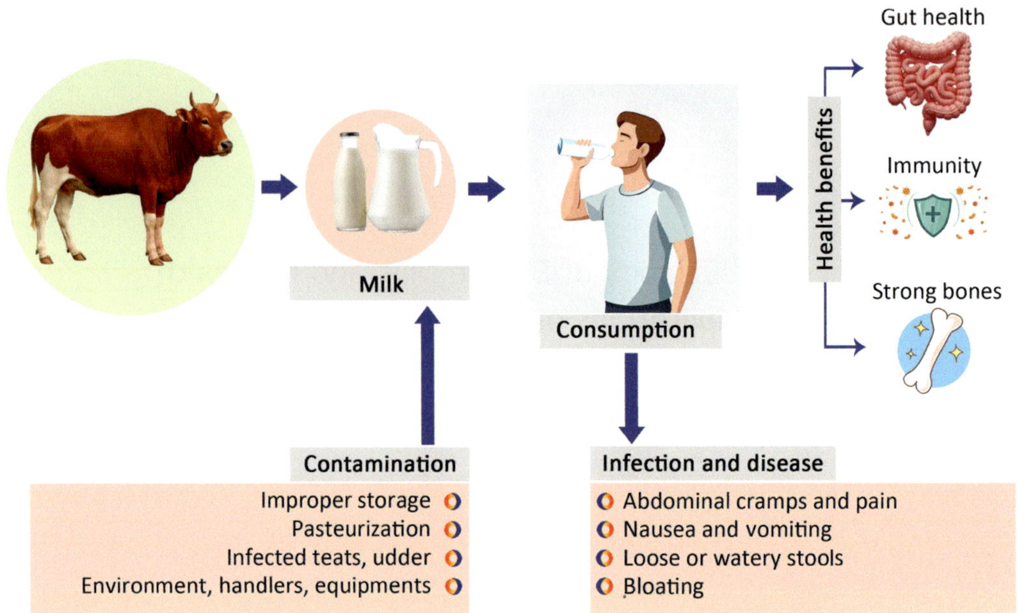

Figure 3.1 Behaviour of raw milk consumption: Risks and benefits.

et al., 2020). The low pH of fermented milk delays gastric emptying and prolongs the gastrointestinal transit time. Fermented milk has been shown to stimulate the Th1 immune response and downregulate the Th2 response, thereby promoting a more balanced and protective immune profile, especially against cancer, viral infections and allergic reactions (LeBlanc et al., 2004; Vinderola et al., 2005).

During fermentation, bacteria and yeasts in milk metabolise lactose to various degradation products depending on the species involved. Certain bacteria can also improve intestinal microbial balance, offering positive health effects in the digestive tract and overall metabolism. Lactobacilli, Streptococci and Bifidobacteria primarily produce lactic acid, acetic acid and monosaccharides (especially galactose), whereas yeasts generate CO_2 and ethanol (Saxelin et al., 2003). Besides glucose and lactose, lactic acid and other organic acids show increased iron absorption, especially when consumed with meals, and create an environment less favourable for the growth of pathogenic bacteria (Branca et al., 2002). Depending on the bacterial strains used in fermentation, products of potentially health-promoting metabolites with distinct flavours and aromas can be obtained.

3.3 SOURCES OF MICROBES IN MILK

Milk in healthy udder cells is generally considered sterile (Tolle et al., 1980). It is colonised by microorganisms soon after secretion through exposure to a variety of sources, including the teat apex, milking equipment, air, water, feed, grass, soil, and other environmental reservoirs (Figure 3.2) (Verdier-Metz et al., 2012). Although maintaining high milk quality is important, certain farming practices negatively affect the characteristic flavours and aromas of dairy products.

Studies reported that feeding practices play a role in the microbial load of cows' milk. The same study also reported higher abundances of *Staphylococcus* spp. in milk from cows that were grazed outdoors than those fed indoors (Hagi et al., 2010). Milking machines may act as reservoirs of microbial contamination, and therefore, equipment hygiene and milking practices strongly influence the microbial population of milk (Michel et al., 2006). Implementing strict hygiene measures in the dairy system can reduce microbial load and ensure milk safety. However, these measures may also unintentionally reduce populations of technologically important microorganisms (Monsallier et al., 2012). As a result, producers may need to compensate for these losses by adding starters and

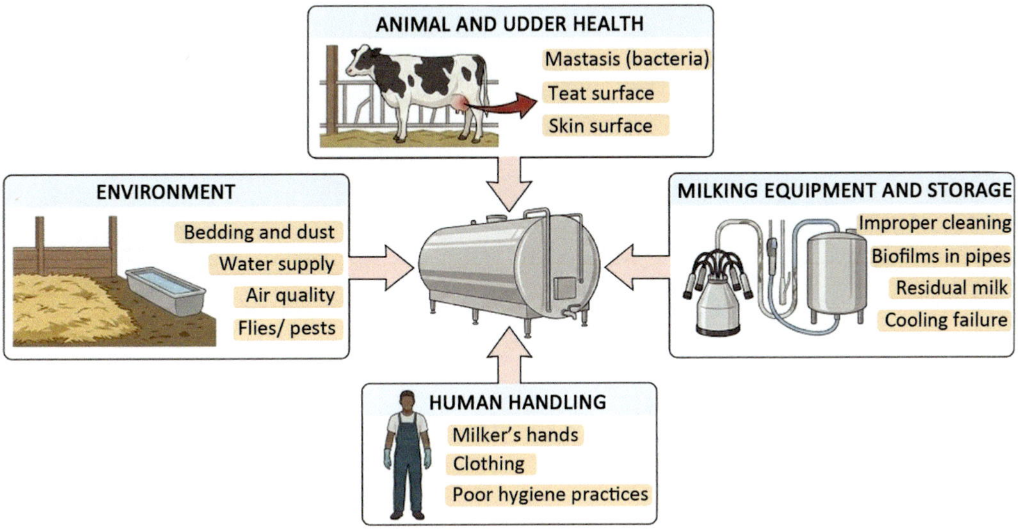

Figure 3.2 Sources of milk contamination.

adjunct cultures to maintain desirable sensory and functional attributes.

3.4 DIVERSITY OF THE BOVINE MILK MICROBIOME

The bovine milk microbiome is a complex and diverse community of microbial species shaped by genetic and physiological factors. Many of these populations subsequently play key roles in natural fermentation processes and are intentionally incorporated as starter or adjunct cultures designed to confer desirable technological and sensory traits to fermented dairy products (Figure 3.3). Their incorporation becomes particularly important when regulatory standards require the use of pasteurised milk, which eliminates native commensal microbiota. Thus, the reintroduction of selected dairy microorganisms can help restore the functional contribution of commensal populations and mitigate associated adverse effects on flavour, texture, and overall product quality (Galli et al., 2022).

3.4.1 Dominant Bacterial Phyla in Raw Bovine Milk

The microbial ecosystem of raw bovine milk is dominated by several major phyla across farms and geographical locations, including Firmicutes, Proteobacteria, Actinobacteria, and Bacteroidetes (Yan et al., 2022; Steinberg et al., 2022; Guo et al., 2019). The relative abundance of these specific phyla reflects microbial ecology of the mammary gland and influences the quality, fermentation and shelf life of dairy products. The phylum Firmicutes encompasses a diverse range of Gram-positive LAB, including Lactobacillus, Lactococcus, Streptococcus, and Enterococcus (Coelho et al., 2022). LAB produces bacteriocins and secondary metabolites that contribute to acidification, fermentation, pathogen exclusion, and flavour development in fermented milk (Quigley et al., 2013; Yuan et al., 2022). However, the phylum also comprises important mastitis-associated pathogens, including *Staphylococcus aureus* and *Streptococcus uberis*, which can trigger inflammatory responses in the mammary gland and alter milk quality (Rešetar et al., 2023; Büthe et al., 2025; Chakroborty et al., 2025). A summary of the dominant microbial species in bovine milk, their ecological roles, and technological relevance is presented in Table 3.1.

Proteobacteria, Actinobacteria, and Bacteroides frequently dominate milk with high microbial loads or poor hygiene practices. Proteobacteria, such as Pseudomonas, Acinetobacter, and Psychrobacter, become particularly dominant during cold storage. They secrete heat-resistant lipases and proteases that can survive pasteurisation, leading to proteolysis, off-flavours, and post-process spoilage (Yuan et al., 2018). Members of the family Enterobacteriaceae (e.g., *Escherichia coli*, Klebsiella, Salmonella) and Bacteroides enter milk

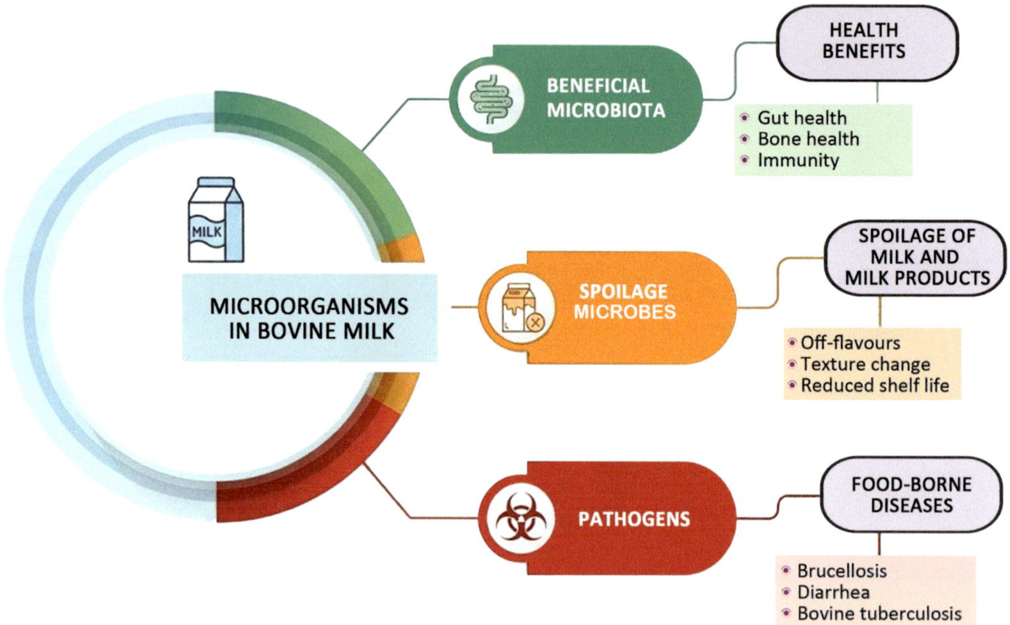

Figure 3.3 Microorganisms in raw milk and their associated behaviour.

through poor hygiene, posing zoonotic risks and reducing the shelf life of dairy products (Tusa et al., 2024).

3.4.2 Role of Beneficial Microorganisms in Raw Bovine Milk

LAB and Bifidobacteria are particularly significant in raw milk due to their probiotic and fermentative properties. According to FAO/WHO (2001), probiotics are live microorganisms that confer health benefits to the hosts when administered in adequate amounts. They contribute to flavour development, pathogen inhibition, and enhancement of the functional and nutritional value of dairy products (Lucey et al., 2015). Most probiotic bacteria used in dairy research today belong to the genera Lactobacillus and Enterococcus. Commonly studied species include *L. acidophilus*, *L. fermentum*, *L. casei*, *L. reuteri*, *L. helveticus*, *L. lactis*, *L. plantarum*, *E. faecalis*, and *E. faecium* (Mulaw et al., 2019). Additionally, LAB can be used to treat several illnesses caused by pathogenic microorganisms that are resistant to drugs, making them a promising alternative for natural biotherapeutic use in both animals and humans.

3.4.2.1 Lactococcus

The genus *Lactococcus* was discovered in the late 19th century by Joseph Lister through his studies on milk fermentation. The organism was first named *Bacterium lactis* and was later reclassified as *Streptococcus lactis* by Orla Jensen before being placed in the genus *Lactococcus*. As of March 2023, the NCBI taxonomy browser describes 22 species of Lactococcus, with 694 genome sequences. Among these, *L. lactis*, *L. cremoris*, *L. raffinolactis*, *L. laudensis*, and *L. hircilactis* have been isolated from dairy environments and animal milk sources (Mahony et al., 2023). Although *Lactococcus garvieae* is historically viewed as a pathogen, it is now recognised as a common inhabitant of dairy environments, with possible probiotic or technological benefits (Dimov et al., 2023). A recent study has shown the predominance of genetically distinct clusters of *L. garvieae* isolates in dairy and aquatic environments; however, lactose utilisation ability was observed only in the dairy isolates. Hence, it is hypothesised that dairy isolates may have gained this metabolic capability through horizontal gene transfer, potentially from related species within the Lactococcus genus (Shahi et al., 2020).

Lactococcus raffinolactis is not typically used as a primary starter culture. However, strain APC 3967,

TABLE 3.1
Some microbial species in bovine milk and their functional roles

Microbial species	Functional/technological role	Health impact	Reference
Lactococcus lactis, L. cremoris	Starter culture, acidification, flavour formation	Improves texture and safety in cheese and yoghurt	Cheng et al., 2022
Lactobacillus helveticus, L. plantarum	Proteolysis and peptide formation	Enhances probiotic activity and digestibility	Zhao et al., 2021
Streptococcus thermophilus	Thermophilic fermentation, EPS production	Improves yoghurt viscosity and gut health	Jurášková et al., 2025
Propionibacterium freudenreichii	Secondary fermentation, vitamin B12 synthesis	Promotes gut health, used in Swiss cheese	Rocchetti et al., 2025
Leuconostoc mesenteroides	Heterofermentation, diacetyl production	Enhances aroma in raw milk cheeses	Sarhir et al., 2023
Enterococcus faecium, E. faecalis	Proteolysis, ester formation	Contributes to flavour but may harbour ARGs	Chaves et al., 2024
Pediococcus pentosaceus	Bacteriocin production	Provides anti-microbial protection	Ghazaei et al., 2023
Pseudomonas fluorescens, P. aeruginosa	Psychrotrophic spoilage	Causes off-flavours during cold storage	Hanim et al., 2025
Staphylococcus aureus	Pathogen, enterotoxin production	Causes mastitis and food-borne illness	Shalaby et al., 2024
Bifidobacterium animalis, B. longum	Gut microbiota modulation	Supports digestive health, used in probiotics	Schöpping et al., 2021

isolated from raw milk, produces a circular bacteriocin raffinocyclicin with broad-spectrum antimicrobial activity, suggesting its application in microbial modulation during cheese ripening (Miceli et al., 2024). Additionally, recent research demonstrated its successful co-fermentation, which can contribute to carbohydrate degradation and flavour development when combined with other lactic acid bacteria (Canon et al., 2020). *L. raffinolactis* strain PO3 enhanced neural marker expression and showed probiotic potential in in vitro models, hinting at broader physiological capabilities beyond dairy use (Khalifa et al., 2023). *L. lactis* subsp. *lactis* and *L. cremoris* strains from raw bovine milk exhibit lower pH values than commercial starter cultures and produce diverse flavour compounds through strong acidification. The product obtained has a stable shelf life during storage (Lee et al., 2020). A recent study on naturally fermented cow milk revealed that strains of *L. lactis* subsp. *cremoris* shows strong acid and bile resistance, high cell surface hydrophobicity, and carries genes for mucosal adhesion and bile salt hydrolase, suggesting promising probiotic potential (Rai et al., 2022).

3.4.2.2 *Lactobacillus*

The genus *Lactobacillus* comprises an exceptionally diverse group of bacteria, now recognised as part of the broader Lactobacillaceae family, which includes over 280 distinct species and subspecies (Walter et al., 2023). Lactobacilli are commonly found in carbohydrate-rich environments, such as plant matter, silage, animal hosts, and raw milk (Parente et al., 2023). Strains of Lactobacillus such as *L. helveticus*, *L. delbrueckii* subsp. *bulgaricus* and *L. delbrueckii* subsp. *lactis* are increasingly utilised in the dairy industry due to their unique technological traits. *Lactobacillus delbrueckii* is frequently detected in raw milk samples and constitutes a dominant population in many traditionally manufactured cheeses (Elean et al., 2023). The widely used industrial strain *L. bulgaricus* 2038 is known for harbouring a specific gene cluster involved in exopolysaccharide biosynthesis, which has been shown to significantly improve the texture and viscosity of yoghurt, especially when cultured in lactose-hydrolysed milk. This trait not only supports better fermentation but also enhances mouthfeel and reduces syneresis in the final product (Yamamoto et al., 2021). *L. helveticus* undergoes rapid autolysis and contributes to reduced bitterness and increased flavour notes in cheese by releasing intracellular enzymes during cheese ripening (Chelladhurai et al., 2023). Their strong proteolytic activity results in the production of bioactive peptides and flavour-enhancing amino acids by protein breakdown (Sıçramaz et al., 2022). In addition, other Lactobacillus species in raw milk can proliferate during dairy processing and often become the dominant member of the microbial community during cheese ripening.

3.4.2.3 *Streptococcus*

The genus *Streptococcus* comprises over 100 recognised species, with the number of subspecies also increasing (Kanungo et al., 2023; Kumar et al., 2022). Several Streptococcus species that are frequently detected in milk and dairy products include *Streptococcus uberis*, *Streptococcus agalactiae*, *Streptococcus dysgalactiae*, *Streptococcus bovis*, and *Streptococcus macedonicus* (Hoque et al., 2024). Among these, *S. macedonicus* displayed desirable characteristics to generate inhibitory compounds against spoilage organisms and pathogens. They lack antibiotic resistance and haemolytic activity and produce peptidases that support their safety profile. Whole-genome sequencing of *S. macedonicus* strains isolated from dairy sources indicates that the species is undergoing progressive genome decay, evidenced by a reduced number of single-copy genes and genomic islands, which is consistent with evolutionary adaptation to a nutrient-rich, stable environment, such as milk (Tarrah et al., 2020). These strains also appear to lack the *pil1* locus, which plays a role in host adhesion and endocarditis in related species, further supporting the safety of this species for use in food applications.

Although many Streptococcus species are pathogenic, *Streptococcus thermophilus* is a thermophilic lactic acid bacterium recognised as "Generally Recognised as Safe" (GRAS) in the United States. The organism was also granted Qualified Presumption of Safety (QPS) status in Europe, owing to its long-standing safe application in food production (Gardan et al., 2024). Many *S. thermophilus* strains are known to produce exopolysaccharides (EPS), giving desirable texture to fermented milk products, particularly yoghurt (Fan et al., 2025). It is commonly isolated from a variety of dairy environments, such as raw milk, artisanal cheeses, and traditional fermented milk products, where it contributes to fermentation and confers technological properties beneficial for dairy processing (Kapse et al., 2024). Its industrial relevance is primarily due to its ability to rapidly catabolise lactose, which drives a swift decline in pH, as well as the synthesis of metabolites that contribute to desirable flavour and texture (Zhao et al., 2024). Whole-genome sequencing studies have also shown that this bacterium lacks several virulence-associated genes typically found in pathogenic Streptococci, including those associated with mucosal adhesion and antibiotic resistance. This absence of harmful traits further supports its recognition as a safe organism under the GRAS status. The study has also provided evidence of horizontal gene transfer (HGT), where genes associated with milk adaptation and EPS production appear to have been acquired from other dairy microorganisms (Hu et al., 2020). These findings reflect how *S. thermophilus* has evolved to thrive in milk and contribute effectively to dairy fermentation.

3.4.2.4 *Propionibacterium*

The genus *Propionibacterium* can be broadly divided into two groups based on their habitats; one group includes strains commonly found on human skin, while the other consists of strains associated with milk and dairy products. Dairy-associated group includes four primary species: *Propionibacterium freudenreichii*, *Acidipropionibacterium acidipropionici*, *Acidipropionibacterium jensenii*, and *Acidipropionibacterium thoenii* (Bücher et al., 2021). These species are widely recognised for their role in cheese fermentation and the production of health-promoting compounds such as vitamin B_{12} and propionic acid (de Assis et al., 2022). They are well adapted to dairy environments and are valued for their technological and probiotic attributes in functional food development and processing (Coronas et al., 2023).

Propionibacterium freudenreichii is widely used as a starter culture in the production of Swiss-style cheeses such as Emmental, where it contributes to eye formation and nutty flavour through propionic acid fermentation (Jeantet et al., 2021). The definite feature of *P. freudenreichii* is its ability to ferment lactate into propionate, acetate, and carbon dioxide, which drives eye formation. Moreover, the distinctive flavours associated with cheese making also arise from the production of fatty acids, through lipolysis, as well as branched-chain acids generated from the catabolism of amino acids (Turgay et al., 2022). Whole-genome sequencing further highlighted its robust genetic potential to withstand diverse environmental stresses, including oxidative stress, bile salt exposure, temperature fluctuations, and even phage attack. These genes are believed to have been acquired through HGT and help *P. freudenreichii* adapt more effectively in different dairy environments (Tarnaud et al., 2020). Another sequencing study has also identified several genes, such as *SlpB*, that encode surface-layer proteins thought to regulate bacterial adhesion and immunomodulatory functions. These surface proteins are highly responsive to environmental conditions and protect gut integrity, thereby reducing inflammation in colitis models (Mantel et al., 2024). The bacterium can synthesise essential nutrients like vitamins B_6, B_9, and B_{12} and mobilise stress-protective compounds trehalose and glycogen to support survival during periods of starvation (Piwowarek et al., 2020). Due to its long-documented safe application in food and the absence of known harmful characteristics, *P. freudenreichii* has been granted GRAS status (de Rezende et al., 2021).

3.4.2.5 Leuconostoc

The genus *Leuconostoc* is taxonomically diverse, comprises 17 recognised species and several subspecies (Raimondi et al., 2022). Leuconostoc species (such as *L. mesenteroides* and *L. pseudomesenteroides*) are frequently associated with plants and fermented foods but are also regularly recovered from milk and dairy-based products. Their presence in milk is often linked to environmental contamination throughout the collection, handling, subsequent storage and processing of milk (Ariute et al., 2025). These bacteria can persist for long periods on surfaces, equipment, and even pasteurisers. They tolerate moderate heat treatments and refrigeration temperatures, which contribute to their survival in dairy environments (Gumustop et al., 2023).

Leuconostoc species have limited proteolytic activity and thus rely on other microorganisms to generate essential amino acids and peptides for optimal growth (Hamdaoui et al., 2023). They act as beneficial NSLAB cultures as they can produce gas (CO_2) through heterolytic fermentation, which is responsible for hole formation and unique texture in raw milk and blue-veined cheese varieties (Sviridenko et al., 2024). Sequencing and cross-analysis of dairy strains of *L. pseudomesenteroides* have also revealed other favourable traits, including genes that support carbohydrate fermentation, protein and amino acid metabolism, and a citrate-derived pathway essential for producing aromatic compounds diacetyl and acetoin. These characteristics underscore their functional relevance in dairy fermentation (Gumustop et al., 2023).

3.4.2.6 Enterococcus

Enterococci are regarded as one of the most controversial food-associated LAB, largely because they can colonise a diverse range of ecological niches, including the GI tracts of both humans and animals. The most detected Enterococcus species found in milk and dairy products are *E. faecalis* and *E. faecium*. However, other species such as *E. durans*, *E. italicus*, and *E. mundtii* are also frequently encountered, depending on the type of dairy product and regional production practices (Terzić-Vidojević et al., 2021; Morandi et al., 2024; Hanzelová et al., 2024). Depending on the strain, it can serve beneficial roles as starter cultures or probiotics or be implicated in food spoilage and serious infections such as nosocomial diseases. This dual nature highlights the need for careful strain-specific evaluation before any application in food systems or as probiotics (Apostolakos et al., 2022; Yordanova et al., 2024). Through their proteolytic activity and fat-hydrolysing activity, Enterococcus species generate flavour compounds such as acetaldehyde, acetoin, and diacetyl. These metabolic functions enhance the sensory profile of fermented milk and cheeses, especially during ripening processes (Terzić-Vidojević et al., 2021).

Due to their physiological adaptability to various growth substrates and environmental conditions, enterococci can withstand both refrigeration and pasteurisation. As a result, they can be found in the microbial ecology of raw and pasteurised milk as well as in the subsequent dairy products. Residual microflora studies further confirm that enterococci are among the most heat-tolerant cocci in pasteurised milk (Sviridenko et al., 2023). Strains of *Enterococcus faecium* isolated from pasteurised milk have been shown to tolerate high temperatures and persist post-pasteurisation (Zhao et al., 2024). The esterase activity of *E. faecalis* has been linked to the release of volatile compounds in milk-based matrices, which enriches the aromatic complexity of dairy products (Vélez et al., 2023). Moreover, a study on *E. faecalis* from white cheese highlights their reduced pathogenic potential, as they possess only a limited set of genes while lacking high-risk determinants such as *esp* and *cylA* (Bozcal et al., 2024). They lack the full complement of pathogenic strains, making them less likely to pose a direct threat to human health.

3.4.2.7 *Pediococcus*

Pediococcus, a lactic acid bacterium, is recognised for its beneficial effects on both human and animal health. Species of *Pediococcus*, such as *P. pentosaceus*, can produce anti-microbial peptides, bacteriocins or bacteriocin-like substances (BLISs) (Ayivi et al., 2020; Jiang et al., 2021). They defend animal health by regulating gut microbes and inhibiting the growth of pathogenic microorganisms. *P. pentosaceus* is a widely distributed and well-characterised probiotic LAB species commonly found in milk and dairy products (Rahman et al., 2024). *P. pentosaceus* has diverse health-promoting functions in humans and animals, including immunomodulatory effects, anti-cancer potential, anti-inflammatory and antioxidant activity, and the ability to reduce blood cholesterol levels (Qi et al., 2021).

3.4.3 Presence of Spoilage and Pathogenic Microorganisms

The raw milk ecosystem supports the growth of both spoilage and harmful microorganisms that can compromise milk quality and pose health risks to humans. Spoilage microorganisms alter basic qualities and characteristics of milk and dairy items. The type and quantity of microorganisms involved in pre- and post-technological processing determine the product defects. Usually, these alterations are connected to a particular unfavourable sensory attribute, such as flavour, consistency, or odour (Samaržija et al., 2007). The most significant pathogens associated with raw milk and bovine mastitis are *Staphylococcus aureus*, *S. uberis*, *E. coli*, *S. dysgalactiae*, *S. agalactiae*, *Pseudomonas aeruginosa*, and *Corynebacterium pyogenes*. Many of these are zoonotic and opportunistic pathogens that are occasionally reported, including *Salmonella* spp., *L. monocytogenes*, *Mycobacterium bovis*, and *Mycobacterium tuberculosis* (Adugna et al., 2015). *S. aureus* is a major cause of food-borne illnesses in humans, primarily due to its ability to produce heat-stable enterotoxins (El-Leboudy et al., 2014). Moreover, *E. coli* and other coliform bacteria are uncommon during mastitis and are not a part of the milk's natural bacterial ecology. They appear as environmental contaminants or as indicators of poor hygiene practices (Pyz-Łukasik et al., 2015).

3.4.3.1 *Psychrotrophs*

Psychrotrophic bacteria are particularly important to the dairy industry because they can grow even at refrigeration temperatures and produce enzymes, such as proteases and lipases. These heat-resistant enzymes, produced during milk storage or processing, can survive pasteurisation or ultra-high-temperature (UHT) treatments, ultimately causing off-flavours, gelation, and a shorter shelf life in dairy products (Samaržija et al., 2012). Psychrotrophic bacteria commonly found in raw milk include species from the genera Pseudomonas, Bacillus, Acinetobacter, Aeromonas, Serratia, Microbacterium, Lactococcus and Staphylococcus (Yuan et al., 2019). Their diversity and abundance vary widely across farms and are strongly influenced by environmental conditions, milking practices, and overall sanitation standards. Poor cleaning and extended milk storage at low temperatures can promote their proliferation and enzyme production. Due to their capacity for biotransfer, most psychrotrophic bacteria create biofilms on a variety of processing and storage devices, serving as persistent reservoirs of contamination and re-entry into the milk supply chain. These biofilms protect bacteria from cleaning and sanitising agents, contributing to recurring spoilage problems in dairy environments (Zarei et al., 2023).

The genus Pseudomonas thrives at refrigeration temperature and is recognised as the primary cause of quality deterioration in both raw and pasteurised milk. The most common spoilage-associated species are *P. aeruginosa*, *P. fragi*, *P. putida*, *P. putrefaciens*, and *P. fluorescens* (De Jonghe et al., 2011). Additionally, post-pasteurisation contamination can occur during handling and packaging at the filling stage, where even minor lapses in hygiene can allow Pseudomonas species to colonise processed milk (Kumaresan et al., 2007). Among the psychrotrophic bacteria, *Listeria monocytogenes* stands out for its resilience and widespread presence in the dairy environment. This Gram-positive, motile, rod-shaped bacterium can cause both sporadic cases and large outbreaks worldwide. *L. monocytogenes* species can multiply at refrigeration temperatures (around 4°C) and survive freezing, enabling them to persist in chilled dairy products. Its presence in milk, cheese, ice cream and other dairy-related products poses a notable risk to consumer health (Kasalica et al., 2011).

3.4.3.2 Staphylococcus aureus

Staphylococcus aureus is a Gram-positive, non-motile, non-spore-forming, facultatively anaerobic bacterium that characteristically appears in grape-like clusters. This catalase-positive member of the Staphylococcaceae family serves as an important mastitis-associated pathogen (Tong et al., 2015). They can be distinguished from other species by their ability to induce plasma clotting through coagulase enzyme activity (Becker et al., 2014). Their growth and persistence are strongly influenced by environmental factors such as temperature, pH, and oxygen availability. *S. aureus* can grow within a broad temperature range of 7–48°C, with optimum growth at 37°C, and tolerates a wide pH range of 4.0–10.0, exhibiting optimum growth between pH 6.0 and 7.0. These growth conditions correspond to the human and host environments as well as those found in most food environments. Milk may become contaminated by *S. aureus* either by direct shedding from the udders of cows with clinical or subclinical mastitis or indirectly through environmental sources during collection, handling or processing (Piccinini et al., 2010). Previous studies have reported that *S. aureus* can grow in moderately acidic or alkaline environments, highlighting its notable environmental adaptability (Kadariya et al., 2014). The organisms can tolerate up to 10% NaCl and hence exhibit remarkable resistance to harsh conditions such as high salinity and osmotic stress. These characteristics, combined with their ability to form biofilms, help them persist in raw milk, dairy environments, and milking equipment (Melchior et al., 2006).

3.4.3.3 Zoonotic Pathogens in Raw Milk

Zoonotic pathogens are microorganisms transmitted from animals to humans, either directly through udder or indirectly via faecal, environmental, or equipment contamination during milking and handling. Members of the genus Brucella are characterised as Gram-negative, non-motile, spore-lacking coccobacilli that cause a highly contagious zoonotic illness called brucellosis (Goldman et al., 1996). *B. suis*, *B. melitensis*, and *B. abortus* are the most common species that infect people; however, *B. canis* also puts humans at risk for brucellosis. The disease can be transmitted through the consumption of raw milk and dairy products manufactured under unregulated conditions. Brucellosis is a frequently underdiagnosed zoonotic disease, largely because of its nonspecific clinical manifestation and limited awareness among healthcare professionals (Islam et al., 2023). Bovine tuberculosis, a chronic disease with a worldwide distribution, is caused by *Mycobacterium bovis* and affects both animals and humans, posing a significant public health risk. Transmission of this disease to humans typically occurs through the consumption of raw or inadequately prepared dairy products, like those prepared without pasteurisation, boiling, or UHT processing (Bolaños et al., 2017).

The genus Campylobacter comprises small, Gram-negative, spiral-shaped rods. Most species exhibit corkscrew-like motility driven by a single, polar, unsheathed flagellum located at one or both ends of the cell (Debruyne et al., 2008). The Campylobacter genus causes "campylobacteriosis", which refers to a broad category of zoonotic diseases. It is the primary cause of bacterial gastroenteritis globally and a major food-borne pathogen in humans. There are nearly 20 species in the genus Campylobacter, and most of the infections are attributed to *C. jejuni* and *C. coli* (El-Zamkan et al., 2016). Salmonella species are Gram-negative, rod-shaped, facultative intracellular bacteria capable of infecting a wide variety of hosts. The genus comprises two species: *Salmonella bongori* and *Salmonella enterica*, which inhabit the intestinal tract of numerous animals, including cattle. They are a leading cause of food-borne illness globally, typically associated with diarrhoea, abdominal cramps, vomiting, and fever (Guibourdenche et al., 2010). Salmonellosis remains one of the most prevalent food-borne bacterial diseases worldwide and continues to be a neglected zoonosis in Egypt and other developing countries. Consumption of raw or inadequately pasteurised milk and dairy products has been identified as a major source of food-borne Salmonella infections (Oliver et al., 2005).

3.4.3.4 Indicator Bacteria in Raw Milk

Indicator bacteria are widely applied to evaluate the sanitary quality of milk and to assess the effectiveness of milking and handling methods. Their presence typically reflects faecal contamination or inadequate equipment sanitation and can signal the possible presence of an enteric pathogen. High levels of coliforms in dairy-based products indicate poor hygiene practices during production. Occasionally, the cheese blows or releases gases after 1–2 days of manufacturing, characterised by large gas holes and a spongy texture (Bintsis et al., 2006). *Escherichia coli* is also an important indicator and opportunistic pathogen that has been increasingly associated with human illness over the past decade. Although most *E. coli* strains are harmless, certain serotypes can cause food-borne infections and GI tract disorders. Among these, enterohemorrhagic *E. coli* (EHEC), particularly the O157:H7 serotype, is considered the most virulent in both the food and dairy industries. *E. coli* O157:H7 has emerged as a major concern for the public, due to its ability to cause severe conditions such as haemorrhagic colitis, haemolytic uremic syndrome and thrombotic thrombocytopenic purpura (Zeinhom et al., 2014).

3.4.4 Bacteriophage

Raw milk is considered the primary reservoir of bacteriophages in the dairy environment, typically containing 10^1 to 10^4 phages per millilitre (Madera et al., 2004). Although phages are naturally present in raw milk, they may also enter the dairy environment through aerosols, personnel, equipment, work surfaces and various dairy byproducts. Lytic bacteriophages act as spoilage agents by infecting bacterial cells through intracellular replication and ultimately lysing them inside the host (Verreault et al., 2011). These phages can withstand a variety of processing treatments and survive at low temperatures, while negatively affecting the quality, safety and economic value of dairy products, particularly cheese. Phages that infect key starter or adjunct species, such as *L. lactis*, *L. delbrueckii*, *L. helveticus* and *S. thermophilus*, are well known to delay or disrupt fermentation processes, leading to slow acidification, poor sensory qualities, or even total batch loss (Marcó et al., 2012).

3.4.5 Yeasts and Moulds

Yeasts and moulds constitute an important component of raw milk microbiota and can significantly influence its quality, shelf life, and downstream dairy fermentations. The prevalence and diversity of yeasts and moulds are influenced by factors such as the cows' physiological status, farm hygiene practices, feeding habits, environmental conditions and seasonal variations. Studies reported the prevalence of fungal communities in raw milk from the teat surface, bedding material, feed and the milking environments (Quigley et al., 2013). Yeasts and moulds such as Kluyveromyces, Candida, Debaryomyces, Penicillium and Geotrichum are frequently detected in raw milk. They assist in shaping the flavour attributes of specific artisanal dairy items but also act as spoilers under poor storage conditions (Garnier et al., 2017; Cenci-Goga et al., 2021). Yeasts can significantly influence dairy fermentations owing to their diverse physiological and biochemical characteristics, including lactose or galactose utilisation, strong proteolytic and lipolytic activities, and their potential to endure low temperatures and high salinity (Kebede et al., 2007). Moulds also contribute significantly to the flavour, aroma, texture and structural characteristics of milk-derived products through extensive proteolytic and lipolytic activities (Ferreira et al., 2012). Understanding the functional roles and metabolic capabilities of dairy-associated yeast and moulds allows the dairy industry to develop novel starter cultures, optimise the performance of existing strains, and strengthen the evidence supporting the "GRAS" status of these technologically significant microorganisms (Ropars et al., 2020). Additionally, several other technologically relevant microorganisms are less abundantly present in raw milk.

3.5 APPROACHES TO DETERMINE MICROBIAL COMMUNITIES OF MILK

Bacteria in raw milk can be identified using two primary approaches: culture-dependent and culture-independent techniques. Culture-dependent methods require viable cells and rely on growing microorganisms on specific media, followed by morphological, biochemical, and physiological characterisation. Although these phenotypic approaches are generally low-tech and inexpensive, they are labour-intensive and time-consuming. In contrast, culture-independent methods analyse bacterial DNA directly, thereby detecting organisms that are difficult to isolate using conventional culturing techniques (Vacheyrou et al., 2011). In recent years, substantial progress has been made in developing rapid, high-throughput DNA-based assays that complement and validate results generated in traditional tests. These approaches potentially avoid the significant underestimation of microbial communities (Quigley et al., 2011).

3.5.1 Culture-Dependent Approaches

Bacterial culturing remains the primary and most reliable method for the bacteriological examination of milk and dairy products. Culture-based methods are used to assess microbial viability, colony morphology and phenotypic characteristics, making them a useful tool in routine diagnosis and quality assessment (Quigley et al., 2013). In laboratory settings, bacteria can be grown successfully when their fundamental, nutritional and environmental requirements are fulfilled. These requirements vary among bacterial species but generally include a carbon and energy source (sugar), a nitrogen source for protein synthesis, a stable pH, suitable aerobic or anaerobic conditions, and an optimal temperature. Culture media used for milk microbiology may be general-purpose, selective, or differential, depending on the target organism. These media often contain specific inhibitors, pH indicators, or chromogenic substrates that help identify and distinguish bacterial species such as coliforms, LAB, or mastitis pathogens (Madigan et al., 1997). For routine culturing, typically, about 10 µl of raw or processed milk is spread onto an agar plate prepared with the necessary nutrients, sometimes with added supplements, and incubated aerobically for 24–48 hours at 37°C. Temperature and atmosphere may be adjusted for psychrotrophs, anaerobes and fastidious organisms. Living bacteria multiply to form discrete visible colonies, known as colony-forming units (CFUs), which serve as a basis for microbial load quantifications. Further identification can be achieved by examining their colony morphology, pigment production, haemolysis pattern, sugar metabolism, Gram-staining reactions, or other biochemical methods. Tests such as catalase, oxidase, or coagulase are used to assess carbohydrate fermentation, enzyme activities, or metabolic pathways for precise identification of milk-associated bacteria (Madigan et al., 2021). These phenotypic methods are often paired with molecular tools to achieve higher taxonomic resolution.

3.5.2 Culture-Independent Methods

The identification and classification of microorganisms have advanced significantly with progress in microbiology, biotechnology, and genomics. Modern rapid detection techniques such as polymerase chain reaction (PCR), biosensors, next-generation sequencing, and immunological assays have transformed the field by providing speed, scalability, and specificity compared to traditional culture-based approaches. These tools facilitate the early detection of spoilage and pathogenic microorganisms, thereby enhancing quality control measures and ensuring improved consumer safety (Guo et al., 2019).

3.5.2.1 Molecular Approaches

Molecular diagnostic assays, particularly nucleic acid–based methods, have recently become the gold standard for assessing microbiota. Advances in high-throughput sequencing technologies have enabled the application of genomics, transcriptomics, and proteomics in animal health. These tools are powerful for understanding the underlying mechanisms of microorganisms in raw milk. PCR forms the backbone of these methods, typically targeting the 23S and 16S ribosomal RNA (rRNA) regions. Conventional PCR amplifies a particular DNA fragment to identify the organism, while reverse transcription PCR (RT-PCR) is used to both detect and quantify specific strains. RT-PCR enhances the safety and cost-effectiveness of PCR testing by eliminating the need for ethidium bromide staining and handling of PCR products. Multiplex PCR has been used to simultaneously identify pathogens in a single milk sample, offering benefits such as high specificity and sensitivity, rapid results, and the potential to detect bacteria even in the presence of antibiotics or preservatives. Though PCR methods are accurate, they cost more compared to immunoassays. These methods often show false positives due to contamination and false negatives due to the presence of milk residue or inhibitors in samples (Algharib et al., 2024).

3.5.2.2 *Amplicon Sequencing*

Amplicon sequencing remains the most widely used culture-independent metagenomic method for targeting specific DNA regions. The 16S rRNA gene is the most used marker for bacterial community profiling, as it is the active fraction of microbiota (Rainard et al., 2017). For Illumina platforms, although the selection of variable regions differs across studies, V3–V4 regions are the most frequently targeted for profiling dairy microbiota. The relatively short length of these amplified regions often restricts taxonomic resolution to the genus level and, less commonly, to the species level. However, species-level or strain-level identification is important in milk microbiome studies because species within the same genus can differ greatly in their technological roles for both the quality and safety of dairy products (Parente et al., 2016). Integrating the amplicon sequencing method with high-resolution non-rRNA genetic markers provides a more comprehensive overview of microbial community structure and enables classification at the species or strain level (Ricciardi et al., 2016).

3.5.2.3 *Shotgun Approaches*

Shotgun metagenomics is a culture-independent approach for characterising the taxonomic and functional composition of a complex microbiome. However, its application in milk and dairy products research is comparatively rare. Meta-transcriptomic studies used for profiling actively expressed microbial genes are even less frequent (Yeluri et al., 2018). High-resolution genome reconstruction and metabolic pathway prediction techniques are likely to gain prominence in dairy microbiome research as sequencing costs decline and bioinformatic pipelines for taxonomic annotation advance (Tilocca et al., 2020). As a cost-effective alternative to whole-genome shotgun sequencing, metagenome inference tools such as PICRUSt and PICRUSt2 can predict functional gene profiles directly from amplicon sequencing data. Although these tools perform relatively well, their application to milk and dairy-associated microbial communities has remained limited (Douglas et al., 2019). Overall, as software tools evolve and sequencing technologies become more accessible, culture-independent omics approaches such as amplicon sequencing, shotgun metagenomics and meta-transcriptomics could transform our understanding of the dairy microbiome. This will provide deeper insights into microbial ecology, food safety, functional properties and their implications in dairy processing.

3.6 MODULATION OF MILK MICROBIOTA BY EXTRINSIC AND INTRINSIC FACTORS

Natural commensal microorganisms of milk not only convert lactose into lactate during fermentation but also produce a variety of compounds that shape the flavour, aroma, and texture of the dairy products we consume (Wouters et al., 2002). At the same time, if milk becomes contaminated with pathogenic bacteria, it can pose a risk to human health by secreting toxins. Therefore, understanding how the composition of milk microbiota changes over time, from milking, transport, storage, and processing, is important as these shifts can influence the final quality and safety of dairy products. Overall, the microbial community in milk is highly responsive to the external environment that can introduce microbes into milk or allow certain populations to grow. An overview of some environmental and management factors affecting milk microbiota and their impact on dairy quality is summarised in Table 3.2. These factors, including farm hygiene, feed, water quality, milking procedure, and storage temperature, ultimately determine the overall microbial load and quality of raw milk.

3.6.1 Farm Environment and Hygiene Practices

The farm environment and the hygiene measures implemented during animal handling and milking greatly influence the microbial load in milk (see Figure 3.2). Pathogenic strains can enter milk in two main ways: directly from an infected animal and indirectly through environmental contamination. Microorganisms may spread between animals within a herd through shared milking equipment, udder hygiene materials and contact with farm personnel (Hogan et al., 1995). A study investigated the transmission of microorganisms to milk through dirt adhering to the teat surface from faeces, bedding, soil, or their combination. The same study also reported that spore concentration in faeces is the primary factor influencing spore concentration in bulk tank milk, due to bacteria capable of producing butyric acid. However, it also showed that effective teat hygiene can significantly lower contamination risks (Vissers et al., 2007). Maintaining high-quality silage, implementing strict cleaning protocols for livestock sheds and

TABLE 3.2

Factors influencing milk microbiota composition and dairy product quality

Factor	Mechanism of action	Effect on microbial community	Impact on product quality	Reference
Farm hygiene	Contamination control via teat and equipment cleaning	Reduces Staphylococcus and Bacillus load	Improves safety and shelf life	Vargova et al., 2023
Feed composition	Alters rumen microbiota and microbial transfer to milk	Modifies Lactobacillus and Prevotella abundance	Enhances flavour profile	Mei et al., 2023
Storage temperature	Psychrotrophic growth under refrigeration	Increases Pseudomonas and Acinetobacter	Causes spoilage and rancidity	Liang et al., 2021
Antibiotic use	Alters microbiota, selects resistant species	Promotes Enterococcus and *E. coli* resistance	Reduces microbial diversity	Vasco et al., 2023
Pasteurisation	Heat inactivation of vegetative cells	Reduces mesophilic flora, activates spores	Improves safety, alters flavour	Dash et al., 2022
Climate variation	Seasonal temperature and humidity shifts	Increases psychrotrophic bacteria in summer	Affects milk stability and safety	Celano et al., 2022
Water quality	Transmission of environmental contaminants	Introduces Enterobacteriaceae	Reduces safety, increases spoilage	Soares et al., 2023
Storage duration	Prolonged cold storage	Accumulation of enzymes and biofilms	Leads to spoilage, texture change	Aouadhi et al., 2025
Milking equipment	Biofilm formation on surfaces	Reservoir for Staphylococcus and Pseudomonas	Reduces hygiene, affects quality	Ryu et al., 2021
Feed silage	Spore contamination from silage	Introduces Clostridium species	Causes off-flavours and defects	Velasco et al., 2025

milking parlours or equipment, and ensuring thorough udder cleaning and teat preparation before milking are essential practices for reducing the entry of anaerobic spore-forming bacteria (ASFB) into bulk tank milk (Doyle et al., 2015). A previous study showed that the most effective way to lower spore levels in milk was to clean the teats with a washable, moist towel followed by drying with a paper towel (Magnusson et al., 2006).

3.6.2 Feed Additives and Antibiotic Use

Silage is a major source of spore contamination in raw milk, with spores commonly originating from grass and maize silage, as well as from concentrated feeds contaminated by aflatoxin B1 (Roussi et al., 2002). Preventive strategies for suppressing aerobic spore outgrowth in silage include the application of LAB or chemical additives. These measures can enhance silage fermentation and substantially reduce the overall spore load entering raw milk (Te Giffel et al., 2002). Some studies have reported predominance of Gram-negative bacteria in animals fed high-forage diets, whereas high-grain diets favour Gram-positive communities, including Lactobacillus. Additionally, cows fed a higher proportion of corn silage in their diet showed increased Prevotella populations and a corresponding decrease in rumen protozoal abundance. These findings further indicate that the composition of rumen microbiota is strongly influenced by diet (Lettat et al., 2013; Hungate et al., 2013). A study conducted in Central Italy reported that supplementation of the marine alga *Ascophyllum nodosum* could influence microbial populations present in the teat canal (Lopez et al., 2016). Since milk-associated microorganisms may originate from the teat-surface litter where cows rest before milking (Zdanowicz et al., 2004), feeding practices also play a significant role in shaping the milk microbiota. Notably, differences have been observed between cows fed indoors and those grazed outdoors, with higher levels of *Staphylococcus* spp. in milk during outdoor feeding (Hagi et al., 2010). Mastitis is one of the reasons for higher antibiotic use on dairy farms, and resistant genes have been isolated from dairy products (Stevens et al., 2016). There is also concern that animals treated with antibiotics may release antibiotic-resistant bacteria (ARB) and antibiotic-resistant genes (ARGs) into the surrounding environment. Moreover, the use of certain anti-microbial agents can diminish the diversity of bovine gut microbiota, potentially altering both animal health and the microbial ecosystem associated with dairy production (Alexander et al., 2008).

3.6.3 Climate-Driven Shifts

Climate change is expected to affect the microbial profile of dairy products through the direct influence of shifting climate variables and seasonal patterns. Climatic changes can promote the proliferation of pathogenic and spoilage microorganisms in raw milk (Van der Spiegel et al., 2012). These climatic variables may include changes in relative humidity, rise in temperature, altered precipitation levels and exposure to sunlight.These are all closely linked to seasonal dynamics and can collectively shape the microbial communities associated with milk and dairy products (López-Gálvez et al., 2018). In addition, climate change can influence the milk microbiome indirectly by inducing heat stress in lactating cows. This results in altered immune responses and increased vulnerability to infections. The anticipated rise in global mean temperature is likely to accelerate microbial growth rates and alter the microbial ecology of both raw and processed milk products (Van der Spiegel et al., 2012). A study across 10 different farms in Shanghai, China, reported that milk samples collected in March, June and July exhibited higher species richness, suggesting an increased risk of bacterial contamination during these months. Seasonal shifts in the milk microbiome are closely aligned with changes in temperature, with lower microbial richness in winter and higher in summer. Notably, during the winter period of November to March, when the average temperature was 8.8°C, Pseudomonas species were more abundant, likely due to their psychrotrophic ability of surviving at lower temperatures (Li et al., 2018).

3.6.4 Impact of Storage Condition

Milk is usually stored under refrigeration to inhibit the growth of most bacteria. However, psychrotolerant microorganisms can withstand and proliferate under refrigeration temperatures, contributing majorly to milk spoilage (De Jonghe et al., 2011). It is important to understand how the microbiology of raw milk changes during storage and subsequent processing treatments. Studies examining seasonal differences in the raw milk microbiome have shown that psychrotolerant bacteria grow readily and show higher protease production in winter milk than summer milk (Marchand et al., 2009). Pseudomonas species, particularly *P. fluorescens*, *P. lundensis*, *P. gessardii*, and *P. fragi*, are common causes of spoilage in milk and cheeses (Mallet et al., 2012). A study examined the impact of 24-hour refrigeration on the microbial content of raw milk and observed proliferation of several bacterial groups, including *L. fermentum*, *L. monocytogenes*, *P. fluorescens*, *S. epidermidis*, *E. durans*, *E. faecium*, *Leuconostoc carnosum*, *S. dysgalactiae*, *K. pneumoniae*, Aeromonas and propionic acid bacteria (Lafarge et al., 2004). A related 48-hour storage period study reported increases in Pseudomonas, Acinetobacter, psychotrophic spore-forming species, and various yeasts and moulds, which are also associated with the spoilage of refrigerated milk (Agarwal et al., 2012).

3.6.5 Impact of Pasteurisation on the Microbiology of Raw Milk

Pasteurisation can reduce the overall microbial load of milk and eliminate spoilage organisms, thereby helping to prevent food-borne illnesses. However, heat treatment can also trigger the activation of spores that may have been dormant in raw milk (Ranieri et al., 2009). The standard pasteurisation method used in the dairy industry is the high-temperature, short-time (HTST) process that involves heating milk to 72°C for 15 seconds. Although this method is very effective, it can diminish the populations of naturally occurring microbes that contribute to the characteristic flavours and aromas of raw milk cheeses. Defined starter cultures are then added to the milk post-pasteurisation to compensate for this loss and generate sensory attributes desirable for the milk (Martin et al., 2011). Heat treatment primarily reduces mesophilic and psychrotropic populations, leaving for consideration those spore-forming thermoduric microorganisms that can enter the product after pasteurisation. Additionally, certain microorganisms may enter a "viable but non-culturable" (VBNC) state after pasteurisation and are underestimated by conventional culture-based detection methods (Paszyńska-Wesołowska et al., 2009).

3.7 APPLICATION AND FUTURE DIRECTION

Dairy-derived microorganisms present significant opportunities for developing next-generation probiotic formulations. Their natural adaptation to the milk environment makes them promising candidates for improving the stability, efficacy, and sensory characteristics of fermented dairy products. The symbiotic association between yeast and bacteria in fermented dairy products confers beneficial effects, including anti-microbial, anti-tumour, anti-carcinogenic, hypoallergenic, and immunomodulating properties (Leite et al., 2013). There are numerous food and animal sources from which probiotic cultures are isolated, but dairy products remain one of the most reliable sources. Certain strains of *Saccharomyces cerevisiae* isolated from kefir have shown strong antagonistic activity against pathogens, including *K. pneumoniae*, *P. aeruginosa*, *S. aureus*, and *B. subtilis* (de Lima et al., 2017). The dairy industry continues to explore novel probiotic and beneficial microorganisms to support the development of functional fermented food products that promote consumer health and well-being. A study in Brazil isolated potential probiotic LAB strains from kefir grains that can produce conjugated linoleic acid (CLA) during the fermentation of whole cow's milk. This organic acid is recognised for its health-promoting properties even at low ingestion rates, including reported reductions in cancer incidence and severity of atherosclerosis. At the same time, insights into the milk microbiome are extending applications to bovine health, particularly to support udder health and improve milk quality (Vieira et al., 2017). Fermented milk products have gained considerable attention since the 1990s, for their health-promoting properties due to the presence of probiotic microorganisms in their composition (10^8–10^9 CFU per day, within a safe and well-established food matrix). Beyond their technological role, important groups such as *Bifidobacteria* spp. are active in deconjugation of bile acid, breakdown of dietary carbohydrates and vitamin synthesis (Admanova et al., 2023). Collectively, these advancements highlight the growing potential of the milk microbiome as a foundation for future functional dairy innovations.

3.8 CONCLUSION

The microbial community in bovine milk plays an essential role in shaping its nutritional value, flavour, and safety. Beneficial microbes such as LAB and Bifidobacteria enrich dairy products by improving taste, texture, and health benefits. At the same time, harmful organisms, including *Staphylococcus aureus* and *Escherichia coli*, reduce milk quality and create health risks if not controlled. Understanding the balance between these groups helps improve both dairy processing and public health outcomes. Modern sequencing and molecular tools have made it easier to study the milk microbiome in greater detail, allowing producers to monitor and guide microbial behaviour more effectively. Future research should combine genetic and ecological methods to design targeted probiotic strains and microbial management systems that enhance milk safety, extend shelf life, and strengthen the nutritional benefits of dairy foods.

REFERENCES

Admanova, G. B., Kuanbay, Z. I., Izimova, R., Keubassova, G. O., & Kozhamzharova, L. S. (2023). Some enzymatic properties of lactic acid bacteria isolated from dairy products. *Fundamental and Experimental Biology*, 112(4), 7–13.

Adugna, M., Asresie, A., & Adigrat, E. (2015). A review on microbiological quality of Ethiopian raw bovine milk. *Food Science and Quality Management*, 35, 17–34.

Agarwal, A., Awasthi, V., Dua, A., Ganguly, S., Garg, V., & Marwaha, S. S. (2012). Microbiological profile of milk: Impact of household practices. *Indian Journal of Public Health*, 56(1), 88–94.

Alexander, T. W., Yanke, L. J., Topp, E., Olson, M. E., Read, R. R., Morck, D. W., & McAllister, T. A. (2008). Effect of subtherapeutic administration of antibiotics on the prevalence of antibiotic-resistant *Escherichia coli* bacteria in feedlot cattle. *Applied and Environmental Microbiology*, 74(14), 4405–4416.

Algharib, S. A., Dawood, A. S., Huang, L., Guo, A., Zhao, G., Zhou, K., ... & Xie, S. (2024). Basic concepts, recent advances, and future perspectives in the diagnosis of bovine mastitis. *Journal of Veterinary Science*, 25(1), e18.

Aouadhi, C. (2025). Comprehending the influence of enzyme sources, environmental parameters, and storage conditions on the proteolytic activity of cow milk. *International Dairy Journal*, 172, 106444.

Apostolakos, I., Tsigkrimani, M., Paramithiotis, S., & Mataragas, M. (2022). Whole-genome sequencing and comparative genomic analysis of Enterococcus spp. isolated from dairy products: Genomic diversity, functional characteristics, and pathogenic potential. *Applied Sciences*, 12(19), 9620.

Ariute, J. C., Coelho-Rocha, N. D., Dantas, C. W. D., de Vasconcelos, L. A. T., Profeta, R., de Jesus Sousa, T., ... & Aburjaile, F. (2025). Probiogenomics of *Leuconostoc mesenteroides* strains F-21 and F-22 isolated from human breast milk reveal beneficial properties. *Probiotics and Antimicrobial Proteins*, 17(2), 500–515.

Ayivi, R. D., Gyawali, R., Krastanov, A., Aljaloud, S. O., Worku, M., Tahergorabi, R., ... & Ibrahim, S. A. (2020). Lactic acid bacteria: Food safety and human health applications. *Dairy*, 1(3), 202–232.

Becker, K., Heilmann, C., & Peters, G. (2014). Coagulase-negative staphylococci. *Clinical Microbiology Reviews*, 27(4), 870–926.

Bintsis, T. (2006). Quality of the brine. In A. Tamime (Ed.), *Brined Cheeses* (pp. 264–301). Blackwell.

Bolaños, C. A. D., Paula, C. L. D., Guerra, S. T., Franco, M. M. J., & Ribeiro, M. G. (2017). Diagnosis of mycobacteria in bovine milk: An overview. *Revista do Instituto de Medicina Tropical de São Paulo*, 59, e40.

Bozcal, E., Tekiner, S., Guler, D. N., & Asal, E. G. (2024). Genomic characterization of multi-drug resistant *Enterococcus faecalis* strain CE_3_5. *Journal of Global Antimicrobial Resistance*, 39, 16–17.

Branca, F., & Rossi, L. (2002). The role of fermented milk in complementary feeding of young children: Lessons from transition countries. *European Journal of Clinical Nutrition*, 56(4), S16–S20.

Bücher, C., Burtscher, J., & Domig, K. J. (2021). Propionic acid bacteria in the food industry: An update on essential traits and detection methods. *Comprehensive Reviews in Food Science and Food Safety*, 20(5), 4299–4323.

Büthe, T., Jessberger, N., Schneider, B., Kreienbrock, L., & Plötz, M. (2025). Occurrence of mastitis pathogens in cow's milk samples from the Hanover region, north-western Germany–An overview of routine laboratory data from 2005 to 2023. *Veterinary Research Communications*, 49(4), 208.

Canon, F., Mariadassou, M., Maillard, M. B., Falentin, H., Parayre, S., Madec, M. N., ... & Gagnaire, V. (2020). Function-driven design of lactic acid bacteria co-cultures to produce new fermented food associating milk and lupin. *Frontiers in Microbiology*, 11, 584163.

Celano, G., Calasso, M., Costantino, G., Vacca, M., Ressa, A., Nikoloudaki, O., ... & De Angelis, M. (2022). Effect of seasonality on microbiological variability of raw cow milk from Apulian dairy farms in Italy. *Microbiology Spectrum*, 10(5), e0051422.

Cenci-Goga, B., Cruciani, D., Crotti, S., Karama, M., Yıldırım, G., Bulut, M., ... & Grispoldi, L. (2021). Diversity of yeasts and moulds in dairy products from Umbria, central Italy. *Journal of Dairy Research*, 88(2), 217–220.

Chakroborty, A., Lungtau, F. L., & Sharma, I. (2025). Antagonistic effects of probiotic *Lactobacillus* strains on multidrug-resistant *Staphylococcus aureus*: Role of mecA/mecC and icaAD gene variants. *Biotechnology and Applied Biochemistry*, 2025, n.pag.

Chaves, C. R. S., Salamandane, A., Vieira, E. J. F., & Salamandane, C. (2024). Antibiotic resistance in fermented foods chain: Evaluating the risks of emergence of *Enterococci* as an emerging pathogen in raw milk cheese. *International Journal of Microbiology*, 2024(1), 2409270.

Chelladhurai, K., Ayyash, M., Turner, M. S., & Kamal-Eldin, A. (2023). *Lactobacillus helveticus*: Health effects, current applications, and future trends in dairy fermentation. *Trends in Food Science & Technology*, 136, 159–168.

Cheng, T., Wang, L., Guo, Z., & Li, B. (2022). Technological characterization and antibacterial activity of *Lactococcus lactis* subsp. cremoris strains for potential use as starter culture for cheddar cheese manufacture. *Food Science and Technology*, 42, e13022.

Coelho, M. C., Malcata, F. X., & Silva, C. C. (2022). Lactic acid bacteria in raw-milk cheeses: From starter cultures to probiotic functions. *Foods*, 11(15), 2276.

Coronas, R., Zara, G., Gallo, A., Rocchetti, G., Lapris, M., Petretto, G. L., ... & Mannazzu, I. (2023). Propionibacteria as promising tools for the production of pro-bioactive scotta: A proof-of-concept study. *Frontiers in Microbiology*, 14, 1223741.

Dash, K. K., Fayaz, U., Dar, A. H., Shams, R., Manzoor, S., Sundarsingh, A., ... & Khan, S. A. (2022). A comprehensive review on heat treatments and related impact on the quality and microbial safety of milk and milk-based products. *Food Chemistry Advances*, 1, 100041.

de Assis, D. A., Machado, C., Matte, C., & Ayub, M. A. Z. (2022). High cell density culture of dairy propionibacterium sp. and acidipropionibacterium sp.: A review for food industry applications. *Food and Bioprocess Technology*, 15(4), 734–749.

De Jonghe, V., Coorevits, A., Van Hoorde, K., Messens, W., Van Landschoot, A., De Vos, P., & Heyndrickx, M. (2011). Influence of storage conditions on the growth of *Pseudomonas* species in refrigerated raw milk. *Applied and Environmental Microbiology*, 77(2), 460–470.

de Lima, M. D. S. F., de Souza, K. M. S., Albuquerque, W. W. C., Teixeira, J. A. C., Cavalcanti, M. T. H., & Porto, A. L. F. (2017). *Saccharomyces cerevisiae* from Brazilian kefir-fermented milk: An in vitro evaluation of probiotic properties. *Microbial Pathogenesis*, 110, 670–677.

de Rezende Rodovalho, V., Rodrigues, D. L. N., Jan, G., Le Loir, Y., de Carvalho Azevedo, V. A., & Guédon, E. (2021). *Propionibacterium freudenreichii*: General characteristics and probiotic traits. In E. F Robles (Ed.), *Prebiotics and Probiotics: From Food to Health*. IntechOpen.

Debruyne, L., Gevers, D., & Vandamme, P. (2008). Taxonomy of the family Campylobacteraceae. *Campylobacter*, 1–25. https://doi.org/10.1128/9781555815554.ch1.

Dimov, S. G. (2023). The controversial nature of some non-starter lactic acid bacteria actively participating in cheese ripening. *BioTech*, 12(4), 63.

Douglas, G. M., Maffei, V. J., Zaneveld, J., Yurgel, S. N., Brown, J. R., Taylor, C. M., ... & Langille, M. G. (2019). PICRUSt2: An improved and extensible approach for metagenome inference. *BioRxiv*, 38(6):672295.

Doyle, C. J., Gleeson, D., Jordan, K., Beresford, T. P., Ross, R. P., Fitzgerald, G. F., & Cotter, P. D. (2015). Anaerobic sporeformers and their significance with respect to milk and dairy products. *International Journal of Food Microbiology*, 197, 77–87.

Elean, M., Albarracin, L., Villena, J., Kitazawa, H., Saavedra, L., & Hebert, E. M. (2023). In silico comparative genomic analysis revealed a highly conserved proteolytic system in *Lactobacillus delbrueckii*. *International Journal of Molecular Sciences*, 24(14), 11309.

El-Leboudy, A. A., Amer, A. A., & El-Mohsen, S. A. (2014). Detection of some pathogenic organisms from dairy farm milk. *Alexandria Journal of Veterinary Sciences*, 44(3), 111–118.

El-Zamkan, M. A., & Hameed, K. G. A. (2016). Prevalence of *Campylobacter jejuni* and *Campylobacter coli* in raw milk and some dairy products. *Veterinary World*, 9(10), 1147.

Fan, Y., Shi, J., Song, X., Wang, G., Xia, Y., Ai, L., & Xiong, Z. (2025). Engineering UDP-sugar synthesis pathways for exopolysaccharide biosynthesis in *Streptococcus thermophilus* AR333. *Journal of the Science of Food and Agriculture*, 105(8), 4657–4664.

FAO/WHO. (2001). Evaluation of Health and Nutritional Properties of Probiotics in Food Including Powder Milk with Live Acid Bacteria. Report of a Joint FAO/WHO Expert Consultation, Córdoba, Argentina.

Ferreira, C. O., Álvarez Martín, P., Flórez García, A. B., Díaz Fernández, J. M., & Mayo Pérez, B. (2012). Biochemical and technological properties of *Penicillium roqueforti* and *Geotrichum candidum* strains isolated from cabrales, a Spanish traditional, blue-veined, starterfree cheese. *Italian Journal of Food Science*, 24, 115–124.

Galli, V., Venturi, M., Mari, E., Guerrini, S., & Granchi, L. (2022). Gamma-aminobutyric acid (GABA) production in fermented milk by lactic acid bacteria isolated from spontaneous raw milk fermentation. *International Dairy Journal*, 127, 105284.

Gardan, R., Honvo-Houeto, E., Mézange, C., Maillot, N. J., Balvay, A., Rabot, S., ... & Juillard, V. (2024). Use of Rgg quorum-sensing machinery to create an innovative recombinant protein expression system in *Streptococcus thermophilus*. *Microbiology*, 170(9), 001487.

Garnier, L., Valence, F., & Mounier, J. (2017). Diversity and control of spoilage fungi in dairy products: An update. *Microorganisms*, 5(3), 42.

Ghazaei, C. (2023). Antibacterial and antibiofilm activity of P. Pentosaceus bacteriocin (pediocin) isolated from cheese and its optimization for the bacteriocin production. *International Journal of Advanced Biological and Biomedical Research*, 11, 134.

Goldman, A. S., Prabhakar, B. S., & Baron, S. (1996). *Medical Microbiology*. University of Texas Medical Branch: Galveston.

Grace, D., Wu, F., & Havelaar, A. H. (2020). MILK symposium review: Foodborne diseases from milk and milk products in developing countries—Review of causes and health and economic implications. *Journal of Dairy Science*, 103(11), 9715–9729.

Guibourdenche, M., Roggentin, P., Mikoleit, M., Fields, P. I., Bockemühl, J., Grimont, P. A., & Weill, F. X. (2010). Supplement 2003–2007 (No. 47) to the white-Kauffmann-Le minor scheme. *Research in Microbiology*, 161(1), 26–29.

Gumustop, I., & Ortakci, F. (2023). Analyzing the genetic diversity and biotechnological potential of Leuconostoc pseudomesenteroides by comparative genomics. *Frontiers in Microbiology*, 13, 1074366.

Guo, W., Liu, S., Khan, M. Z., Wang, J., Chen, T., Alugongo, G. M., ... & Cao, Z. (2024). Bovine milk microbiota: Key players, origins, and potential contributions to early-life gut development. *Journal of Advanced Research*, 59, 49–64.

Guo, Z., Chen, Z., Liu, X., Chen, J., Chen, G., Liu, Z., ... & Lan, Q. (2019). Detection of *Listeria monocytogenes* and *Staphylococcus aureus* via duplex recombinase polymerase amplification. *Journal of Food Safety*, 39(4), e12628.

Hagi, T., Kobayashi, M., & Nomura, M. (2010). Molecular-based analysis of changes in indigenous milk microflora during the grazing period. *Bioscience, Biotechnology, and Biochemistry*, 74(3), 484–487.

Hamdaoui, N., Rokni, Y., Asehraou, A., Mouncif, M., Mennane, Z., Omari, A., ... & Meziane, M. (2023). Technological aptitude and sensitivity of lactic acid bacteria leuconostoc isolated from raw milk of cows: From step-by-step experimental procedure to the results. *Indonesian Journal of Science and Technology*, 8(2), 157–170.

Hanim, S., Ishamri, I., Nurhayati, Y., Abd Ghani, A., & Tang, J. Y. H. (2025). *Pseudomonas fluorescens* and *P. aeruginosa* contaminations of poultry and poultry products: A review on food safety and quality. *International Food Research Journal*, 32(3), 621–640.

Hanzelová, Z., Dudriková, E., Lovayová, V., Výrostková, J., Regecová, I., Zigo, F., & Bartáková, K. (2024). Occurrence of enterococci in the process of artisanal cheesemaking and their antimicrobial resistance. *Life*, 14(7), 890.

Hogan, J. S., Smith, K. L., Todhunter, D. A., & Schoenberger, P. S. (1995). Efficacy of a barrier teat dip containing. 55% chlorhexidine for prevention of bovine mastitis. *Journal of Dairy Science*, 78(11), 2502–2506.

Hoque, M. N., Rahman, M., Sultana, T., Rahman, M. M., Siddique, N., Islam, M. S., ... & Das, Z. C. (2024). Molecular characterization of *Streptococcus* spp. isolated from milk, feces and farm environment of mastitic dairy cows. *Journal of Science and Technology Research*, 6(1), 1–12.

Hu, T., Cui, Y., Zhang, Y., Qu, X., & Zhao, C. (2020). Genome analysis and physiological characterization of four *Streptococcus thermophilus* strains isolated from Chinese traditional fermented milk. *Frontiers in Microbiology*, 11, 184.

Hungate, R. E. (2013). *The Rumen and Its Microbes*. Elsevier.

Islam, M. S., Islam, M. A., Rahman, M. M., Islam, K., Islam, M. M., Kamal, M. M., & Islam, M. N. (2023). Presence of *Brucella* spp. in milk and dairy products: A comprehensive review and its perspectives. *Journal of Food Quality*, 2023(1), 2932883.

Jeantet, R., & Jan, G. (2021). Improving the drying of *Propionibacterium freudenreichii* starter cultures. *Applied Microbiology and Biotechnology*, 105(9), 3485–3494.

Jiang, S., Cai, L., Lv, L., & Li, L. (2021). *Pediococcus pentosaceus*, a future additive or probiotic candidate. *Microbial Cell Factories*, 20(1), 45.

Jurášková, D., Pires, V. C., Ribeiro, S. C., Ferreira, S. S., Bernardo, F., Evtyugin, D., ... & Silva, C. C. (2025). Exopolysaccharide (EPS)-Producing *Streptococcus thermophilus*: Functional and probiotic potential. *Foods*, 14(17), 3013.

Kadariya, J., Smith, T. C., & Thapaliya, D. (2014). *Staphylococcus aureus* and staphylococcal food-borne disease: An ongoing challenge in public health. *BioMed Research International*, 2014(1), 827965.

Kanungo, S., Bonilla-Rosso, G., Sarton-Lohéac, G., Kuffer, M., Hilty, M., Geiser, T., ... & Das, S. (2023). Comparative genomic analysis reveals novel phylogenetically intermediate *Streptococci* with high phenotypic diversity in the human distal lung microbiota. *BioRxiv*, 2023-12.

Kapse, N., Pisu, V., Dhakephalkar, T., Margale, P., Shetty, D., Wagh, S., ... & Dhakephalkar, P. K. (2024). Unveiling the probiotic potential of *Streptococcus thermophilus* MCC0200: Insights from in vitro studies corroborated with genome analysis. *Microorganisms*, 12(2), 347.

Kasalica, A., Vuković, V., Vranješ, A., & Memiši, N. (2011). Listeria monocytogenes in milk and dairy products. *Biotechnology in Animal Husbandry*, 27(3), 1067–1082.

Kebede, A., Viljoen, B. C., Gadaga, H., Narvhus, J. A., & Lourens-Hattingh, A. (2007). The effect of incubation temperature on the survival and growth of yeasts in Sethemi, South African naturally fermented milk. *Food Technology and Biotechnology*, 45(1), 21–26.

Khalifa, A., Ibrahim, H. I. M., Sheikh, A., & Khalil, H. E. (2023). Attenuation of immunogenicity in MOG-induced oligodendrocytes by the probiotic bacterium *Lactococcus* Sp. PO3. *Medicina*, 59(10), 1731.

Kumar S, Bansal K, Sethi SK. (2023). Comparative genomics reconciliations of genus Streptococcus resolves its taxonomy and elucidates biotechnological importance of their constituent species. *Ecological Genetics and Genomics*, 29, 100205.

Kumaresan, G., Annalvilli, R., & Sivakumar, K. (2007). Psychrotrophic spoilage of raw milk at different temperatures of storage. *Journal of Applied Sciences Research*, 3(11), 1383–1387.

Lafarge, V., Ogier, J. C., Girard, V., Maladen, V., Leveau, J. Y., Gruss, A., & Delacroix-Buchet, A. (2004). Raw cow milk bacterial population shifts attributable to refrigeration. *Applied and Environmental Microbiology*, 70(9), 5644–5650.

LeBlanc, J., Fliss, I., & Matar, C. (2004). Induction of a humoral immune response following an *Escherichia coli* O157: H7 infection with an immunomodulatory peptidic fraction derived from *Lactobacillus helveticus*-fermented milk. *Clinical and Vaccine Immunology*, 11(6), 1171–1181.

Lee, H. W., Kim, I. S., Kil, B. J., Seo, E., Park, H., Ham, J. S., ... & Huh, C. S. (2020). Investigation of flavor-forming starter *Lactococcus lactis* subsp. lactis LDTM6802 and *Lactococcus lactis* subsp. cremoris LDTM6803 in miniature gouda-type cheeses. *Journal of Microbiology and Biotechnology*, 30(9), 1404.

Leite, A. M. D. O., Miguel, M. A. L., Peixoto, R. S., Rosado, A. S., Silva, J. T., & Paschoalin, V. M. F. (2013). Microbiological, technological and therapeutic properties of kefir: A natural probiotic beverage. *Brazilian Journal of Microbiology*, 44, 341–349.

Lettat, A., & Benchaar, C. (2013). Diet-induced alterations in total and metabolically active microbes within the rumen of dairy cows. *PLoS One*, 8(4), e60978.

Li, N., Wang, Y., You, C., Ren, J., Chen, W., Zheng, H., & Liu, Z. (2018). Variation in raw milk microbiota throughout 12 months and the impact of weather conditions. *Scientific Reports*, 8(1), 2371.

Liang, C., Zhang, D., Zheng, X., Wen, X., Yan, T., Zhang, Z., & Hou, C. (2021). Effects of different storage temperatures on the physicochemical properties and bacterial community structure of fresh lamb meat. *Food Science of Animal Resources*, 41(3), 509.

Lopez, C. C., Serio, A., Rossi, C., Mazzarrino, G., Marchetti, S., Castellani, F., ... & Martino, G. (2016). Effect of diet supplementation with *Ascophyllum nodosum* on cow milk composition and microbiota. *Journal of Dairy Science*, 99(8), 6285–6297.

López-Gálvez, F., Gil, M. I., & Allende, A. (2018). Impact of relative humidity, inoculum carrier and size, and native microbiota on *Salmonella* ser. Typhimurium survival in baby lettuce. *Food Microbiology*, 70, 155–161.

Lucey, J. A. (2015). Raw milk consumption: Risks and benefits. *Nutrition Today*, 50(4), 189–193.

Madera, C., Monjardín, C., & Suárez, J. E. (2004). Milk contamination and resistance to processing conditions determine the fate of *Lactococcus lactis* bacteriophages in dairies. *Applied and Environmental Microbiology*, 70(12), 7365–7371.

Madigan, M. T., Bender, K. S., Buckley, D. H., Sattley, W. M., & Stahl, D. A. (2021). Ecological diversity of bacteria. *Brock Biology of Microorganisms 16th Edition*; Pearson: Hoboken, NJ, USA, 478–518.

Madigan, M. T., Martinko, J. M., & Parker, J. (1997). *Brock Biology of Microorganisms* (Vol. 11). Upper Saddle River, NJ: Prentice hall.

Magnusson, M., Christiansson, A., Svensson, B., & Kolstrup, C. (2006). Effect of different premilking manual teat-cleaning methods on bacterial spores in milk. *Journal of Dairy Science*, 89(10), 3866–3875.

Mahony, J., Bottacini, F., & van Sinderen, D. (2023). Towards the diversification of lactococcal starter and non-starter species in mesophilic dairy culture systems. *Microbial Biotechnology*, 16(9), 1745–1754.

Mallet, A., Guéguen, M., Kauffmann, F., Chesneau, C., Sesboüé, A., & Desmasures, N. (2012). Quantitative and qualitative microbial analysis of raw milk reveals substantial diversity influenced by herd management practices. *International Dairy Journal*, 27(1–2), 13–21.

Mantel, M., Durand, T., Bessard, A., Pernet, S., Beaudeau, J., Guimaraes-Laguna, J., ... & Rolli-Derkinderen, M. (2024). *Propionibacterium freudenreichii* CIRM-BIA 129 mitigates colitis through S layer protein B-dependent epithelial strengthening. *American Journal of Physiology-Gastrointestinal and Liver Physiology*, 326(2), G163–G175.

Marchand, S., Heylen, K., Messens, W., Coudijzer, K., De Vos, P., Dewettinck, K., ... & Heyndrickx, M. (2009). Seasonal influence on heat-resistant proteolytic capacity of *Pseudomonas lundensis* and *Pseudomonas fragi*, predominant milk spoilers isolated from Belgian raw milk samples. *Environmental Microbiology*, 11(2), 467–482.

Marcó, M. B., Moineau, S., & Quiberoni, A. (2012). Bacteriophages and dairy fermentations. *Bacteriophage*, 2(3), 149–158.

Martin, N. H., Ranieri, M. L., Murphy, S. C., Ralyea, R. D., Wiedmann, M., & Boor, K. J. (2011). Results from raw milk microbiological tests do not predict the shelf-life performance of commercially pasteurized fluid milk. *Journal of Dairy Science*, 94(3), 1211–1222.

Mei, S., He, G., Chen, Z., Zhang, R., Liao, Y., Zhu, M., ... & Chen, C. (2023). Probiotic-fermented distillers grain alters the rumen microbiome, metabolome, and enzyme activity, enhancing the immune status of finishing cattle. *Animals*, 13(24), 3774.

Melchior, M. B., Vaarkamp, H., & Fink-Gremmels, J. (2006). Biofilms: A role in recurrent mastitis infections?. *The Veterinary Journal*, 171(3), 398–407.

Miceli de Farias, F., O'Connor, P. M., Buttimer, C., Kamilari, E., Soria, M. C., Johnson, C. N., ... & Ross, R. P. (2024). Raffinocyclicin is a novel plasmid-encoded circular bacteriocin produced by *Lactococcus raffinolactis* with broad-spectrum activity against many gram-positive food pathogens. *Applied and Environmental Microbiology*, 90(9), e00809–e00824.

Michel, V., Hauwuy, A., & Chamba, J. F. (2006). Gestion de la flore microbienne des laits crus par les pratiques des producteurs. *Rencontres autour des recherches sur les ruminants*, 309–312.

Monsallier, F., Verdier-Metz, I., Agabriel, C., Martin, B., & Montel, M. C. (2012). Variability of microbial teat skin flora in relation to farming practices and individual dairy cow characteristics. *Dairy Science & Technology*, 92(3), 265–278.

Morandi, S., Silvetti, T., Lopreiato, V., Piccioli-Cappelli, F., Trevisi, E., & Brasca, M. (2024). Biodiversity and antibiotic resistance profile provide new evidence for a different origin of enterococci in bovine raw milk and feces. *Food Microbiology*, 120, 104492.

Mulaw, G., Sisay Tessema, T., Muleta, D., & Tesfaye, A. (2019). In vitro evaluation of probiotic properties of lactic acid bacteria isolated from some traditionally fermented Ethiopian food products. *International Journal of Microbiology*, 2019(1), 7179514.

Oliver, S. P., Jayarao, B. M., & Almeida, R. A. (2005). Foodborne pathogens in milk and the dairy farm environment: Food safety and public health implications. *Foodbourne Pathogens & Disease*, 2(2), 115–129.

Parente, E., Guidone, A., Matera, A., De Filippis, F., Mauriello, G., & Ricciardi, A. (2016). Microbial community dynamics in thermophilic undefined milk starter cultures. *International Journal of Food Microbiology*, 217, 59–67.

Parente, E., Zotta, T., Giavalisco, M., & Ricciardi, A. (2023). Metataxonomic insights in the distribution of Lactobacillaceae in foods and food environments. *International Journal of Food Microbiology*, 391, 110124.

Paszyńska-Wesołowska, I., & Bartoszcze, M. (2009). Bacteria in the state of VBNC-a threat to human health. *Medycyna Weterynaryjna*, 65(4), 228–231.

Piccinini, R., Borromeo, V., & Zecconi, A. (2010). Relationship between S. aureus gene pattern and dairy herd mastitis prevalence. *Veterinary Microbiology*, 145(1–2), 100–105.

Piwowarek, K., Lipińska, E., Hać-Szymańczuk, E., Kieliszek, M., & Kot, A. M. (2020). Sequencing and analysis of the genome of *Propionibacterium freudenreichii* T82 strain: Importance for industry. *Biomolecules*, 10(2), 348.

Pyz-Łukasik, R., Paszkiewicz, W., Tatara, M. R., Brodzki, P., & Bełkot, Z. (2015). Microbiological quality of milk sold directly from producers to consumers. *Journal of Dairy Science*, 98(7), 4294–4301.

Qi, Y., Huang, L., Zeng, Y., Li, W., Zhou, D., Xie, J., ... & Yin, J. (2021). *Pediococcus pentosaceus*: Screening and application as probiotics in food processing. *Frontiers in Microbiology*, 12, 762467.

Quigley, L., O'Sullivan, O., Beresford, T. P., Ross, R. P., Fitzgerald, G. F., & Cotter, P. D. (2011). Molecular approaches to analysing the microbial composition of raw milk and raw milk cheese. *International Journal of Food Microbiology*, 150(2–3), 81–94.

Quigley, L., O'Sullivan, O., Stanton, C., Beresford, T. P., Ross, R. P., Fitzgerald, G. F., & Cotter, P. D. (2013).The complex microbiota of raw milk. *FEMS Microbiology Reviews*, 37(5), 664–698.

Rahman, M. M., Siddique, N., Akter, S., Das, Z. C., & Hoque, M. N. (2024). Draft genome sequencing of *Pediococcus pentosaceus strains* isolated from cow milk. *Microbiology Resource Announcements*, 13(5), e0023824.

Rai, R., & Tamang, J. P. (2022). In vitro and genetic screening of probiotic properties of lactic acid bacteria isolated from naturally fermented cow-milk and yak-milk products of Sikkim, India. *World Journal of Microbiology and Biotechnology*, 38(2), 25.

Raimondi, S., Candeliere, F., Amaretti, A., Costa, S., Vertuani, S., Spampinato, G., & Rossi, M. (2022). Phylogenomic analysis of the genus *Leuconostoc*. *Frontiers in Microbiology*, 13, 897656.

Rainard, P. (2017). Mammary microbiota of dairy ruminants: Fact or fiction?. *Veterinary Research*, 48(1), 25.

Ranieri, M. L., Huck, J. R., Sonnen, M., Barbano, D. M., & Boor, K. J. (2009). High temperature, short time pasteurization temperatures inversely affect bacterial numbers during refrigerated storage of pasteurized fluid milk. *Journal of Dairy Science*, 92(10), 4823–4832.

Rešetar Maslov, D., Thomas, F. C., Beletić, A., Kuleš, J., Rubić, I., Benić, M., ... & Mrljak, V. (2023). Distinguishing natural infections of the bovine mammary gland by *Staphylococcus* from *Streptococcus* spp. using quantitative milk proteomics. *Animals*, 13(11), 1829.

Ricciardi, A., De Filippis, F., Zotta, T., Facchiano, A., Ercolini, D., & Parente, E. (2016). Polymorphism of the phosphoserine phosphatase gene in *Streptococcus thermophilus* and its potential use for typing and monitoring of population diversity. *International Journal of Food Microbiology*, 236, 138–147.

Rocchetti, G., Fancello, F., Zara, G., Hatami, F., Coronas, R., Petretto, G. L., ... & Mannazzu, I. (2025). Fermentation of second cheese whey by *Propionibacterium freudenreichii*: Metabolomic profiling and pathway insights for valorization. *Food Research International*, 217, 116838.

Ropars, J., Didiot, E., de La Vega, R. C. R., Bennetot, B., Coton, M., Poirier, E., ... & Giraud, T. (2020). Domestication of the emblematic white cheese-making fungus *Penicillium camemberti* and its diversification into two varieties. *Current Biology*, 30(22), 4441–4453.

Roussi, V., Govaris, A., Varagouli, A., & Botsoglou, N. A. (2002). Occurrence of aflatoxin M1 in raw and market milk commercialized in Greece. *Food Additives & Contaminants*, 19(9), 863–868.

Ryu, S., Shin, M., Yun, B., Lee, W., Choi, H., Kang, M., ... & Kim, Y. (2021). Bacterial quality, prevalence of pathogens, and molecular characterization of biofilm-producing *Staphylococcus aureus* from Korean dairy farm environments. *Animals*, 11(5), 1306.

Samaržija, D., Podoreški, M., Sikora, S., Skelin, A., & Pogačić, T. (2007). Spoilage microorganisms in milk and dairy products. *Mljekarstvo: časopis za unaprjeđenje proizvodnje i prerade mlijeka*, 57(4), 251–273.

Samaržija, D., Zamberlin, Š., & Pogačić, T. (2012). Psychrotrophic bacteria and their negative effects on milk and dairy products quality. *Mljekarstvo: časopis za unaprjeđenje proizvodnje i prerade mlijeka*, 62(2), 77–95.

Sarhir, S. T., Belkhou, R., Bouseta, A., & Hayaloglu, A. A. (2023). Evaluation of techno-functional and biochemical characteristics of selected lactic acid bacteria (*Lactococcus lactis* subsp. lactis and *Leuconostoc mesenteroides* subsp. mesenteroides) used for the production of Moroccan fermented milk: Lben. *International Dairy Journal*, 140, 105592.

Saxelin, M., Korpela, R., & Mäyrä-Mäkinen, A. (2003). Introduction: Classifying functional dairy products. In *Functional Dairy Products* (pp. 1-16). Woodhead Publishing

Schöpping, M., Gaspar, P., Neves, A. R., Franzén, C. J., & Zeidan, A. A. (2021). Identifying the essential nutritional requirements of the probiotic bacteria Bifidobacterium animalis and *Bifidobacterium longum* through genome-scale modeling. *NPJ Systems Biology and Applications*, 7(1), 47.

Shahi, N., & Mallik, S. K. (2020). Emerging bacterial fish pathogen *Lactococcus garvieae* RTCLI04, isolated from rainbow trout (*Oncorhynchus mykiss*): Genomic features and comparative genomics. *Microbial Pathogenesis*, 147, 104368.

Shalaby, M., Reboud, J., Forde, T., Zadoks, R. N., & Busin, V. (2024). Distribution and prevalence of enterotoxigenic *Staphylococcus aureus* and staphylococcal enterotoxins in raw ruminants' milk: A systematic review. *Food Microbiology*, 118, 104405.

Sıçramaz, H., Güven, O. T., Can, A., Ayar, A., & Gül, Y. (2022). Impact of different starter cultures and *Lactobacillus helveticus* on volatile components, chemical and sensory properties of pasta filata cheese. *Current Research in Food Science*, 5, 1009–1016.

Soares, A. S., Miranda, C., Coelho, A. C., & Trindade, H. (2023). Occurrence of coliforms and *Enterococcus* species in drinking water samples obtained from selected dairy cattle farms in Portugal. *Agriculture*, 13(4), 885.

Steinberg, R. S., Silva e Silva, L. C., De Souza, M. R., Reis, R. B., Da Silva, P. C., Lacorte, G. A., ... & Nunes, Á. C. (2022). Changes in bovine milk bacterial microbiome from healthy and subclinical mastitis affected animals of the Girolando, Gyr, Guzera, and Holstein breeds. *International Microbiology*, 25(4), 803–815.

Stevens, M., Piepers, S., & De Vliegher, S. (2016). Mastitis prevention and control practices and mastitis treatment strategies associated with the consumption of (critically important) antimicrobials on dairy herds in Flanders, Belgium. *Journal of Dairy Science*, 99(4), 2896–2903.

Sviridenko, G. M., Komarova, T. V., & Uskova, E. E. (2023). Study of the composition of the residual microflora of milk after pasteurization. *Food Systems*, 5(4), 344–352.

Sviridenko, G. M., Shukhalova, O. M., Vakhrusheva, D. S., & Mamykin, D. S. (2024). Formation of cheese pattern when using monospecies cultures. *Food Systems*, 7(2), 276–281.

Tarnaud, F., Gaucher, F., Do Carmo, F. L. R., Illikoud, N., Jardin, J., Briard-Bion, V., ... & Jan, G. (2020). Differential adaptation of *Propionibacterium freudenreichii* CIRM-BIA129 to cow's milk versus soymilk environments modulates its stress tolerance and proteome. *Frontiers in Microbiology*, 11, 549027.

Tarrah, A., da Silva Duarte, V., Pakroo, S., Corich, V., & Giacomini, A. (2020). Genomic and phenotypic assessments of safety and probiotic properties of *Streptococcus macedonicus* strains of dairy origin. *Food Research International*, 130, 108931.

Te Giffel, M. T., Wagendorp, A., Herrewegh, A., & Driehuis, F. T. (2002). Bacterial spores in silage and raw milk. *Antonie van Leeuwenhoek*, 81(1), 625–630.

Terzić-Vidojević, A., Veljović, K., Popović, N., Tolinacki, M., & Golić, N. (2021). Enterococci from raw-milk cheeses: Current knowledge on safety, technological, and probiotic concerns. *Foods*, 10(11), 2753.

Tilocca, B., Costanzo, N., Morittu, V. M., Spina, A. A., Soggiu, A., Britti, D., ... & Piras, C. (2020). Milk microbiota: Characterization methods and role in cheese production. *Journal of Proteomics*, 210, 103534.

Tolle, A. (1980). The microflora of the udder. *Factors Influencing the Bacteriological Quality of Raw Milk. Document*, 120(4), 4.

Tong, S. Y., Davis, J. S., Eichenberger, E., Holland, T. L., & Fowler Jr, V. G. (2015). *Staphylococcus aureus* infections: Epidemiology, pathophysiology, clinical manifestations, and management. *Clinical Microbiology Reviews*, 28(3), 603–661.

Turgay, M., Falentin, H., Irmler, S., Fröhlich-Wyder, M. T., Meola, M., Oberhaensli, S., & Marchand, H. B. D. G. (2022). Genomic rearrangements in the aspA-dcuA locus of *Propionibacterium freudenreichii* are associated with aspartase activity. *Food Microbiology*, 106, 104030.

Tusa, H., Alemayehu, T., Subussa, B. W., Ayalew, H., & Ali, M. M. (2024). Hygienic practices of vendors and their contribution to Coliform, *Salmonella*, and *Shigella* bacteria of raw milk at Asella Town, Oromia, Ethiopia. *International Journal of Food Science*, 2024(1), 8869022.

Vacheyrou, M., Normand, A. C., Guyot, P., Cassagne, C., Piarroux, R., & Bouton, Y. (2011). Cultivable microbial communities in raw cow milk and potential transfers from stables of sixteen French farms. *International Journal of Food Microbiology*, 146(3), 253–262.

Van der Spiegel, M., Van der Fels-Klerx, H. J., & Marvin, H. J. P. (2012). Effects of climate change on food safety hazards in the dairy production chain. *Food Research International*, 46(1), 201–208.

Vargova, M., Vyrostkova, J., Veszelits Lakticova, K., & Zigo, F. (2023). Effectiveness of sanitation regime in a milking parlour to control microbial contamination of teats and surfaces teat cups'. *Annals of Agricultural and Environmental Medicine*, 30(1), 55–60.

Vasco, K. A., Carbonell, S., Sloup, R. E., Bowcutt, B., Colwell, R. R., Graubics, K., ... & Manning, S. D. (2023). Persistent effects of intramammary ceftiofur treatment on the gut microbiome and antibiotic resistance in dairy cattle. *Animal Microbiome*, 5(1), 56.

Velasco, R., Cabeza, M. C., & Ordóñez, J. A. (2025). Current approaches to minimize the late blowing defect of cheese. *Cogent Food & Agriculture*, 11(1), 2460462.

Vélez, M. A., Wolf, V. I., Espariz, M., Acciarri, G., Magni, C., Hynes, E., & Perotti, M. C. (2023). Study of volatile compounds profiles in milk matrices using *Enterococcus faecalis* EstA and *Rhizomucor miehei* lipase. *Food Research International*, 169, 112861.

Verdier-Metz, I., Gagne, G., Bornes, S., Monsallier, F., Veisseire, P., Delbès-Paus, C., & Montel, M. C. (2012). Cow teat skin, a potential source of diverse microbial populations for cheese production. *Applied and Environmental Microbiology*, 78(2), 326–333.

Verreault, D., Gendron, L., Rousseau, G. M., Veillette, M., Massé, D., Lindsley, W. G., ... & Duchaine, C. (2011). Detection of airborne lactococcal bacteriophages in cheese manufacturing plants. *Applied and Environmental Microbiology*, 77(2), 491–497.

Vieira, C. P., Cabral, C. C., da Costa Lima, B. R., Paschoalin, V. M. F., Leandro, K. C., & Conte-Junior, C. A. (2017). *Lactococcus lactis* ssp. cremoris MRS47, a potential probiotic strain isolated from kefir grains, increases cis-9, trans-11-CLA and PUFA contents in fermented milk. *Journal of Functional Foods*, 31, 172–178.

Vinderola, C. G., Duarte, J., Thangavel, D., Perdigón, G., Farnworth, E., & Matar, C. (2005). Immunomodulating capacity of kefir. *Journal of Dairy Research*, 72(2), 195–202.

Vissers, M. M. M., Driehuis, F., Te Giffel, M. C., De Jong, P., & Lankveld, J. M. G. (2007). Quantification of the transmission of microorganisms to milk via dirt attached to the exterior of teats. *Journal of Dairy Science*, 90(8), 3579–3582.

Walter, J., & O'Toole, P. W. (2023). Microbe profile: The Lactobacillaceae. *Microbiology*, 169(12), 001414.

Wouters, J. T., Ayad, E. H., Hugenholtz, J., & Smit, G. (2002). Microbes from raw milk for fermented dairy products. *International Dairy Journal*, 12(2–3), 91–109.

Yan, H., Du, W., Ji, S., Guo, C., Zhang, Y., Wang, Y., ... & Li, S. (2022). Bacterial composition and interactions in raw milk and teat skin of dairy cows. *Fermentation*, 8(5), 235.

Yamamoto, E., Watanabe, R., Ichimura, T., Ishida, T., & Kimura, K. (2021). Effect of lactose hydrolysis on the milk-fermenting properties of *Lactobacillus delbrueckii* ssp. bulgaricus 2038 and *Streptococcus thermophilus* 1131. *Journal of Dairy Science*, 104(2), 1454–1464.

Yeluri Jonnala, B. R., McSweeney, P. L., Sheehan, J. J., & Cotter, P. D. (2018). Sequencing of the cheese microbiome and its relevance to industry. *Frontiers in Microbiology*, 9, 1020.

Yordanova, R., Hristova, P., & Platikanova, M. (2024). Benefits and risks of using *Enterococcus* species in food or as probiotics. *Journal of IMAB*, 30(2), 5569–5573.

Yuan, H., Han, S., Zhang, S., Xue, Y., Zhang, Y., Lu, H., & Wang, S. (2022). Microbial properties of raw milk throughout the year and their relationships to quality parameters. *Foods*, 11(19), 3077.

Yuan, L., Sadiq, F. A., Burmølle, M., Wang, N. I., & He, G. (2019). Insights into psychrotrophic bacteria in raw milk: A review. *Journal of Food Protection*, 82(7), 1148–1159.

Yuan, L., Sadiq, F. A., Liu, T. J., Li, Y., Gu, J. S., Yang, H. Y., & He, G. Q. (2018). Spoilage potential of psychrotrophic bacteria isolated from raw milk and the thermo-stability of their enzymes. *Journal of Zhejiang University-Science B*, 19(8), 630–642.

Zarei, M., Elmi Anvari, S., Maktabi, S., Saris, P. E. J., & Yousefvand, A. (2023). Identification, proteolytic activity quantification and biofilm-forming characterization of Gram-positive, proteolytic, psychrotrophic bacteria isolated from cold raw milk. *PLoS One*, 18(9), e0290953.

Zdanowicz, M., Shelford, J. A., Tucker, C. B., Weary, D. M., & Von Keyserlingk, M. A. G. (2004). Bacterial populations on teat ends of dairy cows housed in free stalls and bedded with either sand or sawdust. *Journal of Dairy Science*, 87(6), 1694–1701.

Zeinhom, M. M., & Abdel-Latef, G. K. (2014). Public health risk of some milk borne pathogens. *Beni-Suef University Journal of Basic and Applied Sciences*, 3(3), 209–215.

Zhao, L., Xie, Q., Shi, F., Liang, S., Chen, Q., Evivie, S. E., ... & Huo, G. (2021). Proteolytic activities of combined fermentation with *Lactobacillus helveticus* KLDS 1.8701 and *Lactobacillus plantarum* KLDS 1.0386 reduce antigenic response to cow milk proteins. *Journal of Dairy Science*, 104(11), 11499–11508.

Zhao, X., Ge, Y., Yu, X., Liu, C., Li, H., Wang, X., & Yao, S. (2024). Fermentation characteristics of fermented milk with *Streptococcus thermophilus* CICC 6063 and *Lactobacillus helveticus* CICC 6064 and volatile compound dynamic profiles during fermentation and storage. *Molecules*, 29(6), 1257.

CHAPTER FOUR

Role of the Gut Microbiome in Animal Health and Disease

MECHANISTIC INSIGHTS AND APPLICATIONS

Kun Li, Sijia Lu, Qing He, Chang Xu, and Zubair Azhar Nomi

4.1 INTRODUCTION

The gastrointestinal tract of animals has an extraordinarily sophisticated and dynamic assembly of microorganisms that are collectively known as the gut microbiome. This microbial community, which includes bacteria, archaea, fungi, protozoa, and viruses has coevolved with animal hosts and is now seen as part and parcel of animal physiology as opposed to their passive resident (Runge & Rosshart, 2021). The discovery of rapid changes in molecular biology, high-throughput sequencing, and computational biology over the past 20 years caused a profound transformation in the approach to studying the mechanisms of host–microbiome interactions between animals and microorganisms, changing scientific paradigms in animal health, nutrition, disease management (Wilde et al., 2024). The gut microbiome is being thought of as an organ that is functional and acts to enhance metabolic efficiency, immune competence, growth performance, and disease resistance. The challenge of ensuring optimum health without compromising productivity is a major issue in animal production systems, especially in the face of increasing global demand of animal-based foods, limited growth of antibiotic growth promoters and the emerging issues of antimicrobial resistance (Galhano et al., 2021). The existence of these pressures has added to the quest of sustainable, biologically informed substitute to the conventional disease control and growth enhancement approaches. In this context, gut microbiome has become a major area of intervention, which can be used to enhance feed efficiency, immune response and eliminate disease risk by using dietary, microbial and management-based interventions (Barrea et al., 2025). Traditionally, the functions of the gut microorganisms in animals were mainly researched within the framework of digestion and fermentation, in particular, in ruminants, where microbial consortia make it possible to use the fibrous plant materials. Modern studies have however shown that the functional role of the gut microbiome is much greater than degradation of nutrients (Yang et al., 2020). Short-chain fatty acids (SCFAs), secondary bile acids, and bioactive peptides are some examples of microbial metabolites which are signaling molecules that regulate the host gene expression, immune responses, and endocrine pathways. These findings highlight the systemic functionality of the microbiome in animal physiology and the usefulness of the microbiome in a wide range of species, including ruminants, poultry, aquaculture species and companion animals (Forcina et al., 2022). The gut microbiome composition and functioning is predetermined by a plethora

DOI: 10.1201/9781003638568-4

of factors, among which are diet, host genetics, age, environmental conditions, and management practices. The microbial colonization during the early-life stage, especially has a determinant role in the final health patterns (Brodin, 2022). Any type of perturbation at sensitive stages of development, like caused by exposure to antibiotics, malnutrition, or environmental stress agents, may result in lasting changes to microbial communities, thereby placing animals at risk of developing metabolic dysfunction, immune dysfunction, and being predisposed to diseases. The given idea of microbiome programming has a significant impact on animal husbandry where precise interventions in the early stages of growth are necessary (Rizzi et al., 2025). The term dysbiosis, which refers to the lack of balance in the composition or functionality of the gut microbiome, has been more and more involved in the pathogenesis of a broad spectrum of animal diseases. Violations of the homeostasis of microbes have been associated with enteric infections, inflammatory disorders, metabolic syndromes, impaired growth performance, and decreased feed efficiency (Aruwa & Sabiu, 2024). Dysbiosis in livestock systems does not only lead to poor animal welfare but also economic losses as animals become less productive, the cost of veterinary services rises with the mortality rate. It is therefore important to learn the mechanistic associations between microbial imbalance and disease progression to develop the effective prevention and mitigation strategies. The immune system is one of the most vital points of contact between the host and its gut microbiota (Yoo et al., 2020). A large percentage of the animal immune system is concentrated in the gut-associated lymphoid tissue (GALT) where a constant contact with commensal microorganisms influences immunologic development and activity.The beneficial microbial taxa help to maintain immune tolerance, pathogen exclusion and regulation of inflammatory responses, and pathogenic or opportunistic microorganisms may cause an excessive inflammation and tissue destruction (Kogut et al., 2020). Crosstalk between the gut microbiome and immune system has now been established in a two-way direction, as being a foundation of disease resistance and vaccine responsiveness in animals. Despite the fact that this concept has been used extensively on human and rodent models, increasing evidence on animal sciences suggests that microbial communities may influence the stress response, behavior, and overall productivity of livestock (Chen et al., 2021). Changes in gut microbiota composition due to stress can have a negative effect on feed intake, growth rates, and reproductive performance, thus, establishing the connection between the health of microbiomes and animal welfare and economic sustainability. The increasing awareness of microbiome functionality has triggered the realization of microbiome-specific intervention in animal production (Fishbein et al., 2023). Probiotics, prebiotics, synbiotics and postbiotics have become a popular focus as a means to control gut microbial ecosystems toward advantageous taxa and metabolic products. These measures will work to protect gut barrier integrity, cold pathogen microorganisms and increase nutrient use without using antibiotics (Serek & Oleksy-Wawrzyniak, 2021). Nevertheless, interspecies, interproduction system, and environmental context differences in efficacy indicate that a mechanistic grasp of the dynamics of such interventions with host biology and prevailing microbial communities is necessary. The use of technologies in omics such as metagenomics, metatranscriptomics, metaproteomics, and metabolomics has further enlightened the perspective on the functional potential of the gut microbiome (Yang et al., 2025). Combining multiomics data with bioinformatics and artificial intelligence applications is opening the door to accurate microbiome control, where dietary applications and health treatment can be customized to a particular animal group or production objective. Although a great deal has been achieved, there are multiple challenges that remain to limit the implementation of the work on microbiomes into practice (Knight et al., 2018). The great interindividual variation, the lack of standard methods, and microbial behavior in the context contribute to the complexity of the identification of universal biomarkers and intervention measures. Moreover, the various regions have different regulatory systems regarding microbial feed additives and new biotechnological solutions as well, which are another hindrance to their adoption. In order to address such problems, microbiologists, nutritionists, veterinarians, bioinformaticians, and policymakers must collaborate across disciplines (Horvat & Kovačević, 2025). Now is the time and time to have a comprehensive synthesis of the mechanistic knowledge and practice of the gut microbiome in animal health and disease in this shifting scientific landscape. The chapter showcase the basics of microbiome science with applied animal nutrition and animal health management to provide an integrated perspective that can be applied by researchers, industry professionals and policy makers alike.

4.2 COMPOSITION AND DEVELOPMENT OF THE ANIMAL GUT MICROBIOME

The microbiome of the gut of animals is an incredibly structured but dynamic ecosystem, the composition and functional potential of which may be species-specific, developmental, and environmental (Sehnal et al., 2021). Despite the difference in the microbial communities based on their taxonomic characteristics of the animal hosts, they possess functional characteristics which include nutrient metabolism, immune control as well as the intestinal system homeostasis. The compositional logic and developmental dynamics of the gut microbiome are essential to the process of establishing the role of the microbiome in animal health, as well as in the establishment of targeted microbiome-modulating interventions (Ugwu et al., 2024).

4.2.1 Microbial Diversity across Animal Species

4.2.1.1 Ruminants

The ruminant animals including cattle, sheep and goats are among the domesticated species with one of the most complicated and metabolically specialized gut microbiomes. The rumen is a large chamber of fermentation that is nonaerobic and has thick and heterogeneous bacteria, archaea, protozoa and anaerobic fungal populations (Elekwachi et al., 2017). It is the synergistic actions of these microorganisms that allows them to break down structurally complex plant polysaccharides such as cellulose, hemicellulose, and pectin to yield volatile fatty acids (VFAs), predominantly, acetate, propionate, and butyrate, as the major source of energy to the host. The bacterial flora of the rumen are predominantly Firmicutes and Bacteroidetes with genera like Ruminococcus, Fibrobacter, Prevotella and Butyrivibrio playing major roles in the degradation of fibers and carbohydrate fermentation (Cholewińska et al., 2020). Archaea perform the hydrogen disposal process, specifically, the methanogenic species of the genus Methanobrevibacter which influences the effectiveness of fermentation and the environmental sustainability. Protozoa contributes to the sequestration of starch, microbial predatory behavior, and anaerobic fungi contribute to the breakage of plant cell wall so that it can release more fermentable catabolites (Williams et al., 2020). Besides the rumen, microbial communities in the lower gastrointestinal tract of ruminants have a role in protein metabolism and immune signaling and epithelial integrity. Although hindgut microbes are not in such high populations as rumen microbiota, they still influence the process of nitrogen recycling and the immune response of their hosts, which is why microbial communities are so essential to the overall system of ruminant physiology (Liu et al., 2021).

4.2.1.2 Monogastric Animals

The microbiome of the gut in monogastric hosts such as poultry is predominantly located in the distal gastro intestinal tract particularly the cecum and the colon. Monogastrics do not utilize microbial fermentation to a large extent as ruminants do but gut microbes are important in the extraction of nutrients, prevention of pathogens and development of immune systems (Tardiolo et al., 2025). The microbiota of ceca in poultry can be characterized by a great density and variety of bacteria, the most common phyla of which are Firmicutes, Bacteroidetes, and Proteobacteria. Common genera that have been reported are Lactobacillus, Clostridium, Ruminococcus, and Bacteroides all of which have a role in fermenting carbohydrates, in the production of SCFA and gut barrier functionality. The composition of the poultry microbiome is influenced by diet composition, housing conditions, and early-life exposure to microbes particularly, which influence the growth performance and disease resistance (Ruuskanen, 2024).

4.2.1.3 Companion Animals

The microbiomes present in the gut of companion animals, particularly dogs and cats, provide evidence both of their evolutionary origin and diet. The dogs are also omnivorous facultative carnivores and also host a diverse range of different microbial communities, capable of fermenting carbohydrates and proteins that they feed on. Being compulsory carnivores, cats possess a more protein and fat metabolic microbiome (Wu, 2024). Some of the genera that are present in the animals include Clostridium, Blautia and Fusobacterium which are dominant phyla in dogs and help in nutrient metabolism and the health of the gut. It has been discovered that proteolytic bacteria are also enriched in feline gut microbiome, which suggests that the dietary protein consumption is high. Dysbiosis in companion animals has also been associated with inflammatory bowel disease, obesity, and metabolic disorders and suggests that microbiome results can be generalized across populations of animals (Homer et al., 2023).

4.2.2 Developmental Dynamics of the Gut Microbiome

4.2.2.1 Early-Life Microbial Colonization

The intestinal microbiome is established during or shortly after birth and it is a leading indicator of subsequent well-being and physiological functioning. The factors which determine early-life microbial colonization include the maternal microbiota, mode of birth, the intake of colostrum and environmental exposure (Kalbermatter et al., 2021).The neonatal livestock species replace an apparently sterile or low microbial habitat with one of high-rate microbial succession and diversification. Rumen colonization occurs during the early-life activities in ruminants before one starts to consume the solid foods.The niches are initially colonized by facultative anaerobes that consume the rest of the oxygen and create an anaerobic environment that encourages the growth of obligate anaerobic fermenters (André et al., 2021). Gradually adding the amount of solid feed stimulates the growth of fibrolytic and amylolytic microbes that ultimately leads to the creation of a functionally fit rumen microbiome.The conditions in the hatcheries and housing have high effects on the microbial exposure of the poultry and swine at early ages.The commercial production systems are more likely to inhibit the transfer of microbes by mothers hence, the delay or distorted development of the microbiome. This has caused the growing concern of studying early microbial interventions such as ovo-inoculation or probiotic administration at early stages, to promote desirable microbial colonization and immune competence (Shehata et al., 2021).

4.2.2.2 Microbiome Stability and Maturation

Gut microbiome of animals is currently experiencing a maturation process, the result of which is an increasing diversity, functional redundancy and resilience with age of animals. This is closely related to the alterations in the diet, maturation of immune system and physiological maturation (Zhou et al., 2019). Adult microbiome tends to be more stable and resistant to pertussis, both in terms of changing diet or exposure to pathogens. However, the microbial stability may be broken upon physiological stress, leading to dysbiosis because of weaning, transportation, or environment change (Upadhaya & Kim, 2021). Weaning in swine is a very sensitive phase in all aspects where there is a sudden shift in diet, social stress, and immune stress and which is usually marked by short- or long-term shifts in gut microbiota composition. Similarly, when the production in poultry is intense and the growth rate is high, the microbial community dynamics may be subjected to a selective force that can in the long-term influence the health of the animal (Wickramasuriya et al., 2022).

4.2.3 Conditions which Modify the Composition of Gut Microbiome

4.2.3.1 Eating and Nutritional Therapeutics

Diet is one of the best predictors of the gut microbiome composition and functionality of animal species. The nature, quality, and digestibility of the constituents of the feed influence the availability of substrates to be fermented by the microbes hence affecting the structure of the microbial communities (Faniyi et al., 2019). High-protein or high-energy diets stimulate the growth of proteolytic and saccharolytic taxa, whereas fiber diets stimulates the growth of fibrolytic bacteria. The effect of ruminal forageon-to-concentrate ratios on the rumen microbial communities is tremendous, and the implications of the fermentative efficiency, methane production, and metabolic well-being. In monogastric animals, dietary fiber, resistant starch and functional feed additives control hindgut fermenting patterns and profiles of SCFA, but not gut integrity and immune response (Tan et al., 2021).

4.2.3.2 Physiology of the Host and Genetics

Host genetic background also plays a role in the interindividual variation in microbiome composition in the sense that it influences gut morphology, immune responsiveness and metabolic pathways. Genetic selection of production traits may also take place accidentally without the aim of affecting microbial ecologies and the consequences of this may be both health and disease. Other ecological niches within the gastrointestinal tract include physiology of the host which could be gut motility, PH and secretion of mucus (Lindsay et al., 2020).

4.2.3.3 Managing Factors and Environmental

The environmental factors, housing conditions, Hygiene, and exposure to stressors are also important factors which have an effect on the gut microbiota (Singh et al., 2022). The intensive systems are capable of reducing the microbial diversity by raising the biosecurity and application of the antimicrobial, or the extensive systems enlarge the microbial range that animals are exposed. The reduction of stress and promotion of microbiodiversity in the management process are also becoming the popular components of sustainable animal health management (Srivastav et al., 2021).

4.2.4 Microbiome Perturbation and Antibiotics

The extensive use of antibiotics in animals has greatly influenced microbiome composition of the gut, which in most cases results in the reduction of diversity and alteration of functional capacity (Rochegüe et al., 2021). Even though the use of antibiotics may prove effective in control of diseases, it is not selective and disrupts the commensal population, thus, it may induce the emergence of antibiotic-resistant life. The growing constraints on the use of antibiotics have accelerated the move toward microbiome-friendly solutions, which is used to cement the necessity to study how microbiomes develop and become resilient (Pottie et al., 2024). It is founded on the abovementioned compositional and developmental backgrounds that the following will discuss mechanistic functions of the gut microbiome in animal health in terms of metabolic interactions and immune regulation, and host–microbe signaling pathways. Major microbial taxa inhabiting the animal gastrointestinal tract and their functional contributions to host health are summarized in Table 4.1.

4.3 MECHANISTIC ROLES OF THE GUT MICROBIOME IN ANIMAL HEALTH

The gut microbiome is one of the most essential biological interfaces, through which diet, physiology of the host, and external environment interact and control animal health. Gut microbial community is not just a digestive aid, but it also participates in metabolic, immunological and physiological processes influencing growth performance, disease resistance and homeostasis (Wickramasuriya et al., 2022). A multitude of microbial metabolites, structural components and signaling interactions mediate these effects linking microbial activity with host reactions. One of the most significant ways through which the gut microbiome improves the health of animals is through microbial fermentation of foodstuff that cannot be decomposed by the host (Negash, 2022). The process results in SCFAs and other metabolites which serve as energy sources to the intestinal epithelial cells and which regulate the gut motility and epithelial renewal. The metabolites also influence the metabolism of system energy and aid in an effective utilization of nutritional resources therefore the growth and fruitful performance. When this alteration is caused by the microbial fermentation efficiency it can have a direct effect on feed consumption and metabolic health (Lian et al., 2024). The role of the gut microbiome in the integrity of the intestinal barriers is very significant. This is because of positive microbial communities that promote mucus, enhanced tight junction proteins, and pathogen binding through competitive exclusion. These functions protect the host of the excessive microbial translocation and microbial inflammatory stimulants (Cardinale et al., 2020). In microbial balance imbalance, the intestinal permeability and epithelial damage may also be augmented, exposing animals to the local and systemic inflammatory states. Immune control is another major process of microbiome action. The chronic interaction that exists between the commensal microbes and the gut-associated immune system helps in the development of the immune system, tolerance and the balanced immune responses (Shi et al., 2017). The immune cells and cytokines differentiation are also checked by microbial metabolites so that the pathogen immunity and overinflammation could be avoided. On the other hand, a loss of immune competence and susceptibility to disease may occur as a result of dysregulation of host–microbe interactions. The microbiome also has an impact on the systemic cells in the body through metabolic and neuroendocrine signaling mechanisms, in addition to local gut effects (Gribble & Reimann, 2019). Relationships between microbial metabolites and host receptors and hormonal systems mediate stress responses, digestive drive and metabolic productivity. Such interactions relate the wellness of the intestinal microbes to the general animal health and productivity. Collectively, these mechanistic pathways signify the gut microbiome as an animal physiological component (Legan et al., 2022). The insights into such interactions are an indication that there is a biological foundation to microbiome-specific nutritional and management interventions to enhance health, resilience, and sustainable production outcomes. The gut microbiome influences animal health through

TABLE 4.1

Key gut microbial taxa and their functional roles in animal health

Microbial group/taxa	Major functional roles	Primary metabolites / products	Implications for animal health
Firmicutes (e.g., *Lactobacillus*, *Clostridium* spp.)	Carbohydrate fermentation; maintenance of gut barrier integrity; immune modulation	SCFAs (butyrate, acetate); antimicrobial peptides	Enhanced intestinal integrity, reduced inflammation, improved nutrient utilization
Bacteroidetes (e.g., *Bacteroides*, *Prevotella*)	Degradation of complex polysaccharides; nutrient metabolism	Propionate; bioactive oligosaccharides	Improved energy harvest, optimized feed efficiency, metabolic balance
Actinobacteria (e.g., *Bifidobacterium*)	Early-life gut colonization; competitive exclusion of pathogens	Acetate; vitamins	Enhanced immune development, protection against enteric pathogens
Proteobacteria (commensal members)	Nitrogen metabolism; microbial community stabilization	Various metabolic intermediates	Functional microbial diversity; imbalance may indicate dysbiosis
Rumen fibrolytic bacteria (e.g., *Ruminococcus*, *Fibrobacter*)	Plant fiber degradation; fermentation	VFAs (acetate, butyrate)	Efficient fiber utilization; improved growth and milk production
Methanogenic Archaea (e.g., *Methanobrevibacter*)	Hydrogen utilization; fermentation balance	Methane	Stabilization of fermentation; implications for energy efficiency and emissions
Anaerobic fungi	Plant cell wall disruption; fiber accessibility	Fermentation end-products	Enhanced rumen digestion and nutrient availability
Gut-associated immune-interactive microbes	Immune education; regulation of inflammatory responses	Immunomodulatory metabolites	Balanced immune responses; improved disease resistance

integrated metabolic, immunological, and epithelial mechanisms that collectively support physiological homeostasis and disease resistance (Figure 4.1).

4.4 GUT MICROBIOME DYSBIOSIS AND ANIMAL DISEASES

The maintenance of a balanced and healthy gut microbiome is crucial to the dietary health, production and well-being of animals. The disruption of this balance, or so-called gut microbiome dysbiosis, is among the most common causes of the occurrence and further development of a vast array of animal illnesses (Gebrayel et al., 2022). Dysbiosis is not only limited to alteration of the microbial composition but also influences the functional alteration of the microbial metabolic activity and host–microbe interaction. These upheavals can compromise intestinal barrier integrity, disrupt immune responses, and disturb systemic metabolic homeostasis, thereby predisposing animal species to illness (Ali et al., 2025). Gut dysbiosis is typically characterized by a loss of microbial diversity, a reduction in beneficial commensal microbial populations, and an increase

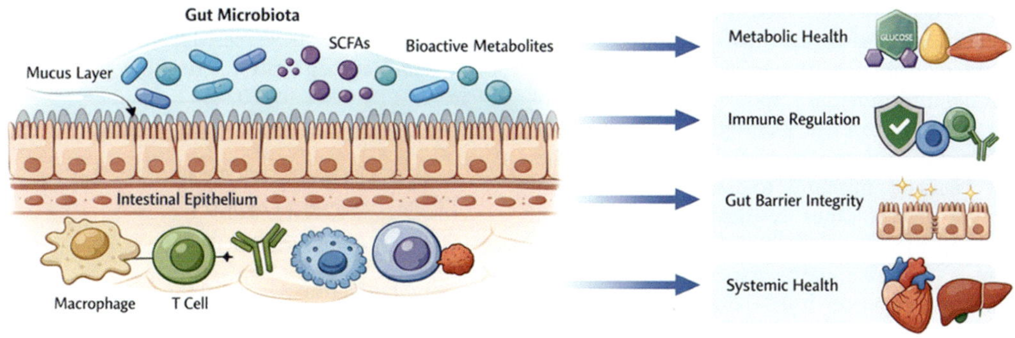

Figure 4.1 Mechanistic pathways linking the gut microbiome to animal health.

in opportunistic or pathogenic microorganisms. These deficiencies may be occasioned by several stressing factors, which may include sudden alterations in the diet, exposure to antibiotics and environmental stresses among others, and improper management mechanisms (Awuchi et al., 2020). Mechanistically, dysbiosis alters the production of key microbial metabolites, particularly SCFAs which play critical roles in epithelial integrity, immune tolerance and anti-inflammatory signaling. Reduction in supply of these metabolites deteriorates the capability of gut boundaries and increases intestinal permeability, which favors the movement of microbial components to systemic circulation causing inflammatory responses (Gasaly et al., 2021). The initial life is a particular susceptible phase, in which dysbiosis can have lasting impacts on the health of animals. Disruption of microbial colonization at such a delicate level of development may result in the fact that immunological growth and metabolicization are disrupted, and animals become prone to the disease later. Ineffective intake of colostrum, low hygiene, or premature exposure to antibiotics can alter normal colonization by microbes in ruminants hence exposing these chickens to neonatal diarrhea, respiratory diseases, and decrease in growth (Robi et al., 2024). Slow and intermittent microbial colonization of poultry is found to lead to poor gut development, absence of immune competency and susceptibility and the importance of early microbial stability is noted. Enteric disease predisposition is the highest acute effect of gut microbiome dysbiosis. Competition between the flora and the pathogens on nutrient and attachment sites, antimicrobial production and activation of the host immune provide colonization resistance by a healthy gut flora (Khan et al., 2021). The dysbiosis compromises such defense mechanisms, and the pathogenic microorganisms can grow. Microbial imbalance has been strongly associated with necrotic enteritis caused by *Clostridium perfringens* as well as increased colonization of pathogenic microorganisms (Salmonella and Campylobacter) caused by food in chickens. These infections not only destroy the health and performance of the animals but also badly affect the health of the people (Ame et al., 2022). Noninfectious gastrointestinal illnesses involving chronic inflammation, slowed digestion and malabsorption of nutrients are also caused by dysbiosis. An alteration in the fermentation patterns and the accumulation of the detrimental microbial metabolites can lead to the damage of the intestinal epithelium and a change in gut motility. Subacute ruminal acidosis may be caused by dysbiosis, particularly that caused by high-concentrate diet, in ruminants and includes reduced feed ratio, hind legs, and inflammation throughout the body (Elmhadi et al., 2022). This demonstrates to the degree to which the microbial imbalance in the intestines can spread its influence to the physiological and production levels. Beside gastrointestinal diseases, dysbiosis in the intestinal microbiome has significant effects on the growth performance and metabolic efficiency. Changes in composition and functionality of microbes may reduce fermentation efficiency, nutrient utilization and morphology of the intestines. Shortening of the intestinal villi, loss of the absorptive surfaces and poor feed ratio and ultimately poor growth rates have been attributed to dysbiosis in poultry (Fathima et al., 2022). The imbalances in the rumen microbial populations lead to disruptions in the voluminal fatty acid production and energy availability in the rumen which negatively affects growth, reproduction, and lactational performance in ruminants. Chronic dysbiosis can also lead to immune-mediated disorders in addition to persistent immune activation and inflammatory responses. Increased microbial tolerance to dietary antigens or commensals may result from the loss of protective microbes that facilitate immune tolerance. Dysbiosis of the gut microbiomes is inextricably associated with inflammatory bowel disease in companion animals, a disease that is associated with chronic inflammatory bowel disease, diarrhea, weight loss, and poor quality of life. Microbial imbalance plays a central role in pathogenesis of diseases despite it being a multifactorial in its origin. The dysbiosis can lead to subclinical inflammation that can influence (Singh et al., 2023) the immune responsiveness and productivity as well as welfare of the production animals despite the fact that it does not manifest itself as a complete disease. The pathology of the dysbiosis of the systemic microbiome of the gut is not gastrointestinal. The increased intestinal permeability allows the remnants of microbes such as lipopolysaccharides to get into the circulation that triggers low-grade systemic inflammation. The metabolic aberrations in response to physiologically challenging periods in ruminants (e.g. early lactation) have been proposed as the source of this inflammatory load and can condition susceptibility to secondary infections (Trevisi et al., 2025). These findings furnish the notion of the gut microbiome as the controller of the whole-body wellness and not a local node of the digestive system. Gut microbiome dysbiosis has been a major contributor of use of antibiotics in production of animals. Even though antibiotics are effective in the management of the disease, their broader spectrum activity is likely to disrupt the commensal microbial population, reducing the diversity of microbial species and their resistance (Huang et al., 2022). The dysbiosis observed in the presence of antibiotics increases the vulnerability of opportunistic infections and fosters the appearance and spread of antimicrobial resistance. The mentioned problems have added to heightened efforts across the globe to reduce the consumption of antibiotics and develop microbiome-based products to avert illnesses and enhance health (Simon et al., 2025). In total, all indicators of dysbiosis of gut microbiome evidence their centrality in the etiology and progression of disease in animals. Diseases caused by dysbiosis compromise the health of animals, reduce their productivity and the cost of production hence the importance of attaining microbial homeostasis is very high. The recognition of dysbiosis as a risk factor can be changed to allow specific intervention regarding nutritional, microbial and management interventions to achieve reestablishment of microbial balance and enhance the health consequences of animals (Attia et al., 2025).

4.5 APPLICATIONS OF GUT MICROBIOME MODULATION IN ANIMAL HEALTH AND PRODUCTION

The heightened importance of the gut microbiome as a prime regulator of animal health has led to the development of certain interventions to manipulate microbial communities to enhance productivity, disease resistance as well as animal welfare. Microbiome modulation might be a fascinating and biologically intelligent concept in the light of an increasing restriction in the use of antibiotics and the need to possess sustainable animal production systems (Beltrán-Velasco & Clemente-Suárez, 2025).

4.5.1 Probiotics

Among intervention interventions that most commonly modify the microbiome in animal production have been the application of probiotics that can be categorized as live microorganisms that confer health-promoting effects to the host when consumed in adequate amounts. The prevalent probiotics taxa in the animal kingdom are Lactobacillus, Bifidobacterium, Enterococcus, Bacillus, and the selection of yeast (Pang et al., 2022). The mechanism of action of the probiotics is positive and is mediated in many different ways. The probiotics win the race against the pathogens in adhesion and nutrients, produce antimicrobial agents, such as organic acids, bacteriocins, and affect host immunity. The use of probiotic supplementation has been revealed to stabilize the rumen fermentation, enhance fiber digestion and feed efficiency in ruminants. It has been known that probiotics enhance gut integrity, reduce pathogen load and growth performance in chicken. However, probiotics are specific to strains, dosage, route, and physiological status of a host (Wahab et al., 2022).

4.5.2 Prebiotics

Prebiotics are nondigestible feed substances, which selectively support the growth and activity of preferred beneficial gut microorganisms. The common prebiotics used in the animal nutrition are oligosaccharides, inulin, fructooligosaccharides, and resistant starches (Morgan, 2023). This is due to the fact that the action of prebiotics is mostly accomplished by provision of fermentable food to the beneficial microbes and an increase in the production of short-chain fatty acid and other bioactive metabolites. These metabolites enhance the ability of the intestinal barrier, regulate the immunologic responses, and repress the pathogenic bacteria (Fu et al., 2023). Prebiotics can have an impact on rumen microbial communities and rumen fermentation performance with the resultant improvement of ruminant metabolic well-being. The targeted nature of prebiotics is both with respect to safety and stability, which is an advantage; however, the use of prebiotics depends on responsive microbial communities in the host gut ecosystem (Yadav et al., 2022).

4.5.3 Synbiotics

Synbiotics are probiotics and prebiotics that are ingested as a combination to synergistically affect the microbiome composition and activity of the gut. Synbiotics are intended to enhance the microbial colonization and resilience of functionality through the maintenance of the presence of both the beneficial and adverse microorganisms as well as the media that aids the growth of microorganisms (Bhatia et al., 2025). Synbiotics have also been found to reduce the colonization and feed intake in chickens. The ruminal fermentation in ruminants could be stabilized and the negative effect of diet change could be compensated by synbiotic preparations. Synbiotic activity can be affected by the compatibility of probiotic and prebiotic strains and data and substrates although they have a potential. It is therefore urgent to develop and justify effectively to succeed in its implementation (Ayesiga et al., 2024).

4.5.4 Postbiotics and Microbial Metabolites

Postbiotics are the nonviable cells or cell fragments or metabolites of microbes which have positive effects on the host. Another advantage of postbiotics is that it does not require microbial viability in comparison to probiotics and therefore, postbiotics is comparatively stable, safe and regulatory acceptable (Guglielmetti et al., 2025). SCFAs, polysaccharides, peptides and cell wall fragments are some of the most crucial postbiotic ingredients. These chemicals can control the immunity mechanism, enhance the wellness of the epithelial lining, and prevent the multiplication of pathogens (Zhou et al., 2025). Postbiotic supplementation of poultry and ruminants has presented an encouraging prospect in enhancing gut health, reduced inflammatory of the gut, and better performance. One recent field of microbiome control is the postbiotics use, particularly when the use of probiotics is challenging in terms of its activation in the environment or when regulatory controls restrict the usage of live microorganisms (Liu et al., 2023).

4.5.5 Phytobiotics and Functional Feed Additives

The reviews of phytobiotics prepared using herbs and spices as well as essential oils have been conducted on the plant-based sources of microbiome modulators. These compounds are antimicrobial, antioxidant, and anti-inflammatory and influence the makeup and functioning of the gut microbes (Oni & Oke, 2025). Phytobiotics have been reported to reduce the concentration of the pathogenic bacteria, increase the activity of the digestive enzymes, and performance in the poultry. There are plant secondary metabolites in ruminants capable of controlling the dynamics of rumen fermentation, reducing methane synthesis, and improving the feed utilization (Kandemir et al., 2025).

4.5.6 Microbiome-Targeted Feeding and Precision Nutrition

Recent findings in microbiome have made it possible to create certain nutrition tactics through dietary plans in order to enhance the functionality of the intestinal microbes (Zhang, 2022). The nutrient use, metabolic strain and performance of the animal can be enhanced through the balance of feed and metabolic capacity of resident microbiota. The microbiome-based diets consider the nature of the fiber, fermentability and functional additives in order to activate the preferred microbial pathways (Feng et al., 2025).

4.5.7 Facilities Management Principals that Promote the Wellness of the Microbiome

Along with nutritional interventions, the management practices play a very significant role in the outcome of the gut microbiome composition and resilience. One of the ways that can be used to ensure a healthy microbiome in the gut would be exposure to microbes at the early age, which will minimize unnecessary antibiotic use, as well as provide a healthy environmental environment. The efficacy of microbiome modulation strategies can be enhanced by food-friendly management strategies along with nutritional interventions (Li et al., 2023).

4.5.8 Problem and Translational Issues

Based on the good outcomes, microbiome modulation strategies have several barriers to implementation. The variation in the findings between the animal groups, nonstandardized assessment processes, and regulation restrictions are very problematic (Bu et al., 2024). Moreover, the interactions of microbiome-focused interventions with the existing management practice should be approached with some care. The additional research study should be focused on the process of identifying loud microbial biomarkers, standardization of the intervention protocols, and efficacy of the intervention under the condition of commercial manufacturing (Simon et al., 2025). Even though microbiome modulation methods have immense potential, the optimization of the methods needs enhanced analytical tools and technologies. Microbiome-targeted nutritional and management interventions modulate gut microbial composition and function, resulting in improved health, productivity, and disease resistance in animals (Figure 4.2).

4.6 EMERGING TOOLS AND TECHNOLOGIES IN ANIMAL MICROBIOME RESEARCH

The recent advancements in technology have played a great role in enhancing the ability of profiling the animal gut microbiome and defining its functional input in disease and health (Rizzi et al., 2025). The profiling of microbial communities has now become possible in a more comprehensive way, both in the terms of taxonomic diversity as well as in terms of functional potential in various production systems and in animal species using high-throughput sequencing technologies. These methods put microbiome research out of descriptive studies into an understanding of how microbes play a role in host physiology (Muralitharan et al., 2023). The functional activity of the microbes, as well as the interactions between the microbes and the host can be understood further with the functional omics technologies, including transcriptomic, proteomic, and metabolomic analyses. Precisely, metabolomics has been helpful in relating the microbial metabolism to the immune response, enteral wellness, and nutritional exploitation (Lagoumintzis & Patrinos, 2023). Approaches to the integration of the multiomics data based on systems biology can be employed to obtain a holistic view of the entire biological networks and identify important microbial pathways that are associated with health and performance. The introduction of bioinformatics, machine learning, and artificial intelligence has also contributed to further development of microbiome research by enabling the analysis of large and complex data and prediction of health and productivity outcome based on microbial signatures (Dakal et al., 2025). Model systems of experimentation and novel diagnostic tools help in causal inference and the timely diagnosis of dysbiosis. All these technologies are accelerating the microbiome science translation into precision, sustainable and health-based animal production systems. Recent advances in sequencing technologies, functional omics, and computational approaches have substantially expanded the analytical toolkit available for animal gut microbiome research (Table 4.2).

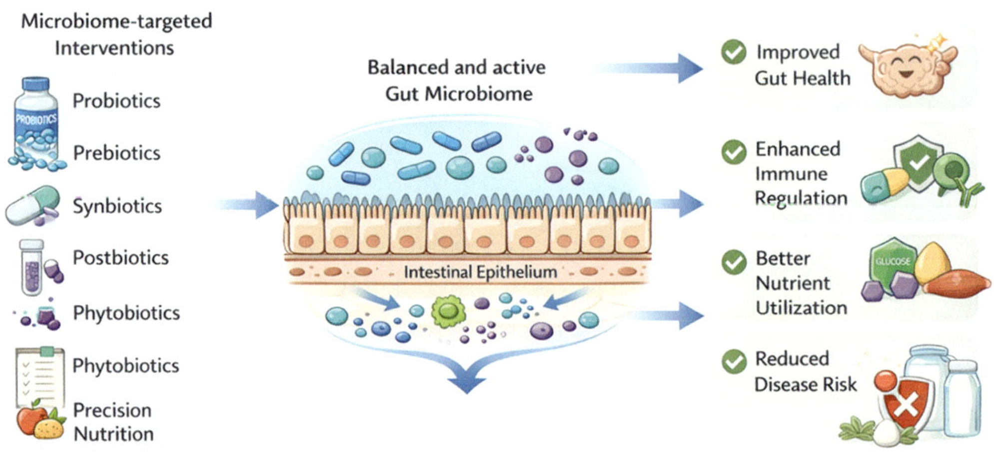

Figure 4.2 Applications of gut microbiome modulation in animal health and production.

TABLE 4.2

Emerging tools and technologies used in animal gut microbiome research

Tool / technology	Principle	Data generated	Key applications in animal microbiome research	Advantages	Limitations / Challenges
16S rRNA gene sequencing	Sequencing of conserved bacterial ribosomal RNA gene regions	Taxonomic profiles of bacterial communities	Characterization of gut microbial diversity; comparison of microbiome structure across diets, health states, and management systems	Cost-effective; high throughput; suitable for large population studies	Limited species-level resolution; no direct functional information
Shotgun metagenomic sequencing	Sequencing of total microbial DNA from samples	Species- and strain-level taxonomic data; functional gene catalogs	Identification of microbial metabolic pathways; analysis of fermentation, immune modulation, and resistance genes	High functional resolution; comprehensive microbial profiling	High cost; complex data analysis; computationally demanding
Metatranscriptomics	Sequencing of total microbial RNA	Actively expressed microbial genes	Assessment of real-time microbial activity during dietary changes, stress, or disease	Captures functional activity; dynamic insights	RNA instability; host RNA contamination; technical complexity
Metaproteomics	Large-scale identification of microbial proteins	Protein expression and enzyme activity profiles	Direct measurement of microbial metabolic and digestive functions	Functional confirmation beyond gene presence	Low throughput; complex sample preparation; limited reference databases
Metabolomics	Profiling of small-molecule metabolites	SCFAs, bile acids, amino acid derivatives, signaling molecules	Linking microbial activity to host metabolism, immunity, and gut integrity	Strong functional relevance; biomarker discovery	Metabolite origin ambiguity (host vs microbe); analytical variability
Multiomics integration	Integration of genomic, transcriptomic, proteomic, and metabolomic data	Systems-level biological networks	Identification of key microbial–host interaction pathways; precision nutrition development	Holistic understanding; predictive modeling	High computational requirements; need for advanced bioinformatics expertise

(Continued)

Table 4.2 (Continued)

Tool / technology	Principle	Data generated	Key applications in animal microbiome research	Advantages	Limitations / Challenges
Bioinformatics pipelines	Computational processing and annotation of sequencing data	Taxonomic, functional, and statistical outputs	Comparative microbiome analysis; microbial network construction	Enables large-scale data interpretation	Methodological variability; lack of standardization
Artificial intelligence & machine learning	Pattern recognition and predictive modeling from large datasets	Predictive microbial signatures	Prediction of disease risk, feed efficiency, and health outcomes	Handles complex, high-dimensional data; precision decision-making	Requires large, high-quality datasets; limited biological interpretability
In vitro fermentation systems	Controlled simulation of gut fermentation	Microbial growth and fermentation outputs	Screening of dietary components and feed additives	Ethical; controlled; cost-effective	Limited physiological realism compared to in vivo systems
Gnotobiotic animal models	Animals with defined or absent microbiota	Causal host–microbe interaction data	Mechanistic validation of microbial functions	Strong causal inference	High cost; ethical and technical constraints
Microbiome transplantation models	Transfer of microbial communities between hosts	Functional microbiome effects	Assessment of microbial community-level impacts on host traits	Demonstrates functional relevance of microbiota	Biosafety and standardization concerns
Microbiome-derived biomarker platforms	Identification of microbial or metabolite indicators	Diagnostic and prognostic biomarkers	Early detection of dysbiosis and disease risk	Supports precision health management	Validation under field conditions required

4.7 CHALLENGES, KNOWLEDGE GAPS, AND FUTURE PERSPECTIVES

Although the gut microbiome is currently making a fast stride, it has several limitations that have continued to limit its full application in the area of animal health and production systems. Microbial composition and microbial process are highly interindividually variable and this makes it hard to define universal biomarkers and leaves it less predictable whether a microbiome-targeted intervention will work (Al-Adham et al., 2025). The irregularities in sampling, sequencing, and data analysis methodology also complicate the comparison of the study outcomes across the research and render them repeatable. Furthermore, many of the published associations between microbial taxa and health outcomes have been merely correlative and thus the need to conduct mechanistic and longitudinal studies to establish causality (Khelfaoui et al., 2025). The knowledge of microbiome research has limited the field level application since interventions have not been validated in the field over the long term and short term due to environmental stressors and changes in management, which are likely to compromise intervention effects. The regulatory uncertainty also exists concerning the microbial products and new biotechnologies that are quite difficult to adopt as well. The second step will be standardized methods, integration of multiomics data, and host and management data and development of accuracy, microbiome-based intervention (Ferrocino et al., 2023). The microbiome-based approaches with enhanced interdisciplinary studies and technological innovations have a potential high possibility of underpinning the sustainable, low-antibiotic, and resilient production systems in animals.

4.8 CONCLUSION

The gut microbiome has established itself as a factor in the health, productivity and disease resistance of animals that is accomplished through complex metabolic, immunological and physiological mechanisms. Recent research of microbiome reported that healthy microbial communities help in efficient utilization of nutrients, intestinal barrier, and immune homeostasis, and dysbiosis is related to massive number of gastrointestinal, metabolic as well as inflammatory illnesses. The increasing environmental restrictions on the use of antibiotics and the growing necessity to seek more sustainable ways to proceed with animal production have merely served to bring microbiome-informed approaches into the limelight. With new technologies, new methods of intervention, including omics technologies, bioinformatics, and precision nutrition methods, it is becoming increasingly possible to look deeper into the mechanisms and intervene more precisely. Still, the path to the transformation of research findings into normal field usage is fraught with grave challenges. More standardization of the methods and long-term validation of them in a commercial environment are required. The combination of microbiome-based solutions can offer a sustainable future of resilient, health-conscious and sustainable animal production systems.

REFERENCES

Al-Adham, I. S., Ali Agha, A. S., Al-Remawi, M., Al-Akayleh, F., Al-Sheikh, A., Aburub, F., & Collier, P. J. (2025). Probiotics, psychobiotics, paraprobiotics, and postbiotics in gut-brain axis modulation: Multi-omics and AI-driven precision nutrition for cognitive health. *Probiotics and Antimicrobial Proteins*, 1–23.

Ali, M., Xu, C., Wang, M., Hina, Q., Ji, Y., Anwar, S., ... & Li, K. (2025). Gut barrier dysfunction and microbiota variations in cryptosporidiosis: A comprehensive review. *Veterinary Sciences*, 12(2), 85.

Ame, N. Y., Mohammed, L. A., & Ame, M. M. (2022). Review on public health importance of salmonellosis of poultry in Ethiopia. *International Journal of Advanced Multidisciplinary*, 9(6), 78–95.

André, A. C., Debande, L., & Marteyn, B. S. (2021). The selective advantage of facultative anaerobes relies on their unique ability to cope with changing oxygen levels during infection. *Cellular Microbiology*, 23(8), e13338.

Aruwa, C. E., & Sabiu, S. (2024). Interplay of poultry–microbiome interactions–influencing factors and microbes in poultry infections and metabolic disorders. *British Poultry Science*, 65(5), 523–537.

Attia, Y. A., Helal, I. E., Nagadi, S. A., Khafaga, A. F., Taha, A. E., Asiry, K. A., ... & Tufarelli, V. (2025). Gut dysbiosis: Causes and prevention strategies–A review. *Annals of Animal Science*, 25(4), 1297–1311.

Awuchi, C. G., Igwe, V. S., & Amagwula, I. O. (2020). Nutritional diseases and nutrient toxicities: A systematic review of the diets and nutrition for prevention and treatment. *International Journal of Advanced Academic Research*, 6(1), 1–46.

Ayesiga, I., Iqbal, S., Kyejjusa, Y., Okoboi, J., Omara, T., Adelina, T., ... & Kahwa, I. (2024). Key regulatory aspects of prebiotics, probiotics, and synbiotics in the management of metabolic disorders. In N. Mishra, S. Ashique, A. Farid, & A. Garg (Eds.), *Synbiotics in Metabolic Disorders* (pp. 214–244). CRC Press.

Barrea, L., Verde, L., Annunziata, G., Chedraui, P., Petraglia, F., Cucalón, G., ... & Muscogiuri, G. (2025). Effectiveness of medical nutrition therapy in the management of patients with obesity and endometriosis: From the Mediterranean diet to the ketogenic diet, through supplementation. The role of the nutritionist in clinical management. *Current Obesity Reports*, 14(1), 68.

Beltrán-Velasco, A. I., & Clemente-Suárez, V. J. (2025). Harnessing gut microbiota for biomimetic innovations in health and biotechnology. *Biomimetics*, 10(2), 73.

Bhatia, A., Sharma, D., Mehta, J., Kumarasamy, V., Begum, M. Y., Siddiqua, A., ... & Mat Rani, N. N. I. (2025). Probiotics and synbiotics: Applications, benefits, and mechanisms for the improvement of human and ecological health. *Journal of Multidisciplinary Healthcare*, 18, 1493–1510.

Brodin, P. (2022). Immune-microbe interactions early in life: A determinant of health and disease long term. *Science*, 376(6596), 945–950.

Bu, L., Wang, C., Bai, J., Song, J., Zhang, Y., Chen, H., & Suo, H. (2024). Gut microbiome-based therapies for alleviating cognitive impairment: State of the field, limitations, and future perspectives. *Food & Function*, 15(3), 1116–1134.

Cardinale, V., Capurso, G., Ianiro, G., Gasbarrini, A., Arcidiacono, P. G., & Alvaro, D. (2020). Intestinal permeability changes with bacterial translocation as key events modulating systemic host immune response to SARS-CoV-2: A working hypothesis. *Digestive and Liver Disease*, 52(12), 1383–1389.

Chen, S., Luo, S., & Yan, C. (2021). Gut microbiota implications for health and welfare in farm animals: A review. *Animals*, 12(1), 93.

Cholewińska, P., Czyż, K., Nowakowski, P., & Wyrostek, A. (2020). The microbiome of the digestive system of ruminants–a review. *Animal Health Research Reviews*, 21(1), 3–14.

Dakal, T. C., Xu, C., & Kumar, A. (2025). Advanced computational tools, artificial intelligence and machine-learning approaches in gut microbiota and biomarker identification. *Frontiers in Medical Technology*, 6, 1434799.

Elekwachi, C. O., Wang, Z., Wu, X., Rabee, A., & Forster, R. J. (2017). Total rRNA-seq analysis gives insight into bacterial, fungal, protozoal and archaeal communities in the rumen using an optimized RNA isolation method. *Frontiers in Microbiology*, 8, 1814.

Elmhadi, M. E., Ali, D. K., Khogali, M. K., & Wang, H. (2022). Subacute ruminal acidosis in dairy herds: Microbiological and nutritional causes, consequences, and prevention strategies. *Animal Nutrition*, 10, 148–155.

Faniyi, T. O., Adegbeye, M. J., Elghandour, M. M. M. Y., Pilego, A. B., Salem, A. Z. M., Olaniyi, T. A., ... & Adewumi, M. K. (2019). Role of diverse fermentative factors towards microbial community shift in ruminants. *Journal of Applied Microbiology*, 127(1), 2–11.

Fathima, S., Shanmugasundaram, R., Adams, D., & Selvaraj, R. K. (2022). Gastrointestinal microbiota and their manipulation for improved growth and performance in chickens. *Foods*, 11(10), 1401.

Feng, Y., Jin, Q., Liu, X., Lin, T., Johnson, A., & Huang, H. (2025). Advances in understanding dietary fiber: Classification, structural characterization, modification, and gut microbiome interactions. *Comprehensive Reviews in Food Science and Food Safety*, 24(1), e70092.

Ferrocino, I., Rantsiou, K., McClure, R., Kostic, T., De Souza, R. S. C., Lange, L., ... & MicrobiomeSupport Consortium. (2023). The need for an integrated multi-OMICs approach in microbiome science in the food system. *Comprehensive Reviews in Food Science and Food Safety*, 22(2), 1082–1103.

Fishbein, S. R., Mahmud, B., & Dantas, G. (2023). Antibiotic perturbations to the gut microbiome. *Nature Reviews Microbiology*, 21(12), 772–788.

Forcina, G., Pérez-Pardal, L., Carvalheira, J., & Beja-Pereira, A. (2022). Gut microbiome studies in livestock: Achievements, challenges, and perspectives. *Animals*, 12(23), 3375.

Fu, Y., Lyu, J., & Wang, S. (2023). The role of intestinal microbes on intestinal barrier function and host immunity from a metabolite perspective. *Frontiers in Immunology*, 14, 1277102.

Galhano, B. S., Ferrari, R. G., Panzenhagen, P., de Jesus, A. C. S., & Conte-Junior, C. A. (2021). Antimicrobial resistance gene detection methods for bacteria in animal-based foods: A brief review of highlights and advantages. *Microorganisms*, 9(5), 923.

Gasaly, N., De Vos, P., & Hermoso, M. A. (2021). Impact of bacterial metabolites on gut barrier function and host immunity: A focus on bacterial metabolism and its relevance for intestinal inflammation. *Frontiers in Immunology*, 12, 658354.

Gebrayel, P., Nicco, C., Al Khodor, S., Bilinski, J., Caselli, E., Comelli, E. M., ... & Edeas, M. (2022). Microbiota medicine: Towards clinical revolution. *Journal of Translational Medicine*, 20(1), 111.

Gribble, F. M., & Reimann, F. (2019). Function and mechanisms of enteroendocrine cells and gut hormones in metabolism. *Nature Reviews Endocrinology*, 15(4), 226–237.

Guglielmetti, S., Boyte, M. E., Smith, C. L., Ouwehand, A. C., Paraskevakos, G., & Younes, J. A. (2025). Commercial and regulatory frameworks for postbiotics: An industry-oriented scientific perspective for non-viable microbial ingredients conferring beneficial physiological effects. *Trends in Food Science & Technology*, 163, 105130.

Homer, B., Judd, J., Mohammadi Dehcheshmeh, M., Ebrahimie, E., & Trott, D. J. (2023). Gut microbiota and behavioural issues in production, performance, and companion animals: A systematic review. *Animals*, 13(9), 1458.

Horvat, O., & Kovačević, Z. (2025). Human and veterinary medicine collaboration: Synergistic approach to address antimicrobial resistance through the lens of planetary health. *Antibiotics*, 14(1), 38.

Huang, C., Feng, S., Huo, F., & Liu, H. (2022). Effects of four antibiotics on the diversity of the intestinal microbiota. *Microbiology Spectrum*, 10(2), e01904-21.

Kalbermatter, C., Fernandez Trigo, N., Christensen, S., & Ganal-Vonarburg, S. C. (2021). Maternal microbiota, early life colonization and breast milk drive immune development in the newborn. *Frontiers in Immunology*, 12, 683022.

Kandemir, A., Bölükbaşı, Ş. C., & Palangi, V. (2025). Inhibition of methane production by plant secondary compounds in ruminants. In *Handbook of Climate Change Mitigation and Adaptation* (pp. 2569–2601). Springer Nature Switzerland.

Khan, I., Bai, Y., Zha, L., Ullah, N., Ullah, H., Shah, S. R. H., ... & Zhang, C. (2021). Mechanism of the gut microbiota colonization resistance and enteric pathogen infection. *Frontiers in Cellular and Infection Microbiology*, 11, 716299.

Khelfaoui, I., Wang, W., Meskher, H., Shehata, A. I., El Basuini, M. F., Abouelenein, M. F., ... & Mahmoud, S. S. (2025). Beyond just correlation: causal machine learning for the microbiome, from prediction to health policy with econometric tools. *Frontiers in Microbiology*, 16, 1691503.

Knight, R., Vrbanac, A., Taylor, B. C., Aksenov, A., Callewaert, C., Debelius, J., ... & Dorrestein, P. C. (2018). Best practices for analysing microbiomes. *Nature Reviews Microbiology*, 16(7), 410–422.

Kogut, M. H., Lee, A., & Santin, E. (2020). Microbiome and pathogen interaction with the immune system. *Poultry Science*, 99(4), 1906–1913.

Lagoumintzis, G., & Patrinos, G. P. (2023). Triangulating nutrigenomics, metabolomics and microbiomics toward personalized nutrition and healthy living. *Human Genomics*, 17(1), 109.

Legan, T. B., Lavoie, B., & Mawe, G. M. (2022). Direct and indirect mechanisms by which the gut microbiota influence host serotonin systems. *Neurogastroenterology & Motility*, 34(10), e14346.

Li, S., He, Y., Zhong, S., Li, Y., Di, Y., Wang, Q., ... & Cao, F. (2023). Antioxidant and anti-aging properties of polyphenol–polysaccharide complex extract from *Hizikia fusiforme*. *Foods*, 12(20), 3725.

Lian, X., Shi, M., Liang, Y., Lin, Q., & Zhang, L. (2024). The effects of unconventional feed fermentation on intestinal oxidative stress in animals. *Antioxidants*, 13(3), 305.

Lindsay, E. C., Metcalfe, N. B., & Llewellyn, M. S. (2020). The potential role of the gut microbiota in shaping host energetics and metabolic rate. *Journal of Animal Ecology*, 89(11), 2415–2426.

Liu, C., Ma, N., Feng, Y., Zhou, M., Li, H., Zhang, X., & Ma, X. (2023). From probiotics to postbiotics: Concepts and applications. *Animal Research and One Health*, 1(1), 92–114.

Liu, K., Zhang, Y., Yu, Z., Xu, Q., Zheng, N., Zhao, S., ... & Wang, J. (2021). Ruminal microbiota–host interaction and its effect on nutrient metabolism. *Animal Nutrition*, 7(1), 49–55.

Morgan, N. K. (2023). Advances in prebiotics for poultry: Role of the caeca and oligosaccharides. *Animal Production Science*, 63(18), 1911–1925.

Muralitharan, R. R., Snelson, M., Meric, G., Coughlan, M. T., & Marques, F. Z. (2023). Guidelines for microbiome studies in renal physiology. *American Journal of Physiology-Renal Physiology*, 325(3), F345–F362.

Negash, A. (2022). Gut microbiota ecology role in animal nutrition and health performance. *Journal of Clinical Microbiology and Antimicrobials*, 6, 1–14.

Oni, A. I., & Oke, O. E. (2025). Gut health modulation in poultry through phytogenics: Mechanisms, benefits, and applications. *Frontiers in Veterinary Science*, 12, 1616734.

Pang, Y., Zhang, H., Wen, H., Wan, H., Wu, H., Chen, Y., ... & Liu, X. (2022). Yeast probiotic and yeast products in enhancing livestock feeds utilization and performance: An overview. *Journal of Fungi*, 8(11), 1191.

Pottie, I., Vázquez Fernández, R., Van de Wiele, T., & Briers, Y. (2024). Phage lysins for intestinal microbiome modulation: Current challenges and enabling techniques. *Gut Microbes*, 16(1), 2387144.

Rizzi, S., Saroglia, G., Kalemi, V., Rimoldi, S., & Terova, G. (2025). Artificial intelligence in microbiome research and beyond: Connecting human health, animal husbandry, and aquaculture. *Applied Sciences*, 15(17), 9781.

Robi, D. T., Mossie, T., & Temteme, S. (2024). A comprehensive review of the common bacterial infections in dairy calves and advanced strategies for health management. *Veterinary Medicine: Research and Reports*, 15, 1–14.

Rochegüe, T., Haenni, M., Mondot, S., Astruc, C., Cazeau, G., Ferry, T., ... & Lupo, A. (2021). Impact of antibiotic therapies on resistance genes dynamic and composition of the animal gut microbiota. *Animals*, 11(11), 3280.

Runge, S., & Rosshart, S. P. (2021). The mammalian metaorganism: A holistic view on how microbes of all kingdoms and niches shape local and systemic immunity. *Frontiers in Immunology*, 12, 702378.

Ruuskanen, S. (2024). Early-life environmental effects on birds: Epigenetics and microbiome as mechanisms underlying long-lasting phenotypic changes. *Journal of Experimental Biology*, 227(Suppl_1), jeb246024.

Sehnal, L., Brammer-Robbins, E., Wormington, A. M., Blaha, L., Bisesi, J., Larkin, I., ... & Adamovsky, O. (2021). Microbiome composition and function in aquatic vertebrates: small organisms making big impacts on aquatic animal health. *Frontiers in Microbiology*, 12, 567408.

Serek, P., & Oleksy-Wawrzyniak, M. (2021). The effect of bacterial infections, probiotics and zonulin on intestinal barrier integrity. *International Journal of Molecular Sciences*, 22(21), 11359.

Shehata, A. M., Paswan, V. K., Attia, Y. A., Abdel-Moneim, A. M. E., Abougabal, M. S., Sharaf, M., ... & Alagawany, M. (2021). Managing gut microbiota through in ovo nutrition influences early-life programming in broiler chickens. *Animals*, 11(12), 3491.

Shi, N., Li, N., Duan, X., & Niu, H. (2017). Interaction between the gut microbiome and mucosal immune system. *Military Medical Research*, 4(1), 14.

Simon, B. O., Nnaji, N. D., Anumudu, C. K., Aleke, J. C., Ekwueme, C. T., Uhegwu, C. C., ... & Onyeaka, H. (2025). Microbiome-based interventions for food safety and environmental health. *Applied Sciences*, 15(9), 5219.

Singh, S., Sharma, P., Pal, N., Kumawat, M., Shubham, S., Sarma, D. K., ... & Nagpal, R. (2022). Impact of environmental pollutants on gut microbiome and mental health via the gut–brain axis. *Microorganisms*, 10(7), 1457.

Singh, S., Sharma, P., Sarma, D. K., Kumawat, M., Tiwari, R., Verma, V., ... & Kumar, M. (2023). Implication of obesity and gut microbiome dysbiosis in the etiology of colorectal cancer. *Cancers*, 15(6), 1913.

Srivastav, A. L., Dhyani, R., Ranjan, M., Madhav, S., & Sillanpää, M. (2021). Climate-resilient strategies for sustainable management of water resources and agriculture. *Environmental Science and Pollution Research*, 28(31), 41576–41595.

Tan, F. P., Beltranena, E., & Zijlstra, R. T. (2021). Resistant starch: Implications of dietary inclusion on gut health and growth in pigs: A review. *Journal of Animal Science and Biotechnology*, 12(1), 124.

Tardiolo, G., La Fauci, D., Riggio, V., Daghio, M., Di Salvo, E., Zumbo, A., & Sutera, A. M. (2025). Gut microbiota of ruminants and monogastric livestock: An overview. *Animals*, 15(5), 758.

Trevisi, E., Cattaneo, L., Piccioli-Cappelli, F., Mezzetti, M., & Minuti, A. (2025). International Symposium on Ruminant Physiology: The immunometabolism of transition dairy cows from dry-off to early lactation: lights and shadows. *Journal of Dairy Science*, 108, 7662–7674

Ugwu, O. P. C., Alum, E. U., Okon, M. B., & Obeagu, E. I. (2024). Mechanisms of microbiota modulation: Implications for health, disease, and therapeutic interventions. *Medicine*, 103(19), e38088.

Upadhaya, S. D., & Kim, I. H. (2021). The impact of weaning stress on gut health and the mechanistic aspects of several feed additives contributing to improved gut health function in weanling piglets—A review. *Animals*, 11(8), 2418.

Wahab, S., Almaghaslah, D., Mahmood, S. E., Ahmad, M. F., Alsayegh, A. A., Abu Haddash, Y. M., ... & Hani, U. (2022). Pharmacological efficacy of probiotics in respiratory viral infections: A comprehensive review. *Journal of Personalized Medicine*, 12(8), 1292.

Wickramasuriya, S. S., Park, I., Lee, K., Lee, Y., Kim, W. H., Nam, H., & Lillehoj, H. S. (2022). Role of physiology, immunity, microbiota, and infectious diseases in the gut health of poultry. *Vaccines*, 10(2), 172.

Wilde, J., Slack, E., & Foster, K. R. (2024). Host control of the microbiome: Mechanisms, evolution, and disease. *Science*, 385(6706), eadi3338.

Williams, C. L., Thomas, B. J., McEwan, N. R., Rees Stevens, P., Creevey, C. J., & Huws, S. A. (2020). Rumen protozoa play a significant role in fungal predation and plant carbohydrate breakdown. *Frontiers in Microbiology*, 11, 720.

Wu, G. (2024). Recent advances in the nutrition and metabolism of dogs and cats. *Adv Exp Med Biol.* 2024;1446:1–14.

Yadav, M. K., Kumari, I., Singh, B., Sharma, K. K., & Tiwari, S. K. (2022). Probiotics, prebiotics and synbiotics: Safe options for next-generation therapeutics. *Applied Microbiology and Biotechnology*, 106(2), 505–521.

Yang, Q., Liang, Q., Balakrishnan, B., Belobrajdic, D. P., Feng, Q. J., & Zhang, W. (2020). Role of dietary nutrients in the modulation of gut microbiota: a narrative review. *Nutrients*, 12(2), 381.

Yang, S. Y., Han, S. M., Lee, J. Y., Kim, K. S., Lee, J. E., & Lee, D. W. (2025). Advancing gut microbiome research: The shift from metagenomics to multi-omics and future perspectives. *Journal of Microbiology and Biotechnology*, 35, e2412001.

Yoo, J. Y., Groer, M., Dutra, S. V. O., Sarkar, A., & McSkimming, D. I. (2020). Gut microbiota and immune system interactions. *Microorganisms*, 8(10), 1587.

Zhang, P. (2022). Influence of foods and nutrition on the gut microbiome and implications for intestinal health. *International Journal of Molecular Sciences*, 23(17), 9588.

Zhou, X., Du, L., Shi, R., Chen, Z., Zhou, Y., & Li, Z. (2019). Early-life food nutrition, microbiota maturation and immune development shape life-long health. *Critical Reviews in Food Science and Nutrition*, 59(sup1), S30–S38.

Zhou, X., Wu, Y., Zhu, Z., Lu, C., Zhang, C., Zeng, L., ... & Zhou, F. (2025). Mucosal immune response in biology, disease prevention and treatment. *Signal Transduction and Targeted Therapy*, 10(1), 7.

CHAPTER FIVE

Livestock Microbiome

Alex Muppasani and Sunil Thomas

5.1 INTRODUCTION

The global population continues to grow rapidly, and within the next few years it is expected to reach nearly 9 billion people. Alongside this growth, rising literacy levels, expanding service sectors, increasing incomes, and evolving lifestyles have significantly altered dietary preferences across the world. As societies become more urbanized and economically stable, consumption patterns are shifting toward protein-rich diets, particularly meat and animal-based products. These trends are placing unprecedented pressure on existing food systems, making it essential to substantially increase food production to meet future global demand (Godber & Wall, 2014; Idamokoro, 2023; Reynolds et al., 2015).

In this context, animal husbandry and livestock rearing play a critical role in ensuring food security while supporting economic development, especially in rural regions. Livestock provide not only meat, milk, and eggs but also employment opportunities, income stability, and a financial safety net for farming communities. For many rural households, livestock represent a form of living capital that can be sold or utilized during times of economic hardship, thereby strengthening resilience and livelihoods (Godber & Wall, 2014; Idamokoro, 2023; Reynolds et al., 2015).

Moreover, livestock husbandry within grassland-based systems offers important environmental and ecological benefits. Properly managed grazing contributes to the conservation of rangeland ecosystems by maintaining plant diversity, preventing shrub encroachment, and improving soil structure. These systems promote the use of land-preserving forages and enable the conversion of grasses and other nonarable vegetation into high-quality food. By producing nutritious food from land unsuitable for crop cultivation, livestock systems enhance overall land-use efficiency while supporting sustainable agricultural practices and long-term environmental stewardship (Godber & Wall, 2014; Idamokoro, 2023).

The conceptual framework underpinning livestock physiology has undergone a fundamental paradigm shift in recent years, evolving from a narrowly host-centric view to a more integrative holobiont model. Within this framework, the animal and its associated symbiotic microbial communities are recognized as a single, coevolving biological entity (Callaway et al., 2025). These microbial assemblages—collectively termed the microbiome—include bacteria, archaea, fungi, protozoa, and viruses that inhabit diverse anatomical niches and collectively contribute genetic, metabolic, and regulatory capacities that vastly exceed those encoded by the host genome alone (Caspani & Swann, 2019; Matijašić et al., 2020). This extensive auxiliary genetic reservoir, often described as the "second genome," encodes a multitude of enzymatic pathways and bioactive metabolites capable of shaping host gene expression, immune system development, neuroendocrine signaling, and adaptive responses to environmental pressures (Bordenstein & Theis, 2015; Callaway et al., 2025).

Far from serving merely as a passive facilitator of digestion, the livestock microbiome functions as a central regulatory interface that maintains

DOI: 10.1201/9781003638568-5

physiological homeostasis and enhances the host's capacity to respond to nutritional variability and environmental stressors (Jandhyala et al., 2015). Its influence extends across multiple biological systems, including immune competence, metabolic efficiency, stress-axis modulation, and even behavioral traits (Farzi et al., 2018). These complex interkingdom interactions are mediated through mechanisms such as epigenetic remodeling, alternative gene splicing, microbial metabolite signaling, and immune crosstalk, collectively shaping animal health, productivity, and resilience across diverse livestock species (Wang & Ibeagha-Awemu, 2021).

The livestock microbiome also occupies a critical intersection between agriculture, public health, and environmental sustainability. In ruminant systems, microbial communities—particularly methanogenic archaea—are primary drivers of enteric methane production, a potent greenhouse gas with significant implications for climate change mitigation strategies (Waters et al., 2025). Simultaneously, the microbiome serves as a reservoir and conduit for antimicrobial resistance (AMR) genes within the "One Health" framework, linking animal, human, and environmental health. Through horizontal gene transfer and ecological connectivity, these resistance determinants can traverse species and ecosystem boundaries, complicating disease control and threatening global public health security (Argudín et al., 2017). Beyond environmental and health dimensions, the microbiome underpins socioeconomic stability in agrarian and resource-limited regions, where livestock such as water buffaloes provide essential nutrition, agricultural labor, and economic resilience (Tong et al., 2022). As global demand for animal-derived protein continues to rise—projected to double by 2050—microbiome-centered strategies offer transformative opportunities to enhance productivity while minimizing ecological footprints and advancing sustainable intensification (Cardoso et al., 2020).

This review provides a comprehensive synthesis of current knowledge on the microbiomes of major livestock species, including cattle, water buffaloes, horses, sheep, goats, and swine. It examines species-specific microbial community structures, functional metabolic networks, and their broader implications for animal health, environmental stewardship, and societal well-being, highlighting the microbiome as a pivotal leverage point for the future of sustainable livestock production.

5.2 GLOBAL LIVESTOCK SYSTEMS

5.2.1 Livestock Production Systems

The domestication of animals represents one of the most consequential transformations of the Anthropocene, fundamentally reconfiguring the biosphere to meet human nutritional, economic, and cultural demands (Cucchi & Arbuckle, 2021). Over millennia, domesticated livestock have become deeply embedded in human societies, shaping settlement patterns, trade networks, and food systems. In the contemporary era, the global livestock sector functions as a vast interface between biological resources and human civilization, occupying a substantial proportion of the planet's ice-free terrestrial surface when rangelands and feed-crop cultivation are considered together. This footprint encompasses extensive grazing systems as well as large areas of arable land devoted to feed crops such as maize, soybean, and barley, effectively reorganizing terrestrial ecosystems around animal-based food production.

The scale of this transformation is underscored by the fact that the total biomass of domesticated livestock now exceeds that of all wild terrestrial mammals combined, reflecting a profound redirection of primary productivity into human-managed food chains (Ritchie & Spooner, 2025). Livestock production exerts strong leverage over global biogeochemical cycles through manure generation, the demand for synthetic fertilizers used in feed-crop cultivation, and greenhouse gas emissions arising from enteric fermentation and manure management (Maree et al., 2025). Ruminant digestion alone accounts for a substantial fraction of anthropogenic methane emissions, while land-use change associated with grazing expansion and feed production contributes significantly to carbon dioxide emissions and biodiversity loss. Despite these environmental pressures, the sector generates approximately 40% of global agricultural value and supports both caloric intake and income for hundreds of millions of rural households worldwide, making livestock indispensable to global food security and rural livelihoods.

5.2.2 Species Distribution and Ecological Adaptation

Although a wide range of vertebrate species are utilized in agriculture, global production of meat, milk, and eggs is dominated by a small number of species—often referred to as the "big five": cattle, pigs, poultry, sheep, and goats—supplemented regionally by water buffalo, camels, and yak (Ritchie et al., 2019). These species differ markedly in digestive physiology, nutritional requirements, reproductive strategies, and climatic tolerance, factors that strongly shape their geographical distribution and functional roles within agroecosystems.

Ruminant species, through foregut fermentation, are uniquely capable of converting fibrous grasses and browse from rangelands, savannas, and shrublands into high-quality animal protein. This allows them to exploit landscapes that are marginal or unsuitable for crop cultivation due to low soil fertility, steep terrain, or limited rainfall (Morgan-Davies et al., 2017). Cattle populations are especially prominent in the Americas and South Asia, while sheep and goats dominate arid and semi-arid regions of North Africa, the Middle East, and Central Asia, where their ability to utilize sparse, low-quality forage underpins pastoral livelihoods and food security (Alders et al., 2021). Water buffalo are concentrated primarily in South and Southeast Asia, where they remain integral to rice-based farming systems, providing milk, meat, and draft power; their physiological and behavioral adaptations make them particularly effective in flooded paddy fields and smallholder production systems.

In contrast, monogastric species such as pigs and poultry depend largely on energy- and protein-rich concentrate feeds and are therefore closely linked to global grain and oilseed markets rather than local forage resources (Bikker & Jansman, 2023). Their production is concentrated in high-density, industrialized systems in East Asia, Europe, and North America, where location is influenced more by infrastructure, market access, and biosecurity regulations than by local agroecological constraints. Across all species, indigenous and locally adapted breeds represent a critical reservoir of genetic diversity, harboring traits such as heat tolerance, disease resistance, and efficiency under low-input conditions—attributes that are increasingly valued as climate variability intensifies and resource constraints grow (Kassahun et al., 2025).

5.2.3 Production Systems

Livestock production systems span a broad continuum, ranging from highly intensive industrial operations to extensive smallholder, mixed crop–livestock, and pastoral systems, with numerous intermediate and hybrid forms (Rust, 2018). In high-income regions and rapidly industrializing economies, production is dominated by large-scale enterprises that rely on specialized, high-yielding breeds, nutritionally optimized feeds, and controlled housing environments designed to maximize biological efficiency, predictability, and output. Increasingly, these systems are incorporating precision livestock farming technologies, including sensor-based health monitoring, automated feeding, and data-driven decision support tools, to optimize animal welfare, growth performance, and feed conversion efficiency at the individual animal level (Papakonstantinou et al., 2024).

While intensive systems achieve high production per unit of land and labor, they are also associated with significant challenges, including animal welfare concerns, localized environmental pollution from nutrient surpluses, and the heightened risk of zoonotic disease transmission and AMR in high-density animal populations (Stel et al., 2022). In contrast, hundreds of millions of smallholder farmers in low- and middle-income countries manage livestock within diversified farming systems in which animals serve multiple functions beyond food production. In these contexts, livestock recycle crop residues, supply manure to maintain soil fertility, provide draft power and transport, and function as mobile assets for savings and risk mitigation, often with minimal reliance on external inputs (Alary et al., 2025). Transhumant and nomadic pastoral systems further distribute grazing pressure across heterogeneous landscapes, and when governed by effective local institutions, they can contribute to rangeland conservation, ecosystem stability, and long-term livelihood resilience.

5.2.4 Socioeconomic Roles

Globally, livestock underpin the livelihoods of an estimated one to more than 1.3 billion people when both on-farm activities and employment across value chains—such as processing, transport, marketing, and retail—are taken into account, making the sector a foundational pillar of rural and periurban economies worldwide (Bilotto et al., 2024). In many low-income, smallholder, and pastoral systems, livestock function as "living capital," embodying a flexible form of wealth that provides savings, insurance, and resilience against shocks including drought, crop failure, market volatility, and household health emergencies (Jensen et al., 2025). Unlike fixed assets, animals can be sold, exchanged, or consumed in response to acute needs, allowing households to stabilize food consumption and income flows and reducing the risk of irreversible descents into poverty following climatic or economic crises (Banda & Tanganyika, 2021).

Ownership, access, and control over livestock assets are deeply shaped by social norms and power relations, particularly gender. Across many regions, men are more likely to control large ruminants and high-value animals, while women often manage poultry, small ruminants, and dairy animals that supply nutrient-dense foods for household consumption and generate frequent, small-scale income (Gitungwa et al., 2021). Evidence consistently shows that income under women's control is disproportionately invested in food security, healthcare, and children's education, amplifying the developmental impacts of livestock beyond direct economic returns (Gitungwa et al., 2021). Consequently, livestock interventions that explicitly address gendered constraints—such as access to veterinary services, credit, markets, and producer organizations—can serve as powerful levers for poverty reduction, social cohesion, and gender-transformative development when designed with these dynamics in mind.

5.2.5 Nutritional Security and One Health

From a nutritional perspective, animal-source foods are uniquely dense in high-quality protein and essential micronutrients, including vitamin B12, bioavailable iron (heme iron), zinc, iodine, and long-chain fatty acids that are often scarce or less bioavailable in predominantly plant-based diets. In low- and middle-income countries, even modest increases in the consumption of milk, meat, or eggs among young children and pregnant or lactating women are associated with reductions in stunting, anemia, and cognitive and physical developmental impairments, particularly during the critical first 1,000 days of life. As a result, livestock-derived foods play a strategic role in addressing "hidden hunger," a condition in which caloric intake may be sufficient while micronutrient deficiencies remain widespread (Adesogan et al., 2025).

At the same time, livestock systems occupy a central position within the One Health framework, which emphasizes the inseparable links between human health, animal health, and environmental integrity (Adesogan et al., 2025). Intensification of production, increasing human–animal contact, and the widespread use of antimicrobials for treatment and prevention can create conditions conducive to the emergence and dissemination of zoonotic pathogens and AMR. These risks are amplified by global trade, dense animal populations, and environmental contamination with pathogens and resistance genes. Addressing them requires integrated approaches that combine improved husbandry, vaccination, nutrition, and biosecurity with coordinated surveillance of pathogens and AMR across human, animal, and environmental compartments. Emerging insights from microbiome science and host–pathogen ecology offer promising avenues for designing livestock systems that simultaneously support nutritional security, disease resilience, and public health protection.

5.2.6 The Sustainability Imperative

Looking ahead, demographic expansion, rapid urbanization, and rising incomes are projected to substantially increase global demand for animal-source foods over the coming decades (Adesogan et al., 2025). Multiple projections suggest that demand for animal-derived protein, particularly meat, could approximately double by 2050 relative to current levels, driven largely by consumption growth in emerging and transitioning economies. If this demand were met through the unmodified expansion of current production practices, it would entail substantial increases in livestock numbers, feed-crop cultivation, and associated greenhouse gas emissions, intensifying pressures on forests, freshwater resources, and biodiversity.

Given that livestock production already occupies a large share of global agricultural land while contributing a minority of total dietary calories, further extensification is neither environmentally sustainable nor economically viable (Ritchie, 2017). The sector is therefore under increasing pressure to achieve sustainable intensification: producing more and higher-quality nutrition per unit of land, water, and emissions while minimizing negative environmental externalities and safeguarding animal welfare. Key strategies include improving feed efficiency and animal health, deploying dietary interventions and breeding approaches to reduce enteric methane emissions, advancing manure management and circular nutrient recycling, and integrating livestock into diversified, agroecological, and landscape-scale planning frameworks (Tseten et al., 2022).

In parallel, broader food system transformations—such as shifts toward more diverse and moderated diets, reductions in food loss and waste, and the emergence of alternative protein sources including plant-based analogues, fermentation-derived proteins, and cultured meat—are likely to interact with livestock trajectories. The balance among these pathways will determine whether livestock remain a major driver of planetary boundary transgression or evolve into a more circular, lower-emission, and socially inclusive component of global food systems. Achieving this transition will require coordinated advances across animal science, microbiome and systems biology, governance, and social policy, ensuring that gains in efficiency are aligned with equity, human health, and long-term ecological integrity.

5.3 THE BOVINE MICROBIOME

Cattle play a multifaceted and indispensable role in human society, providing a wide range of products and services, including milk, meat, leather, and manure used as organic fertilizer, and they have also traditionally served as draught animals in agriculture and transportation. Their economic, nutritional, and cultural importance has made cattle one of the most valuable domesticated animal species worldwide. Beyond their material contributions, cattle have been central to shaping early agricultural systems and sustaining rural livelihoods across diverse environments.

The domestication of cattle was a pivotal development in human history, facilitating the transition from nomadic hunter-gatherer lifestyles to settled, agrarian communities across large parts of Europe, Asia, and Africa. Archaeological and genetic evidence indicates that cattle domestication began more than 10,000 years ago, coinciding with the rise of early farming societies. The ability to rely on cattle for food production, labor, and secondary products supported population growth, land cultivation, and the development of complex societies.

Although the evolutionary relationships among ancestral cattle populations are not yet fully resolved, current evidence suggests that cattle domestication occurred independently on two, and possibly three, separate occasions. These domestication events gave rise to two major cattle lineages: taurine cattle (*Bos taurus*) and indicine cattle (*Bos indicus*). Both lineages share a common wild ancestor, the aurochs (*Bos primigenius*), which is estimated to have diverged approximately 250,000 years ago. Taurine cattle were domesticated in the Fertile Crescent, while indicine cattle originated in the Indus Valley, reflecting adaptation to distinct environmental and climatic conditions (Pitt et al., 2018). Together, these lineages underpin the global diversity of modern cattle breeds and their continued importance in agriculture today.

The gastrointestinal tract (GIT) microbiome of cattle plays a central role in feed digestion, fermentation, and host energy metabolism. Several bacterial genera are particularly important for fiber degradation and metabolic regulation. For instance, Bifidobacterium, Ruminococcus, and Fibrobacter are key fibrolytic microorganisms that contribute to the breakdown of cellulose and other complex plant polysaccharides. In contrast, members of the Succinovibrionaceae family influence hydrogen utilization pathways and are closely associated with methane production and mitigation in ruminants. A major outcome of microbial fermentation in the ruminant gut is the production of volatile fatty acids (VFAs), including acetate, propionate, and butyrate, which serve as the primary energy source for the host animal. Specific taxa contribute to distinct metabolic functions; for example, *Clostridium sticklandii* is involved in ammonia (NH_3) production through amino acid fermentation, while Roseburia species are notable producers of butyrate, a short-chain fatty acid essential for gut health and epithelial integrity (Khan et al., 2023).

Large-scale metagenomic analyses have provided detailed insights into microbial diversity across different components of the dairy production system. In one study, a total of 36 milk samples and 36 fecal samples were collected from three organized dairy farms for metagenomic DNA extraction. Sequencing generated approximately 2.58 billion reads from milk and 2.81 billion reads from feces, corresponding to 200.4 GB and 206.5 GB of data, respectively. The majority of reads mapped to bacterial genomes, revealing the presence of 37 bacterial phyla across the system. Notably, fecal samples exhibited greater phylum-level diversity than milk samples. The dominant phyla identified were Bacillota (53.3%),

Pseudomonadota (24.7%), Bacteroidota (11.0%), and Actinomycetota (8.4%). At the genus level, Clostridium was most abundant in milk, whereas Bifidobacterium predominated in fecal samples, highlighting niche-specific microbial colonization (Amarlapudi et al., 2025).

Spatial variation in microbial composition along the bovine GIT further underscores the functional specialization of different gut compartments. Overall, the GIT microbiome was predominantly bacterial, with Firmicutes (46.5%), Bacteroidetes (33.6%), Proteobacteria (7.5%), and Fibrobacteres being the most abundant phyla. Taxa belonging to Bacteroidetes and Fibrobacteres were dominant in the four-chambered stomach, reflecting their role in primary fiber degradation, whereas Firmicutes and Proteobacteria were more abundant in the postgastric regions, including the small and large intestines. At the genus level, *Prevotella* spp. and *Fibrobacter* spp. were enriched in the forestomach, *Phyllobacterium* spp. and *Eubacterium* spp. were more prevalent in the small intestine, and *Bacteroides* spp. and *Alistipes* spp. dominated the large intestine. Archaeal communities were almost exclusively represented by Euryarchaeota, accounting for 99.5% of archaeal sequences. Among these, Methanobrevibacter spp. were the most prevalent across all GIT regions and were particularly enriched in the postgastric intestine, while *Methanocorpusculum* spp. were more abundant in the large intestine. Eukaryotic taxa were dominated by Nematoda, especially in the postgastric intestine. Ciliate protozoa such as Oxytricha, Stylonychia, and Entodinium were prevalent in the four-chambered stomach, whereas anaerobic fungi including Anaeromyces and Caecomyces (Neocallimastigomycota) were relatively enriched in the large intestine (Lin et al., 2023).

Feed intake is another critical factor shaping the rumen and lower gut microbiomes. Dry matter intake (DMI, kg/day) is commonly used to quantify feed consumption in dairy cows. While cows with higher DMI did not show significant differences in alpha-diversity in either the rumen or lower gut microbiomes compared to cows with lower DMI, beta-diversity analyses revealed a significant association between feed intake levels and microbial community structure in both regions. Rumen fill is a major physiological mechanism regulating feed intake, suggesting that ruminal microorganisms may indirectly influence consumption patterns. Cows with higher DMI exhibited a greater ruminal relative abundance of the Ruminococcus_gauvreauii_group, a fibrolytic taxon that enhances fiber fermentation and accelerates ruminal digesta turnover, thereby supporting increased feed intake. In contrast, higher relative abundance of Howardella was associated with lower DMI. As Howardella is an ureolytic bacterium, its increased abundance may indicate enhanced ruminal urea recycling in cows with reduced feed consumption (Monteiro et al., 2022).

Finally, microbial composition of the lower gut has been linked to production efficiency in dairy cattle. Cows with lower efficiency in milk yield, milk fat, and milk protein production exhibited higher relative abundances of Fusobacterium, Ruminococcaceae UCG-001, and Actinobacteria in the lower gut. Among these, Fusobacterium showed a consistent negative association with all production efficiency traits. Species such as Fusobacterium necrophorum are known to increase in abundance in cattle fed high-grain diets, supporting the observation that less efficient animals may allow greater amounts of readily fermentable carbohydrates to reach the lower gut. This suggests reduced efficiency in nutrient processing in upstream compartments of the GIT, whereas more efficient animals likely utilize fermentable substrates more effectively in the rumen and small intestine, leaving fewer substrates for extensive fermentation in the lower gut (Monteiro et al., 2022).

5.3.1 Mechanisms of Fiber Degradation

Advances in gene-centric metagenomics, metatranscriptomics, and proteomics have revealed that the dominant ruminal cellulolytic bacteria employ fundamentally distinct strategies for plant cell wall deconstruction (Yeoman et al., 2021). Among these, *Fibrobacter succinogenes*, *Ruminococcus flavefaciens*, and *Ruminococcus albus* represent archetypes of divergent evolutionary solutions to the challenge of lignocellulose degradation.

5.3.1.1 Fibrobacter succinogenes

Fibrobacter succinogenes is widely regarded as one of the most effective degraders of crystalline cellulose in the bovine rumen. Unlike many anaerobic fibrolytes, it does not assemble classical multi-enzyme cellulosome complexes. Instead, recent evidence indicates that *F. succinogenes* employs a distinctive outer membrane vesicle (OMV)-mediated strategy (Arntzen et al., 2017). These OMVs are enriched with carbohydrate-active enzymes (CAZymes), particularly endoglucanases, which are delivered directly to the cellulose surface (Arntzen et al., 2017). This localized enzyme deployment minimizes diffusion losses and enhances catalytic efficiency.

Stable attachment to plant fibers is achieved through specialized adhesion mechanisms, including "fibro-slime" proteins and Type IV pili, which anchor the bacterium tightly to the substrate (Burnet et al., 2015). Once bound, glycoside hydrolases from families GH5, GH9, and GH45 cleave cellulose chains into soluble cellodextrins (Ezeilo et al., 2017). These oligomers are rapidly transported into the periplasm, where further hydrolysis occurs, effectively sequestering fermentation products from competing "cheater" microbes and conferring a strong ecological advantage.

5.3.1.2 *Ruminococcus Species*

Ruminococcus flavefaciens exemplifies cellulosome-based degradation. The cellulosome is a highly ordered, multiprotein nanocomplex in which numerous enzymatic subunits are tethered to a central scaffolding protein via high-affinity cohesin–dockerin interactions (Bule et al., 2017). This architecture allows precise spatial organization of enzymes and synergistic degradation of complex plant polymers. The extensively studied strain FD-1 encodes over 223 dockerin-containing proteins, enabling dynamic reconfiguration of its enzymatic repertoire in response to substrate composition and availability (Berg Miller et al., 2009).

By contrast, *Ruminococcus albus*, despite its close phylogenetic relationship, lacks a canonical cellulosomal system. Genomic analyses reveal an absence of cohesin domains, with cellulose attachment instead mediated by a distinctive family of surface-anchored carbohydrate-binding modules (CBM37) (Ezer et al., 2008). This divergence highlights how evolutionary pressures within the rumen ecosystem have driven multiple, functionally equivalent fibrolytic strategies even among closely related taxa.

5.3.1.3 *Prevotella*

Although not a primary degrader of crystalline cellulose, members of the genus *Prevotella*—particularly *P. ruminicola*—are often the most numerically dominant bacteria in the bovine rumen (Betancur-Murillo et al., 2022). These organisms function as metabolic generalists, specializing in the degradation of hemicellulose, pectin, proteins, and starch. Their genomes harbor extensive arrays of polysaccharide utilization loci (PULs), encoding enzymes such as xylanases and arabinofuranosidases (e.g., GH43), which dismantle the hemicellulosic matrix that shields cellulose fibers (Murovec & Accetto, 2024). By removing these structural barriers, Prevotella species facilitate access for primary cellulolytic bacteria, underscoring their keystone role in rumen fiber degradation networks.

5.3.2 Distinct Gut Microbiome Composition and Enhanced Disease Resistance in High-Altitude Weining Cattle

Weining County is located in southwest China at an average altitude of approximately 2,200 m. Weining cattle are an ancient indigenous ruminant breed that has adapted to this high-altitude environment, exhibiting strong hypoxia tolerance, antioxidant capacity, and disease resistance. Although they resemble ordinary cattle with short horns and yellow coats, Weining cattle possess distinctive physiological and metabolic characteristics. Owing to the high nutritional quality of their meat, their physical strength, and their docile nature, they play an important role in local farming systems and regional prosperity. Notably, the incidence of intestinal diseases in Weining cattle is relatively low compared with that in other cattle breeds, suggesting an inherent advantage in gut health (Wang et al., 2023).

At the phylum level, Firmicutes, Bacteroidota, and Desulfobacterota are the dominant bacterial groups in Weining cattle, together accounting for more than 90% of the total gut microbiota. Other phyla, including Proteobacteria, Cyanobacteria, Fibrobacterota, Spirochaetota, Patescibacteria, Verrucomicrobiota, and Campylobacterota, were detected at lower relative abundances. This microbial composition indicates a stable and functionally efficient gut ecosystem optimized for fiber degradation and nutrient utilization (Wang et al., 2023).

At the genus and family levels, several beneficial taxa were markedly enriched in Weining cattle. Lachnospiraceae was the most abundant group, followed by Rikenellaceae RC9 gut group, unclassified [Eubacterium] coprostanoligenes group, Alistipes, and Monoglobus. In addition,

Weining cattle exhibited relatively high levels of Cyanobacteria and Elusimicrobiota. Many of these taxa are known to contribute to short-chain fatty acid production, intestinal barrier maintenance, and immune modulation, thereby supporting overall gut health (Wang et al., 2023).

Comparative analysis revealed distinct differences between Weining and Angus cattle. The relative abundances of Prevotellaceae UCG 003, unclassified Clostridia vadinBB60 group, Campylobacter, Faecalibacterium, Erysipelotrichaceae UCG 009, Parabacteroides, Dorea, Parasutterella, Anaerofustis, Romboutsia, Papillibacter, Blautia, and several other taxa were significantly lower in Weining cattle. In contrast, Anaerosporobacter, unclassified Lachnospiraceae, Lachnospiraceae UCG 001, Ruminobacter, unclassified [Eubacterium] coprostanoligenes group, Ruminococcus, Caproiciproducens, Monoglobus, unclassified Muribaculaceae, and related Clostridia groups were significantly enriched in the Weining group (Wang et al., 2023).

Overall, the abundance of beneficial bacteria—including Lachnospiraceae, Rikenellaceae, Coprostanoligenes, Cyanobacteria, and Muribaculaceae—was significantly higher in Weining cattle than in Angus cattle. In particular, the markedly elevated abundance of Lachnospiraceae in the Weining group may help explain their lower susceptibility to diarrhea. Conversely, potential pathogens such as Campylobacter were more abundant in the Angus group. Collectively, these findings demonstrate that Weining cattle harbor a richer and more favorable gut microbiome structure, which likely contributes to enhanced disease resistance, improved intestinal health, and a reduced risk of diarrheal disorders (Wang et al., 2023).

5.4 METHANOGENESIS AND MITIGATION STRATEGIES

Enteric methane production is an obligate consequence of anaerobic fermentation in the rumen, serving as the primary route for disposal of excess metabolic hydrogen. This process represents a significant energetic loss to the host, accounting for approximately 2–12% of gross energy intake (Xie et al. 2025). Methanogenesis is performed exclusively by archaea, predominantly Methanobrevibacter species, which reduce CO_2 using H_2 as an electron donor (Lyu et al., 2018).

5.4.1 3-Nitrooxypropanol

The development of 3-nitrooxypropanol (3-NOP) marks a major advance in targeted methane mitigation. Unlike antibiotics or broad-spectrum inhibitors, 3-NOP is a structural analogue of methyl-coenzyme M, a key substrate in the terminal step of methanogenesis (Yu et al., 2021). It specifically inhibits methyl-coenzyme M reductase (MCR), the rate-limiting enzyme in methane formation. Structural and biochemical studies show that 3-NOP binds to the MCR active site and oxidizes the essential Ni(I) ion, rendering the enzyme inactive (Duin et al., 2016).

In vivo studies consistently demonstrate methane reductions of 30–40% in dairy and beef cattle. Importantly, inhibition of methanogenesis redirects hydrogen toward alternative fermentation pathways, particularly propionate synthesis. Propionate is an energetically favorable VFA and a major gluconeogenic precursor in ruminants, meaning methane mitigation via 3-NOP can improve feed efficiency rather than compromise it. However, efficacy is diet-dependent, with greater reductions observed in high-concentrate diets. Additionally, while Methanobrevibacter populations decline, other methanogens such as Methanosphaera may persist, indicating partial functional redundancy within the archaeal community (Pitta et al., 2022).

5.4.2 Other Mitigation Agents

A broader portfolio of methane mitigation strategies is under investigation. Halogenated compounds such as bromoform, derived from the red seaweed *Asparagopsis taxiformis*, inhibit cobamide-dependent methyl transferases and can achieve methane reductions of 50–80%, though concerns remain regarding toxicity, residue accumulation, and scalability. Nitrate supplementation acts as a competitive hydrogen sink, reducing methane emissions by approximately 16%, but carries risks of nitrite toxicity and methemoglobinemia if improperly managed. Dietary lipid supplementation can lower methane output by roughly 15% through biohydrogenation of unsaturated fatty acids and suppression of protozoal populations that support methanogens (Rani et al., 2024).

5.5 NITROGEN UTILIZATION EFFICIENCY (NUE) AND BACTERIAL DRIVERS

Ruminant livestock are intrinsically inefficient in their utilization of dietary nitrogen, a limitation rooted in the asynchronous relationship between ruminal protein degradation and microbial nitrogen assimilation. Historically, it has been estimated that only 25–30% of ingested nitrogen is ultimately retained in animal products such as milk and meat, while the remaining 70–75% is excreted, primarily as urinary urea (Harmeyer & Martens, 1980). This inefficiency not only represents a substantial loss of valuable feed nitrogen but also poses significant environmental challenges, as nitrogen excretion contributes to ammonia volatilization, nitrate leaching, and eutrophication of aquatic ecosystems.

5.5.1 Microbial Signatures of High NUE

Emerging evidence indicates that interanimal variation in NUE is driven largely by differences in rumen microbial community structure and function. Recent studies employing stable isotope fractionation and metagenomic profiling have identified distinct microbial signatures associated with high-NUE phenotypes (Li et al., 2022). Cattle with superior NUE consistently harbor increased abundances of bacterial families such as Succinivibrionaceae, Selenomonadaceae, and Acidaminococcaceae. These taxa preferentially channel metabolic hydrogen through alternative fermentation pathways, including the succinate–propionate pathway, which serves as an effective hydrogen sink without generating methane. Importantly, these organisms exhibit enhanced capacity to assimilate ruminal ammonia into microbial protein, thereby improving the flow of high-quality microbial nitrogen to the small intestine for host absorption.

In contrast, animals with low NUE display elevated levels of proteolytic and deaminating bacteria, including members of the Clostridia class and Desulfovibrio species (Rivero et al., 2020). These taxa rapidly hydrolyze dietary protein and amino acids into ammonia at rates that exceed the assimilative capacity of the microbial community, leading to ammonia accumulation, increased absorption into the bloodstream, hepatic urea synthesis, and subsequent urinary excretion. This imbalance highlights the central role of microbial nitrogen partitioning, rather than dietary crude protein alone, in determining whole-animal nitrogen efficiency.

5.6 WATER BUFFALO MICROBIOME

The water buffalo (*Bubalus bubalis*) occupies a pivotal role in agricultural systems across Asia, parts of South America, and the Mediterranean basin, where it supports food production, labor, and rural livelihoods under challenging environmental conditions. Despite routinely consuming lower-quality, high-fiber diets than cattle, water buffaloes consistently exhibit superior body condition, feed efficiency, and productive performance. This apparent paradox is increasingly attributed to the distinctive structure and function of the buffalo rumen microbiome, which enables highly efficient extraction of energy and nutrients from lignocellulosic feeds that are marginal or unsuitable for other livestock species (Chiariotti et al., 2025).

5.6.1 Microbial Community Structure and Functional Coupling of Cellulose Degradation and Methane Metabolism in the Buffalo Digestive Tract

The gut microbiome of buffalo is dominated by a diverse assemblage of bacterial and archaeal taxa that play essential roles in digestion and metabolism. At the phylum level, the bacterial community is primarily composed of Bacteroidota and Firmicutes, which are mainly represented by the classes Bacteroidia and Clostridia, respectively. These dominant phyla are followed in abundance by Verrucomicrobiota, Spirochaetota, and Proteobacteria. At a finer taxonomic resolution, the most prevalent bacterial orders include Bacteroidales and Oscillospirales, while the dominant families are Bacteroidaceae, CAG-272, Rikenellaceae, and UBA932. At the genus level, Alistipes, RC9, and Prevotella are the most abundant bacterial groups, reflecting a microbial community well adapted to complex carbohydrate degradation.

In addition to bacteria, a substantial archaeal community was identified. A total of 74 archaeal metagenome-assembled genomes (MAGs) were recovered, belonging to three phyla: Halobacterota, Thermoplasmatota, and Euryarchaeota. All detected archaeal taxa were methanogens, with Methanocorpusculum, Methanomicrobium, and Methanobrevibacter being the three most abundant genera. These methanogens are key contributors to methane production in the buffalo digestive tract.

Cellulose constitutes the primary structural component of plant cell walls, and its efficient degradation by digestive tract–associated microbes is critical for buffalo nutrition. Traditionally, Fibrobacter, Ruminococcus, and Butyrivibrio have been recognized as the principal cellulolytic bacteria. Notably, Butyrivibrio exhibits broad substrate versatility, degrading not only cellulose but also hemicellulose and proteins. In this study, several taxonomic groupings defined according to the Genome Taxonomy Database (GTDB)—namely Fibrobacter_all (including Fibrobacter and Fibrobacter_A), Ruminococcus_all (including Ruminococcus, Ruminococcus_A, and Ruminococcus_E), and Butyrivibrio_all (including Butyrivibrio and Butyrivibrio_A)—were found to be significantly enriched in the four-chambered (FC) stomach. Their abundances were markedly lower in other digestive tract sections, with Fibrobacter_all being the most abundant among these taxa.

Importantly, the total abundance of Fibrobacter_all was significantly higher in all four FC stomach compartments—the rumen, reticulum, omasum, and abomasum—than in the intestine. Contrary to the long-held view that the rumen harbors the highest abundance of Fibrobacter, our results revealed that Fibrobacter_all peaked in the omasum. This unexpected finding suggests that the omasum may play a more substantial role in cellulolytic digestion than previously recognized.

All archaeal species identified in this study were methanogens and were highly abundant not only in the FC stomach but also in the intestine. This observation challenges the prevailing assumption that methane metabolism occurs primarily in the FC stomach, particularly in the rumen. Instead, our findings highlight a previously underappreciated role of the intestine in methane production. Remarkably, the total abundance of methanogens also peaked in the omasum, mirroring the distribution pattern of Fibrobacter.

Correlation analyses further revealed a strong positive relationship between the abundances of archaea and Fibrobacter across both the FC stomach and intestine. When examining correlations between all bacterial genera and archaeal taxa, Fibrobacter exhibited the highest eco-efficiency correlation with methanogens. These results strongly suggest a functional linkage between Fibrobacter and methane-producing archaea. Previous studies have demonstrated that ruminal methanogens influence cellulolytic microbes by consuming hydrogen, thereby facilitating cellulose degradation. Our findings indicate that similar microbial interactions likely occur in the omasum.

Collectively, these results underscore the critical and previously underestimated roles of the omasum in both cellulose degradation and methane metabolism. They also suggest a close functional coupling between these two processes, providing new insights into the spatial organization and metabolic coordination of the buffalo digestive microbiome (Tong et al., 2022).

5.6.2 Age-Related Dynamics of Rumen Microbial Composition and Function in Dairy Buffaloes

Li et al. (2022) investigated the compositional patterns and functional potential of rumen microorganisms in dairy buffaloes at different stages of life using metagenomic sequencing. Buffalo milk (BM) ranks second globally after cow milk, contributing more than 12% of total world milk production. The nutritional quality of BM is known to be influenced by age, with well-documented effects on total solids, solids-not-fat, lactose, and ash content. Because rumen microorganisms play a central role in feed digestion and energy harvest, age-related changes in rumen microbial communities are of particular importance for understanding productivity and physiological performance in dairy buffaloes.

Rumen microbial colonization begins immediately after birth and is crucial for the development of a fully functional rumen capable of sustaining anaerobic fermentation. To explore how rumen microbial composition and function vary across different stages of physiological development, Li et al. collected a total of 15 rumen samples from dairy buffaloes representing four age groups: juvenile buffaloes aged approximately 1 year (Y group), mature buffaloes aged 3–5 years (M group), early elderly buffaloes aged 6–8 years (E group), and old buffaloes older than 9 years (O group). These groups collectively span the transition from rumen development to peak productivity and eventual aging, allowing for a comprehensive analysis of age-associated microbial dynamics.

Metagenomic analysis revealed that bacteria were the most abundant and dominant microorganisms across all age groups. At the phylum level, the rumen microbiota was largely dominated by Bacteroidetes and Firmicutes, which are commonly associated with carbohydrate degradation and fermentation processes in ruminants. At the genus level, Prevotella was the most dominant taxon across samples, followed by Bacteroides and unclassified bacterial taxa, reflecting the complexity and functional versatility of the rumen microbial ecosystem.

To identify age-specific microbial signatures, differential abundance analysis was performed using linear discriminant analysis (LDA), with taxa showing LDA scores greater than 2.5 and p-values less than 0.05 considered significantly different among groups. This analysis revealed clear age-dependent enrichment patterns. Specifically, six microbial clades were significantly enriched in the Y group, 18 in the M group, five in the E group, and 17 in the O group, indicating substantial restructuring of the rumen microbiome across the buffalo lifespan.

At the phylum level, Actinobacteria, Tenericutes, and Chlamydiae were significantly enriched in the M group, suggesting a microbiome associated with metabolic stability and high productivity during peak adulthood. In contrast, Fibrobacteres, a phylum well known for its role in cellulose degradation, was significantly enriched in the O group, indicating an increased importance of fiber-degrading capacity in older buffaloes.

At the genus level, distinct microbial signatures characterized each age group. The Y group was marked by a significant enrichment of Lactobacillus and Phascolarctobacterium, taxa often associated with early rumen development, fermentation of simple carbohydrates, and gut health, which effectively distinguished juvenile buffaloes from the other age groups. The M group showed significant enrichment of Bifidobacterium, Mycoplasma, Chlamydia, Bacillus, and Burkholderia, reflecting a diverse microbial community potentially linked to enhanced metabolic efficiency and immune interactions during peak productive years. The E group was characterized by a significant enrichment of Stenotrophomonas and Pullulanibacillus, indicating transitional microbial features as animals move toward later life stages. Finally, the O group displayed significant enrichment of Bacteroides, Paludibacter, Parabacteroides, Fibrobacter, and unclassified Bacteroidetes, highlighting a microbial community increasingly specialized for complex polysaccharide degradation in older animals (Li et al., 2022).

Among all identified biomarkers, the bacterial genera showing the greatest degree of enrichment in each age group were Lactobacillus in the Y group, Bifidobacterium in the M group, Stenotrophomonas in the E group, and Bacteroides in the O group. Together, these findings demonstrate that the rumen microbiome of dairy buffaloes undergoes pronounced and systematic shifts with age, reflecting changes in physiological status, digestive requirements, and metabolic function throughout the animal's lifespan (Li et al., 2022).

5.6.3 Buffalo versus Cattle: Microbiological and Physiological Contrasts

Comparative studies between water buffaloes and cattle (Bos taurus/Bos indicus) reveal that buffaloes possess a markedly enhanced capacity for lignocellulose degradation, rooted in both microbiological and physiological adaptations (Xu et al., 2021). First, buffalo rumens harbor significantly higher total microbial biomass, including elevated densities of bacteria and anaerobic fungal zoospores. Anaerobic rumen fungi, particularly genera such as Anaeromyces and Piromyces, play a crucial biomechanical role by penetrating lignified plant tissues with invasive rhizoids, disrupting cell wall architecture and dramatically increasing the surface area available for bacterial enzymatic attack (Chen et al., 2008).

Second, metagenomic and amplicon-based profiling reveal systematic differences in bacterial community composition between species. While Prevotella typically dominates the cattle rumen, the buffalo rumen is enriched with highly specialized cellulolytic and fermentative taxa, including Ruminococcus albus, Succiniclasticum, and Fibrobacter species (Paul et al., 2025). Of particular importance, Ruminococcus albus occurs at significantly higher densities in buffaloes, accelerating cellulose hydrolysis and improving the rate and extent of fiber digestion (Malik et al., 2021). This microbial configuration confers a kinetic advantage that allows buffaloes to extract usable energy from coarse forages with greater efficiency than cattle.

A third critical adaptation is the buffalo's exceptional capacity for endogenous urea recycling. Through enhanced transfer of urea into the rumen via saliva and across the rumen epithelium, buffaloes provide a steady nitrogen supply to ruminal microbes even when dietary protein is scarce. This mechanism sustains robust populations of fibrolytic bacteria belonging to the Fibrobacteres and Firmicutes, preventing microbial collapse under low-protein feeding regimes that would otherwise limit productivity in cattle. Collectively, these microbiological and physiological traits explain the superior performance of buffaloes on nutritionally poor diets.

5.6.4 Socioeconomic Implications of the Buffalo Microbiome

The efficiency and resilience conferred by the buffalo microbiome have far-reaching socioeconomic implications, particularly for smallholder farmers in developing regions. For millions of rural households in Asia and Africa, the water buffalo functions as a "living savings account," providing food security, draft power, and economic buffering against crop failure, market shocks, and climatic stress (Quinzeiro Neto et al., 2010). The animal's ability to convert low- or zero-cost agricultural byproducts—such as rice straw, stubble, and crop residues—into high-value outputs including milk, meat, and labor underpins household resilience and income stability.

By reducing dependence on purchased concentrates and external inputs, the buffalo's microbiome-driven feed efficiency preserves scarce financial resources and enhances the viability of low-input farming systems (Kondaiah, 2025). In many traditional agricultural contexts, women are the primary managers and caretakers of livestock. The buffalo's hardiness, disease tolerance, and efficient nutrient utilization lower labor demands and veterinary costs, strengthening women's control over productive assets and contributing to gender empowerment and intergenerational household stability.

In high-income markets, particularly in southern Europe, the influence of the buffalo microbiome extends beyond primary production into premium dairy value chains. The manufacture of Mozzarella di Bufala Campana PDO relies on a traditional natural whey starter (siero innesto) derived from the previous day's fermentation. This starter culture constitutes a complex, self-renewing microbial ecosystem dominated by *Streptococcus thermophilus* and Lactobacillus species adapted to the BM environment (Bonfanti et al., 2025). During curd maturation, microbial succession favors Lactobacillus dominance, shaping acidification kinetics, texture, and flavor development in the final product (Khongdee et al., 2011). This tight coupling between the animal's microbiome, milk chemistry, and artisanal fermentation processes exemplifies a form of "microbial terroir," in which local microbial ecologies impart distinct sensory and cultural value to regional foods.

5.7 YAK MICROBIOME

The yak (*Bos grunniens*) is the largest ruminant mammal inhabiting the highest altitudes on Earth and represents an ancient lineage that diverged from cattle approximately one to five million years ago, with closer evolutionary ties to bison than to other members of the Bovidae family. As one of the earliest domesticated animals of the Tibetan Plateau, yaks have been integral to human survival in this region for nearly 10,000 years, providing essential resources such as meat, milk, wool, and dung for fuel. Over millennia, the extreme plateau environment—characterized by chronic hypoxia, low temperatures, intense ultraviolet radiation, a short growing season, and prolonged periods of limited, low-quality forage—has imposed strong selective pressures that have shaped both yak biology and their symbiotic gut microbiome (Wang et al., 2024).

To thrive under these harsh conditions, yaks have evolved a suite of distinctive morphological and physiological adaptations. Compared with cattle, yaks possess shorter tongues and stronger, denser conical papillae, anatomical features that enhance their ability to graze and efficiently process coarse, poorly digestible alpine vegetation. Physiologically, yaks are adapted to thin air through enlarged hearts and lungs and smaller red blood cells, which together improve oxygen uptake, transport, and tissue delivery in low-oxygen environments. These integrated evolutionary, anatomical, and physiological traits underpin the yak's remarkable capacity to survive and remain productive at extreme altitudes (Zhao et al., 2022).

Zhao et al. (2022) characterized the yak gut microbiome and identified Bacteroidetes, Firmicutes, Proteobacteria, and Euryarchaeota as the dominant phyla. Among these, Bacteroidetes (53.71%) and Firmicutes (24.74%) were the most abundant. Compared with cattle, yaks exhibited a higher relative abundance of Proteobacteria, Verrucomicrobia, Planctomycetes, Synergistetes, Cyanobacteria, Chloroflexi, and Acidobacteria, whereas Chlamydiae were more prevalent in cattle and Actinobacteria were more abundant in dzomo (yak–cattle hybrids). At the genus level, Bacteroides, Alistipes, Butyrivibrio, Faecalibacterium, and Pseudobutyrivibrio were more dominant in cattle, while Ruminococcus, Selenomonas, Oscillabacter, Treponema, and Fibrobacter were enriched in yaks. Species-level analysis revealed that 19 species—mainly belonging to Bacteroides and

Alistipes—were more abundant in cattle, four species were enriched in dzomo, and eight species, largely from the genus Prevotella, were more predominant in yak.

Functional analysis of the rumen microbiota highlighted clear differences in fermentation pathways among species. Of the 12 identified succinate-producing bacterial species, *Prevotella brevis*, *P. bryantii*, *P. albensis*, *Fibrobacter succinogenes*, *Succinimonas amylolytica*, and *Mitsuokella jalaludinii* were significantly enriched in the yak rumen relative to cattle or dzomo. The succinate-utilizing species *Selenomonas ruminantium* was also more abundant in yaks, suggesting an enhanced capacity for propionate production via the succinate pathway. In contrast, bacterial genera associated with the acrylate pathway for propionate synthesis did not differ significantly among yaks, cattle, and dzomo. Notably, key cellulolytic bacteria—Ruminococcus flavefaciens, Ruminococcus albus, and F. succinogenes—were present at higher relative abundance in yaks, reflecting an enhanced ability to degrade fibrous plant material typical of alpine pastures.

Metabolomic profiling further underscored species-specific rumen adaptations. In total, 185 individual metabolites were identified, and principal component analysis confirmed the stability and robustness of the analytical system. Orthogonal partial least squares discriminant analysis (OPLS-DA) demonstrated distinct rumen metabolite profiles among cattle, dzomo, and yak. After statistical filtering, 27 metabolites were identified as significantly different between pairs of ruminant species. Sixteen metabolites differed between cattle and dzomo: compounds such as caproic acid, benzenepropanoic acid, 2-deoxyerythritol, lithocholic acid, dehydroascorbic acid, phenylacetamide, and uridine 5′-monophosphate were higher in cattle, with lithocholic acid and phenylacetamide showing the greatest upregulation. In contrast, dzomo exhibited higher levels of ethylene glycol, glycine, aminomalonate, chenodeoxycholic acid, 2-picolinic acid, and 3,6-anhydro-D-glucose, with glycine and chenodeoxycholic acid among the most elevated.

When comparing yak and cattle, yak rumen samples showed higher concentrations of ethylene glycol, sebacic acid, 2-picolinic acid, chenodeoxycholic acid, 1,3-dihydroxypyridine, and 3,6-anhydro-D-glucose, but lower levels of serine, lithocholic acid, dehydroascorbic acid, glucose-6-phosphate, uridine 5′-monophosphate, and isochlorogenic acid. Relative to dzomo, yak rumen fluid contained higher valeric acid, boric acid, benzenepropanoic acid, sebacic acid, and 1,3-dihydroxypyridine, while metabolites such as glycine, aminomalonate, pimelic acid, 3-hydroxyphenylacetic acid, and 5′-deoxy-5′-methylthioadenosine were reduced. Together, these microbial and metabolic signatures highlight the yak rumen as a highly specialized ecosystem, optimized for efficient fiber degradation and energy extraction under the extreme conditions of the Tibetan Plateau (Zhao et al., 2022).

5.7.1 Microbiome of Yak on Intensive Feeding System

Intensive feeding systems have been proposed as a strategy to shorten the production cycle and enhance growth performance in yaks. Although improved growth rates have been consistently observed under intensive management, the microbial mechanisms underlying these benefits have only recently begun to be clarified. Rumen metagenomic analyses indicate that intensive feeding markedly reshapes the yak rumen ecosystem, favoring microbial taxa involved in amino acid and carbohydrate metabolism. In particular, increased abundances of Clostridium, Methanobrevibacter, Piromyces, and Anaeromyces were detected in intensively fed yaks, suggesting enhanced fermentative capacity and improved utilization of dietary nutrients (Yang et al., 2024).

Yang et al. (2024) determined that at the bacterial phylum level, the yak rumen microbiome across all samples was dominated by Bacteroidetes (52.29%), Firmicutes (34.81%), and Proteobacteria (5.41%). However, intensive feeding significantly altered this balance. Compared with grazing yaks, intensively fed yaks exhibited a higher relative abundance of Firmicutes (40.85% vs. 28.76%) and a reduced proportion of Bacteroidetes (47.67% vs. 56.90%), a shift often associated with improved energy harvest efficiency. At the genus level, Prevotella (43.96%) was the most dominant taxon, followed by Bacteroides (7.94%), Clostridium (4.46%), Alistipes (3.53%), Fibrobacter (2.77%), and Ruminococcus (2.44%). Species-level discrimination using the LEfSe algorithm revealed that intensively fed yaks were enriched in 11 Clostridium species, 3 Prevotella species, 3 Ruminobacter species, 2 Succinatimonas species, 1 Oscillibacter species, as well as *Succiniclasticum ruminis*, *Ruminococcus bromii*, and *Faecalibacterium prausnitzii*. In contrast, grazing yaks harbored higher abundances of 5 Bacteroides species, 3 Prevotella species, 3 Alistipes species, 3 Fibrobacter species, and *Pseudomonas fluorescens*.

Archaeal communities were also strongly influenced by feeding strategy. Across all samples, Euryarchaeota (92.92%) and Candidatus Heimdallarchaeota (2.53%) were the dominant archaeal phyla. Intensively fed yaks exhibited a higher abundance of Euryarchaeota and lower abundances of Candidatus Heimdallarchaeota, Candidatus Bathyarchaeota, and Candidatus Woesearchaeota. At the genus level, Methanobrevibacter (73.59%) predominated, followed by Methanobacterium (3.49%), Methanosarcina (3.47%), Methanothrix (2.86%), and Methanomassiliicoccus (2.31%). While Methanobrevibacter was more abundant in intensively fed yaks, other dominant methanogenic genera—including Methanosarcina, Methanothrix, Methanomassiliicoccus, Methanosphaera, and Methanolobus—were more prevalent in grazing yaks. At the species level, *Methanobrevibacter curvatus*, *M. filiformis*, *M. cuticularis*, and *Lokiarchaeum* sp. GC14_75 were enriched in the intensive group, whereas grazing yaks showed higher representation of four unidentified Euryarchaeota species, *Methanolobus profundi*, *Methanolobus psychrophilus*, two Candidatus Heimdallarchaeota species, and *Methanosaeta harundinacea*.

The rumen eukaryotic community, particularly anaerobic fungi, also differed substantially between feeding systems. The genera Neocallimastix, Piromyces, and Anaeromyces were among the most abundant eukaryotes and were significantly enriched in intensively fed yaks, consistent with enhanced fiber degradation capacity. Species such as Piromyces sp. E2, *Anaeromyces robustus*, *Neocallimastix californiae*, *Arthrobotrys oligospora*, and *Candida auris* were more abundant under intensive feeding. In contrast, grazing yaks exhibited higher abundances of *Rozella allomycis*, *Rhizophagus irregularis*, *Basidiobolus meristosporus*, *Rhizoclosmatium globosum*, *Allomyces macrogynus*, *Batrachochytrium dendrobatidis*, *Catenaria anguillulae*, *Conidiobolus coronatus*, and *Lobosporangium transversale*.

Rumen viral communities were likewise reshaped by feeding regime. Intensive feeding increased the abundance of viral genera such as C5virus, Cp220virus, and N4virus, whereas grazing yaks showed higher levels of Hokovirus, Alpharetrovirus, Catovirus, Pepy6virus, Bc431virus, and Cervidpoxvirus. At the species level, Geobacillus virus E3, two Campylobacter phages (CP21 and INN35), and Enterobacter phage EcP1 were more abundant in intensively fed yaks. In contrast, grazing yaks were enriched in Hokovirus HKV1, avian myeloblastosis virus, Catovirus CTV1, deerpox virus W-848-83, and Klosneuvirus KNV1.

Finally, species-level network analysis revealed complex microbial interactions across bacteria, archaea, eukaryotes, and viruses. Bacterial species acted as key bridge nodes linking other microbial domains. For example, *Prevotella brevis* and *Prevotella ruminicola* showed positive correlations with *Methanomassiliicoccales archaeon* RumEn M1, *Methanosaeta harundinacea*, Candidatus Heimdallarchaeota archaeon AB_125, and the fungus *Rhizopus microsporus*. In contrast, *Clostridium* sp. CAG-1024 was negatively correlated with Catovirus CTV1 and Hokovirus HKV1. Archaeal species formed distinct positively correlated subunits, including one composed of *Methanobrevibacter olleyae*, *M. ruminantium*, *M. wolinii*, and *M. smithii*, and another comprising *Methanomassiliicoccus luminyensis*, *Methanobrevibacter millerae*, *Candidatus Methanomassiliicoccus intestinalis*, and *Methanobrevibacter* sp. A27. Dominant eukaryotic and viral species also exhibited intricate interaction networks, underscoring the highly integrated and feeding system–dependent nature of the yak rumen microbiome (Yang et al., 2024).

5.8 CAMEL MICROBIOME

Dromedary camels (*Camelus dromedarius*), commonly referred to as one-humped or Arabian camels, are widely distributed across deserts and arid regions stretching from North and northeastern Africa through the Middle East and into the Indian subcontinent. Over evolutionary time, these animals have developed extraordinary physiological, anatomical, and microbial adaptations that allow them to survive and remain productive under extreme environmental pressures. Among their most notable adaptations is an exceptional tolerance to heat and dehydration, enabling them to survive for prolonged periods—up to 40 days—without access to drinking water. Equally remarkable is their ability to subsist on sparse desert vegetation that is largely unsuitable for most other herbivores.

The natural diet of dromedary camels consists primarily of shrubs and woody plants characterized by high levels of structural carbohydrates, particularly lignocellulose, as well as elevated concentrations of anti-nutritional and secondary plant compounds. These include tannins, saponins, phytates, oxalates, and a wide array of phenolic substances that typically reduce palatability and digestibility in other grazing and browsing species. The capacity of camels to efficiently utilize such chemically and structurally complex feed resources underscores the importance of their specialized digestive system and the unique microbial communities it harbors (Al Jassim, 2022; Gharechahi et al., 2023).

Although camels exhibit cud-chewing behavior similar to that of true ruminants such as cattle, sheep, and goats, they are taxonomically classified as pseudo-ruminants and belong to a distinct family, Camelidae. Unlike true ruminants, which possess a four-compartment stomach comprising the rumen, reticulum, omasum, and abomasum, camels have a three-compartment forestomach consisting of C1, C2, and C3. While C1 and C2 function analogously to the rumen and reticulum, camels lack a well-defined omasum-like chamber. Despite this anatomical difference, the camel forestomach provides a highly effective fermentation environment that supports extensive microbial degradation of fibrous plant material (Gharechahi et al., 2023).

Given the highly recalcitrant nature of camel diets and the exceptional capacity of the camel rumen microbiota to degrade lignocellulosic substrates, camels have been used as a model species for in situ rumen incubation experiments. In such studies, different lignocellulosic materials were incubated directly within the camel rumen, followed by whole-metagenome sequencing of substrate-associated microbial communities. These approaches have provided valuable insights into the functional potential, metabolic pathways, and enzymatic repertoire of rumen microorganisms involved in fiber degradation and detoxification processes (Gharechahi et al., 2023).

The camel forestomach hosts a dense and highly diverse consortium of microorganisms, including bacteria, archaea, protozoa, and anaerobic fungi. The relationship between the camel host and its forestomach microbiota is fundamentally symbiotic. The host provides optimal physicochemical conditions—such as temperature, pH, continuous substrate supply, and mechanical mixing—that support microbial growth and activity. In return, the microbiota degrade complex plant polysaccharides, detoxify harmful compounds, synthesize essential nutrients, and produce VFAs that serve as major energy sources for the host. *Streptococcus bovis* is a dominant lactic acid bacterium (LAB) in the camel forestomach. Other LAB commonly detected include *Selenomonas ruminantium*, *Butyrivibrio fibrisolvens*, *Lachnospira pectinoschiza*, and *Prevotella ruminicola* (Al Jassim et al., 2022).

Of particular interest are lactic acid–producing bacteria and those capable of regulating lactate metabolism. Certain isolates of *Selenomonas ruminantium* possess a complex genetic and metabolic machinery that allows them to initially produce L-lactate, subsequently shift to D-lactate production, and ultimately convert lactate to propionate when carbohydrate substrates become limiting (Ghali et al., 2011). These metabolic traits are especially relevant to rumen health, as they may reduce lactate accumulation and thereby lower the risk of ruminal acidosis, highlighting their potential use as probiotic candidates. In addition, *Butyrivibrio fibrisolvens*, which produces substantial amounts of butyric acid, may contribute to the relatively high concentrations of butyrate frequently observed in camel forestomach fluid. Other important LAB include *Lactococcus garvieae* and *Clostridium ramosum*, both recognized as significant L-lactate producers (Al Jassim, 2022).

Metagenomic analyses have further confirmed the extensive diversity of the camel rumen microbiome. Gharechahi et al. (2023) reported that more than 56% of reconstructed metagenome-assembled units (RUIs) were assigned to the phylum Firmicutes_A, followed by Bacteroidota (20%), Firmicutes (7%), and Spirochaetota (6%). Together, these four phyla accounted for over 90% of all RUIs. The remaining RUIs were distributed across 16 additional phyla, including Proteobacteria, Firmicutes_C, Verrucomicrobiota, Desulfobacterota_A, Actinobacteriota, Patescibacteria, Cyanobacteria, Synergistota, Elusimicrobiota, Fibrobacterota, Planctomycetota, Eremiobacterota, Euryarchaeota, Desulfuromonadota, Campylobacterota, and Firmicutes_B, reflecting the remarkable taxonomic breadth of the camel gut ecosystem.

At finer taxonomic resolution, the majority of RUIs that could not be assigned at the genus level belonged to families such as Lachnospiraceae, Anaerovoracaceae, Oscillospiraceae, Treponemataceae, and uncultured clades including CAG-826 and CAG-272. These microbial groups are widely recognized as core members of herbivore gastrointestinal microbiomes and play critical roles in lignocellulose degradation, short-chain fatty acid production, and overall fermentation efficiency in the rumen (Gharechahi et al., 2023).

Studies examining microbial succession under stress conditions further illustrate the dynamic nature of the camel gut microbiome. In original rumen (LG0) and intestinal (CG0) samples, Gammaproteobacteria were the dominant bacterial class. Under IQ stress, an increase in Sphingobacteriia was observed in first-generation rumen (LG1) and intestinal (CG1) samples. In subsequent generations, shifts occurred: Gammaproteobacteria and Alphaproteobacteria became dominant in second- (LG2) and third-generation (LG3) rumen samples, while Gammaproteobacteria and Flavobacteriia dominated second-generation intestinal samples (CG2). By the third generation (CG3), Gammaproteobacteria and Alphaproteobacteria once again emerged as the dominant intestinal flora (Gharechahi et al., 2023).

At the genus level, the original rumen microbiota (LG0) was dominated by Kingella and Moraxella. In first-generation rumen samples (LG1), Sphingobacterium and Pseudomonas prevailed, while Brevundimonas and Pseudomonas dominated in later generations (LG2 and LG3). The original intestinal samples were dominated by Pseudomonadaceae and Escherichia. Under IQ stress, the dominant genera shifted to Sphingobacterium and Stenotrophomonas in CG1, Pseudomonas and Bacillus in CG2, and Ochrobactrum, Pseudomonas, and Bacillus in CG3 (Gharechahi et al., 2023; Lin et al., 2024).

When third-generation rumen (LG3) and intestinal (CG3) contents were incubated with IQ as the sole carbon source, numerous degrading bacteria were isolated. A total of 27 strains from the rumen (L1–L27) and 23 strains from the intestine (C1–C23) were obtained. The dominant genera among rumen isolates were Ochrobactrum, Bacillus, and Pseudomonas, whereas Brevibacillus and Bacillus predominated among intestinal isolates, highlighting the camel gut as a reservoir of metabolically versatile microorganisms (Gharechahi et al., 2023).

In addition to bacteria, protozoa constitute an important component of the camel forestomach microbiota. Rabee et al. (2020) identified two protozoan families, Ophryoscolecidae and Isotrichidae. Seven genera—Diplodinium, Ophryoscolex, Entodinium, Polyplastron, Eudiplodinium, Epidinium, and Trichostomatia—were classified under Ophryoscolecidae, while Dasytricha and Isotricha belonged to Isotrichidae. Although variation in protozoal abundance was observed among individual camels, overall community composition was remarkably consistent, with Diplodinium, Ophryoscolex, and Entodinium consistently emerging as the dominant genera. Collectively, these findings emphasize the extraordinary complexity and functional specialization of the camel forestomach ecosystem, which underpins the animal's ability to thrive in some of the harshest environments on Earth.

5.8.1 Camel Milk Microbiome

Rahmeh et al. (2025) conducted a comprehensive investigation of the microbial community present in raw camel milk using shotgun metagenomic sequencing, a high-resolution approach that enables simultaneous detection of bacteria, bacteriophages, and other microorganisms. Their analysis revealed that raw camel milk harbors a diverse microbiome comprising both beneficial and potentially harmful microbes. Notably, the study identified several technologically important and health-promoting bacteria, including *Lactococcus lactis* and *Streptococcus thermophilus*, which are widely recognized as key starter cultures in the dairy industry due to their roles in fermentation and flavor development. In addition, the presence of bacteriophages capable of infecting a broad spectrum of pathogenic bacteria suggests a potential natural mechanism for microbial regulation within camel milk.

Alongside these beneficial microorganisms, the authors also reported the presence of several pathogenic and opportunistic bacterial species, indicating possible food safety concerns associated with raw camel milk consumption. Among the pathogens detected were *Clostridium botulinum*, *Klebsiella pneumoniae*, *Acinetobacter baumannii*, and *Streptococcus parasuis*, all of which are associated with serious infections in humans and animals. The coexistence of beneficial microbes and pathogens underscores the complex and dynamic nature of the raw camel milk microbiome and highlights the importance of appropriate handling and processing practices.

In milk samples obtained from healthy camels, the bacterial community was dominated at the species level (mean relative abundance ≥1%) by

Acinetobacter baumannii, which accounted for nearly half of the total bacterial population (47.2%). Other abundant species included *Moraxella osloensis* (11.19%), *Klebsiella pneumoniae* (6.01%), *Enterococcus faecium* (5.16%), *Enterococcus lactis* (2.9%), *Enterobacter roggenkampii* (2.49%), *Aeromonas caviae* (2.02%), *Escherichia coli* (1.6%), and *Klebsiella quasipneumoniae* (1.33%). This microbial profile suggests a dominance of environmental and opportunistic bacteria, with relatively lower proportions of classical LAB.

In contrast, camel milk samples from animals affected by mastitis exhibited a markedly different microbial composition. In this group, *Enterococcus faecalis* was the most dominant species, representing 38.6% of the bacterial population, followed by *E. coli* (14.26%) and *Staphylococcus simulans* (3.93%), all of which are commonly associated with mammary gland infections. Additional species detected at notable abundances included *Vibrio kanaloae* (2.34%), *Micromonospora ferruginea* (1.73%), *Conexibacter* sp. SYSU D00693 (1.53%), *Croceimicrobium hydrocarbonivorans* (1.5%), *Klebsiella pneumoniae* (1.44%), and *Sphingomonas paucimobilis* (1.0%).

Several studies have highlighted the remarkable diversity and complexity of LAB populations present in camel milk, underscoring its microbiological and functional significance. Members of the genera Lactobacillus spp., Lactococcus spp., Streptococcus spp., and Enterococcus spp. are consistently reported as core components of the camel milk microbiota. These LAB groups are of particular interest because of their established roles in milk fermentation, preservation, and the promotion of host health (Quilodrán-Vega et al., 2020).

Early culture-based investigations revealed that the most abundant cultivable Gram-positive cocci in camel milk were *Enterococcus faecium* and *Lactococcus lactis*, both of which are commonly associated with dairy fermentations and antimicrobial activity. Within the lactobacilli, dominant species included *Lactobacillus helveticus*, *Lacticaseibacillus casei*, and *Lactiplantibacillus plantarum*. These species are well known for their ability to ferment lactose, tolerate harsh gastrointestinal conditions, and produce bioactive metabolites, making them valuable from both technological and probiotic perspectives (Quilodrán-Vega et al., 2020).

Subsequent studies largely confirmed the predominance of enterococci and lactococci among the cultivable LAB in camel milk, but also reported notable variation in the composition of lactobacilli communities. In particular, differences were observed in the dominant lactobacilli species, with *Lacticaseibacillus paracasei*, other members of the genus Lacticaseibacillus, and *Limosilactobacillus reuteri* frequently detected in later investigations. These variations may reflect differences in geographic origin, feeding practices, milking hygiene, or analytical methods, and they emphasize the dynamic nature of the camel milk microbiome.

Beyond taxonomic characterization, several studies have focused on isolating LAB strains with demonstrated functional and health-promoting properties. For instance, *Lacticaseibacillus casei* TN-2, isolated from fermented camel milk, was shown to produce a bacteriocin capable of inhibiting important foodborne pathogens such as *Escherichia coli* and *Staphylococcus aureus*. Similarly, the *Enterococcus faecium* LCW44 strain isolated from raw camel milk exhibited strong antagonistic activity against *Listeria* spp. and *S. aureus*, highlighting its potential as a natural bioprotective culture (Quilodrán-Vega et al., 2020).

Additional antimicrobial LAB strains have also been identified in camel milk. Lactobacillus acidophilus AA105 demonstrated potent inhibitory effects against *Salmonella paratyphi*, *Shigella* spp., and pathogenic *E. coli*, while *Lactobacillus brevis* CM22 showed notable activity against *Listeria* spp. (Quilodrán-Vega et al., 2020). Collectively, these findings indicate that camel milk is not only a reservoir of diverse LAB but also a promising source of strains with probiotic and biopreservative potential, reinforcing its value for functional food development and food safety applications.

5.9 ALPACA MICROBIOME

South America is home to four native camelid species—the domesticated llama (*Lama glama*) and alpaca (*Vicugna pacos*), and their wild relatives, the guanaco (*Lama guanicoe*) and vicuña (*Vicugna vicugna*)—all of which are closely related to Old World camels but lack humps and are uniquely adapted to the high-altitude, cold, and oxygen-poor environments of the Andes. For thousands of years, these camelids have been central to Andean civilizations, providing essential resources such as high-quality fiber, meat, hides, and, in the case of llamas, reliable transport across rugged mountain terrain. Llamas and guanacos are generally larger and more robust, making them well suited for carrying loads, while alpacas and vicuñas are smaller and prized for their exceptionally fine and valuable wool. Genetic and archaeological evidence indicates that llamas were domesticated from guanacos, whereas alpacas descended from vicuñas, reflecting selective breeding for strength and fiber quality, respectively. Together, these four species form the biological and cultural backbone of traditional Andean life, sustaining livelihoods, economies, and cultural practices across the region for millennia.

Traditionally, alpacas have been maintained in large communal herds that grazed on the high, open plateaus and gentle slopes of the Andes, particularly across southern Peru, western Bolivia, Ecuador, and northern Chile, where cold temperatures, intense solar radiation, and sparse vegetation shaped their management and breeding. For centuries, Andean communities selectively bred alpacas primarily for their wool, which became one of the most important natural fibers in the region due to its warmth, softness, and durability. Over time, this selective breeding led to the development of two distinct modern alpaca breeds, differentiated by their geographic origins and, most notably, by fiber type. The Suri alpaca produces long, silky fibers that hang in elegant, pencil-like locks along the body, while the Huacaya alpaca has shorter, crimped, and dense wool that gives it a fluffy, sheep-like appearance. These differences in fleece structure influence not only the texture and commercial value of the fiber but also the traditional uses of alpaca wool, underscoring the species' enduring economic and cultural significance in Andean pastoral systems (Figueroa et al., 2023).

Alpacas are classified as pseudo-ruminants, possessing a three-compartment foregut that functions in a manner analogous to the multichambered stomach of true ruminants, and they rely heavily on complex gut microbial communities to ferment and degrade plant cell wall carbohydrates such as cellulose and hemicellulose. As in other ruminant and pseudo-ruminant species, the composition of the alpaca microbiome is highly dynamic and varies in response to numerous factors, including diet composition, forage quality, feeding strategies, and husbandry and management practices. These microbial populations play a central role in nutrient extraction, energy metabolism, and overall digestive efficiency, directly influencing animal health, growth, and productivity. Consequently, a detailed understanding of the alpaca gut microbiome is essential for developing evidence-based management and nutritional strategies that promote optimal health and production (Boughey et al., 2025).

The alpaca gastrointestinal microbiome is largely dominated by the bacterial phyla Firmicutes and Bacteroidota, which together comprise the majority of the microbial community. Firmicutes accounted for approximately 53.9–61.5% of the total alpaca biome, reflecting their key role in carbohydrate fermentation and energy harvest, while Bacteroidota contributed between 27.3% and 33.7%, underscoring their importance in polysaccharide breakdown. In contrast, Archaea were present at much lower relative abundance, with the phylum Euryarchaeota representing only about 0.17% of all taxa identified. Within this archaeal phylum, methanogenic classes were unevenly distributed, with Methanobacteriales comprising 0.18% of the total taxa, followed by Methanomassiliicoccales at 0.09% and Methanomicrobiales at 0.01%, indicating a limited but functionally significant methanogenic community in alpacas (Boughey et al., 2025).

Marked spatial variation in microbial composition was observed along the GIT, highlighting the presence of distinct ecological niches. Firmicutes exhibited differential abundance across all body sites, suggesting a broad functional role throughout digestion. Bacteroidetes were particularly enriched in the first gastric compartment (C1) and the distal intestine, while Fibrobacteres and Spirochaetes were most prominent in the C1, consistent with their specialization in fiber degradation. In contrast, Actinobacteria were more abundant in the small intestine, reflecting their adaptation to environments with higher availability of readily digestible nutrients (Carroll et al., 2019).

Dietary inputs further shaped microbial profiles at finer taxonomic resolution. The operational taxonomic unit (OTU) showing significant differences across the greatest number of body sites was a Butyrivibrio OTU, which was consistently more abundant in the ileum, cecum, and large intestine of alpacas fed alfalfa hay compared with those fed grass hay, emphasizing the sensitivity of fermentative bacteria to diet composition. At the genus level, the most abundant taxa in the C1 samples included Prevotella and several unclassified genera within the orders Clostridiales and Bacteroidales, as well as the family Ruminococcaceae. These groups are also dominant in ruminants and other South American camelids, reflecting conserved microbial assemblages that support efficient fermentation and utilization of fibrous plant diets (Carroll et al., 2019).

5.10 EQUINE MICROBIOME

Unlike ruminants, horses are obligate hindgut fermenters, relying on an expansive and highly specialized microbial community housed within the cecum and large colon to extract energy from fibrous plant material. This microbial ecosystem is central to equine health, converting structural carbohydrates into VFAs that fuel metabolism and maintain intestinal integrity. Yet, despite its remarkable efficiency, the equine hindgut is inherently fragile. Small perturbations in diet, management, or stress can destabilize microbial homeostasis, initiating a cascade of local and systemic consequences. Increasingly, research has revealed that disturbances within this microbial milieu extend far beyond the gut, forging pathological connections between the intestine, the hoof, and the brain.

5.10.1 Diet-Driven Modulation of Microbial Composition and Function along the Equine GIT

Raspa et al. (2024) investigated the effects of a high-fiber (HF) versus a high-starch (HS) diet on the microbiota across multiple compartments of the intestinal tract in healthy horses of uniform age and breed. Their findings demonstrated that each section of the equine GIT represents a distinct ecological niche, hosting a unique microbial community shaped by local physiological conditions. Importantly, diet exerted a strong influence on microbial composition throughout the tract. Horses fed the HS diet exhibited a smaller core microbiota and reduced alpha-diversity indices compared with those fed the HF diet, alongside an increased abundance of amylolytic bacteria, reflecting enhanced starch fermentation.

At the phylum level, Firmicutes dominated the microbiota in both dietary groups, surpassing Bacteroidetes, Proteobacteria, and Verrucomicrobia. In the duodenum, comparison of amplicon sequence variants (ASVs) between diets revealed higher relative abundances of Clostridium and Moraxella in horses receiving the HS diet, suggesting an early dietary effect on microbial colonization in the proximal small intestine.

In the jejunum, several taxa associated with carbohydrate fermentation—including Anaerovibrio, Lactococcus, and Turicibacter—were significantly more abundant in horses fed the HS diet. In contrast, Fibrobacter and Lactobacillus, both linked to fiber degradation and gut health, were more prevalent in horses fed the HF diet. Additionally, Enterobacteriaceae were significantly enriched in the jejunum of HS-fed horses, indicating a shift toward facultative anaerobes under high-starch conditions.

Moving to the ileum, Phascolarctobacterium and Barnesiella were more abundant in horses fed the HS diet. Enterobacteriaceae and Ralstonia were also present at higher levels in the ileum of HS-fed horses compared with those on the HF diet, further highlighting the impact of starch-rich feeding on microbial populations in the distal small intestine.

In the caecum, a major site of hindgut fermentation, Fibrobacter was significantly more abundant in horses fed the HF diet, consistent with its role in structural carbohydrate degradation. Conversely, Streptococcus—often associated with rapid starch fermentation—was significantly higher in horses fed the HS diet, suggesting an increased risk of dysbiotic fermentation under high-starch intake.

At the sternal flexure of the colon, Fibrobacter, Paraprevotellaceae, and Prevotella were more abundant in HF-fed horses, reflecting enhanced fiber utilization. In contrast, Anaerovibrio and Erysipelotrichaceae were significantly enriched in HS-fed horses. Methanobrevibacter was also more abundant in the HS group, and Fusobacterium was detected exclusively in horses receiving the HS diet, indicating diet-specific shifts in archaeal and potentially opportunistic bacterial populations.

Analysis of the pelvic flexure revealed higher relative abundances of Ruminococcus, Anaerovibrio, Sarcina, and Roseburia in horses fed the HS diet, whereas Dorea and Oscillospira were more prevalent in HF-fed horses. Succinivibrionaceae were also significantly more abundant in the HS group, reflecting enhanced succinate- and propionate-producing pathways under starch-rich feeding.

In the right dorsal colon, Clostridium and Ruminococcus were present at higher levels in horses fed the HS diet, while Clostridiales were more abundant in HF-fed horses. Prevotella was enriched in the HS diet, whereas Bacteroides were detected only in the HF group. Additionally, Akkermansia and Methanocorpusculum were found at greater levels in horses fed the HF diet, suggesting a more stable and mucin-associated microbial environment under fiber-rich feeding.

Finally, rectal samples showed that horses fed the HF diet harbored higher relative abundances of Lachnospiraceae and Ruminococcaceae, families commonly associated with fiber fermentation and short-chain fatty acid production. In contrast, Treponema and Succinivibrionaceae were more abundant in horses fed the HS diet, further emphasizing the systemic impact of dietary starch on hindgut microbial ecology.

Functional profiling of the microbiota revealed clear metabolic differences between the two dietary regimens. The HS diet was associated with enrichment of genera involved in fermentative pathways, including glycolysis/gluconeogenesis, pyruvate metabolism, methane metabolism, and propanoate metabolism, reflecting intensified carbohydrate fermentation. Conversely, the HF diet was associated with reduced potential microbial functions related to fatty acid degradation, nitrogen metabolism, and sulfur metabolism, highlighting the distinct metabolic landscapes shaped by fiber- versus starch-based feeding strategies in horses (Raspa et al., 2024).

5.10.2 Racehorse Microbiome

Gut microorganisms play a fundamental role in host physiology, influencing metabolism, immunity, and overall performance. However, the extent to which specific microbial functions contribute to the exceptional athletic capacity of racehorses was not completely understood. Li et al. (2025) identified a strong association between gut butyrate-producing bacteria and athletic performance in Thoroughbred racehorses. Both butyrate-producing microbial taxa and genes involved in microbial butyrate biosynthesis were significantly enriched in the racehorse gut, and these findings were further validated by GC–MS analysis, which confirmed elevated butyrate production.

To better understand the microbial contributors underlying this functional enrichment, they examined taxonomic differences at the genus level. Olivibacter and Alicyclobacillus emerged as the most abundant genera in the racehorse gut. In total, 190 MAGs showed significant differential abundance between groups. Among these, MAGs affiliated with Olivibacter, Prevotella, Paludibacteraceae, Saccharofermentans, Ruminococcus, and Colidextribacter were notably enriched in racehorses, with Olivibacter representing the most abundant microbial genome detected.

Functional pathway analysis revealed a striking pattern: among all carbohydrate metabolic pathways, only the butanoate (butyrate) metabolism pathway was significantly enriched in the racehorse gut. This selective enrichment suggests that butanoate metabolism plays a particularly critical role in shaping the gut environment of high-performance horses and may be central to their enhanced athletic capacity.

The elevated abundance and activity of fiber-degrading bacteria—including Olivibacter, Prevotella, Paludibacteraceae, Saccharofermentans, Ruminococcus, and Colidextribacter—likely drive the increased production of butyrate by converting complex dietary carbohydrates into fermentative substrates. This microbial cross-feeding network enhances the local availability of butyrate within the gut and facilitates its systemic distribution.

Once produced, butyrate is absorbed by the host and enters the bloodstream, where it functions as a key bioactive metabolite. In peripheral tissues, particularly skeletal muscle, butyrate influences mitochondrial function, energy metabolism, and muscle efficiency. Through these mechanisms, microbial-derived butyrate may serve as a crucial link between gut microbial composition and the superior athletic performance observed in racehorses (Li et al., 2025).

5.10.3 The Gut–Hoof Axis: Microbial Dysbiosis and the Pathogenesis of Laminitis

Laminitis—a devastating and often career-ending condition characterized by the failure of the lamellar interface between the hoof wall and the pedal bone—is frequently the terminal manifestation of events originating in the hindgut. In many cases, the initiating insult is dietary. The equine digestive tract evolved to process diets rich in structural carbohydrates (fiber), yet modern feeding practices often include high levels of nonstructural carbohydrates in the form of grain-based concentrates or lush pasture grasses rich in fructans. When these carbohydrates exceed the enzymatic capacity of the small intestine, they escape digestion and enter the hindgut intact (Harlow et al., 2016).

This sudden influx of fermentable substrate profoundly alters microbial dynamics. Amylolytic bacteria—most notably Streptococcus bovis, Streptococcus equinus, and Lactobacillus species—rapidly proliferate, outcompeting fiber-fermenting organisms (Milinovich et al., 2006). Their metabolic activity results in excessive lactic acid production, driving a sharp decline in hindgut pH from a physiologic range of 6.5–7.0 to an acidotic state below 6.0, a condition termed hindgut acidosis. This acidic environment is hostile to beneficial fibrolytic microbes, amplifying dysbiosis and perpetuating the cycle.

A critical yet often underappreciated feature of this process is the stereochemistry of lactate production. Under normal conditions, mammalian cells and commensal microbes predominantly generate L-lactate, which is readily metabolized. In contrast, the dysbiotic overgrowth of Lactobacillus and Streptococcus species favors the production of D-lactate (Burgos et al., 2024). Horses possess limited capacity to metabolize this isomer due to inefficient D-lactate dehydrogenase activity. As a result, D-lactate accumulates systemically, contributing to metabolic acidosis and neurologic dysfunction.

Simultaneously, the acidotic hindgut compromises epithelial integrity, leading to increased intestinal permeability, commonly described as "leaky gut." This allows bacterial endotoxins such as lipopolysaccharides (LPS) and vasoactive amines to translocate into the bloodstream (Gossett et al., 1987). Once in circulation, these inflammatory mediators exert profound effects on distant tissues. Within the hoof, they trigger a dysregulated inflammatory response that activates matrix metalloproteinases (MMPs), particularly MMP-2 and MMP-9, within the lamellar tissue (Pollitt, 2004). While MMPs normally regulate extracellular matrix turnover, their excessive activation in laminitis leads to enzymatic degradation of the basement membrane and hemidesmosomal attachments anchoring the hoof wall to the pedal bone. The resulting structural failure permits rotation or sinking of the pedal bone, culminating in severe pain and lameness (Yang & Lopez, 2021).

5.10.4 The Gut–Brain Axis: Microbial Regulation of Behavior, Stress, and Neurochemistry

Parallel to its influence on hoof health, the equine microbiome exerts a powerful effect on the central nervous system through the gut–brain axis—a bidirectional network integrating neural, endocrine, and immune signaling pathways. Emerging evidence supports the assertion that, in horses, gastrointestinal health and mental well-being are inseparable (George, 2023).

Serotonin (5-hydroxytryptamine, 5-HT) is a pivotal neurotransmitter governing mood, anxiety, pain perception, and gastrointestinal motility. Remarkably, the majority of serotonin synthesis occurs in the gut, and this process depends on the availability of the essential amino acid tryptophan. The gut microbiota functions as a metabolic gatekeeper, determining whether tryptophan is utilized for serotonin production or diverted into alternative pathways (Yano et al., 2015). Under conditions of microbial balance (eubiosis), tryptophan remains accessible for serotonin synthesis. However, dysbiosis can redirect tryptophan metabolism toward the kynurenine pathway, reducing central serotonin availability. This shift has been linked to depressive-like states, heightened pain sensitivity, and altered stress responsiveness (Muneer, 2020).

Behavioral manifestations of microbial imbalance are increasingly recognized in equine management. Horses consuming HS, low-fiber diets demonstrate reduced microbial diversity and altered community structure, including a decreased Firmicutes-to-Proteobacteria ratio (Irving et al., 2025). Such microbial profiles correlate strongly with hyperreactivity, heightened "spookiness," and the development of stereotypic behaviors such as crib-biting (aerophagia). Crib-biting is hypothesized to represent a maladaptive coping mechanism, influenced by altered gastric acidity and dysregulation of the gut–brain stress axis involving ghrelin and dopamine signaling (Mach et al., 2020).

Importantly, the gut–brain axis operates in both directions. Psychological and physical stressors can rapidly reshape the microbiome, just as microbial changes influence behavior. Transport stress—a routine but significant challenge for performance horses—has been shown to induce marked shifts in fecal microbiota within hours. As little as two hours of transport can decrease beneficial Bacteroidetes while increasing Proteobacteria, changes likely mediated by cortisol and catecholamine release, which alter gut motility, mucin secretion, and immune function (Perry et al., 2018).This stress-induced dysbiosis transiently compromises gut resilience, increasing susceptibility to colic, inflammation, and opportunistic infections, including enhanced shedding of pathogens such as Salmonella.

5.11 GOAT MICROBIOME

The domestic goat (*Capra hircus*) is one of the most economically significant livestock species worldwide and has long played an indispensable role in human societies. Goats contribute substantially to food security and rural livelihoods by providing meat, milk, fiber, and pelts, particularly in arid, semi-arid, and mountainous regions where other livestock species are less adaptable. Owing to their hardiness, efficient feed utilization, and ability to thrive under diverse environmental conditions, goats are especially important to smallholder and resource-poor farmers. According to the Food and Agriculture Organization (FAO) of the United Nations, the global goat population now exceeds one billion animals, encompassing approximately 1,000 recognized breeds. This remarkable diversity reflects centuries of domestication and selective breeding to meet varied cultural, economic, and ecological needs, underscoring the goat's vital contribution to agricultural sustainability and human well-being worldwide (Cao et al., 2023).

Cao et al. (2023) conducted a comprehensive characterization of the goat gastrointestinal microbiome using MAGs, providing detailed insights into its taxonomic composition, functional potential, and ecological dynamics. At the phylum level, the goat microbiome was dominated by Firmicutes_A and Bacteroidota, followed by Verrucomicrobiota, Proteobacteria, and Spirochaetota. All Firmicutes_A MAGs belonged to the class Clostridia, which was mainly represented by the orders Oscillospirales, 4C28d-15, Lachnospirales, and Lachnospiraceae. In contrast, all Bacteroidota MAGs were assigned to the class Bacteroidia, predominantly comprising the order Bacteroidales, with a small contribution from Flavobacteriales, which included only the family UBA1820.

Members of the phylum Verrucomicrobiota were distributed across three classes: Lentisphaeria, Verrucomicrobiae, and Kiritimatiellae, indicating notable taxonomic diversity. Archaeal MAGs identified in the goat microbiome were exclusively methane-producing taxa and belonged to the phyla Thermoplasmatota, Halobacterota, and Euryarchaeota. Interestingly, when compared with a previous buffalo microbiome study, Halobacterota MAGs were more abundant than Euryarchaeota in both ruminants, though the buffalo microbiome showed a much higher number of Halobacterota MAGs than Euryarchaeota. These observations suggest a higher strain-level diversity within Halobacterota across ruminant species. Moreover, relative abundance analysis revealed that Halobacterota MAGs were detected in only 48.89% of goat samples, whereas Euryarchaeota MAGs occurred in 90.34% of samples. This pattern reinforces the prevailing view that Methanobrevibacter, a member of Euryarchaeota, is a highly prevalent and ecologically important methanogen in ruminants.

Further comparative analysis showed that 56.37% of the goat microbiome MAG catalog (GMMC) exhibited an average nucleotide identity (ANI) of less than 95% compared with genomes in the Ruminant Catalog, indicating substantial novelty. Among these MAGs, 1,241 belonged to Firmicutes (Firmicutes and Firmicutes_A), primarily from Oscillospirales (53.26%), 4C28d-15 (14.75%), and Lachnospirales (14.18%). Notably, Oscillospirales and Lachnospirales are well known for their roles in digestion, while 4C28d-15 is highly abundant in the rumen. These findings demonstrate that the GMMC significantly expands existing public databases by adding previously underrepresented but functionally important microbial taxa.

The authors further investigated microbial dynamics along different sections of the goat GIT. Across all sites, Firmicutes_all (a combined group including Firmicutes, Firmicutes_A, Firmicutes_B, and Firmicutes_C, also referred to as Bacillota) and Bacteroidota were the dominant phyla, accounting for 51.27% and 25.82% of total relative abundance, respectively. Although substantial intersample variation was observed, the Firmicutes-to-Bacteroidota (F/B) ratio remained relatively stable across the four stomach chambers (rumen, reticulum, omasum, and abomasum). In contrast, marked fluctuations in the F/B ratio emerged in the small intestine, particularly in the jejunum, and became more pronounced in downstream GIT regions.

To explore the drivers of these patterns, de novo clustering analysis of F/B ratio dynamics was performed using samples from 23 goats with data from at least seven of nine GIT sites. Two distinct clusters were identified. Cluster 1 exhibited low and relatively constant F/B ratios in the stomach and large intestine, with a sharp increase in the small intestine, especially in the jejunum and ileum. Cluster 2, by contrast, showed a gradual and continuous increase in the F/B ratio along the GIT. Dietary comparison revealed that goats in Cluster 1 were mainly fed silage, whereas those in Cluster 2 predominantly consumed grass. Since Firmicutes and Bacteroidota are often associated with digestive and absorptive functions, respectively, these results suggest that silage is more readily digested and absorbed in the upper intestine, while grass requires prolonged digestion across multiple GIT sites.

Importantly, the two F/B ratio patterns were also associated with distinct disease-related microbial signatures. Comparative genus-level analysis showed that Campylobacter, a known enteritis-associated genus, was significantly enriched in the jejunum of Cluster 1, indicating a potentially elevated disease risk. Conversely, Odoribacter, a genus reported to limit intestinal inflammation, was also enriched, suggesting a compensatory or protective role of the microbiota at this digestion site.

Across the entire GIT, 311 genera displayed significant abundance differences among the four major sections. All four identified methanogenic genera—Methanomethylophilus, Methanocorpusculum, Methanobrevibacter_A, and ISO4-G1—showed differential distributions. Methanomethylophilus was most abundant in the stomach, whereas Methanocorpusculum increased steadily from the stomach to the rectum. The other methanogens exhibited more variable spatial patterns. These findings indicate that methane production is not confined to the stomach but also occurs in downstream GIT regions.

Significant spatial variation was also observed among cellulose-degrading bacteria. Genera such as Prevotella and Fibrobacter were more abundant in the stomach and small intestine, while Faecalibacterium, Acetivibrio_A, Ruminiclostridium_E, and Paraprevotella were enriched in downstream sections. These distributions highlight coordinated host–microbe interactions across the GIT, with different microbial groups contributing to fiber degradation, fermentation, and nutrient absorption at specific locations.

In addition to bacterial and archaeal communities, Cao et al. (2023) annotated viral genomes associated with the goat microbiome using VirusTaxo and Demovir tools. Approximately 75.67% of the viral genomes were assigned to known taxonomic clades. The most abundant viral families were Poxviridae, Mimiviridae, Microviridae, and Siphoviridae.

Finally, given the host specificity of many viruses, the authors predicted viral–host relationships to explore their regulatory roles within the goat microbiome. Using four complementary approaches, including CRISPR-spacer matching, homology-based methods, VirHostMatcher, and binning-based analysis, hosts were assigned to 4,202 viral genomes (58.3% of the total). Although overlap among methods was limited, 5.7% of viral–host links were supported by at least two approaches. In total, 1,321 viruses were classified as specialists with a single host, while the remainder were generalists infecting multiple hosts. Most predicted hosts belonged to Bacteroidota and Firmicutes, with Prevotella, Alistipes, RF16, F082, and Akkermansia being the most frequently targeted genera. Notably, many of these hosts are functionally important for digestion and gut health, suggesting that viruses play key roles in shaping microbial community structure and function.

To assess the potential of viruses in mitigating methane emissions, the authors screened for viruses targeting methanogens and identified 104 such viruses. Among them, 68 were putative lytic viruses capable of infecting methanogens from all four identified genera. Five viral–methanogen associations were supported by multiple prediction methods, highlighting the potential of phage-based strategies to regulate methane-producing archaea in ruminants (Cao et al., 2023).

5.11.1 Variability of Intestinal Microbiota Composition in Goats of Different Breeds

Molotzu et al. (2025) investigated the variability of intestinal microbiota composition in goats of different breeds (Sarda, Maltese, and Alpine) reared in multiple flocks across the Ogliastra region of Sardegna, Italy, with the aim of determining whether breed, farm location, and feeding system (natural pasture grazing–based vs. intensive management) influenced gut bacterial communities. Overall, goat fecal microbiota was dominated by the phyla Firmicutes and Bacteroidetes, with Firmicutes prevailing over Bacteroidetes in all farms except Cardedu; notably, the Baunei farm exhibited a strongly skewed Firmicutes-to-Bacteroidetes ratio in favor of Firmicutes. Across most farms, Proteobacteria, Actinobacteria, and Verrucomicrobia were present at relatively low abundances, whereas the Cardedu farm showed slightly higher proportions of these phyla compared with the others. At the family level, the most abundant taxa included Gracilibacteraceae, Erysipelotrichaceae, Clostridiales Family XI Incertae Sedis, Christensenellaceae, Acidaminococcaceae, Peptostreptococcaceae, Clostridiaceae, Porphyromonadaceae, Prevotellaceae, Flavobacteriaceae, Eubacteriaceae, Ruminococcaceae, Lachnospiraceae, Rikenellaceae, Synergistaceae, and Bacteroidaceae, with marked farm-specific differences. Erysipelotrichaceae was particularly enriched in the Talana, Urzulei, and Perdasdefogu farms, while Acidaminococcaceae and Porphyromonadaceae were especially abundant in Cardedu. Peptostreptococcaceae showed higher relative abundance in Talana and Perdasdefogu but was present at very low levels in Cardedu, whereas Prevotellaceae was more abundant in Perdasdefogu and Villagrande. At the genus level, Ruminococcus, Eubacterium, Roseburia, and Clostridium were the most represented, with Ruminococcus, Roseburia, and Clostridium showing higher abundance in the Talana farm and lower abundance in Cardedu. At the species level, *Ruminococcus faecis*, *Ruminococcus gauvreauii*, *Eubacterium hallii*, *Roseburia faecis*, and *Clostridium lavalense* were predominant, with *R. faecis*, *R. gauvreauii*, and *C. lavalense* being more abundant in Talana and Perdasdefogu and less abundant in Cardedu; notably, *R. faecis* reached its highest abundance in the Talana farm. Collectively, these results demonstrate pronounced farm- and diet-associated variability in the goat intestinal microbiota, underscoring the strong influence of local management practices and feeding systems on gut microbial structure beyond breed differences alone.

5.11.2 Gut Microbiome in Lactating Dairy Goats

Wang et al. (2025) comprehensively characterized the gut microbiome of 134 lactating dairy goats and identified two distinct microbial enterotypes with clear functional and production-related differences. The first enterotype, termed the TR-cluster, was dominated by Turicibacter, Romboutsia, and *Clostridium sensu stricto* 1. The second enterotype, the CO-cluster, showed a higher relative abundance of Christensenellaceae R-7 group and Oscillospiraceae UCG-005. These enterotypes were associated with markedly different metabolic profiles and lactation performances, highlighting the importance of gut microbial structure in dairy goat productivity.

Goats belonging to the TR-cluster exhibited superior production traits, including higher milk yield, fat-corrected milk yield, and milk fat yield, compared with goats in the CO-cluster. This enhanced performance was accompanied by significantly greater concentrations of total VFAs in the gut, particularly acetate, propionate, and butyrate. In addition, TR-cluster goats showed elevated serum glucose and total bile acid levels, suggesting improved energy availability and lipid metabolism. Although microbial diversity in the TR-cluster was lower, functional prediction revealed enrichment of pathways related to energy metabolism, amino acid metabolism, and lipid biosynthesis. In contrast, the CO-cluster microbiome was functionally enriched in pathways associated with membrane transport and genetic information processing, reflecting a different ecological and metabolic strategy.

At the phylum level, Firmicutes and Spirochaetota were significantly enriched in the TR-cluster, whereas Bacteroidota, Actinobacteriota, Patescibacteria, Verrucomicrobiota, and Desulfobacterota were more abundant in the CO-cluster ($P < 0.05$). Consistent patterns were observed at the family level: Peptostreptococcaceae, Erysipelotrichaceae, and Clostridiaceae were overrepresented in the TR-cluster, while Rikenellaceae and several Firmicutes families—such as Oscillospiraceae, Christensenellaceae, and Lachnospiraceae—were more prevalent in the CO-cluster. These taxonomic differences further underscore the contrasting microbial ecosystems represented by the two enterotypes.

Genus-level analysis reinforced these distinctions. The TR-cluster was enriched in Turicibacter, Romboutsia, Clostridium sensu stricto 1, Paeniclostridium, and Treponema, whereas the CO-cluster was characterized by higher abundances of Christensenellaceae R-7 group, Oscillospiraceae UCG-005, unclassified Lachnospiraceae, and the Rikenellaceae RC9 gut group. Correlation analyses revealed that four TR-cluster genera—Turicibacter, Romboutsia, Clostridium sensu stricto 1, and Paeniclostridium—were positively associated with both fat-corrected milk yield and gut butyrate concentration. Moreover, Turicibacter, *Clostridium sensu stricto* 1, and Treponema showed positive correlations with total VFA, acetate, and propionate levels. In contrast, genera enriched in the CO-cluster, including Christensenellaceae R-7 group, Ruminococcus, Monoglobus, Oscillospiraceae NK4A214 group, and Lachnospiraceae NK3A20 group, were negatively correlated with gut VFA concentrations. Collectively, these findings indicate that the TR-cluster microbiome is more metabolically efficient in supporting fermentation end-products and milk fat synthesis, thereby linking specific microbial configurations to improved lactational performance in dairy goats (Wang et al., 2025).

5.12 SHEEP MICROBIOME

Sheep (*Ovis aries*) are domesticated small ruminants that are widely raised across the world for their valuable products, including wool, meat (lamb and mutton), and milk. They play an essential role in agricultural systems, particularly in pastoral and mixed farming communities, due to their adaptability to diverse climates and landscapes ranging from arid regions to temperate grasslands. Beyond their economic importance, sheep are recognized for their intelligence and complex social behavior. They are highly social animals with strong flocking instincts, which provide protection from predators and facilitate group learning and communication.

Sheep exhibit advanced cognitive abilities, such as facial recognition, long-term memory, and social learning, allowing them to identify other sheep and even familiar humans. Their flocking behavior is guided by social hierarchies and cooperative interactions, which help maintain group cohesion and reduce stress. These traits make sheep well suited for managed grazing systems and contribute to their success as one of the earliest domesticated livestock species.

Globally, sheep display remarkable genetic and phenotypic diversity, with approximately 1,200 distinct breeds identified to date (Mehmet et al., 2023). These breeds have been selectively developed to optimize specific traits such as wool quality, meat yield, milk production, disease resistance, and environmental resilience. This extensive diversity highlights the importance of sheep in sustainable agriculture, food security, and rural livelihoods worldwide.

The gastrointestinal microbiome of sheep plays a critical role in nutrient digestion, energy harvest, immune regulation, and adaptation to diverse environments. Advances in next-generation sequencing (NGS) and high-throughput sequencing (HTS) technologies have enabled detailed characterization of microbial communities along different sections of the sheep digestive tract, as well as comparisons across breeds and management systems. Several recent studies highlight both the conserved core microbiota of sheep and the variability driven by husbandry practices, breed genetics, and environmental conditions.

Minozzi et al. (2020) investigated the composition of the sheep hindgut microbiome and assessed whether different husbandry conditions could influence microbial diversity and structure. Using 16S rRNA gene sequencing, the authors found that the hindgut microbiome was dominated by a limited number of major bacterial phyla. Firmicutes constituted the largest proportion (43.6%), followed by Bacteroidetes (30.38%), Proteobacteria (10.14%), and Verrucomicrobia (7.55%). The remaining 13 detected phyla collectively accounted for only 8.3% of the microbial community, indicating a relatively stable and specialized hindgut ecosystem. At deeper taxonomic levels, the most abundant classes were Clostridia (41.5%) and Bacteroidia (30.1%), with corresponding dominance of the orders Clostridiales (41.5%) and Bacteroidales (30.16%). Key families included Ruminococcaceae (28.6%), Rikenellaceae (10.5%), and Campylobacteraceae (7.9%), while the most representative genera were Ruminococcaceae UCG-010 (8.8%) and Campylobacter (7.9%). These results suggest that fiber-degrading and fermentative bacteria form the functional backbone of the sheep hindgut, with management practices exerting measurable but limited effects on overall community structure.

Expanding the focus from the hindgut to the entire GIT, Zhang et al. (2025) identified section-specific microbial signatures along the sheep digestive system. Distinct taxa were associated with different compartments, reflecting functional specialization. Longimicrobiaceae were dominant in the rumen, Candidatus Soleaferrea in the reticulum, Parabacteroides in the omasum, Lactococcus in the abomasum, and Prevotella in the duodenum. Further along the intestine, Anaerotruncus, UCG-002, Pseudobutyrivibrio, Oribacterium, and Moryella were characteristic of the jejunum, ileum, cecum, colon, and rectum, respectively. These findings underscore the strong spatial heterogeneity of the sheep gut microbiome and highlight the coordinated microbial contributions to digestion from foregut fermentation to hindgut nutrient salvage.

Breed-specific differences in the sheep microbiome were examined by Ma et al. (2022), who compared Dorper and Tan sheep. Their results showed that the rumen and hindgut microbiota were largely composed of members of Ruminococcaceae, Rikenellaceae, Prevotellaceae, Christensenellaceae, and Lachnospiraceae, together accounting for approximately 70% of total gut microbes. Firmicutes and Bacteroidetes were the two dominant phyla, comprising nearly 90% of the bacterial community. At the family level, Prevotellaceae predominated in the rumen, whereas Ruminococcaceae were more abundant in the hindgut. Genus-level analysis revealed that Rikenellaceae_RC9_gut_group, Ruminococcaceae_UCG-005, and Ruminococcaceae_UCG-010 were dominant in the hindgut, with higher relative abundances than in the rumen.

To further identify key discriminatory taxa between breeds, Ma et al. (2022) employed LEfSe and random forest analyses. In the rumen, 30 core bacterial taxa were identified, with g_Saccharofermentans and g_Lachnospiraceae_AC2044_group enriched in Tan sheep, while g_Bacteroidales_RF16_group and f_Succinivibrionaceae were more abundant in Dorper sheep. Random forest analysis highlighted f_Enterobacteriaceae, g_Lactobacillus, and s_bacterium_XPB1013 as important differential taxa. In the hindgut, 40 key bacteria were identified, with f_Rikenellaceae and g_Ruminococcaceae_UCG-005 emerging as the most abundant discriminative genera. Joint analysis revealed that f_Enterobacteriaceae, f_Pseudomonadaceae, and s_Lactobacillus_faecis were enriched in both the rumen and hindgut of Tan sheep, whereas f_Marinifilaceae and g_Succinivibrio were more abundant in Dorper sheep. These results demonstrate that host genetics strongly influence microbial community composition in both foregut and hindgut environments.

Environmental adaptation and altitude-related effects on the sheep microbiome were explored by Chang et al. (2020), who analyzed four sheep breeds living on the Qinghai–Tibetan Plateau (QTP): Tibetan sheep (TS), Dorset sheep (DsS), Dorper sheep (DrS), and Small Tail Han sheep (STHS). Using HTS, the authors found that both bacterial abundance and diversity were significantly lower in TS compared to the other three breeds. The gut microbiome composition of TS was also markedly distinct. Although Firmicutes and Bacteroidetes were the dominant phyla across all breeds, TS exhibited higher relative abundances of Spirochaetes, Proteobacteria, and Verrucomicrobia, while Deferribacteres were completely absent. At the genus level, Treponema, Succinivibrio, 5–7 N15, and Prevotella were significantly more abundant in TS than in the other breeds, suggesting microbial adaptations linked to high-altitude environments.

At the family level, Chang et al. (2020) observed no significant differences among the four breeds overall; however, notable trends were evident. Ruminococcaceae and WCHB1–25 were dominant in DsS, DrS, and STHS, whereas families such as Alcaligenaceae, Desulfovibrionaceae, and Barnesiellaceae were nearly absent. In contrast, TS showed higher abundances of Spirochaetaceae, S24-7, Prevotellaceae, Barnesiellaceae, and Succinivibrionaceae, while BS11 and WCHB1–25 were largely absent. Genus-level comparisons revealed more pronounced differences, with TS dominated by Treponema, Succinivibrio, 5–7 N15, and Prevotella, and lacking genera such as Bifidobacterium, Sharpea, and YRC22. Interestingly, while Treponema remained a predominant genus in DsS, DrS, and STHS, other beneficial butyrate-producing genera such as Coprococcus and Roseburia were present at relatively lower abundances. Overall, TS harbored a broader array of relatively abundant taxa, reflecting microbial strategies for energy extraction and survival under extreme plateau conditions.

Collectively, these studies demonstrate that the sheep gastrointestinal microbiome is shaped by gut compartment, breed genetics, husbandry practices, and environmental pressures. While a core community dominated by Firmicutes and Bacteroidetes is conserved, fine-scale taxonomic differences contribute to functional specialization, productivity, and ecological adaptation in sheep.

5.13 MICROBIOME OF PIGS

Domestic pigs (*Sus scrofa domesticus*) are among the most important terrestrial livestock species and represent the largest source of meat consumed worldwide. Their high growth efficiency, adaptability to diverse production systems, and physiological similarities to humans make pigs not only central to global food security but also valuable models for nutritional and biomedical research.

The gut microbiota of pigs plays a pivotal role in regulating host physiology, with profound effects on nutrient digestion, energy metabolism, immune development, and disease resistance.

Intestinal microorganisms contribute to the fermentation of complex carbohydrates, synthesis of bioactive metabolites such as short-chain fatty acids, and modulation of gut barrier integrity. Disruptions in the gut microbial ecosystem have been linked to reduced growth performance, metabolic disorders, and increased susceptibility to enteric infections.

Consequently, detailed characterization of the structure and functional capacity of the pig gut microbiome is essential for advancing our understanding of host–microbe interactions. Insights into microbial composition and metabolic potential can facilitate the development of targeted nutritional strategies, probiotics, and management practices aimed at improving feed efficiency, animal health, and sustainable pig production. Moreover, exploring pig gut microbial resources provides a foundation for uncovering conserved mechanisms of microbiota–host crosstalk relevant to both animal agriculture and human health (Hu et al., 2024).

Lee et al. (2023) investigated the relationship between gut microbiome composition and growth performance in pigs by analyzing age-related and group-specific changes in bacterial communities. Their results demonstrated that the pig gut microbiota undergoes dynamic and stage-dependent shifts as animals mature, with distinct microbial signatures emerging at different growth phases.

When bacterial communities were compared between 70 and 100 days of age, a significant increase in the relative abundance of Corynebacterium was observed at 100 days. Further comparisons between 130 and 160 days of age revealed marked microbial maturation, with significantly higher abundances of Erwinia, Bacteroides, CF231, Ruminococcus, Adlercreutzia, Sphingobacterium, Anaeroplasma, and Methanobrevibacter at 160 days. In contrast, Prevotella was more abundant at 130 days, suggesting a shift from carbohydrate-focused fermentation toward more diverse metabolic functions as pigs aged.

Group-based comparisons further highlighted microbiome differences associated with growth performance. At 160 days of age, pigs in group J exhibited significantly higher abundances of Anaeroplasma, C39, Parabacteroides, Selenomonas, rc4-4, CF231, Anaerovibrio, Bacteroides, Alkalibacterium, Phascolarctobacterium, Oscillospira, Sphingobacterium, and Methanobrevibacter compared with group D. Conversely, group D showed higher relative abundances of Slackia, Roseburia, Lactobacillus, Clostridium, Staphylococcus, Ruminococcus, Collinsella, SMB53, Coprococcus, Sutterella, and Bifidobacterium, indicating distinct microbial strategies associated with differing growth phenotypes.

Age-related changes were also evident within group T. Between 70 and 100 days of age, Ruminococcus, Dialister, and Acholeplasma were significantly enriched at 100 days, while the relative abundance of Coprococcus increased progressively with age. Comparisons between groups D and T at 160 days showed that Clostridium, Adlercreutzia, rc4-4, Lactobacillus, Staphylococcus, Mogibacterium, Bifidobacterium, and Oxalobacter were significantly more abundant in group D, whereas Treponema was significantly enriched in group T.

More detailed temporal analysis within group D revealed pronounced microbiome restructuring across growth stages. At 100 days compared with 70 days, the relative abundances of Trichococcus, Peptococcus, Anaerostipes, Parabacteroides, and Bacillus were significantly higher. From 100 to 130 days of age, increases were observed in Chlamydia, Phascolarctobacterium, and Bilophila, accompanied by a significant decrease in Blautia. Between 130 and 160 days, further maturation occurred, characterized by increased abundances of Turicibacter, Mogibacterium, Bifidobacterium, Dorea, Sutterella, and Fibrobacter, while Lachnospira declined.

Collectively, these findings demonstrate that pig gut microbial communities are highly dynamic and closely associated with age and growth performance. The observed shifts in key bacterial taxa suggest that specific microbial assemblages may contribute to improved nutrient utilization, metabolic efficiency, and overall growth, highlighting the potential of microbiome-informed strategies to enhance pig production (Lee et al., 2023).

5.13.1 Short-Chain Fatty Acid–Producing Microbiota in the Pig Gut

The pig gut microbiome, similar to that of other mammals, plays a vital role in maintaining host health and physiological function. It contributes to protection against pathogen colonization by competing with harmful microorganisms, supports the development and maturation of the immune system, and facilitates the synthesis of essential vitamins and bioactive metabolites that influence digestion and overall metabolism. Through these functions, the gut microbial community is a key determinant of growth performance, disease resistance, and nutritional efficiency in swine.

Despite its importance, a large proportion of microorganisms inhabiting the swine GIT remain poorly characterized. Many of these microbes are difficult or impossible to cultivate using conventional laboratory techniques because they require highly specific, and often unknown, growth conditions. As a result, early studies of the mammalian gut microbiome were largely restricted to a limited set of easily culturable organisms, representing only a small fraction of the true microbial diversity present in the gut ecosystem.

Advances in HTS technologies have transformed this field. In particular, the recovery of MAGs from shotgun metagenomic data has significantly expanded the number of microbial genomes available in reference databases. Although the completeness and accuracy of MAGs can vary, they provide valuable insights into the taxonomic identity and functional potential of previously uncharacterized microbes. This approach allows researchers to link metabolic capabilities and ecological roles to specific microbial species and strains that lack cultured representatives, thereby offering a more comprehensive understanding of the swine gut microbiome and its contributions to host health and productivity (Holman et al., 2022).

Holman et al. (2022) conducted a comprehensive characterization of the short-chain fatty acid (SCFA)–producing microbiome in pigs, with a particular focus on how diet-derived polysaccharides shape microbial composition and metabolic function. After weaning, pigs are commonly transitioned to diets rich in cereal grains such as corn, barley, and wheat. In addition to being high in starch, these feeds contain substantial amounts of complex polysaccharides, including cellulose, hemicellulose (for example, β-mannan, mixed-linkage glucans, xylan, and xyloglucan), and pectin. Because pigs lack the endogenous enzymes required to fully degrade these complex carbohydrates, many of these polysaccharides escape host digestion and reach the hindgut, where they serve as key substrates for microbial fermentation. As a result, the pig gut microbiome harbors an extensive repertoire of genes encoding carbohydrate-active enzymes (CAZymes) that enable the breakdown and utilization of these otherwise indigestible dietary components.

Using genome-resolved metagenomics, the authors reconstructed a large collection of MAGs and classified them using the Genome Taxonomy Database (GTDB). The overwhelming majority of MAGs were bacterial, while only 10 MAGs were assigned to archaea, all of which were identified as methanogens. In total, 19 distinct microbial phyla were represented across the MAG dataset, highlighting the phylogenetic diversity of the swine gut microbiome. Several species were particularly well represented, including Collinsella sp002391315 (22 MAGs), Sodaliphilus sp004557565 (19 MAGs), Prevotella sp000434975 (17 MAGs), and UBA3388 sp004551865 (13 MAGs). Notably, 938 MAGs were assigned to bacterial or archaeal species that currently lack cultured representatives, underscoring the value of metagenomic approaches for revealing previously inaccessible members of the pig gut ecosystem.

Analysis of carbohydrate-degrading potential showed that many of the MAGs encoding the highest numbers of CAZymes and CAZyme families belonged to the Bacteroidaceae family. Prominent species included *Bacteroides fragilis*, *B. heparinolyticus*, *B. stercoris*, *B. thetaiotaomicron*, *B. uniformis*, *B. xylanisolvens*, and *Phocaeicola vulgatus*, all of which are well known for their capacity to degrade a wide range of plant-derived polysaccharides. CAZymes involved in the degradation of arabinose oligosaccharides and mixed-linkage glucans were present in at least 50% of the MAGs. In addition, enzymes predicted to participate in the digestion of amorphous cellulose, arabinan, β-mannan, fucose-containing oligosaccharides, pectin, rhamnose oligosaccharides, starch, xylan, and xyloglucan were encoded by 25% or more of the MAGs. In contrast, mucin-degrading CAZymes were relatively rare, being identified in only 26 MAGs, four of which were classified as *Pauljensenia hyovaginalis* and three as *Tractidigestivibacter* sp004557505.

The functional analysis further revealed a broad potential for SCFA production among the reconstructed genomes. Genes associated with butyrate synthesis via either the butyryl-CoA:acetate CoA-transferase pathway (K01034, K01035) or the butyrate kinase pathway (K00634, K00929) were detected in 18.8% of the MAGs. These included several well-established butyrate-producing species, such as *Anaerostipes hadrus*,

Butyricimonas virosa, *Butyrivibrio crossotus*, *Cloacibacillus porcorum*, *Coprococcus catus*, *Gemmiger formicilis*, *Faecalibacterium prausnitzii*, *Flavonifractor plautii*, and *Megasphaera elsdenii*. Because butyrate is a key energy source for colonocytes and plays an important role in maintaining gut health, the presence of these organisms highlights the functional importance of the pig gut microbiome.

In addition to butyrate, the study identified substantial genetic potential for the production of other fermentation intermediates and SCFAs. Succinate, an important precursor for propionate synthesis, can be produced by several bacterial taxa. Genes encoding fumarate reductase/succinate hydrogenase (K00239, K00240, K00241, K00244, K00245, and K00246) were present in 28.9% of the MAGs. Among the known succinate-producing species carrying these genes were *Akkermansia muciniphila*, *Anaerobiospirillum succiniciproducens*, *Mitsuokella jalaludinii*, *Parabacteroides distasonis*, and *P. vulgatus*. Furthermore, genes encoding lactate dehydrogenase (K00016, K03778), enabling the production of either D-lactate or L-lactate, were detected in 55.3% of the MAGs. A total of 172 MAGs carried genes for the production of both lactate enantiomers, and more than half of these belonged to the Treponema genus or to members of the Lachnospiraceae and Lactobacillaceae families. Together, these findings illustrate the extensive metabolic versatility of the pig gut microbiome and its central role in converting dietary polysaccharides into SCFAs that support host nutrition and gut health.

5.13.2 Pre- versus Postweaning Changes in the Pig Gut Microbiome

Because the MAGs were reconstructed from fecal samples collected both immediately before weaning and seven days after weaning, Holman et al. (2022) were able to examine how the gut microbiome shifts during this critical developmental transition. Weaning represents a major physiological and nutritional change for piglets, as they move from a milk-based diet to solid, plant-based feed. This transition is accompanied by marked alterations in the composition and functional potential of the gut microbial community, which can be detected as changes in the relative abundance of specific MAGs.

A total of 69 MAGs were significantly more abundant in fecal samples collected just prior to weaning. The most strongly enriched preweaning MAGs were classified as *Limousia pullorum*, *Bacteroides fragilis*, *Escherichia coli*, *Pauljensenia hyovaginalis* (referred to as *Schaalia hyovaginalis* in the NCBI taxonomy), and *Phocaeicola vulgatus*. These taxa are commonly associated with milk-fed or early-life gut environments and may be adapted to utilizing host-derived substrates, such as milk components or mucin-associated carbohydrates. Notably, six MAGs with a relative abundance exceeding 0.1% in nursing piglets were completely absent from samples collected from the same animals seven days after weaning. These included *Bacteroides thetaiotaomicron*, *Bulleidia* sp., *Enterococcus faecalis*, *Mediterraneibacter torques*, *Parvimonas* sp., and *P. hyovaginalis*, indicating that certain early-life microbes may be rapidly lost following dietary and environmental changes associated with weaning.

In contrast, a much larger set of MAGs became enriched after weaning, reflecting adaptation to the new, fiber-rich solid diet. Among the 140 MAGs that showed the greatest relative abundance in postweaning samples were genomes assigned to *Copromorpha* sp., *Clostridium* sp000435835, *Fusicatenibacter saccharivorans*, *Intestinibacter* sp., *Oliverpabstia intestinalis*, *Phascolarctobacterium* sp004558595, *Prevotella* sp002251295, *Prevotella* sp004556065, and *Ruminococcus* sp003011855. Many of these taxa are known for their capacity to degrade complex plant polysaccharides and to produce short-chain fatty acids, suggesting a functional reorganization of the microbiome toward efficient utilization of dietary fiber. Collectively, these findings demonstrate that weaning drives rapid and pronounced shifts in the pig gut microbiome, with the loss of milk-adapted microbes and the emergence of taxa specialized for plant-based carbohydrate fermentation.

5.14 CHICKEN MICROBIOME

The chicken (*Gallus gallus domesticus*) is one of the most widely reared and economically important domesticated animals in the world. Chickens are primarily raised for meat and egg production, although many breeds are also kept for ornamental, cultural, and research purposes. As an affordable and high-quality source of animal protein, chicken meat and eggs contribute significantly to global food security and human nutrition, being consumed across diverse cultures and dietary traditions (Mariadassou et al., 2021).

Chickens were domesticated approximately 8,000 years ago in Southeast Asia, a region recognized as the center of origin for modern domestic fowl. Within the genus Gallus, four wild species have been identified: the red junglefowl (*Gallus gallus*), which is considered the primary ancestor of domestic chickens; the gray junglefowl (*Gallus sonneratii*); the Sri Lankan or Lafayette's junglefowl (*Gallus lafayettii*); and the green junglefowl (*Gallus varius*). Genetic and archaeological evidence suggests that domestication involved complex interactions among these wild species, with the red junglefowl playing a dominant role in the ancestry of modern chickens (Mariadassou et al., 2021).

Over thousands of years of domestication and selective breeding, chickens have undergone remarkable diversification. Human-driven selection for traits such as rapid growth, high egg yield, disease resistance, plumage characteristics, and environmental adaptability has resulted in the development of an extraordinary range of chicken breeds. Today, more than one thousand local and indigenous chicken breeds are recognized worldwide, reflecting adaptation to local climates, production systems, and cultural preferences (Mariadassou et al., 2021). This extensive genetic diversity represents an important resource for sustainable poultry production, conservation, and future breeding programs.

The gut microbiota plays a fundamental role in host physiology by regulating digestion, modulating immune responses, and supporting overall health. In recent years, growing attention has been directed toward the chicken gut microbiota due to its critical importance in poultry health, nutrition, and production efficiency. A well-balanced gut microbial community contributes to improved feed utilization, growth performance, and disease resistance, making it a central focus of modern poultry research (Cheng et al., 2025).

Understanding the complex interactions within the chicken gut microbiome is essential not only for optimizing poultry productivity and improving food safety, but also for enhancing host performance and reducing the risk of zoonotic pathogen transmission. The composition and structure of the gut microbiota differ markedly along various segments of the GIT, reflecting differences in pH, nutrient availability, and oxygen levels. Moreover, microbial communities are shaped by multiple factors, including the age and genetic background of the host, diet, management practices, and environmental conditions. Despite considerable spatial and temporal variability, a relatively stable core microbiota is consistently maintained in the chicken gut. This core community consists mainly of commensal bacteria that play key roles in maintaining gut integrity, supporting immune homeostasis, and protecting against pathogenic colonization. The persistence of this core microbiota highlights its importance in sustaining gut health and overall physiological balance in chickens (Cheng et al., 2025).

Burrows et al. (2025) characterized the overall and core GIT microbiota of broiler chickens and evaluated the effects of GIT location, breed, age, and geographical origin on microbial diversity. Across all samples, the microbiota was dominated by Firmicutes, Proteobacteria, and Bacteroidetes, accounting for 80.62%, 7.89%, and 5.91% of total reads, respectively. At the family level, Ruminococcaceae (23.10%), Lactobacillaceae (16.53%), and Lachnospiraceae (10.80%) were most abundant, while Enterobacteriaceae, Oscillospiraceae, and Clostridiaceae each contributed between 2% and 5% of total reads. The dominant genera included Lactobacillus (16.40%), Faecalibacterium (5.10%), Eisenbergiella (3.30%), and Oscillibacter (2.74%), alongside several other genera present at >1% abundance, such as Streptococcus, Bacteroides, Butyricicoccus, and Enterococcus.

Microbial composition varied markedly along the GIT. Cecal samples were enriched in Faecalibacterium, Eisenbergiella, and Oscillibacter, whereas fecal samples were dominated by Lactobacillus, followed by Streptococcus and Enterococcus, indicating limited similarity between cecal and fecal microbiota. Ileal samples were characterized by high abundances of Lactobacillus, Candidatus Arthromitus, and Faecalibacterium, while jejunal samples showed increased proportions of Faecalibacterium and Stenotrophomonas. These findings highlight strong spatial heterogeneity in microbial communities along the chicken GIT.

Age-related shifts in microbial composition were pronounced. During week 1, Lactobacillus strongly dominated, followed by Streptococcus and Eisenbergiella. In week 2, Eisenbergiella and Oscillibacter became more prominent, with intermittent dominance of Lactobacillus. By weeks 3 and 4, the microbiota transitioned toward a more diverse community dominated by Eisenbergiella, Oscillibacter, and Faecalibacterium, with

Lactobacillus becoming less consistently abundant. From week 5 onward, Lactobacillus was no longer a dominant genus, and the community was increasingly composed of Faecalibacterium, Bacteroides, Eisenbergiella, and Oscillibacter. In later stages, particularly after week 6, Bacteroides, Megamonas, and Parabacteroides became dominant, with additional contributions from genera such as Alistipes, Methanocorpusculum, Helicobacter, and Campylobacter. Overall, the study demonstrates a clear age-dependent shift from a Lactobacillus-dominated microbiota in early life toward a more diverse and complex microbial community as broiler chickens mature (Burrows et al., 2025).

5.14.1 Microbiome of Egg-Laying Hen

Eggs are a globally important food source and are central to efficient poultry production. Improving egg safety and laying performance requires a clear understanding of the microbial ecology of the hen reproductive tract. In this context, Ellwood et al. (2025) characterized the microbiome of egg-laying chickens and examined age-related changes in microbial composition and their associations with reproductive performance.

The study identified several bacterial genera known to influence steroid hormone metabolism, including Acinetobacter, Ligilactobacillus, Bacillus, and Akkermansia. Age-related shifts in the abundance of these taxa suggest potential impacts on hormone-regulated reproductive processes. In addition, genera not traditionally linked to reproductive hormone pathways—such as Ruminococcus_torques_group, Mucispirillum, and Fusobacterium—may affect laying efficiency through mechanisms involving mucin degradation, immune modulation, and inflammatory responses. Notably, Turicibacter, newly detected in the ovary, increased in abundance with age and showed a negative correlation with laying performance, highlighting a possible role in bile acid metabolism and stress responses within the reproductive tract.

Across ovarian samples, the 15 most abundant genera included Alistipes, Bacteroides, Clostridia_UCG-014, Clostridia_vadinBB60_group, Faecalibacterium, Fusobacterium, Gastranaerophilales, Lactobacillus, Limosilactobacillus, Mucispirillum, Rikenellaceae_RC9_gut_group, Ruminococcus_torques_group, and Staphylococcus. Eubacterium_ventriosum_group was significantly less abundant in 18-month-old hens compared with 9-month-old hens, whereas Turicibacter, Ligilactobacillus, Cerasicoccus, Candidatus_Saccharimonas, CHKCI001, and Ruminococcus_torques_group were more abundant in older hens. After false discovery rate (FDR) correction, only Turicibacter remained significantly different. Spearman correlation analysis revealed moderate to strong negative associations between laying performance and several ovarian genera, including Ligilactobacillus, CHKCI001, Ruminococcus_torques_group, Turicibacter, UCG-008, and Cerasicoccus.

When the entire reproductive tract was considered, 16 genera showed significant differential abundance between 9- and 18-month-old hens. Genera that significantly decreased with age included Akkermansia, Bacillus, Prevotellaceae_UCG-001, Megasphaera, Fusobacterium, and Mucispirillum. In contrast, Ligilactobacillus, Aeriscardovia, CHKCI001, Acinetobacter, F082, Candidatus_Saccharimonas, UCG-008, NK4A214_group, and Ruminococcus_torques_group increased significantly with age. Correlation analyses showed positive associations between laying performance and Fusobacterium and Megasphaera, whereas Ligilactobacillus, CHKCI001, NK4A214_group, and Ruminococcus_torques_group were negatively associated with egg-laying performance. Collectively, these findings highlight the dynamic nature of the reproductive tract microbiome and its potential influence on age-related declines in laying efficiency (Ellwood et al., 2025).

5.15 CONCLUSION

The livestock microbiome has emerged as a central biological system governing animal health, productivity, and sustainability in modern agricultural systems. Across ruminants, monogastric animals, and hindgut fermenters, microbial communities function as essential metabolic and immunological partners, enabling efficient feed utilization, supporting immune development, and enhancing resilience to disease and environmental stressors. These host–microbe interactions influence not only individual animal performance but also herd-level productivity and the environmental footprint of livestock production. Comparative studies highlight both conserved and species-specific microbial functions that reflect evolutionary adaptations to diet, physiology, and management, underscoring the microbiome's pivotal role in nutrient conversion, gut integrity, and energy metabolism.

Recent advances in HTS, multiomics approaches, and systems biology have substantially improved our understanding of microbiome structure–function relationships and their implications for livestock performance. This growing body of knowledge is opening new avenues for microbiome-informed interventions, including dietary modulation, probiotics and prebiotics, selective breeding, and precision livestock management. Nevertheless, effective translation of these insights into practical, farm-level solutions requires integrated approaches that consider host genetics, nutrition, management practices, and environmental conditions. Ultimately, strategically harnessing the livestock microbiome offers a powerful and sustainable pathway to improve productivity, reduce environmental impacts, and strengthen the resilience of global livestock systems in the face of increasing food demands and environmental challenges.

ACKNOWLEDGMENTS

The author (ST) acknowledges Sharpe-Strumia Research Foundation, W.W. Smith Charitable Trust, and Abraham Thomas Foundation for their generous support in providing the resources that made this work possible.

REFERENCES

Adesogan AT, Mckune SL, Serra R, Miller LC, Bamikole MA, Andrade Laborde JE, Dubeux JCB Jr, Adeoti TM. Animal board invited review: Benefits of livestock and animal-source foods in developing countries. *Animal*. 2025 Nov 19;20:101722.

Al Jassim R. Foregut microbiology of the Arabian camel (Camelus dromedarius). *Anim Front*. 2022 Aug 12;12(4):46–51.

Alary V, Frija A, Abdeladhim M, Sghaier M, Leauthaud C, Farhat M, et al. Manure contribution to rural livelihoods at farm and landscape levels: A systemic approach in semi-arid Central Tunisia. *Agroecol Sustain Food Syst*. 2025 Oct 21;49(9):1489–517.

Alders RG, Campbell A, Costa R, Guèye EF, Ahasanul Hoque M, Perezgrovas-Garza R, et al. Livestock across the world: Diverse animal species with complex roles in human societies and ecosystem services. *Anim Front*. 2021 Oct 1;11(5):20–9.

Amarlapudi MR, Balasubramaniam C, Akash, Singh H, Yadav A, Hirikyathanahalli Vishweswaraiah R, Pradhan D, Kumar N, Dhotre D, Kolte AP, Hogarehalli Mallappa R. Microbiome and antibiotic resistance profile of milk and faeces from cattle in an organized dairy production system. *Int J Antimicrob Agents*. 2025 Dec;66(6):107590.

Argudín MA, Deplano A, Meghraoui A, Dodémont M, Heinrichs A, Denis O, et al. Bacteria from animals as a pool of antimicrobial resistance genes. *Antibiotics (Basel)*. 2017 June 6;6(2):12.

Arntzen MØ, Várnai A, Mackie RI, Eijsink VGH, Pope PB. Outer membrane vesicles from fibrobacter succinogenes S85 contain an array of carbohydrate-active enzymes with versatile polysaccharide-degrading capacity. *Environ Microbiol*. 2017 July;19(7):2701–14.

Banda LJ, Tanganyika J. Livestock provide more than food in smallholder production systems of developing countries. *Anim Front*. 2021 Mar 1;11(2):7–14.

Berg Miller ME, Antonopoulos DA, Rincon MT, Band M, Bari A, Akraiko T, et al. Diversity and strain specificity of plant cell wall degrading enzymes revealed by the draft genome of *Ruminococcus flavefaciens* FD-1. *PLoS One*. 2009 Aug 14;4(8):e6650.

Betancur-Murillo CL, Aguilar-Marín SB, Jovel J. Prevotella: A key player in ruminal metabolism. *Microorganisms*. 2022 Dec 20;11(1):1.

Bikker P, Jansman AJM. Review: Composition and utilisation of feed by monogastric animals in the context of circular food production systems. *Animal*. 2023 July;17:100892.

Bilotto F, Harrison MT, Vibart R, Mackay A, Christie-Whitehead KM, Ferreira CSS, et al. Towards resilient, inclusive, sustainable livestock farming systems. *Trends Food Sci Technol*. 2024 Oct 1;152:104668.

Bonfanti A, Silvestri R, Novellino E, Tenore GC, Schiano E, Iannuzzo F, et al. Microbial profiling of buffalo mozzarella whey and ricotta exhausted whey: Insights into potential probiotic subdominant strains. *Microorganisms*. 2025 Aug;13(8):1804.

Bordenstein SR, Theis KR. Host biology in light of the microbiome: Ten principles of holobionts and hologenomes. *PLoS Biol*. 2015 Aug 18;13(8):e1002226.

Boughey I, Samsing F, Hall E, Rodney R, Bush R. Characterisation of the faecal microbiome of alpacas raised in South Eastern Australia. *Animals (Basel)*. 2025 Jun 13;15(12):1748.

Bule P, Alves VD, Israeli-Ruimy V, Carvalho AL, Ferreira LMA, Smith SP, et al. Assembly of *Ruminococcus flavefaciens* cellulosome revealed by structures of two cohesin-dockerin complexes. *Sci Rep*. 2017 Apr 7;7:759.

Burgos RA, Manosalva C, Alarcón P, Navarro M, Quiroga J, Morán G, et al. The D-lactate enigma: Exploring the inflammatory influence of D-lactate in cattle. *Front Vet Sci*. 2024 Dec 18;11.

Burnet MC, Dohnalkova AC, Neumann AP, Lipton MS, Smith RD, Suen G, et al. Evaluating models of cellulose degradation by *Fibrobacter succinogenes* S85. *PLOS One*. 2015 Dec 2;10(12):e0143809.

Burrows PB, Godoy-Santos F, Lawther K, Richmond A, Corcionivoschi N, Huws SA. Decoding the chicken gastrointestinal microbiome. *BMC Microbiol*. 2025 Jan 20;25(1):35.

Callaway T, Perez HG, Corcionivoschi N, Bu D, Fluharty FL. International symposium on ruminant physiology: The holobiont concept in ruminant physiology—More of the same, or something new and meaningful to food quality, food security, and animal health?. *J Dairy Sci*. 2025 July 1;108(7):7567–75.

Cao Y, Feng T, Wu Y, Xu Y, Du L, Wang T, Luo Y, Wang Y, Li Z, Xuan Z, Chen S, Yao N, Gao NL, Xiao Q, Huang K, Wang X, Cui K, Rehman SU, Tang X, Liu D, Han H, Li Y, Chen WH, Liu Q. The multi-kingdom microbiome of the goat gastrointestinal tract. *Microbiome*. 2023 Oct 2;11(1):219.

Cardoso ADS, Barbero RP, Romanzini EP, Teobaldo RW, Ongaratto F, Fernandes MHMDR, et al. Intensification: A key strategy to achieve great animal and environmental beef cattle production sustainability in Brachiaria grasslands. *Sustainability*. 2020 Aug 18;12(16):6656.

Carroll C, Olsen KD, Ricks NJ, Dill-McFarland KA, Suen G, Robinson TF, Chaston JM. Bacterial communities in the alpaca gastrointestinal tract vary with diet and body site. *Front Microbiol*. 2019 Jan 18;9:3334.

Caspani G, Swann J. Small talk: Microbial metabolites involved in the signaling from microbiota to brain. *Curr Opin Pharmacol*. 2019 Oct 1;48:99–106.

Chang J, Yao X, Zuo C, Qi Y, Chen D, Ma W. The gut bacterial diversity of sheep associated with different breeds in Qinghai province. *BMC Vet Res*. 2020 Jul 23;16(1):254.

Chen XL, Wang JK, Wu YM, Liu JX. Effects of chemical treatments of rice straw on rumen fermentation characteristics, fibrolytic enzyme activities and populations of liquid- and solid-associated ruminal microbes in vitro. *Anim Feed Sci Technol*. 2008 Mar 1;141(1):1–14.

Cheng YC, Krieger M, Korves AM, Camarinha-Silva A. The chicken gut microbiome in conventional and alternative production systems. *J Anim Sci Biotechnol*. 2025 Nov 19;16(1):153.

Chiariotti A, Borghese A, Boselli C, Barile VL. Water buffalo's adaptability to different environments and farming systems: A review. *Animals (Basel)*. 2025 May 24;15(11):1538.

Cucchi T, Arbuckle B. Animal domestication: From distant past to current development and issues. *Anim Front*. 2021 June 19;11(3):6–9.

Duin EC, Wagner T, Shima S, Prakash D, Cronin B, Yáñez-Ruiz DR, et al. Mode of action uncovered for the specific reduction of methane emissions from ruminants by the small molecule 3-nitrooxypropanol. *Proc Natl Acad Sci*. 2016 May 31;113(22):6172–7.

Ellwood KM, Kramer AE, Dutta A. Age-driven changes in the layer hen reproductive microbiome are associated with lay performance. *Poult Sci*. 2025 Nov 1;104(11):105703.

Ezeilo UR, Zakaria II, Huyop F, Wahab RA. Enzymatic breakdown of lignocellulosic biomass: The role of glycosyl hydrolases and lytic polysaccharide monooxygenases. *Biotechnol Biotechnol Equip*. 2017 May 16;31:1–16.

Ezer A, Matalon E, Jindou S, Borovok I, Atamna N, Yu Z, et al. Cell surface enzyme attachment is mediated by Family 37 carbohydrate-binding modules, unique to *Ruminococcus albus*. *J Bacteriol*. 2008 Dec 15;1S90(24):8220–2.

Farzi A, Fröhlich EE, Holzer P. Gut microbiota and the neuroendocrine system. *Neurotherapeutics*. 2018 Jan;15(1):5–22.

Figueroa D, Corredor FA, Mamani-Cato RH, Gallegos-Acero RF, Condori-Rojas N, Estrada R, Heredia L, Salazar W, Quilcate C, Arbizu CI. Microsatellite-based genetic diversity and population structure of Huacaya alpacas (Vicugna pacos) in Southern Peru. *Animals (Basel)*. 2023 May 5;13(9):1552.

George. *The Relationship between the Equine Gut and Brain*. Succeed Vet – New. 2023.

Ghali, MB, Scott PT, Alhadrami GA, and Al Jassim RAM. Identification and characterisation of the predominant lactic acid producing and utilising bacteria in the foregut of the feral camel (Camelus dromedarius) in Australia. *Anim. Prod. Sci*. 2011;51:597–604.

Gharechahi J, Vahidi MF, Sharifi G, Ariaeenejad S, Ding XZ, Han JL, et al. Lignocellulose degradation by rumen bacterial communities: New insights from metagenome analyses. *Environ Res*. 2023 July 15;229:115925.

Gitungwa H, Gustafson CR, Jimenez EY, Peterson EW, Mwanzalila M, Makweta A, et al. Female and male-controlled livestock holdings impact pastoralist food security and women's dietary diversity. *One Health Outlook*. 2021 Dec;3(1):3.

Godber OF, Wall R. Livestock and food security: Vulnerability to population growth and climate change. *Glob Chang Biol*. 2014 Oct;20(10):3092–102.

Gossett KA, Cleghorn B, Adams R, Church GE, McCoy DJ, Carakostas MC, et al. Contribution of whole blood L-lactate, pyruvate, D-lactate, acetoacetate, and 3-hydroxybutyrate concentrations to the plasma anion gap in horses with intestinal disorders. *Am J Vet Res*. 1987 Jan;48(1):72–5.

Harlow BE, Lawrence LM, Hayes SH, Crum A, Flythe MD. Effect of dietary starch source and concentration on equine fecal microbiota. *PLoS One*. 2016 Apr 29;11(4):e0154037.

Harmeyer J, Martens H. Aspects of urea metabolism in ruminants with reference to the goat. *J Dairy Sci*. 1980 Oct;63(10):1707–28.

Holman DB, Kommadath A, Tingley JP, Abbott DW, 2022. Novel insights into the pig gut microbiome using metagenome-assembled genomes. *Microbiol Spectr* 10:e02380-22.

Hu J, Chen J, Ma L, Hou Q, Zhang Y, Kong X, Huang X, Tang Z, Wei H, Wang X, Yan X. Characterizing core microbiota and regulatory functions of the pig gut microbiome. *ISME J*. 2024 Jan 8;18(1):wrad037.

Idamokoro EM. The relevance of livestock husbandry in the context of food security: A bibliometric outlook of research studies from 1938 to 2020. *Front. Sustain. Food Syst*. 2023; 7:1204221.

Irving J, Pineau V, Shultz S, Ter Woort F, Julien F, Lambey S, et al. Impact of low-starch dietary modifications on faecal microbiota composition and gastric disease scores in performance horses. *Animals*. 2025 June 28;15(13):1908.

Jandhyala SM, Talukdar R, Subramanyam C, Vuyyuru H, Sasikala M, Reddy DN. Role of the normal gut microbiota. *World J Gastroenterol*. 2015 Aug 7;21(29):8787–803.

Jensen N, Barrett C, Mude A, Banerjee R, Noritomo Y, Shikuku K, et al. Index-based Livestock insurance to support pastoralists against droughts. *Food Policy*. 2025 Nov 1;137:102909.

Kassahun D, Tilahun M, Tesema Z, Meseret M, Lakew E, Tesfa A, et al. Climate-resilient genetic selection in cattle: Insights from resilience phenotypes of Fogera cattle under climate variabilities. *Vet Anim Sci*. 2025 Dec 1;30:100538.

Khan FA, Pandupuspitasari NS, Huang C, Negara W, Ahmed B, Putri EM, Lestari P, Priyatno TP, Prima A, Restitrisnani V, Surachman M, Akhadiarto S, Darmawan IWA, Wahyuni DS, Herdis H. Unlocking gut microbiota potential of dairy cows in varied environmental conditions using shotgun metagenomic approach. *BMC Microbiol*. 2023 Nov 16;23(1):344.

Khongdee T, Sripoon S, Vajrabukka C. The effects of high temperature and wallow on physiological responses of swamp buffaloes (Bubalus bubalis) during winter season in Thailand. *J Therm Biol*. 2011 Oct 1;36(7):417–21.

Kondaiah N. Buffalo meat production and export. In: Chatli MK, Kumar PSGH, Singh I, editors. *Buffalo (Bubalus bubalis) Meat: Production, Processing and Quality*. Singapore: Springer Nature; 2025. pp. 33–68.

Lee JH, Kim S, Kim ES, Keum GB, Doo H, Kwak J, Pandey S, Cho JH, Ryu S, Song M, Cho JH, Kim S, Kim HB. Comparative analysis of the pig gut microbiome associated with the pig growth performance. *J Anim Sci Technol*. 2023 Jul;65(4):856–864.

Li C, Li X, Liu K, Xu J, Yu J, Liu Z, Mach N, Ni W, Liu C, Zhou P, Wang L, Hu S. Multiomic analysis of different horse breeds reveals that gut microbial butyrate enhances racehorse athletic performance. *NPJ Biofilms Microbiomes*. 2025 May 24;11(1):87.

Li M, Zhong H, Li M, Zheng N, Wang J, Zhao S. Contribution of ruminal bacteriome to the individual variation of nitrogen utilization efficiency of dairy cows. *Front Microbiol*. 2022 Mar 18;13:815225.

Lin J, Zeng C, Li X, Tang Q, Liao J, Jiang Y, Zeng X. Microorganisms in the rumen and intestine of camels have the ability to degrade 2-amino-3-methylimidazo[4, 5-f]quinoline. *Food Sci Nutr*. 2024 Mar 18;12(7):4667–4679.

Lin L, Lai Z, Zhang J, Zhu W, Mao S. The gastrointestinal microbiome in dairy cattle is constrained by the deterministic driver of the region and the modified effect of diet. *Microbiome*. 2023 Jan 20;11(1):10.

Lyu Z, Shao N, Akinyemi T, Whitman WB. Methanogenesis. *Curr Biol*. 2018 July 9;28(13):R727–32.

Ma Y, Yang X, Hua G, Deng X, Xia T, Li X, Feng D, Deng X. Contribution of gut microbiomes and their metabolomes to the performance of Dorper and Tan sheep. *Front Microbiol*. 2022 Nov 28;13:1047744.

Mach N, Ruet A, Clark A, Bars-Cortina D, Ramayo-Caldas Y, Crisci E, et al. Priming for welfare: Gut microbiota is associated with equitation conditions and behavior in horse athletes. *Sci Rep*. 2020 May 20;10:8311.

Malik PK, Trivedi S, Kolte AP, Sejian V, Mech A, Biswas S, et al. Comparative assessment of rumen microbial diversity in cattle and buffaloes. In: Sejian V, Chauhan SS, Devaraj C, Malik PK, Bhatta R, editors. *Climate Change and Livestock Production: Recent Advances and Future Perspectives*. Singapore: Springer; 2021. pp. 223–38.

Maree E, Blignaut J, Gilliland J, Lee MRF, Manzano P, McCosker T, et al. Ruminant livestock farmers and industry are leading innovation to deliver human nutrition and improved environmental outcomes through sector lifecycle collaboration: A review of case studies. *Anim Front*. 2025 Feb 1;15(1):55–71.

Mariadassou M, Suez M, Sathyakumar S, Vignal A, Arca M, Nicolas P, Faraut T, Esquerré D, Nishibori M, Vieaud A, Chen CF, Manh Pham H, Roman Y, Hospital F, Zerjal T, Rognon X, Tixier-Boichard M. Unraveling the history of the genus *Gallus* through whole genome sequencing. *Mol Phylogenet Evol*. 2021 May;158:107044.

Matijašić M, Meštrović T, Čipčić Paljetak H, Perić M, Barešić A, Verbanac D. Gut Microbiota beyond bacteria—mycobiome, virome, archaeome, and eukaryotic parasites in IBD. *Int J Mol Sci*. 2020 Apr 11;21(8):2668.

Mehmet, K., Arzik, Y., Behrem, S., White, S.N., Ulas Cinar, M. Comparative genomic characterization of indigenous fat-tailed Akkaraman sheep with local and transboundary sheep breeds. *Food and Energy Secur*. 2023;13(1). Article e508.

Milinovich GJ, Trott DJ, Burrell PC, Van Eps AW, Thoefner MB, Blackall LL, et al. Changes in equine hindgut bacterial populations during oligofructose-induced laminitis. *Environ Microbiol*. 2006 May;8(5):885–98.

Minozzi G, Biscarini F, Dalla Costa E, Chincarini M, Ferri N, Palestrini C, Minero M, Mazzola S, Piccinini R, Vignola G, Cannas S. Analysis of hindgut microbiome of sheep and effect of different husbandry conditions. *Animals (Basel)*. 2020 Dec 22;11(1):4.

Molotzu MR, Cabras PA, Di Marcantonio L, Atzeni R, Macciotta NPP and Canu A Metagenomic analysis of goat feces from Ogliastra (Sardinia, Italy). *Front. Microbiomes* 2025;3:1474497.

Monteiro, H.F., Zhou, Z., Gomes, M.S. et al. Rumen and lower gut microbiomes relationship with feed efficiency and production traits throughout the lactation of Holstein dairy cows. *Sci Rep* 2022; 12, 4904.

Morgan-Davies C, Wilson R, Waterhouse T. Impacts of farmers' management styles on income and labour under alternative extensive land use scenarios. *Agric Syst*. 2017 July 1;155:168–78.

Muneer A. Kynurenine Pathway of Tryptophan Metabolism in Neuropsychiatric Disorders: Pathophysiologic and Therapeutic Considerations. *Clin Psychopharmacol Neurosci*. 2020 Nov 30;18(4):507–26.

Murovec U, Accetto T. Transcriptomic analysis of polysaccharide utilization loci reveals substrate preferences in ruminal generalists *Segatella bryantii* TF1-3 and *Xylanibacter ruminicola* KHP1. *BMC Genomics*. 2024 May 20;25(1):495.

Papakonstantinou GI, Voulgarakis N, Terzidou G, Fotos L, Giamouri E, Papatsiros VG, et al. Precision livestock farming technology: Applications and challenges of animal welfare and climate change. *Agriculture*. 2024 Apr 15;14(4): 620.

Paul SS, Dey A, Baro D, Andonissamy J, Paul J, Punia BS. Rumen microbiota in cattle and buffaloes: Insights into host-specific bacterial diversity. *Biology (Basel)*. 2025 Sept 1;14(9):1166.

Perry E, Cross TWL, Francis JM, Holscher HD, Clark SD, Swanson KS. Effect of road transport on the equine cecal microbiota. *J Equine Vet Sci*. 2018 Sept 1;68:12–20.

Pitt D, Sevane N, Nicolazzi EL, MacHugh DE, Park SDE, Colli L, Martinez R, Bruford MW, Orozco-terWengel P. Domestication of cattle: Two or three events? *Evol Appl*. 2018 Jul 23;12(1):123–136.

Pitta DW, Indugu N, Melgar A, Hristov A, Challa K, Vecchiarelli B, et al. The effect of 3-nitrooxypropanol, a potent methane inhibitor, on ruminal microbial gene expression profiles in dairy cows. *Microbiome*. 2022 Sept 13;10(1):146.

Pollitt CC. Equine laminitis. *Clin Tech Equine Pract*. 2004 Mar 1;3(1):34–44.

Quilodrán-Vega S, Albarracin L, Mansilla F, Arce L, Zhou B, Islam MA, Tomokiyo M, Al Kassaa I, Suda Y, Kitazawa H, Villena J. Functional and genomic characterization of *Ligilactobacillus salivarius* TUCO-L2 isolated from *Lama glama* milk: A promising immunobiotic strain to combat infections. *Front Microbiol*. 2020 Dec 8;11:608752.

Quinzeiro NetoT, Garcia AR; Santos J. et al. *Socioeconomical Importance of Buffalo Breeding to Small Farm Agriculture at Xingu River Floodplains*, Pará, Brazil. Revista Veterinária, 2010; v. 21, Sup. 1.

Rabee, A.E., Forster R., Elekwachi C., Sabra E., and Lamara M. Comparative analysis of the metabolically active microbial communities in the rumen of dromedary camels under different feeding systems using total rRNA sequencing. *PeerJ*. 2020; 8:e10184.

Rahmeh R, Jordamovic NB, Akbar A, Almutairi B, Kishk M, et al. Shotgun sequencing of camel milk microbiome. *Appl Food Res* 2025; 5: 101324.

Rani A, Pundir A, Verma M, Joshi S, Verma G, Andjelković S, et al. Methanotrophy: A biological method to mitigate global methane emission. *Microbiol Res*. 2024 June;15(2):634–54.

Raspa F, Chessa S, Bergero D, Sacchi P, Ferrocino I, Cocolin L, Corvaglia MR, Moretti R, Cavallini D, Valle E. Microbiota characterization throughout the digestive tract of horses fed a high-fiber vs. a high-starch diet. *Front Vet Sci*. 2024 May 14;11:1386135.

Reynolds LP, Wulster-Radcliffe MC, Aaron DK, Davis TA. Importance of animals in agricultural sustainability and food security. *J Nutr*. 2015 Jul;145(7):1377–9.

Ritchie H. How much of the world's land would we need in order to feed the global population with the average diet of a given country? *Our World in Data*. 2017 Oct 3.

Ritchie H, Rosado P, Roser M. Meat and dairy production. *Our World in Data*. 2019 Nov 1.

Ritchie H, Spooner F. Almost all of the world's mammal biomass is humans and livestock. *Our World in Data*. 2025 Dec 1.

Rivero MJ, Keim JP, Balocchi OA, Lee MRF. In vitro fermentation patterns and methane output of perennial ryegrass differing in water-soluble carbohydrate and nitrogen concentrations. *Animals*. 2020 June;10(6):1076.

Rust JM. The impact of climate change on extensive and intensive livestock production systems. *Anim Front*. 2018 Oct 26;9(1):20–5.

Stel M, Eggers J, Alonso WJ. Mitigating zoonotic risks in intensive farming: Solutions for a sustainable change. *Ecohealth*. 2022;19(3):324–8.

Tong, F., Wang, T., Gao, N.L. et al. The microbiome of the buffalo digestive tract. *Nat Commun* 2022; 13, 823.

Tseten T, Sanjorjo RA, Kwon M, Kim SW. Strategies to mitigate enteric methane emissions from ruminant animals. *J Microbiol Biotechnol*. 2022 Mar 28;32(3):269–77.

Wang D, Yin Q, Tang G, Sun L, Yu J, Cao Y, Yao J. Gut microbiome enterotypes drive divergent lactation performance in dairy goats through host-microbe metabolic cross-talk. *Microbiol Spectr*. 2025 Nov 4;13(11):e0136725.

Wang L, Wu D, Zhang Y, Li K, Wang M and Ma J. Dynamic distribution of gut microbiota in cattle at different breeds and health states. *Front. Microbiol*. 2023; 14:1113730.

Wang M, Ibeagha-Awemu EM. Impacts of epigenetic processes on the health and productivity of livestock. *Front Genet*. 2021 Feb 23;11:613636.

Wang R, Bai B, Huang Y, Degen A, Mi J, Xue Y, Hao L. Yaks are dependent on gut microbiota for survival in the environment of the Qinghai Tibet Plateau. *Microorganisms*. 2024 May 31;12(6):1122.

Waters SM, Roskam E, Smith PE, Kenny DA, Popova M, Eugène M, et al. International symposium on ruminant physiology: The role of rumen microbiome in the development of methane mitigation strategies for ruminant livestock. *J Dairy Sci*. 2025 July 1;108(7):7591–606.

Xie X, Cao Y, Li Q, Li Q, Yang X, Wang R, Zhang X, Tan Z, Lin B, Wang M. Mitigating enteric methane emissions: An overview of methanogenesis, inhibitors and future prospects. *Anim Nutr*. 2025 Feb 8;21:84–96.

Xu Q, Zhong H, Zhou J, Wu Y, Ma Z, Yang L, et al. Lignin degradation by water buffalo. *Trop Anim Health Prod*. 2021 June 5;53(3):344.

Yang Q, Lopez MJ. The equine hoof: Laminitis, progenitor (stem) cells, and therapy development. *Toxicol Pathol*. 2021 Oct 1;49(7):1294–307.

Yang, S., Zheng, J., Mao, H. et al. Multiomics of yaks reveals significant contribution of microbiome into host metabolism. *NPJ Biofilms Microbiomes* 2024; 10: 133.

Yano JM, Yu K, Donaldson GP, Shastri GG, Ann P, Ma L, et al. Indigenous bacteria from the gut microbiota regulate host serotonin biosynthesis. *Cell*. 2015 Apr 9;161(2):264–76.

Yeoman CJ, Fields CJ, Lepercq P, Ruiz P, Forano E, White BA, et al. In vivo competitions between *Fibrobacter succinogenes*, *Ruminococcus flavefaciens*, and *Ruminoccus albus* in a gnotobiotic sheep model revealed by multiomic analyses. *mBio*. 2021 Mar 3;12(2):e03533–20.

Yu G, Beauchemin KA, Dong R. A Review of 3-Nitrooxypropanol for enteric methane mitigation from ruminant livestock. *Animals (Basel)*. 2021 Dec 13;11(12):3540.

Zhang Y, Cheng J, Lin C, Li F, Zhang X, Li C, Zhang D, Yang X, Xu D, Zhao Y, Zhao L, Li X, Tian H, Weng X, Wang W. Spatial heterogeneity determines the gastrointestinal microbiome signatures and ecological processes that govern bacterial community assembly in sheep. *Microbiol Spectr*. 2025 Feb 4;13(2):e0111024.

Zhao C, Wang L, Ke S, Chen X, Kenéz Á, Xu W, Wang D, Zhang F, Li Y, Cui Z, Qiao Y, Wang J, Sun W, Zhao J, Yao J, Yu Z, Cao Y. Yak rumen microbiome elevates fiber degradation ability and alters rumen fermentation pattern to increase feed efficiency. *Anim Nutr*. 2022 Aug 9;11:201–214.

CHAPTER SIX

Birds, Microbes, and Disease

A REVIEW OF THE AVIAN MICROBIOME AND ITS IMPLICATIONS

Zi Yan, Huiyan Xu, Jinfeng Zhao, and Longxian Zhang

6.1 INTRODUCTION

Microbiome consist of bacteria, archaea, fungi, algae, and small protists. The microbiome is a key interface connecting the host with its environment and plays a central role in the host's health, adaptation and evolution. For instance, host-associated microbiomes contribute in many ways to the homeostasis of the organism (Peixoto et al., 2021). The microbiome's contributions range from helping to provide nutrition and aiding growth, development, and behavior to protecting against pathogens and toxic compounds. Notably, the fungal component (mycobiome), while less explored, constitutes a critical part of this ecosystem. Birds can harbor diverse fungi, including commensals and potential pathogens such as Aspergillus and Candida species, which may impact host health and contribute to environmental spore dispersal.

Birds possess a specialized intestinal structure, and their gut microbiota demonstrates notable instability and high plasticity, which can be attributed to their diverse living environments—such as forests, deserts, and urban areas—as well as their varied dietary lifestyles, including herbivores and omnivores (Bodawatta et al., 2021; Ruuskanen, 2024). Factors in both the external and internal environments of the host can influence the composition of the gut microbiota. For instance, birds' complex life histories and varied dietary patterns, combined with the selective pressures associated with flight, such as high energy demands and physiological adaptations can result in even more intricate changes in their gut microbiota. Other anthropogenic influences such as urbanization and domestication also play noticeable roles in manipulating microbiota assemblage of avians (Teyssier et al., 2018). Over long-term coexistence and coevolution with their avian hosts, the gut microbiota has come to play a vital role in host defense, immune regulation, digestion and absorption, vitamin synthesis, and other essential functions (Grond et al., 2018; Knutie, 2020; Kohl, 2012). On the other hand, microalternation of gut microbiota assemblage potentially impacts their various behavior and recognition abilities (Davidson et al., 2020).

Animal hosts harbor diverse and complex microbial communities on their surfaces and within their bodies—a subject of growing interest as metagenomic technologies advance our understanding of these intricate ecosystems. While animal microbiomes present valuable opportunities for discovery, such as novel enzymes, ecological insights, and probiotic candidates, they also pose significant challenges

DOI: 10.1201/9781003638568-6

to public and animal health (Benskin et al., 2009; Jourdain et al., 2007). Both domesticated and free-living birds play a significant role in the dissemination of commensal bacteria and pathogens. A particular concern is their contribution to the spread of ARGs. Birds can acquire and harbor ARGs in their gut microbiomes from environmental sources, acting as mobile reservoirs that facilitate the long-distance dispersal of resistance traits. This capacity poses a substantial threat to global health. Through long-distance migration, frequent social interactions, and proximity to human habitats, birds facilitate the widespread transmission of microbes, including those harboring ARGs, across geographical and species boundaries, thereby contributing to the emergence of infectious diseases and the escalation of antimicrobial resistance. (Ayala et al., 2019; Smith et al., 2020).

Collectively, research on the avian metagenome offers novel insights and potential strategies for the poultry industry (e.g., developing antibiotic alternatives and enhancing feed efficiency), for the conservation of endangered species (e.g., health assessment and optimized reintroduction programs), and for public health (e.g., early detection of zoonotic disease transmission and ARG dissemination). This chapter reviews existing studies on the avian gut microbiota, with a focused discussion on the composition of the gastrointestinal microbial community, factors shaping its dynamics, and the role of birds in transmitting zoonotic pathogens along with antibiotic resistance genes. Special attention is given to interactions between the gut microbiota and host health, illustrated, where appropriate, with examples from domestic chickens.

6.2 CORE GUT MICROBIOTA

6.2.1 Diversity of Gut Microbiota Composition

Birds, as flying vertebrates, have evolved a highly efficient and streamlined digestive system with distinct functional compartments. Extending from the beak to the cloaca, this system is dedicated to digesting food, absorbing nutrients, and expelling waste with minimal weight—thereby supporting their high metabolic rate and continuous energy demands for flight. Like other vertebrates, the avian digestive tract consists of multiple specialized organs. Notably, microbial communities, particularly bacteria, extensively colonize the stomach and intestines, playing essential roles in digestion and host health. Using scanning electron microscopy, researchers observed microbial populations on the duodenal, ileal, and cecal mucosal surfaces of the chicks' intestinal epithelia. The distribution of bacteria in this case, reflecting the role of them in digestion in the GI tract. The gastrointestinal microbiota of young chicks fed a corn-based diet varies by section. In the duodenum and ileum, facultative anaerobes (e.g., Streptococcus, Staphylococcus, Lactobacillus, and Escherichia coli) are the dominant flora, alongside substantial populations of obligate anaerobes like Eubacterium, Propionibacterium, Clostridium, Gemmiger, and Fusobacterium. The cecal environment, however, is almost exclusively composed of strict anaerobes, including anaerobic Gram-positive cocci, Eubacterium, Clostridium, Gemmiger, Fusobacterium, and Bacteroides (Salanitro et al., 1978). The distinct spatial distribution of gut microbiota underscores their specialized roles in the digestive process. In emus, nutrient digestion and absorption occur mainly in the small intestine, whereas the cecum acts as the primary site for plant fiber fermentation. Unlike young chicks, free-range adult emus receiving a barley-alfalfa-canola–supplemented diet exhibit lower interindividual similarity and higher individuality in gut microbial composition. Firmicutes and Proteobacteria were identified as the dominant bacterial phyla in the small intestine. The small intestinal microbiota, particularly in the jejunum, appears to contribute to the detoxification of plant materials and environmental chemicals. In contrast, the cecal microbiome contains more genes related to defense mechanisms and immune responses against enteric pathogens. Microbial digestion and fermentation are predominantly carried out in the jejunum and cecum. It is also noteworthy that while young chicks maintain a stable core microbiota with clearly dominant genera such as Lactobacillus and Streptococcus, only two core OTUs were shared across the entire small intestine of emus (Kim et al., 2023). Moreover, in Muscovy ducks, the ileum, which is primarily responsible for nutrient absorption and transport, hosts a microbial community dominated by the phyla Bacteroidetes, Firmicutes, and Proteobacteria (Z. Fu et al., 2022) (Table 6.1).

TABLE 6.1

Composition of gut microbiota succession the GI tract

Host lifestyle	Species	Sampling site	Predominant genera	Ref.
Poultry farm	Chicken and Turkey	Cecum	Clostridium, Ruminococcus, Lactobacillus, and Bacteroides	Wei et al., 2013
Poultry farm	Broiler chickens	Duodenum, Jejunum, Ileum, Colon	**Firmicutes (>60%)** Bacteroidetes, Proteobacteria, Actinobacteria, Cyanobacteria	Xiao et al., 2017
		Cecum	**Bacteroidetes (50%)** Firmicutes, Proteobacteria, Actinobacteria, Cyanobacteria	
Zoo	Birds	Intestine	Overall: Firmicutes (50.58%), Proteobacteria (29.95%) Bacteroidetes (7.82%), Fusobacteria (4.22%), Actinobacteria (4.01%), Campylobacteria (1.15%) terrestrial birds: Proteobacteria(16.94%),Fusobacteria(0.11%), Campylobacteria (0.02%) waterfowl bird: Proteobacteria(37.06%),Fusobacteriota(10.03%) ,Campilobacterota(2.75%) arboreal birds: Proteobacteria(33.55%), Fusobacteriota(0.28%), Campilobacterota(0.05%)	Wang et al., 2022
Poultry farm	Domestic duck (*Anas platyrhyncho*)	Small, Large Intestine	Firmicutes, Actinobacteria, and Bacteroidetes.	Susanti et al., 2023

6.2.2 Major Bacterial Phyla in Birds

To adapt to their specific ecological niches, birds have evolved diverse digestive tract structures. These distinct gut morphologies select for specialized gut microbial communities capable of efficiently breaking down the host's particular diet, thereby maximizing energy assimilation (Bodawatta et al., 2022). The avian gut microbiota is consistently dominated by four main bacterial phyla: Firmicutes, Bacteroidetes, Proteobacteria and Actinobacteria. This composition is observed across species with vastly different diets, from carnivores such as penguins and vultures to herbivores including kakapo, turkey, duck, goose, and emu (Waite & Taylor, 2015).

6.2.3 Firmicutes

The intestinal Firmicutes represent a vast and phylogenetically diverse phylum, many members of which remain poorly characterized due to their strict anaerobic growth requirements (Vesth et al., 2013). Known for their role in degrading insoluble dietary fibers (Berry, 2016), Firmicutes are key producers of short-chain fatty acids (SCFAs) through fermentation. These SCFAs are subsequently absorbed by the host intestinal epithelium and serve as an important energy source (Grond et al., 2018; Svihus et al., 2013). Members of the phylum Firmicutes play important roles in metabolism, digestion, and absorption of protein and other nutrients (Bernini et al., 2016; Berry, 2016). For example, Firmicutes are associated with breakdown of fatty acids, and certain members produce substantial butyrate—a health-promoting molecule known to increase insulin sensitivity, regulate energy metabolism, and enhance leptin gene expression.

On the other hand, Firmicutes include species that catabolize complex carbohydrates, polysaccharides, sugars, and fatty acids to provide an energy source for the host (Flint et al., 2008). A positive correlation has been observed between Firmicutes abundance and host mass gain and immune competence, suggesting functional conservation of this bacterial group in energy harvest and health across both mammalian and avian species (Zhao et al., 2015). This is further supported by studies showing that dietary supplementation with Firmicutes as probiotics can enhance nutrient absorption and improve metabolic efficiency in chickens (Li et al., 2015).

6.2.4 Bacteroidetes

In parallel, Bacteroidetes are also widely detected in the gastrointestinal tracts of both domesticated and wild birds, including chickens (Rychlik, 2020), turkeys (Scupham et al., 2008), geese (Xu et al., 2017), and ostriches (P. Li et al., 2020). Bacteroidetes, particularly members of the genus Bacteroides, are highly specialized commensals in the distal gut, where they play critical roles as degraders of complex carbohydrates and producers of SCFAs (Fan et al., 2023). The relationship between Bacteroidetes and their animal hosts is now recognized as mutualistic rather than commensal, as both partners experience increased fitness (Comstock, 2009). Key beneficial contributions of Bacteroidetes to host health include modulation of the immune system, such as through activation of T-cell–mediated responses (Mazmanian et al., 2008), and inhibition of gastrointestinal colonization by potential pathogens (Mazmanian, 2008). For example, in week-old chicks, high abundance of Bacteroides has been associated with reduced markers of intestinal inflammation. A major functional output of gut Bacteroidetes is the production of butyrate, a key end-product of colonic fermentation. Butyrate contributes to colonic health through its antineoplastic properties and roles in maintaining epithelial integrity (Kim & Milner, 2007). Furthermore, Bacteroidetes are integral to bile acid metabolism and the biotransformation of toxic or mutagenic compounds, thereby supporting metabolic homeostasis and detoxification (Smith et al., 2006).

6.2.5 Proteobacteria

Birds generally harbor a higher relative abundance of Proteobacteria in their gastrointestinal tracts compared to mammals and domestic chickens. This phylum encompasses a diverse range of genera, including several known opportunistic pathogens such as Escherichia, Helicobacter, Salmonella, and Vibrio, all of which have been isolated from various avian hosts (Antunes et al., 2016; Ayala & Ogbunugafor, 2023; Ewers et al., 2003; Samir & Zaher, 2025). Beyond their pathogenic potential, certain Proteobacteria are also implicated in functional roles such as cellulose degradation (Fan et al., 2023). Additionally, a substantial presence of Fusobacteria is frequently observed in the gut microbiota of carnivorous and omnivorous birds (Waite & Taylor, 2014).

6.3 IMPACT OF DIFFERENT FACTORS ON THE AVIAN GUT MICROBIOTA

6.3.1 Intrinsic Factors

6.3.1.1 Gender

In domestic ducks, the duodenal Chao index at the genus level was significantly higher in male ducks than in females, indicating greater microbial richness in the duodenum of males at 10-week age. Although Firmicutes, Proteobacteria, Bacteroidetes, and Actinobacteria constituted the consistent core microbiota across developmental stages in domestic ducks, distinct sex-based differences emerged at finer taxonomic and functional levels. The jejunal microbiota composition shows significant difference. Male ducks exhibited significantly higher abundances of Escherichia-Shigella, Pseudomonas, *Clostridium sensu stricto_1*, Sphingomonas, and Desulfovibrio, whereas females showed a notably greater abundance of *Rothia* (X. Chen et al., 2021).

6.3.1.2 Mating Behavior

In avian species, sexual transmission is a significant pathway for disease dissemination, though its mechanisms in wild populations remain poorly characterized. The cloaca, a common orifice for excretion and reproduction, acts as a critical site for this exchange, serving as a reservoir that allows both gastrointestinal pathogens and endosymbionts to integrate into ejaculates during copulation. The influential role of sexual transmission on cloacal microbiota is supported by experimental evidence. For instance, suppressing insemination leads to progressively divergent microbial communities between mates. Females in such pairs exhibit reduced cloacal diversity and a decline in bacteria shared with their partners, indicating a reversion to a precopulatory state. This resilience underscores the concept of the cloaca as a stable microbial ecosystem (White et al., 2010). Furthermore, theory suggests that sexual transmission may not be solely negative; if the probability of transmitting beneficial microbes exceeds that of pathogens, and their positive effects on host fitness outweigh the negatives, this dynamic could encourage increased copulatory rates (Lombardo et al., 1999).

The composition of the cloacal microbiota is also shaped by sex-specific factors, particularly during the breeding season. A study of free-living rufous-collared sparrows (*Zonotrichia capensis*) revealed that males exhibited a significant increase in bacterial diversity at the onset of reproduction, followed by a decrease as they transitioned to a nonbreeding condition. In contrast, females showed no significant changes in microbial diversity across reproductive stages. These findings demonstrate that sex-linked physiological or behavioral factors selectively shape microbiota assembly in wild bird populations (Escallón et al., 2019).

In female Gentoo penguins, the gut microbiota undergoes a noticeable reconstruction from egg-laying until chicks leave the nest. This period is marked by a continuous decrease in microbial diversity, characterized by a rise in Fusobacteria and pathogenicity and a decline in Bacteroidetes and genes related to carbohydrate metabolism (Tian et al., 2024). Conversely, a study of African penguins revealed that more productive females—those with higher egg production—exhibited a higher relative abundance of potentially beneficial bacteria like Alphaproteobacteria, Rhizobiales, and Bradyrhizobium. In contrast, less productive penguins showed an increased proportion of opportunistic pathogens such as Klebsiella and Plesiomonas (Jiang et al., 2024).

Together, these studies demonstrate that reproductive processes profoundly influence the avian microbiota, but the specific effects are species-specific and may be indicative of the individual's health and reproductive success.

6.3.1.3 Development Stage

Development stage is a key determinant of microbial community differences between nestlings and adult birds. Multiple studies have demonstrated significant age-dependent differentiation in gastrointestinal microbiota across both domesticated (J. Gong et al., 2008) and wild bird species (Y. Liu et al., 2021). Specifically, house sparrow nestlings exhibit a characteristic microbial succession pattern during development: around postnatal day 9, the relative abundance of Proteobacteria in the intestinal lumen begins to decline significantly, while Firmicutes increase correspondingly (Kohl et al., 2019). The gut microbiota at this developmental stage displays two notable characteristics: first, overall diversity remains at relatively low levels; second, the community composition is dominated by transient opportunistic species that gradually transition toward a more stable adult-type microbiota during development. Remarkably, despite sharing identical nest environments with their parents and having only food source through regurgitation feeding, nestlings and adults show exceptionally low operational taxonomic unit (OTU) overlap (van Dongen et al., 2013). Beyond the gut ecosystem, similar age-dependent microbial patterns have been documented at three distinct respiratory sites (nasal cavity, trachea, and lungs) in domestic chickens across developmental stages (2 days, 3 weeks, and 30 months), indicating systemic age-driven microbial assembly patterns (Glendinning et al., 2017). Similarly, samples of domestic duck from the duodenum, jejunum, and ileum were all clustered into two distinct phases, 2–4 and 6–10 weeks, based on developmental time, indicating a stage-specific shift in the gut microbiota structure (Zhu et al., 2020). This phenomenon suggests that host developmental stages may exert stronger selective pressure on microbiota assembly than environmental factors.

6.3.1.4 Migration

The long-distance migrations of wild birds may exert substantial selective pressures on their gut microbiota composition, subjecting them to diverse environmental influences and fostering distinct microbial community architectures (Li et al., 2024). Furthermore, environmental influences during migration may drive species-specific variation in gut microbiota composition across migratory populations, suggesting the microbiota's critical role in modulating immune responses.

Comparative analyses demonstrate significant divergence in microbial profiles between migratory and resident populations of barn swallows (*Hirundo rustica*), even when these populations are conspecific and sympatric (Turjeman et al., 2020). Investigations of Blackpoll warblers (*Setophaga striata*) further reveal dynamic alterations in gut microbiota across migratory phases (Trevelline et al., 2023). Notably, these studies document a progressive decline in microbiome phylogenetic diversity and reduced within-group variation throughout migration, suggesting microbial community convergence among individuals during prolonged migratory periods. This phenomenon was primarily attributable to a marked elevation in Proteobacteria abundance, particularly a single amplicon sequence variant (ASV) from the Enterobacteriaceae family. Collectively, these findings demonstrate that migratory behavior drives specific gut microbial assemblage patterns, likely reflecting adaptive responses to distinct physiological and metabolic requirements across different life-history stages. The study on red-necked stint (*Calidris ruficollis*) shows limited environmental effects on microbiota composition due to constrained microbial acquisition from its surroundings through foraging behavior (Risely et al., 2017). Specifically, the study found only marginal compositional differences between stint flocks occupying ecologically distinct sites, with migrating populations demonstrating comparable microbial diversity and taxonomic structure to year-round residents at the same location. This pattern not only indicates resistance of the gut microbiota to invasion from ingested microbes but also highlights its remarkable stability despite environmental variation. As ecological transitions during migration consistently expose avian hosts to novel microbial communities, they potentially increase susceptibility to enteric infections while simultaneously creating opportunities for cross-geographic pathogen transmission (Altizer et al., 2011; Leung & Koprivnikar, 2016).

Consequently, given these dual pressures of microbial exposure and disease risk, the adaptive resilience of a migratory bird's gut microbiota to newly encountered microbes may represent a critical determinant of disease resistance capacity, particularly when facing repeated environmental challenges across migratory routes. Other combined pressures, namely dietary shifts, environmental factors, and physiological stress together lead to substantial seasonal variation in the gut microbiota of migratory birds. Furthermore, migration pressure also results from different times of traveling. For instance, the gut microbiota diversity rises during the spring season that is dominated by *Clostridium_sensu_stricto_13* and Brochothrix where in the autumn migration dominated by potential pathogens such as Fusobacterium and Escherichia-Shigella (Ouyang et al., 2025).

6.3.2 Extrinsic Factors

6.3.2.1 Environment

Unlike mammals that consistently demonstrate phylogeny-dependent gut microbiome (GM) variation, avian species exhibit stronger environmental determinants of GM composition along with greater developmental plasticity. Notably, passerine gut microbiota displays remarkable plasticity that is predominantly shaped by environmental factors rather than dietary influences (Schmiedová et al., 2023). Recent studies reveal that while seasonal variation drives significant compositional shifts in tropical bird GM between dry and rainy seasons (Schmiedová et al., 2023) as well as migratory species like white-tailed eagles additionally show substantial GM fluctuations that appear mediated by combined effects of dietary shifts, environmental factors, and physiological migration stress (Ouyang et al., 2025). Furthermore, hydrological extremes exert particularly strong impacts, as severe drought conditions not only reduce bacterial diversity but also demonstrate that interannual water fluctuations affect microbial community structure more profoundly than monthly variations, suggesting adaptive responses in drought-exposed species like common cranes (Wang et al., 2024). Beyond these established patterns, habitat–thermal interactions constitute an additional influential factor shaping avian gut microbiota dynamics. Mixed-forest habitats demonstrate enhanced capacity to support winter GM diversity, while ambient temperature exerts more subtle yet statistically significant compositional effects (Liukkonen et al., 2024). Thermal conditions particularly influence host–parasite interactions through their impact on infestation susceptibility (Albert et al., 2023). Experimental evidence reveals contrasting thermal effects across species: swallow nests with elevated temperatures showed reduced parasite loads compared to control nests, whereas bluebird nests under identical thermal manipulation exhibited increased parasite abundance. These opposing outcomes suggest that climate change may generate complex, system-specific net effects on parasite fitness and host health across multihost–parasite interaction networks (Scharsack et al., 2016).

6.3.2.2 Human Behaviors

Human-induced factors, particularly urbanization, represent a primary driver of rapid habitat change globally. The process of urbanization alters food availability, intensifies environmental stressors, and can heighten disease susceptibility in wildlife. These selective pressures can drive adaptive divergence in behavioral, physiological, and morphological traits among local avian populations. Consequently, urban environments are likely to induce species-specific alterations in the gut microbiota, generating uncertainty regarding the precise impact of these habitats on the avian GM.

For instance, while some avian species may exhibit higher gut bacterial richness in urban settings, certain songbirds demonstrate a contrasting pattern. Both house sparrows (*Passer domesticus*) (Teyssier et al., 2018) and Darwin's finches (*Geospiza fuliginosa* and *Geospiza fortis*) (Knutie, 2020) display reduced bacterial richness in more urbanized areas compared to their rural counterparts. Moreover, urban populations of these birds have been found to harbor a significantly higher relative abundance of Campylobacteraceae, a bacterial family that includes known avian and zoonotic pathogens representing potential public health threat for urban residents (Berlow et al., 2021).

6.3.2.3 Food Source

Human activities, particularly through the provision of food sources, significantly influence the gut microbiota of avian species. Taxonomic analyses indicate that the gut microbiota of bar-headed geese is predominantly composed of the phyla Firmicutes, Proteobacteria, Actinobacteria, and Bacteroidetes. Notably, the relative abundance of Bacteroidetes was significantly higher in artificially reared individuals compared to their wild counterparts (W. Wang et al., 2017). Functional profiling further revealed that the GM of captive bar-headed geese was enriched for bacterial genes associated with carbohydrate transport and metabolism, energy production, and coenzyme transport and metabolism, suggesting a metabolic adaptation to an anthropogenic diet.

This shift in microbial community structure due to diet can also have pathogenic consequences. For instance, domestic mallards (*Anas platyrhynchos*) fed a different diet than wild populations exhibited a greater abundance of potential pathogenic microbes (He et al., 2023).

However, the impact of human provisioning is not uniform and appears to be moderated by rearing conditions. In farmed Swan geese (*Anser cygnoides*), the gut microbiota demonstrated a limited degree of differentiation from that of wild geese. Both groups were dominated by Firmicutes, followed by Proteobacteria, Bacteroidetes, and Actinobacteria. This minimal divergence is likely attributable to the farmed environment, which simulated natural conditions by allowing the geese to forage freely and providing a mixed diet of natural and artificial foods (W. Wang et al., 2016) (Figure 6.1).

6.4 MYCOBIOME, AVIAN FITNESS AND ZOONOTIC POTENTIAL

Birds harbor a highly complex fungal system, namely mycobiome. Similar to the bacterial microbiome, the diversity and composition of fungi carried by birds are influenced by factors such as environment and dietary habits (W. Li et al., 2024). Fungi are distributed throughout the bird's body, inhabiting areas ranging from feathers and skin to digestive tract. The fungi carried by birds can generally be categorized into three types: commensal fungi, opportunistic pathogens, and pathogenic fungi. Commensal fungi constitute part of the normal microbiota and play a role in maintaining physiological balance in birds. Pathogenic fungi, often acquired from the external environment, can directly cause diseases in birds. There is no clear-cut distinction between opportunistic and pathogenic fungi. Typically, opportunistic fungi do not affect healthy birds but can induce various disease responses when the immune system is compromised (Chege, 2016). Additionally, birds may carry environmental fungi that are harmless to them but may potentially induce zoonotic disease in other mammals and humans. This article will focus on fungi that pose direct threats to avian health, particularly Aspergillus and Candida, while also providing a brief overview of the role of commensal fungal communities in maintaining health and balance of avians.

Fungi represent a ubiquitous and diverse kingdom of eukaryotic organisms, with over 300 species identified as human pathogens. In avian species, the primary fungal pathogens are Aspergillus and various yeasts, which predominantly cause opportunistic infections in

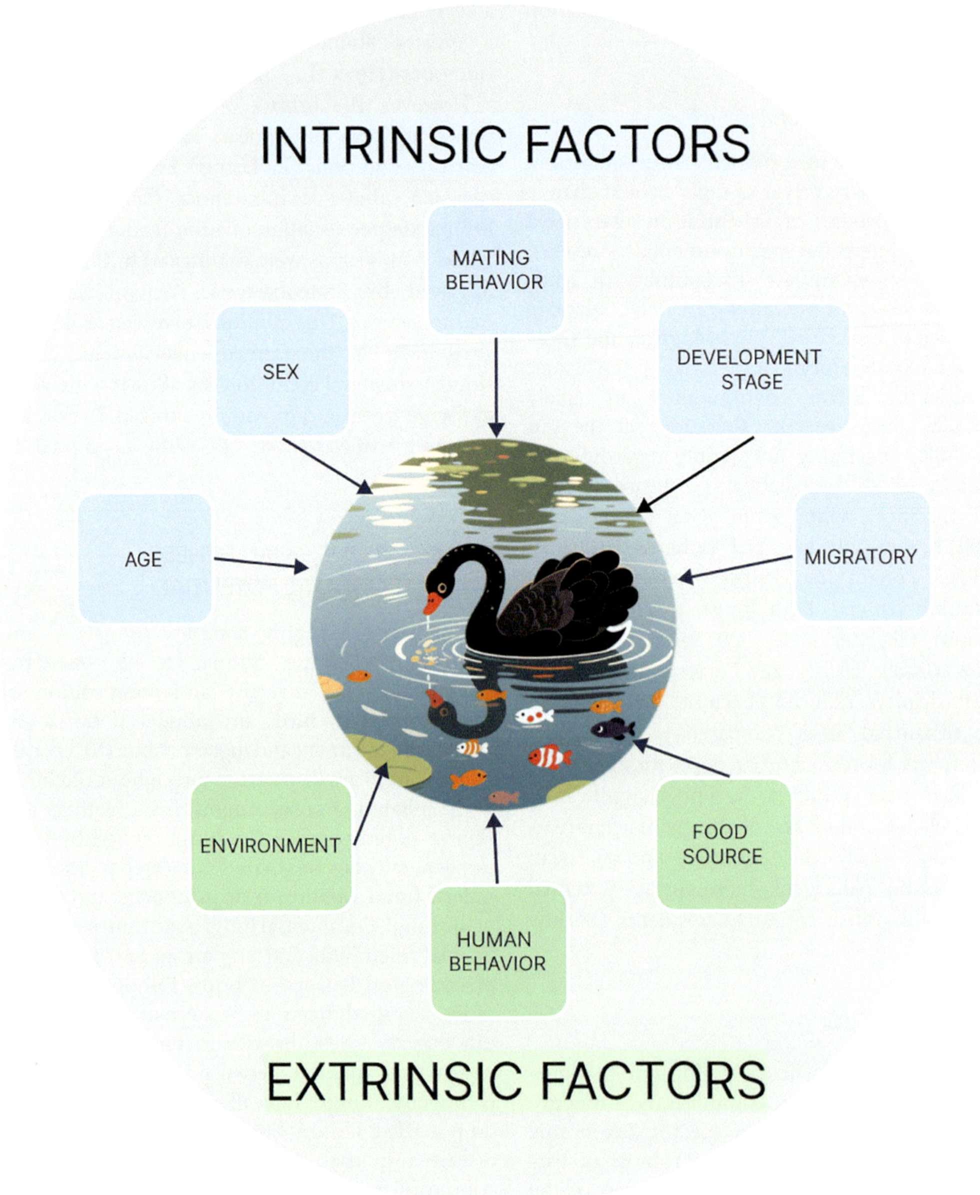

Figure 6.1 The different factors influencing the composition of gut microbiota.

immunocompromised juveniles and debilitated adults. The composition and dynamics of the avian mycobiome, the fungal community within and on a host, are shaped by a complex interplay of factors, analogous to those governing the bacterial microbiome. Key determinants include host physiological traits such as body size and dietary niche, as well as environmental variables like seasonality and precipitation, which collectively influence microbial diversity and pathogen emergence (Woodhams et al., 2020).

6.4.1 Aspergillosis

Avian aspergillosis, especially *Aspergillus fumigatus*, remains a major respiratory pathogen in birds. Other species like *A. flavus*, *A. niger*, *A. nidulans*, and *A. terreus* may also be isolated from avian cases of aspergillosis (sometimes in mixed infections) but much less frequently than *A. fumigatus* (Arné et al.,

2021). The pathogen primarily originates in the respiratory tract. However, the infection can disseminate systemically, causing a range of acute or chronic clinical manifestations in various organs. All avian species are considered susceptible to *A. fumigatus*, a fungus implicated in numerous mortality outbreaks across domestic and wild birds in both captive and free-living settings (Deem, 2003; Garg & Sethi, 1969; Seyedmousavi et al., 2018; Simi et al., 2019). Moreover, the economic impact of aspergillosis is most significant in poultry production including goose (Morishita, 2004; de Jong et al., 2025), duck (de Jong et al., 2025), broiler chicken (Cruciani et al., 2022) and ostrich (Fotouh et al., 2025), as the disease typically occurs during the later stages of the growth cycle and treatments for infected poultry are nonexistent; therefore, prevention is the only way to reduce economic loss (Arné et al., 2011).

Aspergillosis is predominantly observed in animals subjected to poor hygiene or malnutrition (Alhassani et al., 2025; Asfaw, 2017; Cacciuttolo et al., 2009a). This occurs because *A. fumigatus* proliferates in contaminated bedding or feed, and its prolific sporulation generates abundant, easily aerosolized conidia (Seyedmousavi et al., 2015). Consequently, in poorly ventilated and dusty farm environments, these spores are inhaled and deposited deep in the respiratory tract. Acute cases, which are characterized by high morbidity and mortality, typically occur in young birds following massive spore inhalation, while the chronic form is more sporadic and generally affects older birds. *A. fumigatus* has been documented in wild birds since the early 19th century. Significant die-offs in free-ranging species like waterfowl, gulls, and corvids have been linked to the dumping of moldy waste grains in feeding areas (Sacristán et al., 2014). The pathogen is also prevalent in captive settings, affecting birds of prey (Deem, 2003), penguins (Alvarez-Perez et al., 2010; Debergh et al., 2023), and parrots (Hock et al., 2023; Souza et al., 2024; Tanase et al., 2024). A report also documented the sickness of large, precious waterfowl in a zoo after they consumed fresh fish contaminated with *A. fumigatus*. This outbreak resulted from a combination of factors: abundant rainfall, limited sunlight, mild temperatures suitable for fungal growth, lush vegetation in the breeding area, and inadequate routine disinfection of the fish-rearing ponds (Feng et al., 2019). In wildlife rehabilitation centers, birds experiencing stress from trauma, captivity, starvation, or prolonged antibiotic therapy are particularly susceptible (Deem, 2003), and a high incidence in these debilitated individuals often adversely affects rehabilitation success. This susceptibility to stress is also evident in game birds, such as the red-legged partridge, which are used for restocking and can be negatively affected by stressors like captivity in aviaries, transport in cages, release in unsuitable environments, and malnutrition (Cacciuttolo et al., 2009b).

6.4.2 Candidiasis

Candidiasis, commonly referred to as thrush or sour crop, is a fungal disease predominantly caused by the yeast *Candida albicans*. This organism is an opportunistic pathogen that is frequently isolated from the droppings of healthy birds in a commensal state but can undergo a commensal-to-pathogen transition, causing primary disease in stressed or immunocompromised individuals (Jacobsen, 2023). The condition primarily affects the upper gastrointestinal tract, with clinical manifestations including gastrointestinal stasis, anorexia, and oral lesions. This pathogenic overgrowth is facilitated by a range of predisposing factors. Host age is a critical determinant, as young birds with incomplete immunocompetence exhibit significantly higher susceptibility; developing immune systems mount a less balanced response compared to more resistant adults (Talazadeh et al., 2022). Furthermore, extrinsic and physiological stressors such as prolonged antibiotic use, concurrent malnutrition, other immunosuppressive conditions, and poor environmental hygiene collectively compromise the host's health, increasing the risk of symptomatic infection (Shrubsole-Cockwill et al., 2010; Talazadeh et al., 2022). The epidemiological relevance of avian candidiasis is underscored by its detection in a diverse host range. The fungus has been isolated from domesticated avians such as geese, broiler chickens, and pigeons; pet birds including parrots; and captive species in zoological and rehabilitation settings, such as peacocks, flamingos, and various birds of prey (Al Rubaiee et al., 2017; Deem, 2003; Magalhães Pinto et al., 2019; Proença et al., 2014).

6.4.3 Cryptococcus

Birds are key carriers of Cryptococcus, an opportunistic pathogen that causes cryptococcosis in humans and various animals. Although they rarely contract the disease themselves, they facilitate environmental contamination by shedding the fungus in their feces (Cafarchia et al., 2006; Hagen et al., 2015; Kielstein et al., 2000; Raso et al., 2004). The fungus thrives in natural settings, particularly in soils enriched with bird droppings. Hence, mammals and humans exposed to the contaminated material are susceptible to the disease.

Cryptococcus has been detected in a wide range of bird species. These include wild birds of prey such as *Aegypius monachus*, *Gyps fulvus*, and *Neophron percnopterus* (Cafarchia et al., 2006; Pitarch et al., 2017) as well as domesticated pet birds like parrots (González-Hein et al., 2010). It is acquired through inhalation of infectious propagules, often resulting in meningitis and meningoencephalitis (Siqueira et al., 2022). Histological examinations have revealed cryptococcal cells in numerous tissues, including the beak, choana, sinuses, lungs, air sacs, heart, liver, spleen, kidneys, intestines, and central nervous system (Raso et al., 2004) demonstrating systemic distribution despite the absence of clinical disease.

The pathogenesis of avian cryptococcosis demonstrates two distinct disease patterns, dictated by the specific Cryptococcus species and the host involved. For instance, *Cryptococcus bacillisporus* exhibits characteristics of a primary pathogen in seemingly immunocompetent parrots, rather than acting as an opportunistic infection. It causes localized, invasive disease of the upper respiratory tract, with clinical signs referable to mycotic rhinitis. The infection can extend to involve adjacent structures such as the sinuses, beak, and palate. In contrast, *Cryptococcus neoformans* var. *grubii* in pigeons typically behaves as an opportunistic pathogen in immunodeficient hosts. The disease in these birds is not confined to superficial sites like the upper respiratory tract or subcutis; instead, it often penetrates the lower respiratory tract or disseminates widely to various internal organs. Furthermore, *C. bacillisporus* strains were also isolated from captive kiwi birds in Australia, all of which succumbed to severe diffuse cryptococcal pneumonia or widely disseminated disease. This aligns with a previously reported case in a kiwi from New Zealand, indicating that this Cryptococcus species is highly pathogenic to kiwis.

Furthermore, the role of pet birds in the zoonotic transmission of this pathogen is of particular concern. A study in Thuringia, Germany, detected a high prevalence of *Cryptococcus neoformans* var. *neoformans* in pooled fecal samples from pet bird flocks. Crucially, the predominant strain was serovar A, which is the most common strain responsible for human cryptococcosis cases in Germany. This direct match indicates that pet birds act as a reservoir for this pathogen, posing a direct zoonotic risk to their owners and handlers (Kielstein et al., 2000). On the other hand, keepers and caregivers of avians working at rehabilitation center zoo are facing greater risk of acquiring human cryptococcosis since birds of prey as potential carriers and spreaders of *C. neoforman* and other yeasts of importance in human infections (Cafarchia et al., 2006; Pitarch et al., 2017). In contrast, a separate study of pigeon excreta in Innsbruck, Austria, isolated only the less pathogenic *Cryptococcus albidus* and found no *C. neoformans* (Kielstein et al., 2000). This disparity underscores that the zoonotic risk is not uniform across all avian populations but is significantly influenced by the specific microbial ecology of the bird group, with pet bird flocks presenting a notably higher risk.

6.5 AVIANS AS VECTORS OF ANTIBIOTIC RESISTANCE

Avians, encompassing domestic, wild, and captive populations, are critical in the global transmission network of antibiotic resistance genes (ARGs) (Figure 6.2). Their ecological features and mobility underpin their capacity to act as reservoirs, amplifiers, and vectors. Consequently, antibiotic-resistant bacteria (ARB) carrying these ARGs are frequently isolated from their intestinal tracts and external surfaces. Notably, members of the Enterobacteriaceae family, such as *E. coli* and *Salmonella* spp., along with *C. jejuni* and Enterococcus species, are among the most commonly reported ARB in birds, carrying a wide array of ARGs including those conferring resistance to beta-lactams (blaCTX-M, blaTEM), tetracyclines (tetA, tetB), fluoroquinolones (qnr), and colistin (mcr-1) (Begum et al., 2024). ARGs are primarily transmitted through horizontal gene transfer (HGT) mediated by mobile genetic elements such as plasmids, transposons, and integrons, and vertical gene transfer (VGT), spreading through agricultural soils, crops, and food chains to humans, thereby reducing the efficacy of antibiotic treatments (Fang et al., 2018; Michaelis & Grohmann, 2023). If left unaddressed, the annual global death toll due to antibiotic resistance is expected to reach 10 million by 2050 (Huemer et al., 2020).

TABLE 6.2
Global summary of ARB and ARGs in birds

ARB	Country of sampling	Host	ARG	Resistance to drugs	Ref.
E. coli	China	Migratory birds	blaCTX-M, blaTEM-1, tet(A), tet(B), tet(M), sul1, sul2, sul3, cmlA, floR, and intI1	β-lactams, tetracycline	(Yuan et al., 2021)
	China	Aquatic bird(egret)	mcr-1	-	(Wu et al., 2018)
	China	Migratory Birds	$bla_{CTX-M-27}$ and $bla_{TEM-176}$	-	(D. Wang et al., 2024)
	Poland	Wild birds	*sul2*, bla_{TEM}, *irp-2*, *ompT*	Tetracycline, ciprofloxacin, gentamicin, and ampicillin	(Nowaczek et al., 2021)
	Poland	Free-living birds	$bla_{CTX-M-15}$, bla_{CMY-2}, bla_{SHV-12}, bla_{TEM-1B}, *qnrS1*, *qnrB19*, *mcr-1*, *fosA7*, *aac(3)-IIa*, *ant(3")-Ia*, *aph(6)-Id*, chromosomal *gyrA*, *parC*, *parE* mutations	cephalosporins, quinolones, polymyxins, aminoglycosides, fosfomycin	(Skarżyńska et al., 2021)
	Australia	Silver gull	blaIMP-4, blaCTX-M, blaOXA, qnrS	meropenem, cefotaxime,ciprofloxacin	(Wyrsch et al., 2022)
	Alaska	sympatric gulls and bald eagles	bla_{CTX-M} and bla_{CMY}	-	(Ahlstrom et al., 2018)
	Germany	semi-wild mallard ducks	$Bla_{CTX-M-1}$ and $bla_{CTX-M-32}$	-	(Dreyer et al., 2022)
	Tunisia	Wild birds	*tet(M)*, *tet(L)*, *erm(B)*, *erm(C)*, *aac(6′)-Ie–aph(2″)-Ia*, *cat*, *vanA*, *vanB2*	Tetracycline, Erythromycin. Ciprofloxacin, Trimethoprim-sulfamethoxazole, Kanamycin, Chloramphenicol,Gentamicin, Streptomycin, Vancomycin	(Ben Yahia et al., 2018)
Salmonella spp.	Bangladesh	Migratory and captive wild birds	blaTEM, blaSHV, sul1, agfA	tetracycline, ampicillin streptomycin, ceftriaxone, colistin sulphate	(Begum et al., 2024)
	Spain	wild birds	-	ciprofloxacin, nalidixic acid and colistin.	(Martín-Maldonado et al., 2020)

(Continued)

Table 6.2 (Continued)

ARB	Country of sampling	Host	ARG	Resistance to drugs	Ref.
	Korea	Broiler Chicken	blaCTX-M-15	-	(Shang et al., 2021)
	United States	poultry	-	Tetracycline, Streptomycin, Nalidixic acid	(Velasquez et al., 2018)
	Canada	poultry	-	Ceftiofur, Gentamicin, Amoxicillin,Tetracycline	(St. Amand et al., 2013)
	Bangladesh	Migratory birds	gyrB and/or parC	Tetracycline, ampicillin, chloramphenicol, sulfamethoxazole, trimethoprim, ciprofloxacin.	(Card et al., 2023)
Campylobacter	Canada	Broiler chicken	-	Tetracycline, Quinolone	(Dramé et al., 2020)
	Spain	Urban wild birds	-	ciprofloxacin, tetracycline, nalidixic acid streptomycin	(Mencía-Gutiérrez et al., 2024)
	Egypt	Migratory bird , broiler chicken	tetA , BlaOXA-61	doxycycline	(Tawakol et al., 2023)
	Spain	Birds of prey	-	nalidixic acid, ciprofloxacin, tetracycline, streptomycin.	(Mencía-Gutiérrez et al., 2021)
Staphylococcus spp.	Atlantic Ocean	Arctic tern (*Sterna paradisaea*)		β-lactam and quinolone	(Begum et al., 2024)

ARG pollution in poultry farming environments has reached critical levels, largely driven by the extensive use of antibiotics in animal husbandry. Prolonged antibiotic exposure has led to the emergence and dissemination of ARGs and ARB through both horizontal and vertical gene transfer (Y. Chen et al., 2025). The complexity of ARG pollution in this setting is influenced by intensive farming practices, host physiological traits, and inefficient waste management systems, with manure, wastewater, and airborne particulates serving as major pathways for environmental dispersion (Korver, 2023). Common ARBs in poultry farms include Salmonella, Campylobacter, and Gram-negative bacilli such as *E. coli* (D. Liu et al., 2012; Miles et al., 2006). These bacteria employ diverse resistance mechanisms, including the production of inactivating enzymes (e.g., β-lactamases and aminoglycoside-modifying enzymes) and active efflux systems (Islam et al., 2024). A notable example is a case of respiratory infection in ostriches in Romania caused by multidrug-resistant *E. coli*. Isolates from the lungs and bone marrow exhibited resistance to enrofloxacin, doxycycline, and amoxicillin, underscoring the severity of antimicrobial resistance in veterinary settings (Iorgoni et al., 2025). Similarly, avian pathogenic *E. coli* (APEC) isolates from Egyptian broiler farms showed high resistance to ampicillin, tetracycline, nalidixic acid, and chloramphenicol, with 29.3% (34/116) carrying integrons (Awad et al., 2016). The situation is particularly severe in eastern China, where multidrug resistance is widespread among APEC isolates. All tested strains (100%) were resistant to three or more antimicrobial classes, with exceptionally high resistance rates (>90%) to erythromycin, enrofloxacin, tetracycline, doxycycline, lincomycin, and streptomycin. In contrast, carbapenems and polymyxins remained largely effective, with resistance rates as low as 1.7% for imipenem and 0.4% for polymyxin B (Afayibo et al., 2022).

Furthermore, the high density and operational intensity of poultry farms significantly increase the risk of ARG transmission to humans—especially farm workers—and the broader environment. Studies have detected plasmid-mediated quinolone resistance (PMQR) genes, including mcr-1.1, in *E. coli* from healthy humans, chickens, and farm environments. The same gene has been identified in isolates from poultry manure and drinking water, indicating a persistent cycle of cross-contamination within the farm ecosystem. The horizontal transfer of PMQR genes across human–poultry–environment interfaces represent a serious public health threat and facilitates the ongoing dissemination of antimicrobial resistance.

Wildlife inhabiting anthropogenically influenced environments are frequently exposed to and can acquire ARB (Marcelino et al., 2019; J. Wang et al., 2017). Synanthropic species, which live in close association with human habitats and forage in urban settings, landfills, or intensive agricultural areas, exhibit higher carriage rates of ARB in their gut microbiomes compared to those in undisturbed environments. These birds often carry clinically relevant ARGs, such as *bla*CTX-M-15, *van*A and mcr-1, which are associated with resistance to last-resort antibiotics in human medicine (Ben Yahia et al., 2018; Wyrsch et al., 2022; Wu et al., 2018). Over the years, wild animals have been increasingly recognized as potential vectors and secondary reservoirs of ARB, posing transmission risks to humans and livestock (Elsohaby et al., 2021; Fu et al., 2023; Łopucki et al., 2024; Tardón et al., 2021; Wu et al., 2018).

Among them, wild birds are of particular concern due to their high mobility and long-distance migratory behavior, which enables the cross-regional and even intercontinental spread of antibiotic resistance (Esposito et al., 2024; Russo et al., 2022; Sacristán et al., 2014; Tardón et al., 2021). A compelling example comes from a study on egrets, which act as mobile vectors transporting antibiotic resistance from polluted waterways to terrestrial ecosystems (Wu et al., 2018). Soil analysis revealed significantly higher levels of antimicrobial resistance in egret-frequented parks compared to remote control sites. Furthermore, the observed decline in the abundance of ARGs and mobile genetic elements (MGEs) along the river–egret–park pathway suggests that egrets facilitate resistance dissemination primarily through HGT to native soil bacteria, rather than via clonal expansion of specific bacterial strains. Driven by food scarcity, climate change, and overpopulation, wild birds are increasingly foraging in urban areas, feeding on human leftovers in landfills. This behavior has turned them into significant reservoirs of Salmonella. Of particular concern is that the most common Salmonella types found are zoonotic and often resistant to critically important antibiotics like ciprofloxacin and colistin. Consequently, these birds act as a critical link, spreading antimicrobial resistance from environmental reservoirs into human populations (Martín-Maldonado et al., 2020).

Although direct evidence of ARG transmission between avians and mammals or humans is still limited (Łopucki et al., 2024), several epidemiological clues suggest nonnegligible spillover risks. Nevertheless, the elevated prevalence of integrons and specific antibiotic resistance (e.g., to sulfisoxazole and trimethoprim/sulfamethoxazole) in humans associated with small-scale production birds implies a potential spillover of resistance from poultry to humans (Aworh et al., 2023; Moser et al., 2018) (Figure 6.2). Future studies combining genomic surveillance with ecological modeling are needed to clarify the directionality and frequency of such cross-species transmission events.

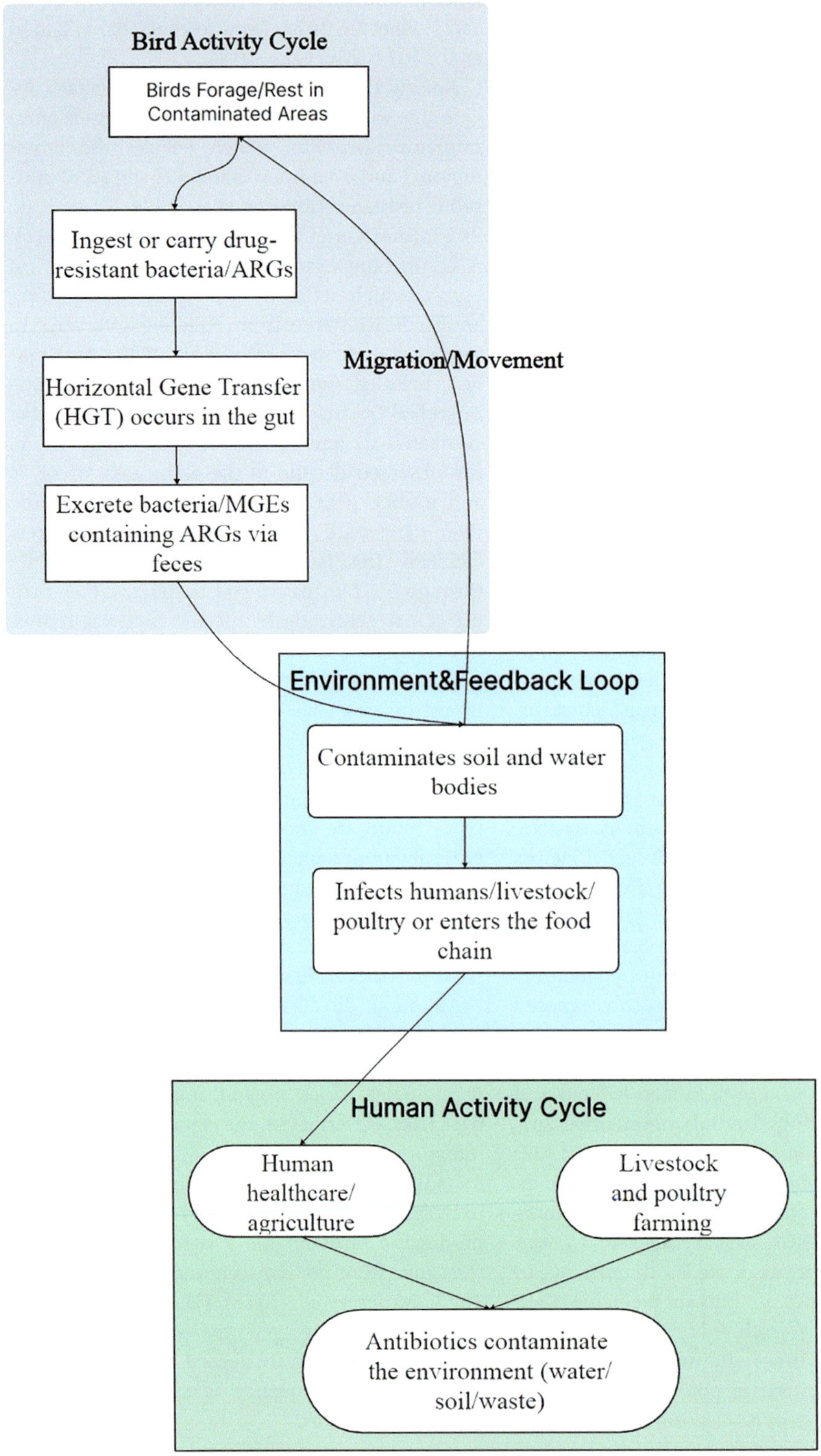

Figure 6.2 Acquisition and dissemination of AMR in wild birds posing threat to public health system.

6.6 CONCLUSION AND FUTURE PERSPECTIVES

This paper reviews the avian GM, with a specific focus on its microbial composition, the factors shaping its community structure, and the role of birds as reservoirs of zoonotic pathogens relevant to public health. As one of the most widespread and diverse groups of vertebrates, birds occupy nearly all terrestrial and aquatic niches beyond the deep ocean—including forests, deserts, grasslands, and urban environments. Consequently, avian health, and particularly the gut microbiota that modulates it, represents a critical area for scientific research. As highlighted throughout this review, studying the avian GM not only advances fundamental knowledge in microbial ecology and host–microbe interactions, but also offers actionable insights for applications in poultry production, conservation biology, and public health surveillance. Nevertheless, significant challenges—such as constraints in sample accessibility and the difficulty of culturing key bacterial taxa under standard laboratory conditions—continue to hinder a comprehensive understanding of the functional roles and mechanistic contributions of the avian gut microbiota. Genomic studies have further strengthened this link; for instance, the identification of identical *bla*CTX-M-15–bearing plasmids in *E. coli* from gulls, sewage, and clinical isolates worldwide demonstrates the potential for avian-mediated ARG spread across ecosystems and into healthcare settings. Overall, further research is warranted to elucidate the functional capabilities of the avian gut microbiota and the underlying mechanisms governing its complex interactions with the host.

REFERENCES

Afayibo, D. J. A., Zhu, H., Zhang, B., Yao, L., Abdelgawad, H. A., Tian, M., Qi, J., Liu, Y., & Wang, S. (2022). Isolation, molecular characterization, and antibiotic resistance of avian pathogenic *Escherichia coli* in Eastern China. *Veterinary Sciences*, 9(7), 319. https://doi.org/10.3390/vetsci9070319

Ahlstrom, C. A., Bonnedahl, J., Woksepp, H., Hernandez, J., Olsen, B., & Ramey, A. M. (2018). Acquisition and dissemination of cephalosporin-resistant *E. coli* in migratory birds sampled at an Alaska landfill as inferred through genomic analysis. *Scientific Reports*, 8(1), 7361. https://doi.org/10.1038/s41598-018-25474-w

Al Rubaiee, Z., Al Murayati, H., Nielsen, J. T., & Møller, A. P. (2017). Fungi, feather damage, and risk of predation. *Ecology and Evolution*, 7(24), 10797–10803. https://doi.org/10.1002/ece3.3582

Albert, L., Rumschlag, S., Parker, A., Vaziri, G., & Knutie, S. A. (2023). Elevated nest temperature has opposing effects on host species infested with parasitic nest flies. *Oecologia*, 201(4), 877–886. https://doi.org/10.1007/s00442-023-05343-8

Alhassani, A. N. A., Ahmed, A. T., Sanghvi, G., Ganesan, S., Al-Hetty, H. R. A. K., Sapaev, I. B., Singh, A., Sudan, P., Mustafa, Y. F., & Gholami-Ahangaran, M. (2025). A review on Aspergillosis in Turkey: As a main fungal disease in Poultry. *Veterinary Medicine and Science*, 11(6), e70605. https://doi.org/10.1002/vms3.70605

Altizer, S., Bartel, R., & Han, B. A. (2011). *Animal Migration and Infectious Disease Risk | Science*. www.science.org/doi/abs/10.1126/science.1194694?casa_token=pEkCvQqtipEAAAAA:MWXmibqFp-79eCNL0IZYwTraqlGJFbYk_RnoVTSeexMNtxpE-AFMVQYhFbhe2yJTOLf-rlphIcCLvRA&casa_token=4apnOfg7VhcAAAAA:FnrbsN7mMM9MtR6GirR9hOe_5PbGifhzosl1YgdT1LGIx-PXUQnw45nUR14PBAFHsbK5yMHhLQaz35c

Alvarez-Perez, S., Mateos, A., Dominguez, L., Martinez-Nevado, E., Blanco, J. L., & Garcia, M. E. (2010). Polyclonal *Aspergillus fumigatus* infection in captive penguins. *Veterinary Microbiology*, 144(3), 444–449. https://doi.org/10.1016/j.vetmic.2010.02.026

Antunes, P., Mourão, J., Campos, J., & Peixe, L. (2016). Salmonellosis: The role of poultry meat. *Clinical Microbiology and Infection*, 22(2), 110–121. https://doi.org/10.1016/j.cmi.2015.12.004

Arné, P., Risco-Castillo, V., Jouvion, G., Le Barzic, C., & Guillot, J. (2021). Aspergillosis in wild birds. *Journal of Fungi*, 7(3), 241. https://doi.org/10.3390/jof7030241

Arné, P., Thierry, S., Wang, D., Deville, M., Le Loc'h, G., Desoutter, A., Féménia, F., Nieguitsila, A., Huang, W., Chermette, R., & Guillot, J. (2011). *Aspergillus fumigatus* in Poultry. *International Journal of Microbiology*, 2011(1), 746356. https://doi.org/10.1155/2011/746356

Asfaw, M. (2017). *Review on Major Fungal Disease of Poultry*. https://doi.org/10.5829/IDOSI.BJPS.2017.16.25

Awad, A., Arafat, N., & Elhadidy, M. (2016). Genetic elements associated with antimicrobial resistance among avian pathogenic *Escherichia coli*. *Annals of Clinical Microbiology and Antimicrobials*, 15(1), 59. https://doi.org/10.1186/s12941-016-0174-9

Aworh, M. K., Kwaga, J. K. P., Hendriksen, R. S., Okolocha, E. C., Harrell, E., & Thakur, S. (2023). Quinolone-resistant *Escherichia coli* at the interface between humans, poultry and their shared environment- a potential public health risk. *One Health Outlook*, 5(1), 2. https://doi.org/10.1186/s42522-023-00079-0

Ayala, A. J., & Ogbunugafor, C. B. (2023). When vibrios take flight: A meta-analysis of pathogenic vibrio species in wild and domestic birds. In S. Almagro-Moreno & S. Pukatzki (Eds.), *Vibrio spp. Infections* (pp. 295–336). Springer International Publishing. https://doi.org/10.1007/978-3-031-22997-8_15

Ayala, D. I., Cook, P. W., Franco, J. G., Bugarel, M., Kottapalli, K. R., Loneragan, G. H., Brashears, M. M., & Nightingale, K. K. (2019). A systematic approach to identify and characterize the effectiveness and safety of novel probiotic strains to control foodborne pathogens. *Frontiers in Microbiology*, 10, 1108. https://doi.org/10.3389/fmicb.2019.01108

Begum, R., Asha, N. A., Dipu, D. C. C., Roy, M., Rahman, A., Chowdhury, Md. S. R., Hossain, H., Islam, Md. R., Uddin, M. B., Rahman, Md. M., & Hossain, Md. M. (2024). Virulence and antimicrobial resistance patterns of *Salmonella spp.* recovered from migratory and captive wild birds. *Veterinary Medicine and Science*, 10(6), e70102. https://doi.org/10.1002/vms3.70102

Ben Yahia, H., Chairat, S., Hamdi, N., Gharsa, H., Ben Sallem, R., Ceballos, S., Torres, C., & Ben Slama, K. (2018). Antimicrobial resistance and genetic lineages of faecal enterococci of wild birds: Emergence of *vanA* and *vanB2* harbouring *Enterococcus faecalis*. *International Journal of Antimicrobial Agents*, 52(6), 936–941. https://doi.org/10.1016/j.ijantimicag.2018.05.005

Benskin, C. McW. H., Wilson, K., Jones, K., & Hartley, I. R. (2009). Bacterial pathogens in wild birds: A review of the frequency and effects of infection. *Biological Reviews*, 84(3), 349–373. https://doi.org/10.1111/j.1469-185X.2008.00076.x

Berlow, M., Phillips, J. N., & Derryberry, E. P. (2021). Effects of urbanization and landscape on gut microbiomes in white-crowned sparrows. *Microbial Ecology*, 81(1), 253–266. www.jstor.org/stable/27291092

Bernini, L. J., Simão, A. N., Alfieri, D. F., Lozovoy, M. A., Mari, N. L., de Souza, C. H., Dichi, I., & Costa, G. N. (2016). Beneficial effects of Bifidobacterium lactis on lipid profile and cytokines in patients with metabolic syndrome: A randomized trial. Effects of probiotics on metabolic syndrome. *Nutrition*, 32(6), 716–719. https://doi.org/10.1016/j.nut.2015.11.001

Berry, D. (2016, February 4). *The emerging view of Firmicutes as key fibre degraders in the human gut—Berry—2016—Environmental Microbiology—Wiley Online Library*. https://enviromicro-journals.onlinelibrary.wiley.com/doi/10.1111/1462-2920.13225

Bodawatta, K. H., Freiberga, I., Puzejova, K., Sam, K., & Poulsen, M. (2021, February 18). *Flexibility and resilience of great tit (Parus major) gut microbiomes to changing diets | Animal Microbiome*. https://link.springer.com/article/10.1186/s42523-021-00076-6

Bodawatta, K. H., Hird, S. M., Grond, K., Poulsen, M., & Jønsson, K. A. (2022). Avian gut microbiomes taking flight. *Trends in Microbiology*, 30(3), 268–280. https://doi.org/10.1016/j.tim.2021.07.003

Cacciuttolo, E., Rossi, G., Nardoni, S., Legrottaglie, R., & Mani, P. (2009a). Anatomopathological aspects of avian aspergillosis. *Veterinary Research Communications*, 33(6), 521–527. https://doi.org/10.1007/s11259-008-9199-7

Cacciuttolo, E., Rossi, G., Nardoni, S., Legrottaglie, R., & Mani, P. (2009b). Anatomopathological aspects of avian aspergillosis. *Veterinary Research Communications*, 33(6), 521–527. https://doi.org/10.1007/s11259-008-9199-7

Cafarchia, C., Romito, D., Iatta, R., Camarda, A., Montagna, M. T., & Otranto, D. (2006). Role of birds of prey as carriers and spreaders of *Cryptococcus neoformans* and other zoonotic yeasts. *Medical Mycology*, 44(6), 485–492. https://doi.org/10.1080/13693780600735452

Card, R. M., Chisnall, T., Begum, R., Sarker, M. S., Hossain, M. S., Sagor, M. S., Mahmud, M. A., Uddin, A. S. M. A., Karim, M. R., Lindahl, J. F., & Samad, M. A. (2023). Multidrug-resistant non-typhoidal Salmonella of public health significance recovered from migratory birds in Bangladesh. *Frontiers in Microbiology*, 14, 1162657. https://doi.org/10.3389/fmicb.2023.1162657

Chege, S. M. (2016). Opportunistic infections of avians. In C. J. Hurst (Ed.), *The Rasputin Effect: When Commensals and Symbionts Become Parasitic* (pp. 221–260). Springer International Publishing. https://doi.org/10.1007/978-3-319-28170-4_9

Chen, X., Hu, B., Huang, L., Cheng, L., Liu, H., Hu, J., Hu, S., Han, C., He, H., kang, B., Xu, H., Zhang, R., Wang, J., & Li, L. (2021). The differences in intestinal growth and microorganisms between male and female ducks. *Poultry Science*, 100(2), 1167–1177. https://doi.org/10.1016/j.psj.2020.10.051

Chen, Y., Liu, Y., Zhao, C., Ma, J., & Guo, J. (2025). Antibiotic resistance gene pollution in poultry farming environments and approaches for mitigation: A system review. *Poultry Science*, 104(3), 104858. https://doi.org/10.1016/j.psj.2025.104858

Comstock, L. E. (2009). Importance of glycans to the host-bacteroides mutualism in the mammalian intestine. *Cell Host & Microbe*, 5(6), 522–526. https://doi.org/10.1016/j.chom.2009.05.010

Cruciani, D., Crotti, S., Maresca, C., Pecorelli, I., Verdini, E., Rodolfi, M., Scoccia, E., Spina, S., Valentini, A., & Agnetti, F. (2022). Preliminary investigation about

Aspergillus spp. spread in umbrian avian farms. *Journal of Fungi*, 8(11), 1213. https://doi.org/10.3390/jof8111213

Davidson, G. L., Wiley, N., Cooke, A. C., Johnson, C. N., Fouhy, F., Reichert, M. S., de la Hera, I., Crane, J. M. S., Kulahci, I. G., Ross, R. P., Stanton, C., & Quinn, J. L. (2020). Diet induces parallel changes to the gut microbiota and problem solving performance in a wild bird. *Scientific Reports*, 10(1), 20783. https://doi.org/10.1038/s41598-020-77256-y

Debergh, H., Becker, P., Vercammen, F., Lagrou, K., Haesendonck, R., Saegerman, C., & Packeu, A. (2023). Pulmonary aspergillosis in humboldt penguins—susceptibility patterns and molecular epidemiology of clinical and environmental *Aspergillus fumigatus* isolates from a Belgian zoo, 2017–2022. *Antibiotics*, 12(3), 584. https://doi.org/10.3390/antibiotics12030584

Deem, S. L. (2003). Fungal diseases of birds of prey. *Veterinary Clinics of North America: Exotic Animal Practice*, 6(2), 363–376. https://doi.org/10.1016/S1094-9194(03)00004-5

de Jong, J. E., Heuvelink, A. E., Dieste Pérez, L., & Holstege, M. M. C. (2025). *Aspergillus spp*., aspergillosis and azole usage in animal species in Europe: Results from a multisectoral survey and review of recent literature. *Medical Mycology*, 63(2), myaf009. https://doi.org/10.1093/mmy/myaf009

Dramé, O., Leclair, D., Parmley, E. J., Deckert, A., Ouattara, B., Daignault, D., & Ravel, A. (2020). Antimicrobial resistance of campylobacter in broiler chicken along the food chain in Canada. *Foodborne Pathogens and Disease*, 17(8), 512–520. https://doi.org/10.1089/fpd.2019.2752

Dreyer, S., Globig, A., Bachmann, L., Schütz, A. K., Schaufler, K., & Homeier-Bachmann, T. (2022). Longitudinal study on extended-spectrum beta-lactamase-E. *coli* in sentinel mallard ducks in an important baltic stop-over site for migratory ducks in Germany. *Microorganisms*, 10(10), 1968. https://doi.org/10.3390/microorganisms10101968

Elsohaby, I., Samy, A., Elmoslemany, A., Alorabi, M., Alkafafy, M., Aldoweriej, A., Al-Marri, T., Elbehiry, A., & Fayez, M. (2021). Migratory wild birds as a potential disseminator of antimicrobial-resistant bacteria around Al-Asfar Lake, Eastern Saudi Arabia. *Antibiotics*, 10(3), 260. https://doi.org/10.3390/antibiotics10030260

Escallón, C., Belden, L. K., & Moore, I. T. (2019). The cloacal microbiome changes with the breeding season in a wild bird. *Integrative Organismal Biology*, 1(1), oby009. https://doi.org/10.1093/iob/oby009

Esposito, E., Scarpellini, R., Celli, G., Marliani, G., Zaghini, A., Mondo, E., Rossi, G., & Piva, S. (2024). Wild birds as potential bioindicators of environmental antimicrobial resistance: A preliminary investigation. *Research in Veterinary Science*, 180, 105424. https://doi.org/10.1016/j.rvsc.2024.105424

Ewers, C., Janssen, T., & Wieler, L. H. (2003). [Avian pathogenic *Escherichia coli* (APEC)]. *Berliner und Munchener tierarztliche Wochenschrift*, 116(9–10), 381–395.

Fan, Y., Ju, T., Bhardwaj, T., Korver, D. R., & Willing, B. P. (2023). Week-old chicks with high bacteroides abundance have increased short-chain fatty acids and reduced markers of gut inflammation. *Microbiology Spectrum*, 11(2), e03616–e03622. https://doi.org/10.1128/spectrum.03616-22

Fang, T., Wang, H., Cui, Q., Rogers, M., & Dong, P. (2018). Diversity of potential antibiotic-resistant bacterial pathogens and the effect of suspended particles on the spread of antibiotic resistance in urban recreational water. *Water Research*, 145, 541–551. https://doi.org/10.1016/j.watres.2018.08.042

Feng, H. J., Yang, L., & Lu, B. B. (2019). Diagnosis, treatment, and control of fungal diseases in Oriental storks and red-crowned cranes. *Guangxi Journal of Animal Husbandry and Veterinary Medicine*, (2), 3.

Flint, H. J., Bayer, E. A., Rincon, M. T., Lamed, R., & White, B. A. (2008). Polysaccharide utilization by gut bacteria: Potential for new insights from genomic analysis. *Nature Reviews Microbiology*, 6(2), 121–131. https://doi.org/10.1038/nrmicro1817

Fotouh, A., Hamed, A. M. A.-E., Elbarbary, N. K., Almasoudi, S. H., Alsyaad, K. M., Abdulaziz, O., Ahmed, A. E., & Elsayed, A. K. (2025). Pathological investigation of late embryonic death in ostrich hatcheries and their economic impact in Egypt. *American Journal of Veterinary Research*, 1(aop), 1–9. https://doi.org/10.2460/ajvr.25.07.0261

Fu, Y., Zhang, K., Shan, F., Li, J., Wang, Y., Li, X., Xu, H., Qin, Z., & Zhang, L. (2023). Metagenomic analysis of gut microbiome and resistome of Whooper and Black Swans: A one health perspective. *BMC Genomics*, 24(1), 635. https://doi.org/10.1186/s12864-023-09742-2

Fu, Z., Yang, H., Xiao, Y., Wang, X., Yang, C., Lu, L., Wang, W., & Lyu, W. (2022). Ileal microbiota alters the immunity statues to affect body weight in muscovy ducks. *Frontiers in Immunology*, 13, 844102. https://doi.org/10.3389/fimmu.2022.844102

Garg, D. N., & Sethi, M. S. (1969). Fungal flora of the avian respiratory tract. *Poultry Science*, 48(1), 339–341. https://doi.org/10.3382/ps.0480339

Glendinning, L., McLachlan, G., & Vervelde, L. (2017). Age-related differences in the respiratory microbiota of chickens. *Plos One*, 12(11), e0188455. https://doi.org/10.1371/journal.pone.0188455

González-Hein, G., González-Hein, J., & Jarabrán, M. C. D. (2010). Isolation of *Cryptococcus neoformans* in dry droppings of captive birds in Santiago, Chile. *Journal of Avian Medicine and Surgery*, 24(3), 227–236. https://doi.org/10.1647/2009-003.1

Grond, K., Sandercock, B. K., Jumpponen, A., & Zeglin, L. H. (2018). The avian gut microbiota: Community, physiology and function in wild birds. *Journal of Avian Biology*, 49(11), e01788. https://doi.org/10.1111/jav.01788

Hagen, F., Khayhan, K., Theelen, B., Kolecka, A., Polacheck, I., Sionov, E., Falk, R., Parnmen, S., Lumbsch, H. T., & Boekhout, T. (2015). Recognition of seven species in the *Cryptococcus gattii/Cryptococcus neoformans* species complex. *Fungal Genetics and Biology*, 78, 16–48. https://doi.org/10.1016/j.fgb.2015.02.009

He, Y., Zhang, M., Dai, C., & Yu, L. (2023). Comparison of the gut microbial communities of domestic and wild mallards (Anas platyrhynchos) based on high-throughput sequencing technology. *Animals*, 13(18), Article 18. https://doi.org/10.3390/ani13182956

Hock, L., Herold, M., Walczak, C., Schoos, A., Penny, C., Cauchie, H.-M., & Ragimbeau, C. (2023). Environmental dynamics of Campylobacter jejuni genotypes circulating in Luxembourg: What is the role of wild birds? *Microbial Genomics*, 9(6), 001031. https://doi.org/10.1099/mgen.0.001031

Huemer, M., Mairpady Shambat, S., Brugger, S. D., & Zinkernagel, A. S. (2020). Antibiotic resistance and persistence—Implications for human health and treatment perspectives. *EMBO Reports*, 21(12), e51034. https://doi.org/10.15252/embr.202051034

Iorgoni, V., Iancu, I., Popa, I., Gligor, A., Orghici, G., Sicoe, B., Dreghiciu, C., Purec, D., Nistor, P., Florea, B., & Herman, V. (2025). First report of respiratory infection caused by multidrug-resistant *Escherichia coli* in an Ostrich in Romania. *Antibiotics*, 14(4), 354. https://doi.org/10.3390/antibiotics14040354

Islam, R., Ferdous, F. B., Hoque, M. N., Asif, N. A., Rana, Md. L., Siddique, M. P., & Rahman, Md. T. (2024). Characterization of β-lactamase and virulence genes in Pseudomonas aeruginosa isolated from clinical, environmental and poultry sources in Bangladesh. *Plos One*, 19(4), e0296542. https://doi.org/10.1371/journal.ponc.0296542

Gong, J., Yu, H., Liu, T., Gill, J. J., & Chambers, J. R. (2008). *Effects of zinc bacitracin, bird age and access to range on bacterial microbiota in the ileum and caeca of broiler chickens | Journal of Applied Microbiology | Oxford Academic*. https://academic.oup.com/jambio/article-abstract/104/5/1372/6718816?login=false

J Scupham, A., Patton, T. G., Bent, E., & Bayles, D. O. (2008). Comparison of the cecal microbiota of domestic and wild Turkeys. *Microbial Ecology*, 56(2), 322–331. https://doi.org/10.1007/s00248-007-9349-4

Jacobsen, I. D. (2023). The role of host and fungal factors in the commensal-to-pathogen transition of *Candida albicans*. *Current Clinical Microbiology Reports*, 10(2), 55–65. https://doi.org/10.1007/s40588-023-00190-w

Jiang, J., Hu, D., & Pei, E. (2024). Integrated omics analysis reveals a correlation between gut microbiota and egg production in captive African penguins (Spheniscus demersus). *Animal Reproduction Science*, 263, 107448. https://doi.org/10.1016/j.anireprosci.2024.107448

Jourdain, E., Gauthier-Clerc, M., Bicout, D., & Sabatier, P. (2007). Bird migration routes and risk for pathogen dispersion into western mediterranean wetlands. *Emerging Infectious Diseases*, 13(3), 365–372. https://doi.org/10.3201/eid1303.060301

Kielstein, H., Schmalreck, K., & Glawischnig. (2000). Occurrence of *Cryptococcus spp.* in excreta of pigeons and pet birds. *Mycoses*, 43(1–2), 7–15. https://doi.org/10.1046/j.1439-0507.2000.00534.x

Kim, Y. S., & Milner, J. A. (2007). Dietary modulation of colon cancer risk. *The Journal of Nutrition*, 137(11 Suppl), 2576S–2579S. https://doi.org/10.1093/jn/137.11.2576S

Kim, J. E., Tun, H. M., Bennett, D. C., Leung, F. C., & Cheng, K. M. (2023). Microbial diversity and metabolic function in duodenum, jejunum and ileum of emu (Dromaius novaehollandiae). *Scientific Reports*, 13(1), 4488. https://doi.org/10.1038/s41598-023-31684-8

Knutie, S. A. (2020). Food supplementation affects gut microbiota and immunological resistance to parasites in a wild bird species. *Journal of Applied Ecology*, 57(3), 536–547. https://doi.org/10.1111/1365-2664.13567

Kohl, K. D. (2012). Diversity and function of the avian gut microbiota. *Journal of Comparative Physiology B*, 182(5), 591–602. https://doi.org/10.1007/s00360-012-0645-z

Kohl, K. D., Brun, A., Caviedes-Vidal, E., & Karasov, W. H. (2019). Age-related changes in the gut microbiota of wild house sparrow nestlings. *Ibis*, 161(1), 184–191. https://doi.org/10.1111/ibi.12618

Korver, D. R. (2023). Review: Current challenges in poultry nutrition, health, and welfare. *Animal*, 17, 100755. https://doi.org/10.1016/j.animal.2023.100755

Leung, T. L. F., & Koprivnikar, J. (2016). *Nematode parasite diversity in birds: The role of host ecology, life history and migration—Leung—2016—Journal of Animal Ecology—Wiley Online Library*. https://besjournals.onlinelibrary.wiley.com/doi/full/10.1111/1365-2656.12581

Li, Y., Zhang, H., Chen, Y. P., Yang, M. X., Zhang, L. L., Lu, Z. X., Zhou, Y. M., & Wang, T. (2015). Bacillus amyloliquefaciens supplementation alleviates immunological stress in lipopolysaccharide-challenged broilers at early age. *Poultry Science*, 94(7), 1504–1511. https://doi.org/10.3382/ps/pev124

Li, P., Zhang, Y., Yang, J., Zhang, X., Ye, L., Li, B., Luo, Q., & Wang, J. (2020). *Analysis of colon microbial diversity of African Ostriches at different ages*. In Review. https://doi.org/10.21203/rs.3.rs-125467/v1

Li, W., Cheng, L., He, X., He, G., Liu, Y., Sang, Z., Wang, Y., Shao, M., Xiong, T., Xu, H., & Zhao, J. (2024). Gut fungi of black-necked cranes (Grus nigricollis) respond to dietary changes during wintering. *BMC Microbiology*, 24(1), 232. https://doi.org/10.1186/s12866-024-03396-0

Liu, D., Chai, T., Xia, X., Gao, Y., Cai, Y., Li, X., Miao, Z., Sun, L., Hao, H., Roesler, U., & Wang, J. (2012). Formation and transmission of *Staphylococcus aureus* (including MRSA) aerosols carrying antibiotic-resistant genes in a poultry farming environment. *Science of The Total Environment*, 426, 139–145. https://doi.org/10.1016/j.scitotenv.2012.03.060

Liu, Y., Yan, T., Ren, Z., & Yang, X. (2021). Age-associated changes in caecal microbiome and their apparent correlations with growth performances of layer pullets. *Animal Nutrition*, 7(3), 841–848. https://doi.org/10.1016/j.aninu.2020.11.019

Liukkonen, M., Muriel, J., Martínez-Padilla, J., Nord, A., Pakanen, V.-M., Rosivall, B., Tilgar, V., van Oers, K., Grond, K., & Ruuskanen, S. (2024). Seasonal and environmental factors contribute to the variation in the gut microbiome: A large-scale study of a small bird. *Journal of Animal Ecology*, 93(10), 1475–1492. https://doi.org/10.1111/1365-2656.14153

Lombardo, M. P., Thorpe, P. A., & Power, H. W. (1999). The beneficial sexually transmitted microbe hypothesis of avian copulation. *Behavioral Ecology*, 10(3), 333–337. https://doi.org/10.1093/beheco/10.3.333

Łopucki, R., Stępień-Pyśniak, D., Christensen, H., Kubiński, K., Lenarczyk, E., Martinez-de-Tejada, G., Kitowski, I., & Masłyk, M. (2024). Interspecies transmission of antimicrobial-resistant bacteria between wild birds and mammals in urban environment. *Veterinary Microbiology*, 294, 110130. https://doi.org/10.1016/j.vetmic.2024.110130

Magalhães Pinto, L., de Assis Bezerra Neto, F., Araújo Paulo de Medeiros, M., Zuza Alves, D. L., & Maranhão Chaves, G. (2019). *Candida* species isolated from pigeon (*Columbia livia*) droppings may express virulence factors and resistance to azoles. *Veterinary Microbiology*, 235, 43–52. https://doi.org/10.1016/j.vetmic.2019.05.022

Marcelino, V. R., Wille, M., Hurt, A. C., González-Acuña, D., Klaassen, M., Schlub, T. E., Eden, J.-S., Shi, M., Iredell, J. R., Sorrell, T. C., & Holmes, E. C. (2019). Meta-transcriptomics reveals a diverse antibiotic resistance gene pool in avian microbiomes. *BMC Biology*, 17(1), 31. https://doi.org/10.1186/s12915-019-0649-1

Martín-Maldonado, B., Vega, S., Mencía-Gutiérrez, A., Lorenzo-Rebenaque, L., de Frutos, C., González, F., Revuelta, L., & Marin, C. (2020). Urban birds: An important source of antimicrobial resistant *Salmonella* strains in Central Spain. *Comparative Immunology, Microbiology and Infectious Diseases*, 72, 101519. https://doi.org/10.1016/j.cimid.2020.101519

Mazmanian, S. K., Round, J. L., & Kasper, D. L. (2008). A microbial symbiosis factor prevents intestinal inflammatory disease. *Nature*, 453(7195), 620–625. https://doi.org/10.1038/nature07008

Mencía-Gutiérrez, A., García-Peña, F. J., González, F., Pastor-Tiburón, N., Pérez-Cobo, I., Marín, M., & Martín-Maldonado, B. (2024). Exploring the prevalence and resistance of campylobacter in urban bird populations. *Veterinary Sciences*, 11(5), 210. https://doi.org/10.3390/vetsci11050210

Mencía-Gutiérrez, A., Martín-Maldonado, B., Pastor-Tiburón, N., Moraleda, V., González, F., García-Peña, F. J., Pérez-Cobo, I., Revuelta, L., & Marín, M. (2021). Prevalence and antimicrobial resistance of *Campylobacter* from wild birds of prey in Spain. *Comparative Immunology, Microbiology and Infectious Diseases*, 79, 101712. https://doi.org/10.1016/j.cimid.2021.101712

Michaelis, C., & Grohmann, E. (2023). Horizontal gene transfer of antibiotic resistance genes in biofilms. *Antibiotics*, 12(2), 328. https://doi.org/10.3390/antibiotics12020328

Miles, T. D., McLaughlin, W., & Brown, P. D. (2006). Antimicrobial resistance of *Escherichia coli* isolates from broiler chickens and humans. *BMC Veterinary Research*, 2(1), 7. https://doi.org/10.1186/1746-6148-2-7

Morishita, T. Y. (2004). Common diseases in backyard ducks and geese. *Seminars in Avian and Exotic Pet Medicine*, 13(4), 191–196. https://doi.org/10.1053/j.saep.2004.04.004

Moser, K. A., Zhang, L., Spicknall, I., Braykov, N. P., Levy, K., Marrs, C. F., Foxman, B., Trueba, G., Cevallos, W., Goldstick, J., Trostle, J., & Eisenberg, J. N. S. (2018). The role of mobile genetic elements in the spread of antimicrobial-resistant *Escherichia coli* from chickens to humans in small-scale production poultry operations in rural ecuador. *American Journal of Epidemiology*, 187(3), 558–567. https://doi.org/10.1093/aje/kwx286

Nowaczek, A., Dec, M., Stępień-Pyśniak, D., Urban-Chmiel, R., Marek, A., & Różański, P. (2021). Antibiotic resistance and virulence profiles of *Escherichia coli* strains isolated from wild birds in Poland. *Pathogens*, 10(8), 1059. https://doi.org/10.3390/pathogens10081059

Ouyang, X., Guan, Y., Pei, J., Ge, J., Wang, H., & Bao, L. (2025). Seasonal variation in gut microbiota of migratory wild raptors: A case study in white-tailed eagles. *Animal Microbiome*, 7(1), 37. https://doi.org/10.1186/s42523-025-00406-y

Peixoto, R. S., Harkins, D. M., & Nelson, K. E. (2021). Advances in microbiome research for animal health. *Annual Review of Animal Biosciences*, 9(1), 289–311. https://doi.org/10.1146/annurev-animal-091020-075907

Pitarch, A., Gil, C., & Blanco, G. (2017). Oral mycoses in avian scavengers exposed to antibiotics from livestock farming. *Science of The Total Environment*, 605–606, 139–146. https://doi.org/10.1016/j.scitotenv.2017.06.144

Proença, L. M., Mayer, J., Schnellbacher, R., Sanchez, S., Huang, C.-T., Brown, H., Jiménez, D., Stelmach, D., & Divers, S. J. (2014). Antemortem diagnosis and successful treatment of pulmonary candidiasis in a sun conure (Aratinga solstitialis). *Journal of Avian Medicine and Surgery*, 28(4), 316–321. https://doi.org/10.1647/2013-073

Raso, T. F., Werther, K., Miranda, E. T., & Mendes-Giannini, M. J. S. (2004). Cryptococcosis outbreak in psittacine birds in Brazil. *Medical Mycology*, 42(4), 355–362. https://doi.org/10.1080/13693780410001712061

Risely, A., Waite, D., Ujvari, B., Klaassen, M., & Hoye, B. (2017). Gut microbiota of a long-distance migrant demonstrates resistance against environmental microbe incursions. *Molecular Ecology*, 26(20), 5842–5854. https://doi.org/10.1111/mec.14326

Russo, T. P., Minichino, A., Gargiulo, A., Varriale, L., Borrelli, L., Pace, A., Santaniello, A., Pompameo, M., Fioretti, A., & Dipineto, L. (2022). Prevalence and phenotypic antimicrobial resistance among ESKAPE Bacteria and enterobacterales strains in wild birds. *Antibiotics*, 11(12), 1825. https://doi.org/10.3390/antibiotics11121825

Ruuskanen, S. (2024). Early-life environmental effects on birds: Epigenetics and microbiome as mechanisms underlying long-lasting phenotypic changes. *Journal of Experimental Biology*, 227(Suppl_1), jeb246024 https://doi.org/10.1242/jeb.246024

Rychlik, I. (2020). Composition and function of chicken gut microbiota. *Animals*, 10(1), 103. https://doi.org/10.3390/ani10010103

Sacristán, C., Esperón, F., Herrera-León, S., Iglesias, I., Neves, E., Nogal, V., Muñoz, M. J., & De La Torre, A. (2014). Virulence genes, antibiotic resistance and integrons in *Escherichia coli* strains isolated from synanthropic birds from Spain. *Avian Pathology*, 43(2), 172–175. https://doi.org/10.1080/03079457.2014.897683

Salanitro, J. P., Blake, I. G., Muirehead, P. A., Maglio, M., & Goodman, J. R. (1978). Bacteria isolated from the duodenum, ileum, and cecum of young chicks. *Applied and Environmental Microbiology*, 35(4), 782–790. https://doi.org/10.1128/aem.35.4.782-790.1978

Samir, A., & Zaher, H. M. (2025). Chicken as a carrier of emerging virulent Helicobacter species: A potential zoonotic risk. *Gut Pathogens*, 17(1), 31. https://doi.org/10.1186/s13099-025-00707-2

Scharsack, J. P., Franke, F., Erin, N. I., Kuske, A., & Büscher, J. (2016). *Effects of environmental variation on host–parasite interaction in three-spined sticklebacks (Gasterosteus aculeatus)—ScienceDirect*. www.sciencedirect.com/science/article/pii/S0944200616300332

Schmiedová, L., Kreisinger, J., Kubovčiak, J., Těšický, M., Martin, J.-F., Tomášek, O., Kauzálová, T., Sedláček, O., & Albrecht, T. (2023). Gut microbiota variation between climatic zones and due to migration strategy in passerine birds. *Frontiers in Microbiology*, 14, 1080017. https://doi.org/10.3389/fmicb.2023.1080017

Seyedmousavi, S., Bosco, S. de M. G., de Hoog, S., Ebel, F., Elad, D., Gomes, R. R., Jacobsen, I. D., Jensen, H. E., Martel, A., Mignon, B., Pasmans, F., Piecková, E., Rodrigues, A. M., Singh, K., Vicente, V. A., Wibbelt, G., Wiederhold, N. P., & Guillot, J. (2018). Fungal infections in animals: A patchwork of different situations. *Medical Mycology*, 56(suppl_1), S165–S187. https://doi.org/10.1093/mmy/myx104

Seyedmousavi, S., Guillot, J., Arné, P., de Hoog, G. S., Mouton, J. W., Melchers, W. J. G., & Verweij, P. E. (2015). Aspergillus and aspergilloses in wild and domestic animals: A global health concern with parallels to human disease. *Medical Mycology*, 53(8), 765–797. https://doi.org/10.1093/mmy/myv067

Shang, K., Wei, B., Cha, S.-Y., Zhang, J.-F., Park, J.-Y., Lee, Y.-J., Jang, H.-K., & Kang, M. (2021). The occurrence of antimicrobial-resistant *Salmonella enterica* in hatcheries and dissemination in an integrated broiler chicken operation in Korea. *Animals*, 11(1), 154. https://doi.org/10.3390/ani11010154

Shrubsole-Cockwill, A. N., Millins, C., Jardine, C., Kachur, K., & Parker, D. L. (2010). Avian pox infection with secondary *Candida albicans* encephalitis in a juvenile golden eagle (Aquila chrysaetos). *Journal of Avian Medicine and Surgery*, 24(1), 64–71. https://doi.org/10.1647/2009-010R.1

Simi, W. B., Leite-Jr, D. P., Paula, C. R., Hoffmann-Santos, H. D., Takahara, D. T., & Hahn, R. C. (2019). Yeasts and filamentous fungi in psittacidae and birds of prey droppings in midwest region of Brazil: A potential hazard to human health. *Brazilian Journal of Biology*, 79, 414–422. https://doi.org/10.1590/1519-6984.181192

Siqueira, N. P., Favalessa, O. C., Maruyama, F. H., Dutra, V., Nakazato, L., Hagen, F., & Hahn, R. C. (2022). Domestic birds as source of cryptococcus deuterogattii (AFLP6/VGII): Potential risk for cryptococcosis. *Mycopathologia*, 187(1), 103–111. https://doi.org/10.1007/s11046-021-00601-w

Skarżyńska, M., Zając, M., Bomba, A., Bocian, Ł., Kozdruń, W., Polak, M., Wiącek, J., & Wasyl, D. (2021). Antimicrobial resistance glides in the sky—Free-living birds as a reservoir of resistant Escherichia coli with zoonotic potential. *Frontiers in Microbiology*, 12. https://doi.org/10.3389/fmicb.2021.656223

Smith, C. J., Rocha, E. R., & Paster, B. J. (2006). The medically important Bacteroides spp. in health and disease. In *The Prokaryotes* (pp. 381–427). Springer. https://doi.org/10.1007/0-387-30747-8_14

Smith, O. M., Snyder, W. E., & Owen, J. P. (2020). Are we overestimating risk of enteric pathogen spillover from wild birds to humans? *Biological Reviews*, 95(3), 652–679. https://doi.org/10.1111/brv.12581

Souza, L. R., Santana, C. H., Ribeiro, L. N., Sousa, D. R., Gomes, K. M. C., Barroca, J. M. J. F., Silva, M. B. G., & Santos, R. L. (2024). Pneumonia, airsacculitis and osteomyelitis caused by *Aspergillus fumigatus* in an African grey parrot (*Psittacus erithacus*)—Case report. *Arquivo Brasileiro de Medicina Veterinária e Zootecnia*, 76, 269–274. https://doi.org/10.1590/1678-4162-13094

St. Amand, J. A., Otto, S. J. G., Cassis, R., & Annett Christianson, C. B. (2013). Antimicrobial resistance of *Salmonella enterica* serovar Heidelberg isolated from poultry in Alberta. *Avian Pathology*, 42(4), 379–386. https://doi.org/10.1080/03079457.2013.811465

Susanti, R., Utami, N. R., & Yuniastuti, A. (2023). Characterization of microbiota and secretory Ig-A in the domestic duck (Anas platyrhynchos) small and large intestine. *Biodiversitas Journal of Biological Diversity*, 24(4). https://doi.org/10.13057/biodiv/d240461

Svihus, B., Choct, M., & Classen, H. L. (2013). Function and nutritional roles of the avian caeca: A review. *World's Poultry Science Journal*, 69(2), 249–264. https://doi.org/10.1017/S0043933913000287

Talazadeh, F., Ghorbanpoor, M., & Shahriyari, A. (2022). Candidiasis in birds (*Galliformes, Anseriformes, Psittaciformes, Passeriformes*, and *Columbiformes*): A focus on antifungal susceptibility pattern of *Candida albicans* and *Non-albicans* isolates in avian clinical specimens. *Topics in Companion Animal Medicine*, 46, 100598. https://doi.org/10.1016/j.tcam.2021.100598

Tanase, O. I., Pavel, G., Hritcu, O. M., Dascalu, M. A., Bratuleanu, B. E., Rimbu, C. M., & Bocaneti, F. D. (2024). Disseminated infection with *Aspergillus fumigatus* in a scarlet macaw parrot (Ara macao)—A case report. *Animals: An Open Access Journal from MDPI*, 14(15), 2282. https://doi.org/10.3390/ani14152282

Tardón, A., Bataller, E., Llobat, L., & Jiménez-Trigos, E. (2021). Bacteria and antibiotic resistance detection in fractures of wild birds from wildlife rehabilitation centres in Spain. *Comparative Immunology, Microbiology and Infectious Diseases*, 74, 101575. https://doi.org/10.1016/j.cimid.2020.101575

Tawakol, M. M., Nabil, N. M., Samir, A., M., H. H., Yonis, A. E., Shahein, M. A., & Elsayed, M. M. (2023). The potential role of migratory birds in the transmission of pathogenic Campylobacter species to broiler chickens in broiler poultry farms and live bird markets. *BMC Microbiology*, 23(1), 66. https://doi.org/10.1186/s12866-023-02794-0

Teyssier, A., Rouffaer, L. O., Saleh Hudin, N., Strubbe, D., Matthysen, E., Lens, L., & White, J. (2018). Inside the guts of the city: Urban-induced alterations of the gut microbiota in a wild passerine. *Science of The Total Environment*, 612, 1276–1286. https://doi.org/10.1016/j.scitotenv.2017.09.035

Tian, J., Sanganyado, E., Li, D., Li, Y., Gao, X., Lu, Z., & Liu, W. (2024). Changes of gut microbiota diversity and stability over the reproduction process in captive female Gentoo penguins (Pygoscelis papua). *Polar Biology*, 47(7), 651–662. https://doi.org/10.1007/s00300-024-03260-x

Trevelline, B. K., Sprockett, D., DeLuca, W. V., Andreadis, C. R., Moeller, A. H., & Tonra, C. M. (2023). Convergent remodelling of the gut microbiome is associated with host energetic condition over long-distance migration. *Functional Ecology*, 37(11), 2840–2854. https://doi.org/10.1111/1365-2435.14430

Turjeman, S., Corl, A., Wolfenden, A., Tsalyuk, M., Lublin, A., Choi, O., Kamath, P. L., Getz, W. M., Bowie, R. C. K., & Nathan, R. (2020). Migration, pathogens and the avian microbiome: A comparative study in sympatric migrants and residents. *Molecular Ecology*, 29(23), 4706–4720. https://doi.org/10.1111/mec.15660

van Dongen, W. F., White, J., Brandl, H. B., Moodley, Y., Merkling, T., Leclaire, S., Blanchard, P., Danchin, É., Hatch, S. A., & Wagner, R. H. (2013). Age-related differences in the cloacal microbiota of a wild bird species. *BMC Ecology*, 13(1), 11. https://doi.org/10.1186/1472-6785-13-11

Velasquez, C. G., Macklin, K. S., Kumar, S., Bailey, M., Ebner, P. E., Oliver, H. F., Martin-Gonzalez, F. S., & Singh, M. (2018). Prevalence and antimicrobial resistance patterns of *Salmonella* isolated from poultry farms in southeastern United States. *Poultry Science*, 97(6), 2144–2152. https://doi.org/10.3382/ps/pex449

Vesth, T., Ozen, A., Andersen, S. C., Kaas, R. S., Lukjancenko, O., Bohlin, J., Nookaew, I., Wassenaar, T. M., & Ussery, D. W. (2013). Veillonella, firmicutes: Microbes disguised as Gram negatives. *Standards in Genomic Sciences*, 9(2), 431–448. https://doi.org/10.4056/sigs.2981345

Waite, D. W., & Taylor, M. W. (2014). Characterizing the avian gut microbiota: Membership, driving influences, and potential function. *Frontiers in Microbiology*, 5, 223. https://doi.org/10.3389/fmicb.2014.00223

Waite, D. W., & Taylor, M. W. (2015). Exploring the avian gut microbiota: Current trends and future directions. *Frontiers in Microbiology*, 6, 673. https://doi.org/10.3389/fmicb.2015.00673

Wang, D., Ji, X., Jiang, B., Yuan, Y., Liang, B., Sun, S., Zhu, L., Liu, J., Guo, X., Yin, Y., & Sun, Y. (2024). Prevalence of antibiotic resistance and virulence genes in *Escherichia coli* carried by migratory birds on the inner Mongolia Plateau of Northern China from 2018 to 2023. *Microorganisms*, 12(6), 1076. https://doi.org/10.3390/microorganisms12061076

Wang, J., Hong, M., Long, J., Yin, Y., & Xie, J. (2022). Differences in intestinal microflora of birds among different ecological types. *Frontiers in Ecology and Evolution*, 10. https://doi.org/10.3389/fevo.2022.920869

Wang, J., Ma, Z.-B., Zeng, Z.-L., Yang, X.-W., Huang, Y., & Liu, J.-H. (2017). The role of wildlife (wild birds) in the global transmission of antimicrobial resistance genes. *Zoological Research*, 38(2), 55

Wang, W., Cao, J., Yang, F., Wang, X., Zheng, S., Sharshov, K., & Li, L. (2016). High-throughput sequencing reveals the core gut microbiome of Bar-headed goose (Anser indicus) in different wintering areas in Tibet. *MicrobiologyOpen*, 5(2), 287–295. https://doi.org/10.1002/mbo3.327

Wang, W., Zheng, S., Sharshov, K., Sun, H., Yang, F., Wang, X., Li, L., & Xiao, Z. (2017). Metagenomic profiling of gut microbial communities in both wild and artificially reared Bar-headed goose (Anser indicus). *MicrobiologyOpen*, 6(2), e00429. https://doi.org/10.1002/mbo3.429

Wei, S., Morrison, M., & Yu, Z. (2013). Bacterial census of poultry intestinal microbiome. *Poultry Science*, 92(3), 671–683. https://doi.org/10.3382/ps.2012-02822

White, J., Mirleau, P., Danchin, E., Mulard, H., Hatch, S. A., Heeb, P., & Wagner, R. H. (2010). Sexually transmitted bacteria affect female cloacal assemblages in a wild bird. *Ecology Letters*, 13(12), 1515–1524. https://doi.org/10.1111/j.1461-0248.2010.01542.x

Woodhams, D. C., Bletz, M. C., Becker, C. G., Bender, H. A., Buitrago-Rosas, D., Diebboll, H., Huynh, R., Kearns, P. J., Kueneman, J., Kurosawa, E., LaBumbard, B. C., Lyons, C., McNally, K., Schliep, K., Shankar, N., Tokash-Peters, A. G., Vences, M., & Whetstone, R. (2020). Host-associated microbiomes are predicted by immune system complexity and climate. *Genome Biology*, 21(1), 23. https://doi.org/10.1186/s13059-019-1908-8

Wu, J., Huang, Y., Rao, D., Zhang, Y., & Yang, K. (2018). Evidence for environmental dissemination of antibiotic resistance mediated by wild birds. *Frontiers in Microbiology*, 9, 745. https://doi.org/10.3389/fmicb.2018.00745

Wyrsch, E. R., Nesporova, K., Tarabai, H., Jamborova, I., Bitar, I., Literak, I., Dolejska, M., & Djordjevic, S. P. (2022). Urban wildlife crisis: Australian silver gull is a bystander host to widespread clinical antibiotic resistance. *mSystems*, 7(3), e00158–22. https://doi.org/10.1128/msystems.00158-22

Xu, Q., Yuan, X., Gu, T., Li, Y., Dai, W., Shen, X., Song, Y., Zhang, Y., Zhao, W., Chang, G., & Chen, G. (2017). Comparative characterization of bacterial communities in geese fed all-grass or high-grain diets. *Plos One*, 12(10), e0185590. https://doi.org/10.1371/journal.pone.0185590

Yuan, Y., Liang, B., Jiang, B., Zhu, L., Wang, T., Li, Y., Liu, J., Guo, X., Ji, X., & Sun, Y. (2021). Migratory wild birds carrying multidrug-resistant *Escherichia coli* as potential transmitters of antimicrobial resistance in China. *Plos One*, 16(12), e0261444. https://doi.org/10.1371/journal.pone.0261444

Zhao, W., Wang, Y., Liu, S., Huang, J., & Zhai, Z. (2015, February 17). *The Dynamic Distribution of Porcine Microbiota across Different Ages and Gastrointestinal Tract Segments | PLOS One*. https://journals.plos.org/plosone/article?id=10.1371/journal.pone.0117441

Zhu, C., Song, W., Tao, Z., Liu, H., Zhang, S., Xu, W., & Li, H. (2020). Analysis of microbial diversity and composition in small intestine during different development times in ducks. *Poultry Science*, 99(2), 1096–1106. https://doi.org/10.1016/j.psj.2019.12.030

CHAPTER SEVEN

An Overview of Poultry Litter Microbial Ecology and Application of Microbiome Analyses

Steven C. Ricke, Elena G. Olson, Carson T. Priddle, Maxim A. Peckenschneider, and Brett M. Hale

7.1 INTRODUCTION: COMPOSITION, SOURCES, AND MANAGEMENT OF POULTRY LITTER

The poultry industry generates three types of waste during the production cycle: litter from broilers, manure from laying hens, and dead birds (Edwards & Daniels, 1992). Poultry litter is a complex matrix composed of excreta, feed, feathers, broken eggs, feed waste, and bedding materials (Nahm, 2005; Wang et al., 2025). While poultry litter originates from bedding material, it is distinct from unused bedding. Bedding refers to fresh material applied to the floor of the poultry house before day-old birds arrive (Figure 7.1). Litter, in contrast, is a bedding material that subsequently includes poultry waste products, such as feces, feed particles, and moisture (Leigsalz et al., 2025). Used litter primarily consists of water and carbon, with lesser amounts of nitrogen, phosphorus, and potassium, and minor levels of various elements, such as chlorine, manganese, iron, magnesium, sodium, and arsenic (Edwards & Daniels, 1992). In a commercial production system, litter is generally classified as single use, partial reuse, or multiuse, depending on how many production cycles it remains in the house (Bolan et al., 2010).

Poultry litter materials can originate from a variety of sources, including straw, wood shavings, sawdust, straw pellets, and seed husks (Bodie et al., 2019; Liegsalz et al., 2025). Choosing a litter source may depend on local availability. For example, rice is an important cereal crop in the state of Arkansas (Bodie et al., 2019). Therefore, rice husks gained regional popularity as a poultry litter material due to their ready availability as a rice processing byproduct (Bodie et al., 2019). Given the immense number of birds produced in the United States, it is not surprising that equally large quantities of used litter are continuously generated throughout the production cycle (Wang et al., 2025).

Managing these vast amounts of poultry waste presents a significant disposal problem for the poultry industry. A variety of reuse and disposal strategies have been explored, including feeding litter to ruminants, biogas production, and composting prior to land application (Edwards & Daniels, 1992). In practice, litter is generally reprocessed for ruminant feed, and its composition is less suitable for biogas production (Edwards & Daniels, 1992). Ultimately, the primary outlet for dispensing poultry waste and manure is agricultural land (Bolan et al., 2010). When applied to soil, these materials can serve as valuable sources of nitrogen, phosphorus, and potassium, improving soil fertility (Bolan et al., 2010; Wang et al., 2025). Soil improvement through poultry waste amendment extends beyond supplying nutrients. These materials can also contribute micronutrients, increase organic matter content, and enhance water-holding capacity, among other benefits that support agricultural crop production (Bolan et al., 2010). While this chapter does not aim to provide an exhaustive review of

DOI: 10.1201/9781003638568-7

Figure 7.1 Fresh poultry litter versus used poultry litter.

the published literature in this field of research, it offers an overview of poultry litter management practices, associated environmental concerns, microbial ecology of litter, and the application of microbiome-based analytical approaches, with key observations emphasized throughout.

7.2 ENVIRONMENTAL IMPLICATIONS OF POULTRY LITTER USE

Environmental issues can arise from poor management, particularly when land near water bodies is overloaded with poultry litter (Wang et al., 2025). Such practices can contribute to environmental air pollution by producing odorous nitrogen- and sulfur-based gases, as well as by volatilizing ammonia and generating greenhouse gases during the decomposition of poultry waste (Chavez et al., 2004; Bolan et al., 2010). Additionally, other organic compounds and micronutrients, including nitrogen, phosphorus, arsenic, copper, and zinc, can also contribute to environmental pollution of ground and surface water (Jones, 2007; Bolan et al., 2010). However, the contribution of arsenic to soil ecosystems remains controversial. Pathogenic organisms associated with litter create another environmental concern. The microorganisms in poultry litter can not only cause disease in the poultry house but also pose ecological and public health risks. During heavy rainfall events, litter-associated bacteria can be transported via soil runoff, contributing to the dissemination of antibiotic-resistant pathogens (Yang et al., 2021; Wang et al., 2025). If human pathogens are among the mobilized microorganisms, the risk extends beyond agricultural settings and becomes a broader public health concern. However, knowledge of the microbial composition of poultry litter and the factors that influence the occurrence of specific organisms is limited. Microbial identification has evolved substantially with the advent of next-generation sequencing technologies, enabling far more comprehensive assessments of poultry litter microbiology than previously possible. With the addition of multiomic tools, including metagenomics, metatranscriptomics, and metabolomics, researchers can move beyond simple taxonomic profiles to examine functional enzymes, metabolic activity, and nutrient transformations. These approaches provide new insight into how microbial communities cycle key nutrients and how these nutrient-driven processes influence microbial succession, pathogen dynamics, and overall litter quality. Together, these technologies offer a detailed understanding of litter ecology and their relevance to poultry production and soil fertility.

7.3 PATHOGENS ASSOCIATED WITH POULTRY LITTER

Poultry litter and bedding materials are exposed to a diverse array of microorganisms throughout the poultry production cycle (Figure 7.2). In addition to those contributed directly by the birds, environmental sources, such as insects, rodents, aerosols, dust, and wild animals, continually introduce new microbial populations into the litter matrix. During the lifespan of litter, microbial metabolism drives the breakdown and mineralization of organic compounds, ultimately determining the final chemical composition of the waste prior to land application (Edwards & Daniels, 1992; Nahm, 2005). While most microorganisms are not pathogenic, others can cause disease in birds and/or foodborne illness. Some of the pathogenic

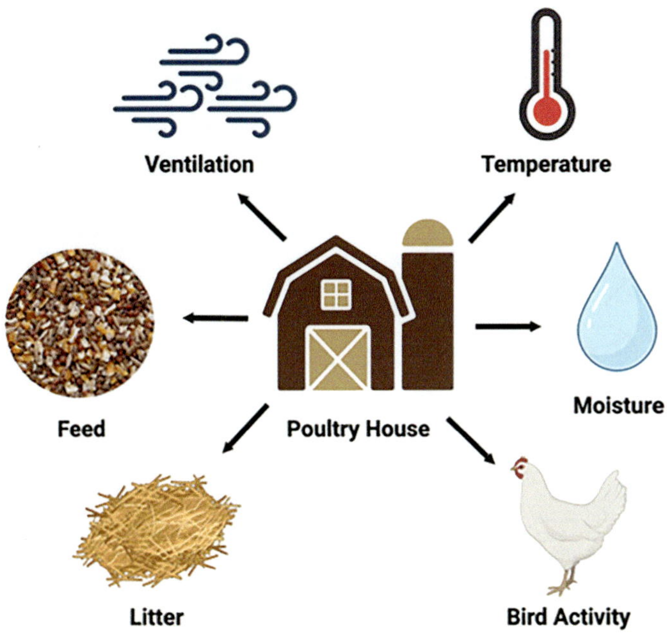

Figure 7.2 Potential contamination routes for poultry litter. Created in BioRender. Olson, E. (2025) https://BioRender.com/1xhqj9l.

and other microorganisms identified in poultry litter include Salmonella, pathogenic *Escherichia coli*, Staphylococcus, Streptococcus, *Listeria monocytogenes*, coliforms, Mycobacterium, Clostridium, Bordetella, and Campylobacter (Terzich et al., 2000; Locatelli et al., 2017; Wang et al., 2025). Salmonella is among the most frequently associated with litter and the most studied foodborne pathogen (Wang et al., 2025). Its presence in litter poses a risk to the flock and the surrounding environment, as Salmonella can be transmitted to water and land environments, where it may persist for extended periods of time (Gu et al., 2019; Wang et al., 2025). For example, Gu et al. (2019) reported that Salmonella could survive for 4–6 months in soil fertilized with poultry litter. This notable environmental persistence, despite often suboptimal conditions, underscores the need for effective detection, surveillance, and management strategies for Salmonella and other litter-associated pathogens.

7.4 LITTER MANAGEMENT STRATEGIES AND INTERVENTION APPROACHES

Limiting the spread of pathogens from litter before land application is an important management consideration. Composting combined with aeration is a standard method used to decrease overall litter volume and odors, minimize flies, reduce weed viability, and reduce pathogens (Edwards & Daniels, 1992). In general, composting proceeds through several distinct stages: an initial mesophilic phase, a transition to thermophilic temperatures, and cooling and maturation. This progression yields a uniform, stable, and nutrient-dense soil additive (Wang et al., 2025). Combining composting principles with in-house windrowing is a common strategy for managing poultry litter. In this approach, poultry litter is piled into rows spanning the length of the poultry house, with aeration achieved either through mechanical turning or forced-air systems (Edwards & Daniels, 1992; Dittoe et al., 2018). Although the two processes share similar goals, they differ in time required to complete the cycle, the temperature profiles achieved, and the characteristics of the residual material at the end of treatment (Dittoe et al., 2018). The thermophilic conditions generated during composting are generally assumed to inactivate many pathogens present in litter (Edwards & Daniels, 1992). Additionally, applying water to windrowed litter via a sprinkler system can enhance microbial heating and help the litter reach core and peripheral temperatures sufficient to inactivate many pathogens when maintained for an adequate duration (Dittoe et al., 2018). Litter can also be amended directly in the

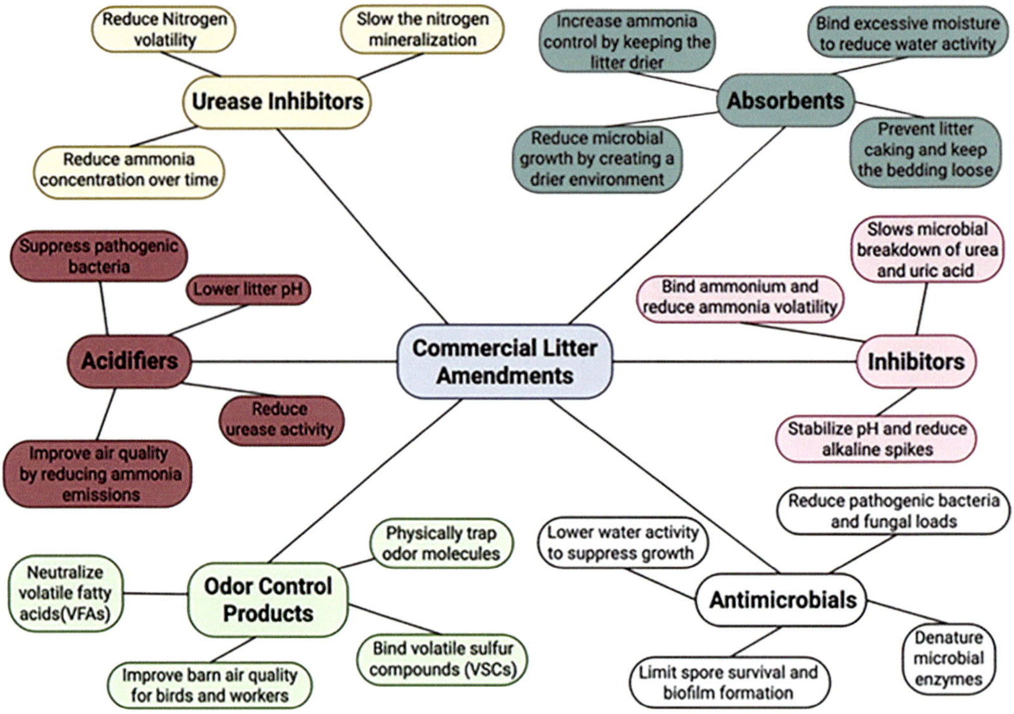

Figure 7.3 Impact of commercial litter amendments on commercial poultry production. Created in BioRender. Olson, E. (2025) https://BioRender.com/8rwnuxh.

poultry house before composting (Figure 7.3). Most amendment strategies aim to reduce elevated ammonia levels in poultry litter and include ventilation, competitive exclusion cultures, enzymes, and chemical acidifiers (Wang et al., 2025). Several commercial acidifiers, generally formulated with aluminum sulfate, sodium bisulfate, and acidified clay, are implemented in the poultry industry to convert volatile ammonia to nonvolatile ammonium (Moore et al., 2025; Wang et al., 2025). In addition to controlling ammonia emissions, these acidifiers can also reduce pathogens and inhibit microbial processes associated with ammonia production (Wang et al., 2025). Since the majority of ammonia emissions arise from microbial degradation of nitrogenous compounds excreted by poultry, particularly uric acid and urea, some mitigation strategies have focused on directly inhibiting the microorganisms responsible for these initial catabolic activities. Approaches include the use of nitro-containing compounds (nitroethanol, nitropropanol, and nitropropionic acid), supplementation with specific minerals such as copper, or dietary inclusions of egg-derived antibodies targeting uricase activity (Hunde et al., 2012; Kim et al., 2013). Future research may reveal that dietary management strategies can influence not only the microbial composition of poultry litter but also the metabolic pathways and function of the organisms that shape these ecosystems.

7.5 MICROBIAL ECOLOGY AND FUNCTIONAL DYNAMICS OF POULTRY LITTER

Given the diverse components of poultry litter, it is not surprising that its microbial ecology is highly complex. The sources of microorganisms are inherently variable, and as litter transitions from fresh bedding to used litter and eventually to composting, the associated microbial communities are expected to shift in composition and metabolic activities as substrate availability changes (see Figure 7.2). These dynamics become even more complex when multiple flocks are raised sequentially on the same litter. Nodar et al. (1990) conducted one of the early comprehensive studies of poultry litter microbiology by combining chemical and physical analyses with culture-based enumeration methods, including liquid most probable number (MPN) assays and plating to determine colony-forming units (CFUs) per gram of dry litter. Over 14 weeks in pine-sawdust litter, they identified numerous functional microbial groups, such as denitrifiers, acidophiles, aerobic spore-forming bacteria, actinomycetes, fungi, and several anaerobic groups, including nitrogen-fixing organisms, cellulolytics, amylolytics, and pectinolytics. Overall, their findings indicated a decrease in bacterial density, accompanied by rapid mineralization of organic nitrogen to ammonium, whereas carbon mineralization was relatively low, particularly from cellulose. These shifts in nitrogen and carbon mineral profiles would be expected to have an impact on the fertilizer value of poultry litter when applied to agronomic soils.

7.6 POULTRY LITTER MICROBIOMES: ADVANCES IN SEQUENCING AND SAMPLING CONSIDERATIONS

Culture-based methods quantify viable microorganisms within an ecosystem and allow isolation of pure species for further metabolic and biochemical characterization. Combining enumeration with metabolic profiling can provide additional insights into community structure, functional potential, and the overall microbial ecology. However, attaining a comprehensive profile of microorganisms within an ecosystem is limited by the choice of media and the risk of missing viable and nonculturable species. In addition, pathogen recovery from poultry litter can suffer from considerable variation (Plumblee Lawrence et al., 2022). In recent years, substantial progress has been made in next-generation sequencing of microbial communities using the 16S rRNA gene to identify and catalogue microbial taxa (Olson et al., 2025). Sequencing approaches now include not only short-read targeting of hypervariable regions but also emerging long-read technologies that enable resolved taxonomic identification of microbial diversity within and between ecosystems (Olson et al., 2025). These advances, coupled with improvements in bioinformatic pipelines for processing and statistically analyzing raw sequence data, have greatly expanded the scope and utility of microbiome research applications (Olson et al., 2025).

While most microbiome research has focused on gastrointestinal and fecal-associated microbiota, poultry litter presents a more complex, soil-like matrix with distinct spatial and physicochemical heterogeneity. As a result, applying microbiome analyses to litter requires additional consideration of how to capture representative samples (Figure 7.4). Locatelli et al. (2017) compared fecal and litter microbial compositions from different locations within a commercial broiler barn housing approximately 25,000 birds. Samples were collected near cooling pads, water/feed lines, bulk litter between water and feeders, bulk litter areas along the sides of the poultry barn, and exhaust fans. The authors observed that litter microbial composition was drastically different among sample sites, whereas fecal samples were not different by location within the barn. Additionally, pooled samples did not differ significantly from the litter microbial composition but were substantially different from fecal samples. These findings indicate that litter behaves as a spatially structured ecosystem and that sampling location strongly influences observed microbial profiles. Consequently, representative litter sampling requires attention to within-house variability, whereas fecal samples tend to be more homogeneous.

Furthermore, litter management may affect the development of intestinal microbiota in chickens throughout their production cycle. Wang et al. (2016) examined the relationship between the broiler gut and litter microbiota across six consecutive 6-week growth cycles, with a 2-week downtime between each cycle. Fresh and reused litter, ileal mucosa, and cecal digesta were compared. The gut microbiota in both gut compartments was affected by the bird's age and litter management. More specifically, reused poultry litter was associated with greater cecal richness than fresh litter. These findings suggest that reused litter can accelerate microbial colonization in young chicks and play a critical role in shaping gut microbial communities. Thus, characterizing litter microbial composition before bird placement is crucial for understanding its potential influence on early gut colonization and subsequent poultry health outcomes. These observations also suggest that litter microbial populations mirror the age-related shifts occurring in the poultry gut microbiota as the bird progresses throughout the production cycle. For example, Danzeisen et al. (2015) reported that litter microbiomes in turkey houses shifted over time, mirroring changes in gut microbiota, particularly in ileal microbiomes. As research advances, additional relationships between litter microbiomes and gut microbial populations will likely emerge, including potential health risks associated with specific litter

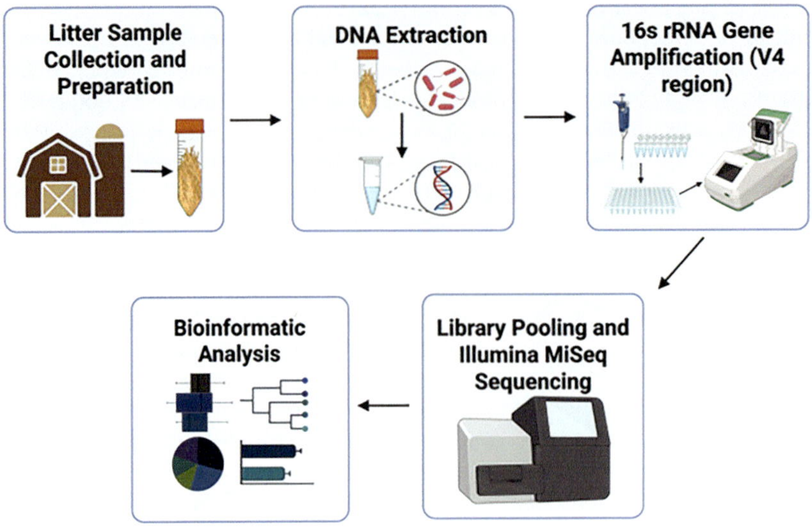

Figure 7.4 Microbiome sequencing basic steps for poultry litter samples. Created in BioRender. Olson, E. (2025) https://BioRender.com/mof3438.

management practices. Moreover, the impact of litter on the underlying soil should be considered, as certain microorganisms may persist in subsoils following litter cleanouts and contribute to long-term microbial reservoirs within the production system (Crippen et al., 2019).

7.7 DOWNSTREAM IMPACTS: POULTRY LITTER INFLUENCE ON SOIL MICROBIOMES

While poultry litter microbial communities can influence poultry gut microbiomes, the impact of litter on soil microbial ecology also warrants deliberation. Microorganisms in poultry litter can persist in soils where it is applied. While some can be beneficial, others may be pathogenic and pose public health risks. For example, Gu et al. (2019) observed that Salmonella serovars were primarily detected in fresh poultry litter and in field soils amended with it. However, more recent metagenomic studies suggest that litter-associated microbiota and antibiotic-resistant (AMR) genes exert minimal long-term impact on indigenous soil and water microbial ecosystems. Durán-Visera et al. (2024) reported that litter-associated taxa were either outcompeted by environmental microbial communities or introduced at too low concentrations to persist. Further research will be needed using whole-genome sequencing to precisely identify individual pathogen strains and/or other organisms originating in the poultry house, track their potential routes from the litter to soil and associated water microbial ecosystems, and assess their ability to persist in these environments. Importantly, these observations demonstrate that taxonomic identification or the detection of AMR genes alone does not entirely describe the pathogenic potential of a microbial ecosystem. The presence of Salmonella may not necessarily indicate expression of virulence factors, survival capacity, or the ability to colonize hosts. Similarly, identifying AMR genes in a metagenomic dataset does not reveal whether these genes are expressed, mobilized, or confer phenotypic resistance in situ. Further research employing whole-genome sequencing and strain-resolved metagenomics will be needed to precisely track pathogen lineages originating in poultry houses and follow their potential movement. Such approaches will be critical for evaluating how these microorganisms retain virulence, fitness, and persistence when introduced into new environments.

Litter serves as an agronomic amendment, enriching soils with nutrients that promote the growth of indigenous microorganisms (Yang et al., 2019). To evaluate these effects in pasture systems, Yang et al. (2019) conducted long-term field studies examining soil taxonomic diversity after 13 years of conservation management with or without poultry litter application. Microbial populations from control pasture sites, which received neither poultry litter nor cattle manure, were taxonomically distinct and exhibited lower richness and diversity than those from amended sites. The authors concluded that continuous cattle grazing and repeated poultry litter application contribute to increasing soil microbial diversity. In a follow-up study, Yang et al. (2021) reported that watershed bacterial richness also increased after poultry litter amendment. As sequencing technologies continue to advance, more comprehensive taxonomic analyses of soils before and after poultry litter application will provide deeper insights into how these systems respond to amendment. Additional factors, such as litter type, amendment use, bird type, flock density, and litter age, are also likely to influence soil microbial outcomes. Likewise, the choice of cover crop may modulate the soil microbial diversity response to poultry litter, particularly when nitrogen-fixing legumes or other plants with distinct rhizosphere communities are present.

7.8 CONCLUSIONS: INTEGRATING MICROBIAL ECOLOGY, MANAGEMENT, AND ENVIRONMENTAL OUTCOMES

Poultry litter remains a significant area of scientific and practical importance. While it plays a central role in maintaining poultry flock health and performance, it also serves as an agronomic fertilizer, bringing both benefits and challenges for environmental and public health issues. Although numerous poultry litter management strategies and chemical amendments have been employed to mitigate issues such as ammonia volatilization and nutrient runoff, the underlying microbial ecology of poultry litter has historically received far less attention. While the mineralization of litter components by certain bacterial groups is well documented, much less is known about broader microbial interactions, successional dynamics, and the ecological processes that shape litter communities over time. The introduction of 16S rRNA gene sequencing has greatly expanded our understanding by revealing relationships between litter microbiome, soil microbial diversity, and even the poultry gut microbiota, suggesting previously unrecognized ecological linkages among these systems. As sequencing technologies, whole-genome approaches, and bioinformatic capabilities continue to advance, many of these potential relationships will become clearer. Likewise, the introduction of analytical tools such as metabolomics and other functional assays will further advance our understanding of the metabolic pathways that drive litter transformation and nutrient cycling. Collectively, these tools will enable a comprehensive view of poultry litter as an interconnected microbial ecosystem, from the introduction of bedding, through its evolution during production, to its eventual landscape application. These insights will be critical for developing sustainable litter management strategies that support animal health and environmental stewardship.

REFERENCES

Bodie AR, Micciche AC, Atungulu GG, Rothrock, Jr MJ, Ricke SC (2019) Current trends of rice milling byproducts for agricultural applications and alternative food production systems. *Front. Sustain. Food Syst.* 3:47. doi: 10.3389/fsufs.2019.00047.

Bolan NS, Szogi AA, Chuasavathi T, Seshadri B, Rothrock Jr, MJ, Panneerselvam (2010) Uses and management of poultry litter. *World's Poult Sci J.* 66: 673–698. doi:10.1017/S0043910000656.

Chavez C, Coufal CD, Carey JB, Lacey RE, Beier RC, Zahn A (2004) The impact of supplemental dietary methionine sources on volatile compound concentrations in broiler excreta. *Poultry Sci.* 83: 901–910.

Crippen TL, Sheffield CL, Singh B, Byrd JA, Beier RC. (2019) How management practices within a poultry house during successive flock rotations change the structure of the soil microbiome. *Fronts. Microbiol.* 10: 2100. doi.3389/fmicb.2019.02100.

Danzeisen JL, Clayton JB, Huang H, Knights D, McComb B, Hayer SS, Johnson TJ (2015) Temporal relationships exist between cecum, ileum, and litter microbiomes in a commercial turkey flock, and subtherapeutic penicillin treatment impacts ileum bacterial community establishment. *Fronts. Vet. Sci.* 2:56. doi:10.3389/fvets.2015.00056.

Dittoe, DK, McDaniel CD, Tabler T, Kiess AS (2018) Windrowing poultry litter after a broiler house has been sprinkled with water. *J. Appl. Poultry Res.* 27: 1–15. doi.org/10.3382/japr/pfx034.

Durán-Visera A, Lindner BG, Hatt JK, Lai A, Wallace R, Ginn O, Brown J, Konstantinidis KT (2024) Metagenomic insights into the impact of litter from poultry concentrated animal feeding operations (CAFOs) to adjacent soil and water microbial communities. *Science Total Environ.* 920: 170772. doi.org/10.1016/j.scitotenv.2024.170772.

Edwards DR, Daniel TC (1992) Environmental impacts of on-farm poultry waste disposal – A review. *Bioresource Technol.* 41: 9–33. doi.org/10.1016/0960-8524(92)90094-E.

Gu G, Strawn LK, Zheng J, Reed EA, Rideout SL (2019) Diversity and dynamics of *Salmonella enterica* in water sources, poultry litters, and field soils amended with poultry litter in major agricultural area of Virginia. *Front. Microbiol.* 10: 2868. doi: 10.3389/fmmicb.2019.02868.

Hunde AR, Patterson PH, Ricke SC, Kim WK. (2012) Supplementation of poultry feeds with dietary zinc and other minerals and compounds to mitigate nitrogen emissions – A review. *Biol. Trace Element Res.* 147: 386–394. doi. 10.1007/s12011-011-9310-8

Jones FT (2007) A broad view of arsenic. *Poult Sci.* 86: 2–14.

Kim WK, Patterson PH, Rodriguez-Lecompte JC, Ricke SC (2013) The potential to reduce poultry nitrogen emissions with specific uricase egg yolk feed grade antibodies. *World's Poult Sci. J.* 69: 45–55. doi:10.1017/S0043933913000056.

Liegsalz TM, Blechschmidt S, Heide, M, Blum U, Lehnhardt F, Gasti M, Steinhoff-Wagner J (2025) Characterizing framework for broiler chicken's bedding materials for focusing on preventative health. *BMC Poult Sci Manag.* 2: 6. d.org/10.1186/s44364-025-00009-y.

Locatelli A, Hiett KL, Caudill AC, Rothrock Jr MJ (2017) Do fecal and litter microbiomes vary within the major areas of a commercial poultry house, and does this affect sampling strategies for whole-house microbiome studies? *J. Appl. Poult Res.* 26: 325–336. doi.org/10.3382/japr/pfw076.

Moore Jr PA, Anderson K, Martin J, Li H, Pursell JL (2025). Development and testing of an indoor air scrubber within poultry houses. *J Appl Poult Res.* 34: 100599. doi.org/10.1016/j.japr.2025.100599.

Nahm KH (2005) Factors influencing nitrogen mineralization during poultry litter composting and calculations for available nitrogen. *World's Poult Sci. J.* 61: 238–255. doi: 10.1079/wps200455.

Nodar R, Acea MJ, Carballas T (1990) Microbial populations of poultry pine-dust litter. *Biol. Wastes* 33: 295–306.

Olson EG, Hale, BM, Chatman CC, Mantovani HC, E. Majumder EL-W, Ricke SC (2025) Gut microbiota and advances in microbiome sequencing-based technologies: Opportunities for potential biologics discovery in meat animals. *Meat and Muscle Biol.* 9(1): 1–21. doi.org/10.22175/mmb.20022.

Plumblee Lawrence JR, Cudnik D, Oladeinde (2022) Bacterial detection and recovery from poultry litter. *Fronts Microbiol.* 12: 803150. doi: 10.3389/fmicb.2021.803150.

Terzich M, Pope MJ, Cherry TE, Hollinger J (2000) Survey of pathogens in poultry litter in the United States. *J. Appl. Poult Res.* 9: 287–291.

Wang H, Jiang X, Wang J. 2025. Chapter 15. Managing poultry litter to improve safety and reduce environmental impact. In: S.C. Ricke, (Ed.) 2025. *Achieving Sustainable Production of Poultry Meat- Volume 1 Safety, Quality and Sustainability*, 2nd Edition. Burleigh Dodd Publishing, Cambridge, UK. pp. 367–402. doi.org/10.19103/AS.2024.0146.19.

Wang L, Lilburn M, Yu Z (2016) Intestinal microbiota of broiler chickens as affected by litter management regimens. *Fronts. Microbiol.* 7: 593. doi: 10.3389/fmicb.2016.00593.

Yang Y, Ashworth AJ, DeBruyn JM, Willet C, Durso LM, Cook K, Moore Jr PA, Owens PR (2019) Soil bacterial diversity is driven by long-term pasture management, poultry litter, and cattle manure inputs. *PeerJ* 7: e7839. doi.org/10.7717/peerj.7839.

Yang Y, Ashworth AJ, Durso LM, Savin M, DeBruyn JM, Cook K, Moore Jr PA, Owens PR (2021) Do long-term conservation pasture management practices influence microbial diversity and antimicrobial resistant genes in runoff? *Fronts. Microbiol.* 12: 617066. doi: 10.3389/fmicb.2021.617066.

CHAPTER EIGHT

The Microbiome of Companion Animals

Alex Muppasani and Sunil Thomas

8.1 INTRODUCTION

Dogs have been associated with humans for more than 15,000 years (Galibert et al., 2011; Tancredi & Cardinali, 2023) while cats have been closely associated with humans for more than 9,000 years (Driscoll et al., 2009). There are around 360 breeds of dogs (Federation Cynologique Internationale; www.fci.be/en/) and 73 breeds of cats (The International Cat Association; https://tica.org/ticas-breeds/browse-all-breeds/). Traditionally many breeds of dogs were used for hunting and retrieving or for guarding livestock from predators, while cats were used to control rodent pests in agricultural environment.

Companion animals, such as dogs and cats, play an increasingly important role in human society, providing not only companionship but also contributing to human health and wellbeing (Alessandri et al., 2020; Abdolghanizadeh et al., 2024). The human–animal bond has strengthened significantly in recent decades, and pets are now regarded as integral members of many households across the globe. Millions of dogs and cats are kept as pets in houses all over the world. This widespread companionship reflects not only emotional attachment but also the recognition of the multiple benefits that pets bring to human lives. Beyond companionship, pet ownership has been associated with a wide range of positive physical and psychological outcomes. Numerous studies have demonstrated that interacting with pets can reduce stress, alleviate symptoms of depression, mental illness, lower blood pressure, and decrease the risk of cardiovascular disease (Dotson & Hyatt, 2008; Wisdom et al., 2009; Levine et al., 2013). Furthermore, pets contribute to improved self-esteem, social interaction, and overall happiness among their owners (Scoresby et al., 2021). Evidence also suggests that pet ownership during early life may help reduce the risk of certain chronic conditions, including asthma, and allergies (Hesselmar et al., 1999; Litonjua et al., 2002; Black, 2012).

Recent research has begun to explore the mechanisms underlying these health benefits, with particular attention to the role of the microbiota—the diverse community of microorganisms that inhabit the bodies of both humans and animals. The mutual exchange of microbiota between pets and their owners may play a key role in influencing health outcomes (Trinh et al., 2018). Factors such as environmental exposure, diet, host genetics, medication use, and age can also shape the composition and diversity of the microbiota in both humans and animals (Jarett et al., 2021). Understanding these microbial interactions is crucial, as imbalances in the microbiota, known as dysbiosis, have been linked to various diseases. Therefore, future strategies for preventing and controlling infections and metabolic or immune disorders may benefit from incorporating knowledge about microbiota dynamics.

Research from the 1980s demonstrated that pets and their owners share common intestinal bacteria (Caugant et al., 1984). Studies in infants and young children have shown that early-life exposure to household furry pets increases richness and diversity of the human gut microbiome (Azad

DOI: 10.1201/9781003638568-8

et al., 2013; Tun et al., 2017). By characterizing and comparing the microbiota of pets and their owners, researchers can gain valuable insights into how microbial sharing influences both human and animal wellbeing. Moreover, maintaining the health of companion animals requires a detailed understanding of their microbial communities, as these can serve as reservoirs for potential zoonotic pathogens—microbes capable of transmitting between animals and humans (Wu et al., 2024).

Several studies have already begun to investigate the extent and implications of microbiota transmission between companion animals and their human owners. These findings underscore the importance of examining the microbiome as a critical link between animal and human health. Accordingly, this chapter reviews current knowledge on the microbiota of companion animals, with particular focus on their composition, factors influencing their diversity, and the potential pathways of microbiota sharing between pets and humans.

8.2 PETS IN THE HUMAN ECOSYSTEM

In modern society, domestic dogs and cats are the most prevalent and culturally significant companion animals. They occupy a central position within the human domestic environment, reflecting a long history of coevolution and social integration between humans and animals (Hernandez et al., 2022). In the United States alone, approximately 38.4% of households include at least one dog, while 25.4% include at least one cat (Nugent & Daugherty, 2022). This widespread cohabitation creates a shared living space in which humans and their pets exist in continuous and often intimate contact, forming an ideal environment for the bidirectional exchange of microorganisms (Song et al., 2013).

The human–animal bond extends far beyond mere cohabitation. For many individuals, companion animals occupy a deeply emotional and social role within the household. Surveys indicate that between 90% and 97% of pet owners regard their animals as full members of the family (Nugent & Daugherty, 2022; Brown, 2023).This attachment provides substantial emotional support and companionship, often resembling the social dynamics of human familial relationships (Korukorikoes et al., 2019). Such relationships can have measurable psychological and physiological benefits, including reductions in stress and loneliness and increases in perceived social connectedness.

From a scientific standpoint, the concept of a "companion animal" is defined not only by species identity but by this unique sociobehavioral relationship between humans and animals. A central mechanism underlying the depth of this relationship is anthropomorphism—the attribution of human-like emotions, intentions, and cognitive processes to nonhuman species (Serpell, 2002). Anthropomorphism allows pet owners to interpret and respond to animal behavior through a human lens, reinforcing emotional attachment and social reciprocity (Serpell, 2002). As a result, companion animals become integrated into the human social world, functioning as both social partners and household members.

This deep social and physical integration also carries biological implications. Close human–animal interactions facilitate the continuous exchange of microbial communities through touch, shared spaces, and even airborne particles (Song et al., 2013). Such microbial sharing contributes to the development of a "shared household microbiome," a complex ecological system comprising both human and animal microorganisms (Serpell, 2002). The concept of the human body as a "superorganism"—a composite entity formed by mammalian and microbial cells (Turnbaugh et al., 2007)—can thus be extended to include household pets. In pet-owning households, the boundaries between the microbiomes of humans, animals, and the domestic environment become increasingly porous.

Cohabitation and physical intimacy merge humans and their companion animals into an integrated ecological and microbial unit—a multispecies superorganism. Within this unit, microbial communities are constantly exchanged, adapted, and comanaged through daily activities such as petting, feeding, and sharing living spaces. This ongoing microbial transfer represents a dynamic process that not only shapes the composition of the household microbiome but also reflects the profound interconnectedness of biological and social life in modern human–animal relationships.

8.3 CANINE MICROBIOME

Across their bodies, dogs host diverse and site-specific microbial communities, each adapted to the unique physiological and environmental conditions of different anatomical regions. Distinct microbiota has been characterized on the skin, within the ear canal, on the conjunctiva of the eyes, and throughout the respiratory, genitourinary, and gastrointestinal tracts. Over the past several years, these microbial ecosystems have become a major focus of veterinary and comparative microbiome research, driven by growing recognition of their roles in maintaining health and preventing disease. While there appears to be a core microbiota—a set of bacterial taxa commonly shared among dogs—this core composition varies substantially across body sites and individuals. Such variation is shaped by a complex interplay of intrinsic and extrinsic factors. Intrinsic factors include age, which influences immune function and skin barrier properties; breed, which can affect physiology and predisposition to certain dermatological or metabolic conditions; and sex, which may alter hormone-related microbial patterns. In addition to host-specific factors, environmental and lifestyle influences play a crucial role in shaping the canine microbiome. For instance, dogs living in urban environments are often exposed to different microbial reservoirs and pollutants than those in rural settings, leading to differences in microbial diversity and composition. Similarly, diet—whether commercial kibble, raw, or fresh formulations—has been shown to significantly alter microbial communities, particularly within the gut and skin, by affecting nutrient availability and metabolic interactions among microbes (Pereira & Clemente, 2021).

8.3.1 Canine Gut Microbiome

The gastrointestinal tract (GIT) of a healthy dog hosts a highly complex and dynamic microbial ecosystem that plays a fundamental role in maintaining physiological balance and overall health (Kim et al., 2025). This diverse microbial community—collectively known as the gut microbiota—is intimately involved in numerous vital processes, including nutrient digestion and absorption, energy metabolism, immune system modulation, and even neurobehavioral regulation through the gut–brain axis (Kim et al., 2025; Sun et al., 2025). In recent years, advances in high-throughput sequencing and metagenomic analyses have deepened our understanding of the canine gut microbiome, revealing it to be a stable yet adaptable system that interacts intricately with both host genetics and environmental factors.

The gut microbiome of dogs is typically dominated by five major bacterial phyla: Firmicutes, Bacteroidetes, Fusobacteria, Proteobacteria, and Actinobacteria (Kim et al., 2025; Garrigues et al., 2022). While these phyla are also prominent in other mammalian species, the canine gut microbiome exhibits a distinctive feature—its notably high abundance of Fusobacteria in healthy adults (Garrigues et al., 2022). In humans, members of this phylum are often linked to gastrointestinal disorders; however, in dogs, Fusobacterium species are considered normal and beneficial commensals, typically comprising around 20% of the total microbial community (Garrigues et al., 2022). This dominance likely reflects an evolutionary adaptation to the dog's ancestral carnivorous diet, as Fusobacteria are highly efficient at degrading dietary proteins into amino acids and peptides—processes central to energy extraction and nitrogen balance in a meat-based diet (Garrigues et al., 2022).

Other dominant taxa perform essential metabolic roles that are conserved across mammalian hosts. Members of Firmicutes, particularly within the class Clostridia, are key producers of short-chain fatty acids (SCFAs) such as butyrate, which serves as the primary energy source for colonocytes and supports intestinal barrier integrity (Garrigues et al., 2022). Bacteroidetes exhibit remarkable metabolic flexibility, capable of fermenting diverse substrates, including complex polysaccharides, proteins, and host-derived glycans, thereby contributing to nutrient salvage and microbial cross-feeding interactions within the gut ecosystem (Garrigues et al., 2022). Proteobacteria and Actinobacteria, though present in smaller proportions, play important roles in mucosal immune modulation and the maintenance of redox balance in the intestinal lumen.

The composition and density of the microbial community vary markedly along the gastrointestinal tract, influenced by local physicochemical conditions such as pH, oxygen availability, nutrient gradients, and transit time (Kim et al., 2025; Huang et al., 2020). The upper GIT, the duodenum and jejunum, generally exhibits lower bacterial diversity due to higher oxygen tension and faster digesta flow. The duodenum is typically enriched in Firmicutes (46.4%) and Proteobacteria (26.6%) (Kim et al., 2025). In contrast, the colon and rectum provide a largely anaerobic environment conducive to the proliferation of strict anaerobes, leading to a higher overall microbial diversity dominated by Firmicutes and Bacteroidetes (Huang et al., 2020).

To assess and monitor intestinal health in dogs, researchers have developed quantitative molecular tools such as the Canine Microbiota Dysbiosis Index (CMDI) (AlShawaqfeh et al., 2017). The CMDI is a quantitative PCR (qPCR)-based assay that measures the relative abundance of seven bacterial taxa associated with dysbiosis and summarizes the results into a single numerical value. This standardized index provides a robust, objective measure of microbial imbalance, facilitating both clinical diagnosis and longitudinal monitoring of therapeutic interventions (Kim et al., 2025; AlShawaqfeh et al., 2017).

The persistent presence of Fusobacterium and other protein-fermenting taxa as core members of the healthy canine gut underscores an evolutionary optimization for a high-protein, meat-based diet (Garrigues et al., 2022). However, the dietary landscape of modern companion dogs has shifted dramatically. Contemporary commercial pet foods frequently contain substantial quantities of carbohydrates and plant-derived fibers as cost-effective energy sources and structural components (Li et al., 2017). This dietary transition represents an evolutionary mismatch between the dog's ancestral gut microbiome and its present-day nutritional environment. Such a mismatch may disrupt microbial equilibrium, predisposing dogs to chronic, low-grade dysbiosis—an imbalance increasingly implicated in metabolic dysfunctions, inflammatory bowel diseases, obesity, and other chronic conditions in companion animals (Huang et al., 2020; Garrigues et al., 2022).

Overall, the canine gut microbiome represents a finely tuned evolutionary system that reflects both the dietary history and ecological adaptations of the species. Modern feeding practices, while nutritionally adequate by traditional standards, may inadvertently shift this microbial ecosystem away from its evolutionary norm, with profound and long-term consequences for canine health. Understanding these microbial–dietary interactions is therefore essential for developing nutritionally and ecologically informed approaches to canine nutrition and disease prevention.

8.3.2 Breed-Specific Differences in the Canine Fecal Microbiome

Reddy et al. (2019) investigated the fecal microbiome composition in three dog breeds—Maltese, Miniature Schnauzer, and Poodle—of the same age, fed identical diets, and housed under the same conditions. The study revealed that breed significantly influences the composition of the fecal microbiome, highlighting the role of host genetic factors in shaping gut microbial communities.

At the phylum level, Firmicutes was the most dominant in all three breeds, although its relative abundance was significantly lower in the Maltese compared to the Miniature Schnauzer and Poodle. Within Firmicutes, the genus Lactobacillus was the most abundant across all breeds, showing the greatest prevalence in the Poodle, though differences among breeds were not statistically significant. Other Firmicutes genera, including Megamonas, Streptococcus, Blautia, and Catenibacterium, were also highly abundant. Streptococcus showed significant variation, being most abundant in Miniature Schnauzer, moderate in Poodle, and least abundant in Maltese fecal samples. Turicibacter abundance was significantly higher in Maltese and Poodle compared to Miniature Schnauzer. Blautia was enriched in Poodle and Miniature Schnauzer relative to Maltese, while Megamonas was highest in Miniature Schnauzer with minor differences between the other two breeds. Several Firmicutes genera—including Allobaculum, Phascolarctobacterium, Faecalibacterium, Peptococcus, and Clostridium—showed no significant differences among the breeds.

The second most dominant phylum, Bacteroidetes, included three known genera and two unclassified families. Prevotella was the most abundant genus, with the Maltese showing the highest abundance relative to Miniature Schnauzer and Poodle. Other genera, including Bacteroides and Paraprevotella, were present at 2.83% and 0.22%, respectively. The unclassified families Prevotellaceae and Bacteroidaceae did not differ significantly among the breeds.

Within Proteobacteria, genera such as Sutterella, Escherichia, Succinivibrio, and Anaerobiospirillum, along with an unclassified Sutterellaceae family, were detected. Notable differences included Succinivibrio, which was significantly higher in Miniature Schnauzer, and Anaerobiospirillum, which was highest in Poodle, moderate in Maltese, and lowest in Miniature Schnauzer.

Actinobacteria was the third most dominant phylum. Genera such as Bifidobacterium and Collinsella showed no significant variation across the breeds. In contrast, within Fusobacteria, the genus Fusobacterium was significantly more abundant in Maltese dogs than in Poodle or Miniature Schnauzer breeds.

This study underscores that fecal microbiome composition varies among dog breeds, even when diet and environmental factors are standardized. These differences suggest that host genetics play a key role in shaping gut microbial communities. Understanding breed-specific microbial patterns is essential for optimizing nutritional strategies and developing tailored interventions to maintain a healthy gut microbiome, which may ultimately support overall canine health and disease prevention.

You and Kim (2021) conducted a comparative analysis of the fecal microbiome across eight dog breeds: Dachshund, Maltese, Bichon, Yorkshire Terrier, Chihuahua, Pomeranian, Poodle, and Bulldog. Across all breeds, the major microbial phyla identified were Firmicutes, Bacteroidetes, Fusobacteria, Proteobacteria, and Actinobacteria.

At the phylum level, breed-specific differences were most notable for Fusobacteria, which showed significant variation among breeds. For instance, Maltese dogs exhibited a high abundance of Fusobacteria, whereas Poodles were enriched in Firmicutes and Actinobacteria. At the genus and species level, the most predominant species across all breeds included *Fusobacterium perfoetens*, *Mediterranea massiliensis*, *Prevotella copri*, and *Romboutsia timonensis*. Microbial richness, a measure of the total number of distinct taxa, was highest in Bulldogs, while microbial evenness, assessed by the Shannon diversity index, did not differ significantly among breeds.

The study also explored associations between the gut microbiome and body condition. Compared to normal-weight dogs, the overweight group exhibited higher abundances of Actinobacteria and Deferribacteres. At the species level, overweight dogs had increased levels of *Enterococcus cecorum*, *Peptostreptococcus russellii*, *Bacteroides fragilis*, *Bacteroides massiliensis*, *Bacteroides thetaiotaomicron*, and *Corynebacterium lactis*, whereas *Anaerobiospirillum succiniciproducens*, *Sutterella stercoricanis*, and *Mucispirillum schaedleri* were reduced.

Comparisons between lean and normal-weight dogs revealed that the lean group had higher abundances of Streptococcus equinus, Peptostreptococcus russellii, and Beduini massiliensis. Similarly, in lean versus overweight dogs, the lean group showed elevated levels of Anaerobiospirillum succiniciproducens, Beduini massiliensis, Romboutsia timonensis, and Streptococcus equinus, whereas the overweight group was enriched in Peptostreptococcus russellii, Schaalia cardiffensis, and Enterococcus cecorum. Additional distinctions included Fenollaria being more abundant in the lean group and Cellulosilyticum ruminicola in the normal group.

Of particular note, Fusobacterium perfoetens have been identified as an aging-related or opportunistic pathogen, frequently observed in older dogs, highlighting its potential role in age-associated microbiome shifts and disease susceptibility.

These findings underscore that both host genetics (breed) and physiological status (body condition) are significant determinants of the canine gut microbiome. Recognizing these patterns may help guide breed-specific nutrition and management strategies to support gut health and prevent obesity-related dysbiosis.

8.3.3 Working Canine Microbiome

Working dogs are vital contributors to societies worldwide, fulfilling a myriad of tasks in distinct environments that shape both their physical health and their microbiomes (Serpell, 2002). Historically, dogs have aided humans through hunting, herding, protection, and specialized work such as sled pulling in the Arctic, which remains a hallmark of high-latitude regions like Siberia, Canada, and Alaska (Smith et al., 2024). Sled dogs living in these regions endure extreme cold and physically demanding labor, consuming meat-based, high-protein diets due to ancestral feeding practices and the demands of their environment (Freeman, 2015). Recent research indicates that Alaskan and Inuit sled dogs possess gut microbiomes dominated by the phyla Firmicutes, Bacteroidota, and Fusobacteria, with highly abundant families such as Clostridiaceae and Fusobacteriaceae. These microbial communities support nutrient digestion and metabolic health. Factors like glucosamine supplementation and physical activity also influence the diversity and abundance of specific gut taxa in these athletic dogs (Wang et al., 2024).

Search and rescue dogs regularly operate in diverse, often challenging environments. These canines experience frequent exposure to various soils, moisture conditions, and environmental microbial sources, which results in a highly diverse nasal microbiome. In healthy dogs, the nasal cavity is dominated by the phylum Proteobacteria—particularly the family Moraxellaceae, with the genus Moraxella constituting most of the microbial community. Other genera, such as Streptococcus and Pseudomonas, are present but at considerably lower relative abundances (Tress et al., 2017). Unlike dogs in urban Western societies, who have skin microbiomes dominated by human-associated genera such as Streptococcus and Corynebacterium, subway environments reveal that human skin bacteria—particularly Cutibacterium, Corynebacterium, Staphylococcus, and Streptococcus—are consistently prevalent on surfaces due to frequent human contact and sanitized spaces. Research from multiple cities worldwide shows these same human-related taxa not only reside on subway surfaces, but that subway passengers' skin and the surfaces rapidly exchange these microbiota through physical contact, resulting in both skin and environmental flora closely resembling that of people and their household settings (Peimbert & Alcaraz, 2023).

Military and working dogs' gut microbiota exhibits notable adaptability as their environments and duties change. In a controlled cohort of military dogs, the gut microbiome underwent rapid changes during early growth and gradually stabilized as dogs matured and were exposed to shared external environments and similar diets. The microbiota's core was composed of resilient genera like Bacteroides, Peptoclostridium, Fusobacterium, Lactobacillus, and Blautia. Shifts in diversity and taxonomic composition reflected dietary transitions and the onset of field training, showing that the canine microbiome is sensitive to both environmental exposure and workload. Notably, factors such as stress, infection, and environment significantly shaped the structure and diversity of the microbiome, illustrating its dynamic and adaptive character in working dogs (An et al., 2025).

Breed is a central factor that determines the nature of the microbiome. Scandinavian sled dogs (such as Alaskan Huskies and Malamutes) maintain higher Fusobacterium levels than North American Labradors or German Shepherds. Reddy et al. (2019) showed that Maltese, Poodles, and Schnauzers fed identical diets and housed together still possessed distinct gut microbiome signatures, which implies genetics and regional climates shape microbial communities. Bai et al. (2023) extended this finding to Labradors and Poodles in disease settings, where breed-specific differences influenced recovery from gastrointestinal disturbance.

Diet remains another principal driver behind microbiome variation. Sled dogs and dogs in resource-limited Arctic and rural environments often eat fresh protein from natural sources, boosting microbial diversity and favoring protein-fermenting taxa such as Fusobacterium and *Clostridium hiranonis*. Urban working dogs on commercial diets show reduced richness, with more Firmicutes, Bacteroidetes, and Staphylococcus on the skin (Leverett et al., 2022; Pilla & Suchodolski, 2021). A fresh diet supports more SCFA-producing bacteria—important for energy, inflammation control, and gut health.

Environment and work type strongly influence microbial health and zoonotic risk. Search and rescue dogs actively transport soil and environmental bacteria into homes and vehicles, sharing Actinobacteria, Staphylococcus, and Corynebacterium with handlers. Song et al. (2013) documented how cohabiting humans have skin microbiomes more similar to their dogs than to other dogs, emphasizing frequent microbial exchange. Dogs working in rural Africa or South America, exposed to livestock and wildlife, have skin communities with higher Enterobacteriaceae, while those in humid tropics and Southeast Asia show increased Staphylococcus and risk for skin dysbiosis, especially under stressful work conditions (Sigudu et al., 2024)

Microbiome diversity proves essential to immune defense and disease resilience. Working dogs exposed to environmental extremes and stress often develop adaptive skin and gut ecosystems, but are also at risk for dysbiosis and zoonotic transmission. Atopic dermatitis, chronic GI disorders, and MRSA infection risk are all increased during periods of low microbial diversity or heavy antibiotic use (Kim et al., 2025). Bradley et al. (2016) found disease states in dogs are defined by the collapse of beneficial Firmicutes and overgrowth of opportunists such as *Staphylococcus pseudintermedius*. Antibiotics further induce persistent reductions in microbial richness and disrupt host health (Pilla et al., 2020).

The microbiota of dogs is strongly shaped by their living environment. Dogs residing in urban areas, which typically feature higher levels of sanitation and more standardized diets, tend to exhibit lower microbial diversity compared to their rural counterparts. These urban dogs display a gut microbiota with less richness and evenness, but it is relatively stable and often characterized by an increased abundance of certain bacterial groups, including Firmicutes. In contrast, rural dogs are exposed to a broader spectrum of environmental microbes, supporting a more diverse microbiota enriched with taxa such as Actinobacteria, Escherichia, Collinsella, and Holdemanella. Such differences highlight the significant impact of lifestyle and environment on shaping the composition and diversity of the canine gut microbiota, paralleling similar patterns seen in humans living in cities versus villages (Pavlíčková et al., 2025).

Globally, working dogs serve as ecological sentinels, reflecting diet, region, task, and environment in their microbiome signatures. From the subarctic sled dogs of Siberia to guide dogs in Tokyo, police K9s in urban Paris, and herding dogs in rural South America, every combination of breed, lifestyle, and geography produces a unique microbial profile. This adaptation is both protective and vulnerable, determining each dog's resistance to disease, efficiency of digestion, and ability to thrive amid the demands of human society (Garrigues et al., 2022; Leverett et al., 2022; Reddy et al., 2019).

8.3.4 The Canine Skin Microbiota

Throughout their bodies, dogs host distinct microbial communities that vary according to specific anatomical locations such as the skin, ear canal, conjunctiva, respiratory tract, genitourinary tract, and gastrointestinal tract. These site-specific microbiomes have become an important focus of research in recent years, as scientists strive to understand their complex roles in maintaining health and contributing to disease. While evidence suggests the existence of a core microbiota shared among dogs across various body sites, this composition is far from static. It is influenced by a combination of intrinsic factors—including age, breed, and sex—and extrinsic factors such as environmental exposure, lifestyle (urban versus rural settings), and diet (Pereira & Clemente, 2021).

Among these anatomical sites, the skin stands out as the dog's largest organ and serves as the primary interface between the host and the external environment. Far beyond acting as a simple physical barrier, the skin harbors a rich, dynamic, and diverse community of microorganisms—including bacteria, fungi, archaea, and viruses—that coexist in a delicate ecological balance (Štempelová et al., 2025; Gallo, 2017). This complex microbial ecosystem, collectively known as the skin microbiome, plays a fundamental role in maintaining cutaneous homeostasis and overall health.

Recent advances in high-throughput sequencing technologies have revolutionized our understanding of this ecosystem, revealing that the diversity and functional potential of the canine skin microbiota far surpass what was previously uncovered through traditional culture-based studies (Štempelová et al., 2025). Beyond serving as a passive collection of microorganisms, the skin microbiome performs vital functions in sustaining epidermal barrier integrity, regulating immune responses, and providing colonization resistance against opportunistic and pathogenic microbes (Pierezan et al., 2016; Flowers & Grice, 2020; Štempelová et al., 2025).

In healthy dogs, the skin bacterial community is primarily composed of four dominant phyla: Bacillota (formerly Firmicutes), Actinomycetota (Actinobacteria), Pseudomonadota (Proteobacteria), and Bacteroidota (Bacteroidetes) (Štempelová et al., 2025). At finer taxonomic levels, the composition varies widely among individuals and anatomical sites, but certain genera consistently appear as core members of the canine skin microbiome. Among these, Streptococcus is often the most abundant, followed by Curtobacterium, Bacteroides, and Corynebacterium (Štempelová et al., 2025). Each of these taxa fulfills distinct ecological roles—from Corynebacterium species that contribute to lipid metabolism and skin pH regulation to Streptococcus species that modulate immune signaling and occupy protective ecological niches on the epidermis.

The composition and structure of the canine skin microbiome are not uniform across the body surface but vary significantly by anatomical location. Different regions, such as the ear canal, interdigital spaces, groin, and muzzle, harbor distinct microbial communities shaped by variations in microenvironmental conditions including moisture, temperature, sebaceous secretion, and exposure to external elements (Štempelová et al., 2025). For instance, moist and occluded areas tend to support higher abundances of Proteobacteria and Bacteroidota, while drier sites are often dominated by Actinomycetota and Bacillota. These site-specific communities form discrete microbial niches that collectively contribute to the skin's overall ecological resilience and defense capacity.

A particularly striking observation from beta-diversity analyses, which assess differences in microbial community structure among samples, is that the individual dog itself exerts the strongest influence on the composition of its skin microbiome (Štempelová et al., 2025). Microbial samples collected from different body sites on the same dog are more similar to one another than samples from the same body site across different dogs. This pattern indicates the presence of a host-specific microbial fingerprint, a unique and stable signature reflecting the complex interplay of host genetics, immune function, physiological traits, and long-term environmental exposures (Rodrigues Hoffmann et al., 2014; Štempelová et al., 2025). Factors such as breed, age, sex, diet, grooming practices, and even the household environment further modulate this individual microbial identity.

This high level of interindividual variability has important implications for understanding human–animal microbial exchange. When microbial transfer occurs between a dog and its owner—through touch, shared environments, or close physical contact—it does not represent a generic or uniform exchange of "dog microbes." Instead, what is transferred is a highly personalized microbial consortium, specific to the dog's individual microbiome profile. Consequently, microbial sharing between humans and their pets reflects a personalized microbial dialogue, influenced by the unique microbiological identities of both host species.

This individuality in microbial composition may help explain why the health outcomes associated with pet ownership—such as the modulation of immune function, allergy development, or protection against atopic diseases—vary widely among individuals (Song et al., 2013). Each instance of human–animal cohabitation represents a distinct ecological partnership, where the microbial communities of both species interact in unique and dynamic ways. Understanding these personalized microbial exchanges offers valuable insight into the broader concept of the shared human–animal microbiome and its potential role in shaping host health and disease susceptibility in multispecies households.

Cusco et al. (2017) conducted a comprehensive analysis of the variability of the canine skin microbiota across eight distinct skin sites in a homogeneous, well-controlled cohort of Golden–Labrador Retriever crossbred dogs cohabiting in the same kennel. The study aimed to understand microbial diversity and site-specific differences in the skin microbiota of healthy dogs under uniform environmental conditions. The authors collected samples from eight defined skin regions: the inner pinna, chin, nasal skin, dorsal back, axilla, abdomen, interdigital region, and perianal area. Their investigation revealed a remarkable microbial diversity comprising 2,216 bacterial operational taxonomic units (OTUs), which were taxonomically classified into 17 phyla, 41 classes, 62 orders, 128 families, and 242 genera. The dominant phyla inhabiting the skin of these healthy dogs were Proteobacteria, Firmicutes, Bacteroidetes, Actinobacteria, Cyanobacteria, and Fusobacteria, followed by less abundant taxa such as TM7 and Tenericutes.

Among the various skin sites, the inner pinna harbored a notably higher abundance of Proteobacteria compared to other areas. Within this phylum, the major representative classes were Gammaproteobacteria, Alphaproteobacteria, and Betaproteobacteria. The phyla Firmicutes and Bacteroidetes were also well represented through the classes Bacilli and Flavobacteriia, respectively, occurring in similar abundances to Proteobacteria. Interestingly, the inner pinna exhibited a broad range of less abundant bacterial taxa, indicating it may be a particularly diverse microenvironment on the canine skin.

The chin and nasal skin regions both showed enrichment in Gammaproteobacteria, though their dominant bacterial families differed. The chin microbiota was primarily represented by the Pseudomonadaceae family, while the nasal skin was dominated by Pasteurellaceae. The differential distribution of these families highlights the distinct ecological niches created by local environmental factors, such as moisture, exposure, and glandular secretions, within each site.

The dorsal back and axilla shared similar overall taxonomic profiles. Both regions were dominated by Gammaproteobacteria, with Moraxellaceae being the predominant family, followed by Bacilli, notably Lactobacillaceae. However, one key difference distinguished the axilla: a markedly higher abundance of Staphylococcaceae (class Bacilli), which was also differentially distributed compared with other skin sites. This observation suggests that the axilla, being a more occluded and humid region, supports the proliferation of Staphylococcus species commonly associated with moist skin environments.

Likewise, the abdomen and interdigital regions shared comparable microbial communities. Both were dominated by Gammaproteobacteria, particularly members of Enterobacteriaceae, Moraxellaceae, and Pseudomonadaceae, along with the Cyanobacteria family Xenococcaceae. Notably, Planococcaceae (phylum Firmicutes) was detected exclusively in the abdomen and absent in the interdigital region, implying subtle yet important site-specific differences even between anatomically similar regions. These findings demonstrate that bacterial composition on canine skin is not only taxonomically complex but also spatially organized according to microenvironmental conditions.

Among all sampled areas, the perianal region exhibited the most distinct microbiota composition. Here, Firmicutes, particularly class Bacilli, was the dominant phylum, followed by Actinobacteria. Numerous bacterial families were differentially distributed in this region, indicating a highly specialized microbial niche. Prominent families included Erysipelotrichaceae, Lachnospiraceae, Lactobacillaceae, and Veillonellaceae (Firmicutes); Corynebacteriaceae (Actinobacteria); and Bacteroidaceae (Bacteroidetes). The perianal region also showed an enrichment of Fusobacteriaceae, although this was not statistically different compared to other sites. Collectively, these results underscore that the perianal area is the most divergent site in terms of bacterial composition and diversity on canine skin.

In addition to site-specific variation, Cusco et al. (2017) observed that host sex contributed to differences in microbiota composition. The authors found that in the abdomen, sex explained 11% of the microbiome variation according to the weighted UniFrac analysis, surpassing temporal influences. Across the three skin sites significantly affected by sex (abdomen, back, and axilla), males demonstrated an overrepresentation of bacteria from the phylum Fusobacteria, particularly the genera Sneathia and Fusobacterium. Other genera enriched in males included Actinomycetospora, Gemella, Parvimonas, Brevundimonas, and members of the candidate phylum SR1. Conversely, females exhibited an overrepresentation of the family Enterobacteriaceae. These findings suggest that sex-linked physiological or hormonal factors may modulate microbial colonization patterns on the skin, influencing both bacterial diversity and site-specific community structure.

8.3.5 Influence of Dry and Fresh Foods on the Canine Skin Microbiome

Leverett et al. (2022) conducted a study to examine whether diet, specifically fresh versus dry food, affects the skin microbiome composition of domestic dogs. The researchers analyzed bacterial communities across different skin sites and dietary groups to determine how diet influences microbial diversity and abundance.

At the phyla level, dogs consuming a dry-food diet exhibited increased relative abundances of Actinobacteria, Bacteroidetes, and Proteobacteria, while showing decreases in Planctomycetes, Firmicutes, and Chloroflexi compared to dogs fed a fresh-food diet. These shifts suggest that dietary composition can substantially alter the dominant bacterial groups residing on the skin.

When considering the three skin sites sampled—the groin, ear, and paw—site-specific differences were also observed. The groin region harbored a higher proportion of Firmicutes than the other areas, indicating that local microenvironmental factors such as moisture and sebum levels may favor certain bacterial taxa. In contrast, the paw displayed elevated proportions of Cyanobacteria and Planctomycetes, possibly reflecting greater environmental exposure and contact with surfaces.

Across all skin sites and diets, the most prevalent bacterial genera were Staphylococcus, Porphyromonas, Streptococcus, Corynebacterium, and Conchiformibius. Among these, Staphylococcus was particularly abundant in the groin compared to the ear and paw, highlighting its prominence as a core component of the canine skin microbiota. An overall dietary effect on the skin microbiome was detected, independent of sampling site. Specifically, dogs on a dry-food diet exhibited a significant decrease in alpha diversity relative to those on a fresh-food diet. This reduction in microbial richness suggests that fresh diets may support a more diverse and potentially more resilient skin microbiome.

In terms of taxa contributing most strongly to differences between diets, Staphylococcus, Porphyromonas, Streptococcus, Corynebacterium, Conchiformibius, and Pseudomonas accounted for the largest percentage of dissimilarity between the fresh- and dry-food groups. Furthermore, Pedobacter, Porphyromonas, Hymenobacter, Spirosoma, Corynebacterium, and Bacteroides were identified as the most effective discriminators, as they exhibited the highest ratio of average dissimilarity to standard deviation.

When individual genera were compared, the dry-food diet was associated with increased relative abundances of Hymenobacter, Acinetobacter, Neisseria, Stenotrophomonas, and Janthinobacterium. Conversely, it led to reduced levels of Actinomycetospora, Massilia, Bacteroides, Spirosoma, Mycoplasma, Jonesia, DA101, Sporosarcina, and Actinotelluria compared with the fresh-food diet. These findings collectively suggest that dietary formulation not only alters the overall diversity of the canine skin microbiome but also selectively enriches or suppresses particular bacterial taxa, potentially influencing skin health and microbial balance.

8.3.6 Effect of Diet on the Canine Microbiome

Extruded commercial diets are typically high in carbohydrates, though high-protein, low-carbohydrate alternatives are available. Raw diets, either frozen or freeze-dried, are increasingly popular and generally consist of meat with very low carbohydrate content. A smaller but growing segment of owners prepare homemade diets, either cooked or raw. Studies have demonstrated that gut microbiome profiles in different species closely reflect diet composition, particularly when there are large macronutrient differences, such as between carnivorous and herbivorous diets. In omnivores, including humans, even short-term consumption of diets composed entirely of animal or plant products can alter microbial community structure, overriding interindividual differences in microbial gene expression. In humans, an animal-based diet, which increases fat and protein intake while reducing fiber to near zero, results in an increase in bile-tolerant microorganisms and a decrease in Firmicutes, which are important for metabolizing plant polysaccharides (Pilla & Suchodolski, 2021).

In dogs, overall macronutrient content appears more influential than specific ingredients. One study found that an extruded diet made exclusively from plant protein did not significantly alter the gut microbiome compared with a traditional extruded diet containing both plant and animal proteins, provided the macronutrient content was similar. Major shifts in macronutrient composition were evaluated in a study of healthy dogs fed four different dry prescription diets formulated for weight loss, renal disease, low-fat content, or hypoallergenic purposes. The weight-loss diet, which had the highest protein and fiber content, produced the largest changes in microbiome composition. Higher protein intake was associated with increased Fusobacteria, while the abundance of SCFA producers such as Bacteroides, Prevotella, and Faecalibacterium increased in dogs fed the weight-loss diet. Faecalibacterium also increased in dogs fed the low-fat diet. Both the weight-loss (28.1% fiber) and low-fat (8.6% fiber) diets contained soluble and insoluble fibers, including beet pulp, fructooligosaccharides (FOS), and psyllium, which likely contributed to the observed increases in SCFA-producing genera (Pilla & Suchodolski, 2021).

Different types of dietary fiber have been studied for their prebiotic effects, with most enriching fiber-fermenting, SCFA-producing Firmicutes. Inulin-type fructans, for example, increase fecal SCFA levels—including acetate, butyrate, and propionate—as well as total bile acids, while reducing Enterobacteriaceae abundance. Raw, meat-based diets are markedly different in macronutrient composition from control extruded diets, containing higher protein and lower carbohydrate and fiber levels. Dogs fed raw diets generally show decreases in Firmicutes and Bacteroidetes, reflecting reduced carbohydrate fermentation, while Proteobacteria, Fusobacteria, and protein-associated genera increase in abundance. One study of six dogs fed a raw diet for over a year found increased microbiome diversity along with elevated levels of *Clostridium perfringens* and *Fusobacterium varium*, both of which produce butyrate from protein, suggesting microbiome adaptation to high-protein diets (Pilla & Suchodolski, 2021).

Although some members of the Clostridiaceae family can be associated with gastrointestinal disease, their increase in protein-rich diets may not be detrimental. Instead, these bacteria appear to contribute to protein digestion, as their abundance correlates positively with protein digestibility and firmer fecal scores, and negatively with undigested protein in feces. *Clostridium hiranonis*, a species involved in bile acid metabolism, has been shown to increase in dogs fed high-protein diets and is often decreased in dogs with gastrointestinal disease or after antibiotic treatment. Studies report that bile acid metabolism remains normal in healthy dogs fed raw diets, similar to kibble-fed controls. Proper bile acid metabolism is crucial not only for lipid digestion but also for the regulation of intestinal inflammation and is often disrupted in chronic gastrointestinal disorders (Pilla & Suchodolski, 2021).

Bifidobacterium, Lactobacillus, and Faecalibacterium are considered beneficial in omnivores due to their role in fermenting carbohydrates to produce butyrate, a key energy source for colonocytes. However, SCFAs such as butyrate are present in feces of all mammals, even those consuming minimal carbohydrates, suggesting alternative pathways for SCFA production. Supporting this, studies comparing high-fat and high-starch diets in dogs found no differences in acetate, butyrate, or propionate levels, indicating that SCFA production is not solely dependent on carbohydrate intake. Another study showed that adding minced meat to a dry-food diet slightly increased fecal butyrate and isovalerate levels, further highlighting the microbiome's metabolic adaptability to dietary protein (Pilla & Suchodolski, 2021).

8.3.7 The Modulators of Canine Microbial Ecology

The composition of the canine microbiome is not fixed; rather, it represents a dynamic and adaptive ecosystem shaped by a complex interplay of intrinsic and extrinsic factors. Among the most influential of these determinants are diet, age, and host genetics or breed. Each of these factors exerts distinct pressures that modulate microbial structure and function, thereby influencing overall health, metabolism, and disease susceptibility in the domestic dog.

Diet is widely recognized as one of the most potent and rapidly acting modulators of the gut microbiome (Pilla & Suchodolski, 2021; Huang et al., 2020). The macronutrient composition of a dog's diet, particularly the relative proportions of protein, fat, fiber, and carbohydrates, can induce substantial shifts in microbial community structure within days (Pilla & Suchodolski, 2021). High-protein, low-carbohydrate (HPLC) diets typically promote the proliferation of Firmicutes, resulting in a decreased Bacteroidetes-to-Firmicutes ratio and enrichment of specific proteolytic bacteria such as *Clostridium hiranonis* and *Clostridium perfringens* (Li et al., 2017). In contrast, carbohydrate-rich diets tend to favor saccharolytic taxa, including *Bacteroides uniformis* and other Bacteroidota members that specialize in fermenting complex polysaccharides (Li et al., 2017).

Dietary format and processing also play a major role in shaping microbial diversity and functionality. Comparative studies have shown that dogs fed raw, meat-based diets exhibit higher microbial richness and alpha diversity than those consuming extruded kibble-based diets (Kim et al., 2017). This greater diversity may reflect the exposure to a wider range of microbial substrates and environmental microbes in raw foods. However, it also carries potential health tradeoffs, including increased abundance of opportunistic pathogens such as *C. perfringens* (Kim et al., 2017). These findings underscore that diet does not merely feed the host—it also nourishes and selects for specific microbial communities, influencing metabolic outputs such as SCFA production, bile acid metabolism, and immune signaling.

Age represents another critical determinant of microbial composition and stability. The canine gut microbiome undergoes a well-defined process of maturation across the lifespan (Kim et al., 2025; Garrigues et al., 2022). During early development, microbial colonization begins shortly after birth, shaped by maternal contact, nursing, and environmental exposure. As puppies transition from milk to solid food, microbial diversity increases dramatically, leading to the establishment of a relatively stable adult microbiome (Garrigues et al., 2022). In senior dogs, however, this stability often declines. Aging is associated with reduced microbial diversity, diminished abundance of beneficial Firmicutes, including lactic acid-producing Lactobacillus species, and an increased prevalence of Proteobacteria, a phylum often linked to gut inflammation and dysbiosis (Kim et al., 2025; Masuoka et al., 2017). These compositional changes can impair digestive efficiency, nutrient absorption, and immune competence, leaving older dogs more vulnerable to metabolic disturbances and gastrointestinal disorders.

Finally, breed and host genetics have emerged as important, though sometimes underappreciated, influences on the canine gut microbiome (Bai et al., 2023). Controlled studies have demonstrated that distinct breeds harbor unique microbial profiles even when maintained on similar diets and under comparable environmental conditions (Bai et al., 2023). For instance, healthy Poodles and Labrador Retrievers exhibit significant differences in microbial diversity and taxonomic composition, suggesting that breed-specific genetic factors, such as immune regulation, gut physiology, or metabolic rate, shape the microbial ecosystem (Bai et al., 2023). Moreover, these breed-related microbial signatures influence not only baseline gut health but also the host's response to disease. During episodes of acute diarrhea, Poodles and Labradors display distinct patterns of microbiome disruption and recovery, along with differential shifts in microbial metabolites (Bai et al., 2023). Such findings reinforce the notion that breed must be considered a key biological variable in both clinical diagnosis and nutritional management of gastrointestinal disorders in dogs.

Taken together, these factors—diet, age, and breed—highlight the intricate interdependence between the canine host and its microbiota. The gut microbiome is not merely a passive reflection of physiological state; it is an adaptive biological system that coevolves with the host's life history, environment, and genetic background. Understanding these dynamic interactions is essential for developing targeted nutritional strategies and precision microbiome interventions that support optimal health throughout a dog's lifespan.

8.4 THE FELINE MICROBIOME

8.4.1 Feline Gut Microbiome

The gastrointestinal microbiome of the domestic cat represents a highly specialized and finely balanced microbial ecosystem that reflects the cat's unique dietary and evolutionary adaptations. Similar to other mammals, the feline gut microbiome is primarily composed of four dominant bacterial phyla—Firmicutes, Bacteroidetes, Proteobacteria, and Actinobacteria (Lyu et al., 2020; Suchodolski et al., 2015). However, unlike omnivorous species such as humans or dogs, the composition and functionality of this microbial community are distinctly shaped by the cat's obligate carnivorous lifestyle. Over evolutionary time, cats have adapted to thrive on a diet composed almost entirely of animal-derived proteins and fats, with minimal carbohydrate intake. This dietary specialization has influenced both their host metabolism and the ecological dynamics of their gut microbiota (Lyu et al., 2020).

Because of their protein- and fat-rich diet, cats rely far less on microbial fermentation of carbohydrates for energy production than omnivores do. In species like humans, fermentation of dietary fibers produces SCFAs, an important energy source. In contrast, the feline gut microbiota has evolved to function efficiently in an environment low in fermentable substrates, instead focusing on protein fermentation and amino acid metabolism (Lyu et al., 2020). Despite this reduced role in carbohydrate fermentation, the microbiome continues to play a vital role in maintaining gastrointestinal homeostasis. It contributes to key physiological functions, including vitamin synthesis (e.g., vitamin K_2 by *Escherichia coli* and *Bacteroides* spp., and folate by *Enterococcus* spp.), bile acid metabolism, and the protection of the gut epithelium from pathogenic invasion (Lyu et al., 2020).

Recent studies aimed at defining the characteristics of a "healthy" feline microbiome have provided valuable insights into its core structure. Ganz et al. (2022) identified a core set of approximately 30 bacterial genera that are consistently found across a large cohort of healthy North American cats. Among the most prevalent taxa are Prevotella, Bacteroides, Collinsella, Catenibacterium, Blautia, Faecalibacterium, and Megasphaera. These taxa represent the foundational microbial community that supports digestive stability and metabolic balance.

Understanding this core microbial composition is essential for recognizing dysbiosis—disruptions in the normal gut microbial community—that have been linked to various feline gastrointestinal and systemic diseases (Ganz et al., 2022; Pilla & Suchodolski, 2021). By establishing a clear baseline for what constitutes a healthy gut microbiome, researchers and veterinarians can better diagnose, prevent, and manage disorders such as inflammatory bowel disease, obesity, and chronic enteropathies in domestic cats. This emerging field of feline microbiome research thus holds great promise for advancing both veterinary medicine and comparative microbiology.

8.4.2 Aging and Diet Type Influence the Feline Microbiome

Bermingham et al. (2018) investigated how aging and diet type—specifically kibbled (dry) versus canned (wet) food—affect the gut microbiome composition of domestic cats across different life stages. The study aimed to understand how age-related physiological changes interact with diet to influence microbial community structure in feline gastrointestinal health.

Across all age groups and diets, Peptostreptococcaceae emerged as a prominent bacterial family, demonstrating notable variability in abundance depending on diet. This group represented an average of 8.1% of the total microbial community in cats fed kibbled diets and increased substantially to 33.2% in cats fed the canned diet, suggesting that diet moisture and nutrient composition may significantly shape gut microbial populations.

Among young cats fed the kibbled diet, unclassified Peptostreptococcaceae were the most relatively abundant taxa, averaging 22.5% of the microbial community. However, this dominance shifted with aging: in older cats maintained on the same kibbled diet, Prevotella became the predominant genus, accounting for an average relative abundance of 29.7%. Additionally, Lactobacillus, which was present at moderate levels in the feces of young cats consuming kibbled diets, declined sharply in abundance in older cats on the same diet. This pattern suggests potential age-related reductions in beneficial lactic acid bacteria under dry feeding conditions.

In contrast, cats consuming canned diets also exhibited a strong presence of unclassified Peptostreptococcaceae across all age groups. Their relative abundance increased from 16% at 8 weeks of age to approximately 30% at 17 and 104 weeks, before declining slightly to 23.4% at 260 weeks. Other major bacterial taxa identified in canned-fed cats included Prevotella (ranging from 6.8% to 16%) and unclassified Prevotellaceae (ranging from 7.9% to 10.3%). Furthermore, Clostridium species were relatively abundant in younger cats fed canned diets (4.2% at 8 weeks and 10.0% at 17 weeks), but their prevalence decreased with age, dropping to 1.9% and 3.2% in cats aged 104 and 260 weeks, respectively.

When comparing the overall effects of diet, unclassified Peptostreptococcaceae demonstrated higher relative abundances in cats fed canned diets (25.7%) than in those fed kibbled diets (17.0%). Conversely, Prevotella was more abundant in cats consuming kibbled diets, averaging 20.0%, compared to 10.8% in cats on canned diets. Similarly, Megasphaera (11.8% vs. 0.2%) and Blautia (9.6% vs. 5.7%) were both more prevalent in kibbled-fed cats, indicating that dry diets may favor these fermentative genera. On the other hand, unclassified Prevotellaceae (8.1% vs. 1.2%) and Bacteroides (7.3% vs. 1.8%) were more abundant in cats fed canned diets, suggesting that wet diets promote distinct microbial niches rich in carbohydrate-degrading taxa.

Overall, Bermingham et al. (2018) demonstrated that both age and diet type exert significant influences on the feline gut microbiome. While the canned diet generally promoted higher levels of Peptostreptococcaceae and Bacteroides, the kibbled diet was associated with greater microbial diversity among genera such as Prevotella, Megasphaera, and Blautia. These findings highlight the dynamic nature of the gut microbiota in response to dietary composition and aging, emphasizing the importance of tailored nutritional strategies for maintaining gut health across a cat's lifespan.

8.4.3 The Feline Skin Microbiome

The skin of the domestic cat hosts a remarkably rich and intricate microbial ecosystem—one that far exceeds what was once understood from traditional culture-based methods (Meason-Smith et al., 2017; Older et al., 2023). Advances in next-generation sequencing have revolutionized our understanding of this microbiome, revealing that the feline skin surface supports a diverse bacterial community dominated by the phyla Proteobacteria, Bacteroidetes, Firmicutes, and Actinobacteria (Meason-Smith et al., 2017; Older et al., 2023). Alongside these bacteria resides a complex fungal population, known collectively as the skin mycobiome, which is primarily composed of members of the phylum Ascomycota—a group that includes numerous environmental and soil-associated fungi (Meason-Smith et al., 2017). Together, these microbial populations form an essential biological interface that influences skin health, immunity, and disease resistance.

A variety of factors determine the composition and stability of the feline skin microbiome, with host genetics and behavior emerging as two of the most influential (Older et al., 2019). Breed-specific differences, in particular, have been shown to significantly shape both microbial diversity and community structure. For example, cats such as the nearly hairless Sphynx or the short-coated Devon Rex exhibit distinct microbial signatures compared to long-haired or dense-coated breeds (Older et al., 2019). These variations are largely driven by physiological and environmental differences at the skin surface—such as pH, temperature, hydration level, and lipid secretion—which create unique microhabitats that favor the growth of specific microbial taxa. Thus, the diversity of feline breeds offers a natural experiment in how anatomy and physiology influence microbial ecology.

Among the most intriguing and distinctive influences on the feline skin microbiome is the cat's meticulous self-grooming behavior (Meason-Smith et al., 2017). Unlike most mammals, cats spend a significant portion of their waking hours grooming, a process that continuously transfers saliva to the fur and skin. This behavior establishes a dynamic interconnection between the oral and cutaneous microbiomes, as grooming acts both as a source of microbial seeding and a mechanism of microbial removal (Meason-Smith et al., 2017). Saliva introduces oral microbes to the skin surface while simultaneously cleansing and redistributing microbial populations across the body.

This self-grooming is therefore far more than a hygienic act—it represents a sophisticated form of microbial homeostasis. By continually reseeding the skin with microbes from the relatively stable oral microbiota, cats may be employing an evolutionary strategy of microbial regulation that maintains a consistent and protective cutaneous barrier. Such regulation helps defend against pathogenic invasion and supports the immune system's capacity to distinguish between harmless commensals and potential threats. Importantly, this tight oral–skin microbial connection also suggests that disturbances in one environment, such as oral dysbiosis caused by dental disease, could have downstream effects on skin health. Understanding these cross-site microbial interactions offers valuable insights into feline biology and provides a foundation for developing microbiome-based approaches to managing skin disorders in cats.

8.4.4 Breed, Environment, and Host Factors Shaping the Feline Skin Microbiome

Older et al. (2019) conducted a comprehensive analysis of the skin microbiome in multiple cat breeds, including Bengal, Cornish Rex, Devon Rex, Siberian, Sphynx, and indoor domestic cats (mixed genotype). Samples were collected from several body sites, including the axilla, dorsum, ear canal, nostril, and oral cavity, to assess the skin microbiome across diverse niches. The study also compared environmental influences by including indoor and outdoor domestic cats.

Alpha diversity, which measures both the richness (number of taxa) and evenness (distribution of taxa), differed significantly between cat breeds. Devon Rex cats exhibited the lowest median alpha diversity, indicating a relatively simplified microbial community, while Bengal cats had the highest diversity. The most pronounced breed-associated differences were observed on the dorsum and ear canal, with Sphynx cats displaying the most diverse microbial populations on the dorsum. Hair length did not significantly affect bacterial alpha diversity, nor did the indoor versus outdoor environment, regardless of the metric or body site analyzed.

The major bacterial phyla across all breeds were Proteobacteria (44.03%), Firmicutes (21.04%), Bacteroidetes (16.65%), and Actinobacteria (10.38%). Among the most abundant taxa were the family Pasteurellaceae (11.14%) and the genera Porphyromonas (7.40%) and Staphylococcus (4.79%). The Veillonellaceae family, typically associated with the gut microbiota, varied significantly between breeds, with higher relative abundance in the ear canal of Sphynx cats. Other taxa, including *Porphyromonas* spp. and *Lactobacillus* spp., also displayed breed-specific differences.

Environmental influences were also evident: many bacterial taxa differed in abundance between indoor and outdoor cats, particularly in nostril samples. For instance, *Corynebacterium* spp. was more abundant in indoor cats across all body sites.

As with bacteria, alpha diversity of fungal communities varied significantly among cat breeds. Sphynx and Bengal cats exhibited the highest fungal diversity across all body sites. Site-specific differences were observed on the dorsum, ear canal, and nostril, whereas the indoor versus outdoor environment had no significant effect on fungal alpha diversity.

Hair length influenced fungal diversity: short-haired breeds (DSH) and very short-haired breeds, including Cornish Rex, Devon Rex, and Sphynx, exhibited more diverse fungal communities than medium- or long-haired cats. The composition of fungal communities was more variable than that of bacteria. Dominant genera included *Cladosporium* spp. and *Malassezia* spp., with many other taxa, such as *Alternaria* spp. and *Aspergillus* spp., showing breed-specific differences. Among body sites, the dorsum exhibited the most pronounced taxonomic shifts, followed by the nostril and groin, particularly in the relative abundance of *Malassezia* spp. and *Alternaria* spp.

Environmental influences were evident for certain fungal taxa. Ustilaginomycetes and Ustilaginales, primarily plant-associated fungi, were more abundant on outdoor cats, especially on the dorsum, and were generally absent in the oral cavity. In nostril samples, two phyla, Ascomycota and Basidiomycota, along with unassigned sequences, were differentially abundant. Analyses of indoor domestic cats revealed host-specific influences on the microbiome. Female cats exhibited greater fungal diversity in the oral cavity and on the skin compared to males. Older cats displayed increased bacterial and fungal diversity on the skin relative to younger adult cats, suggesting that host age may contribute to the maturation and complexity of the skin microbiome over time.

8.4.5 Effect of Diet on Feline Microbiome

HPLC diets, including both extruded and raw formulations, have been extensively studied in cats as alternatives to traditional moderate-protein extruded diets. In addition to dry HPLC diets, canned diets provide a convenient higher-protein option and are frequently fed alone or in combination with extruded foods. The benefits of moist foods in cats extend beyond macronutrient content; research indicates that feeding canned diets promotes increased water intake, reduces voluntary energy intake, and lowers urine specific gravity, which may provide protective effects for the urinary tract and support overall health, particularly in cats prone to urolithiasis or chronic kidney disease (Pilla & Suchodolski, 2021).

Microbiome studies in kittens weaned onto HPLC extruded diets have revealed notable effects on gut microbial composition and function. HPLC diets increase species diversity compared with medium-protein, medium-carbohydrate (MP/MC) diets, and selectively enrich five genera known for butyrate production: Clostridium and Eubacterium, which can generate butyrate from carbohydrates via the butyryl-CoA:acetate CoA-transferase (*but*) pathway or from protein via the butyrate kinase (*buk*) pathway, and Faecalibacterium, Ruminococcus, and Blautia, which produce butyrate through the *but* pathway. Dietary differences influenced 194 metabolic pathways, including those related to amino acid biosynthesis and metabolism, highlighting the profound impact of the dietary protein-to-carbohydrate ratio on microbiome functionality. While the significance of these metabolite alterations for overall feline health is not fully understood, these findings underscore the microbiome's responsiveness to macronutrient composition in early-life nutrition.

Research on raw meat–based diets in cats is more limited but suggests similar trends. Feeding raw 1- to 3-day-old chicks to kittens or raw beef to adult cats increased the abundance of genera associated with SCFA production. While absolute fecal butyrate concentrations did not rise with raw beef feeding, the molar ratio of butyrate increased, indicating a shift in SCFA composition that could affect colonic health and metabolism. The addition of plant fiber, such as inulin and cellulose at 2% of feed, partially mitigated these effects, producing a microbiome profile more similar to cats fed standard extruded diets and normalizing fecal SCFA ratios, suggesting that small amounts of fermentable fiber can modulate microbial activity even in high-protein diets.

Canned diets, similar to raw meat diets, are typically high in protein and fat and low in carbohydrates. Studies in adult cats and kittens show that these diets promote greater microbial diversity in the gut, and many of the enriched genera are associated with butyrate production, potentially supporting intestinal health. Historically, hindgut fermentation in cats was considered negligible due to their strict carnivorous nature. However, as in dogs, supplementation with prebiotic fibers has been shown to increase SCFA-producing bacteria, including Bifidobacterium. Combining fructooligosaccharides (FOS) and galactooligosaccharides further elevates concentrations of total SCFAs, including butyrate and valerate, demonstrating that even obligate carnivores possess the capacity for diet-modulated hindgut fermentation. Collectively, these findings highlight the capacity of dietary composition—both macronutrient ratios and inclusion of fermentable fibers—to shape the feline gut microbiome, enhance microbial diversity, and support functional metabolites that contribute to intestinal health (Pilla & Suchodolski, 2021)..

8.4.6 Environmental Determinants of Feline Microbiome

The environment in which a cat resides exerts a profound influence on the composition and functionality of its microbiome, shaping microbial communities across the skin, respiratory tract, and gastrointestinal system. Lifestyle differences—whether indoor, outdoor, or feral—create distinct microbial signatures that reflect varying exposures to diet, pathogens, and environmental microbes. Studies comparing these populations have revealed a nuanced and body site–specific picture, where the effects of environmental exposure differ depending on both the microbial community being examined and the diversity metrics used. For example, research on the nasal microbiome has shown that healthy indoor cats actually exhibit greater species richness and microbial diversity than cats with outdoor access (Dorn et al., 2017). This seemingly counterintuitive finding may result from the continuous and concentrated exposure to the complex microbial environment of indoor spaces, which are heavily influenced by the microbial communities associated with human cohabitants and household materials (Dorn et al., 2017).

In contrast, studies of the gut microbiome reveal an even more striking interplay between environmental exposure, diet, and microbial function. When comparing domestic (typically urban and indoor) cats with feral cats, researchers have found that although overall taxonomic diversity may not differ dramatically, the functional capacity of the microbiome diverges significantly (Aizpurua et al., 2025). Feral cats possess gut microbiomes that are enriched with genes responsible for the degradation of amino acids and lipids, reflecting their diet of hunted prey—rich in raw protein and fat. By contrast, domestic cats consuming commercially prepared diets show a microbial profile more adapted to the digestion of processed nutrients. These dietary differences are mirrored in taxonomic composition: feral cats often exhibit higher abundances of bacteria from the phyla Bacillota and Fusobacteriota, whereas domestic cats display greater representation of Pseudomonadota and Bacteroidota (Aizpurua et al., 2025).

This divergence illustrates a broader ecological process: domestication and urbanization appear to drive microbial homogenization and functional simplification. Feral and rural cats, with their variable diets and exposure to dynamic outdoor environments, exhibit greater interindividual microbiome variability, suggesting more adaptable and resilient microbial ecosystems (Páez-Triana et al., 2025; Aizpurua et al., 2025). In contrast, domestic cats, constrained by standardized diets and sanitized living conditions, possess more uniform and less metabolically diverse microbiomes. This pattern closely parallels that observed in industrialized human populations, where modernization has led to a measurable loss of microbial diversity due to decreased contact with natural environments and reliance on processed foods (Reese et al., 2021).

The implications of this environmental influence are far-reaching. Reduced microbial diversity and functional capacity may weaken the symbiotic resilience of domestic cats, potentially increasing their susceptibility to metabolic disorders, inflammatory bowel disease, and immune dysregulation. The microbiome functions as an adaptive interface between the host and its environment—processing nutrients, training the immune system, and resisting pathogens. Thus, as cats transition from the ecological richness of the outdoors to the controlled environments of modern homes, they may be trading microbial adaptability for stability. Understanding these environmental effects not only informs feline health management but also provides a valuable comparative model for studying how domestication, diet, and environment shape the microbiomes of mammals, including humans.

8.5 DYSBIOSIS AND DISEASE IN COMPANION ANIMALS

Dysbiosis, defined as the disruption or imbalance of microbial communities that normally coexist symbiotically with the host, is increasingly recognized as a key driver in the development of both acute and chronic diseases in companion animals. In healthy states, the microbiome performs critical functions such as maintaining epithelial integrity, aiding digestion, modulating immune responses, and preventing colonization by pathogens. However, when this delicate equilibrium is disturbed—whether through dietary changes, antibiotic use, infection, or environmental stressors—it can trigger a cascade of adverse physiological effects. These include altered metabolic function, immune dysregulation, and chronic inflammation, which collectively contribute to disorders affecting the gastrointestinal system, skin, and systemic metabolic health. Importantly, the dysbiotic patterns observed in dogs and cats show striking parallels to those documented in human diseases, emphasizing the evolutionary conservation of host–microbiome interactions and their importance in maintaining health across mammalian species.

8.5.1 Inflammatory Bowel Disease

Inflammatory bowel disease (IBD) represents one of the most thoroughly studied conditions associated with dysbiosis in companion animals. It refers to a group of chronic enteropathies characterized by persistent gastrointestinal inflammation and is a leading cause of chronic vomiting, diarrhea, and weight loss in both dogs and cats (Honneffer et al., 2014; Cornell University, 2018). The pathogenesis of IBD is multifactorial, involving a complex interplay among host genetics, immune dysregulation, environmental triggers, and gut microbial imbalances (Honneffer et al., 2014; Jergens, 2012). Mounting evidence indicates that gut dysbiosis is not merely a secondary effect but a central mechanism driving both the initiation and persistence of intestinal inflammation in IBD (Honneffer et al., 2014).

A consistent microbial pattern has emerged across species and studies: a reduction in microbial diversity accompanied by a shift in bacterial composition at the phylum level. This dysbiotic signature is typically defined by a decrease in beneficial Firmicutes, particularly butyrate-producing bacteria such as Faecalibacterium and members of Ruminococcaceae (Clostridium clusters IV and XIVa), alongside an increase in proinflammatory Proteobacteria, including Escherichia coli and other Enterobacteriaceae (Honneffer et al., 2014). The decline in butyrate-producing taxa is especially detrimental, as butyrate serves as a primary energy source for colonocytes, reinforces intestinal barrier integrity, and exerts anti-inflammatory effects by modulating regulatory T-cell function (Garrigues et al., 2022). Thus, the depletion of these beneficial bacteria contributes directly to epithelial damage and immune hyperactivation—hallmarks of IBD in both animals and humans.

In cats, IBD is often associated with an increased bacterial load adherent to the duodenal mucosa, particularly members of the Enterobacteriaceae family, whose abundance correlates strongly with the severity of mucosal inflammation (Jergens, 2012). Additional microbial shifts include a decrease in beneficial taxa such as Bifidobacterium and an increase in sulfate-reducing bacteria like Desulfovibrio, both of which can disrupt mucosal integrity and promote oxidative stress (Honneffer et al., 2014). Clinically, feline IBD has become a major focus of veterinary gastroenterology not only because of its chronic, relapsing nature but also due to its potential progression to low-grade intestinal T-cell lymphoma (LGITL), suggesting a possible inflammatory–neoplastic continuum (Marsilio, 2021).

In dogs, similar microbial alterations have been documented in chronic enteropathies. A marked depletion of butyrate-producing Firmicutes and an overgrowth of opportunistic pathogens such as *E. coli* and *Clostridium perfringens* have been consistently reported (Garrigues et al., 2022; Honneffer et al., 2014). These dysbiotic changes result in metabolic shifts toward the production of inflammatory metabolites, weakening the mucosal barrier and perpetuating immune activation.

The consistent dysbiotic signatures across cats, dogs, and humans suggest that IBD and related inflammatory disorders share a conserved pathological mechanism rooted in microbial imbalance. The loss of beneficial, immune-regulating commensal bacteria disrupts intestinal homeostasis, facilitating chronic inflammation and tissue damage. Understanding these microbial dynamics opens new avenues for therapeutic interventions, including microbiome restoration strategies such as probiotics, prebiotics, dietary modulation, and fecal microbiota transplantation (FMT). Continued comparative research in companion animals not only improves veterinary treatment but also enhances our understanding of the microbiome's role in inflammatory diseases across species, reinforcing the interconnectedness of animal and human health under the One Health paradigm.

8.5.2 Metabolic Disorders

The gut microbiome functions as a metabolically active organ that plays a central role in the regulation of host energy harvest, storage, and expenditure. Through its vast repertoire of enzymes and metabolic pathways, the gut microbiota profoundly influences nutrient processing, lipid metabolism, and glucose regulation. Consequently, alterations in microbial composition or function—collectively termed dysbiosis—are increasingly recognized as important contributors to metabolic disorders such as obesity and diabetes mellitus (DM) (Turnbaugh et al., 2008; Brunkwall & Orho-Melander, 2017). These disorders are now prevalent among companion animals, paralleling trends in human populations, and highlight the shared metabolic vulnerabilities of mammals living in modern, nutrient-rich environments.

8.5.3 Obesity

Obesity has become one of the most widespread nutritional and metabolic disorders in both dogs and cats, mirroring a global epidemic also observed in humans. It is associated with a constellation of metabolic disturbances, including insulin resistance, chronic low-grade inflammation, and altered lipid metabolism. Increasing evidence implicates the gut microbiome as a central mediator of these processes (Li et al., 2017; Kieler et al., 2016).

In dogs, obesity correlates with distinct alterations in the gut microbial community, though the direction and magnitude of these changes may depend on dietary composition. Obese dogs fed high-protein diets exhibit a pronounced dysbiotic profile characterized by an increase in Firmicutes and a decrease in Bacteroidetes, a pattern reminiscent of that found in obese humans (Li et al., 2017). This microbial shift is thought to enhance caloric extraction from food and promote fat storage. In contrast, other studies have identified elevated levels of Proteobacteria and Fusobacteriota and a reduction in Firmicutes in obese Golden Retrievers, suggesting that breed-specific physiology and diet can modulate the microbiome's response to metabolic stress (Cai et al., 2024). Such variability underscores the context-dependent nature of microbial dysbiosis in obesity and the need for species- and breed-specific research.

The situation in cats presents a striking metabolic paradox. Unlike humans, mice, and dogs—where obesity is often marked by an increased Firmicutes-to-Bacteroidetes (F/B) ratio—obese cats display the opposite trend, with a lower F/B ratio and a significant decrease in Firmicutes abundance (Ma et al., 2022). Furthermore, feline obesity is associated with a reduction in overall microbial diversity, suggesting a loss of ecological stability in the gut. This inversion of the classic F/B pattern reflects the cat's unique obligate carnivorous physiology. In cats, obesity-linked dysbiosis may not stem from enhanced energy extraction from carbohydrates, as their diets and metabolism are protein- and fat-dominant. Instead, feline obesity likely involves altered microbial metabolism of amino acids and fatty acids, alongside inflammatory signaling pathways distinct from those in omnivores (Ma et al., 2022; Yeon et al., 2024). These findings suggest that the microbiome–metabolism axis in carnivores operates under fundamentally different ecological and biochemical constraints.

8.5.4 Diabetes Mellitus

DM in companion animals, as in humans, is a complex metabolic disease marked by impaired insulin production or function, leading to chronic hyperglycemia. Emerging research suggests that gut dysbiosis may contribute to the onset and progression of diabetes by inducing systemic low-grade inflammation, disrupting gut barrier integrity, and altering the production of microbial metabolites that modulate glucose metabolism (Brunkwall & Orho-Melander, 2017).

In dogs, DM is one of the most common endocrinopathies and has been linked to measurable microbial shifts. Dogs with diabetes, particularly those experiencing hyperlactatemia, show an increased abundance of lactate-producing bacteria, including Enterococcus and Bifidobacterium (Kim et al., 2023). These changes may influence intestinal pH, enhance lactic acid accumulation, and alter SCFA dynamics, potentially exacerbating insulin resistance (Laia et al., 2022).

In cats, the pathophysiology of DM closely parallels human Type 2 DM (T2DM), with insulin resistance as a central feature (Slingerland et al., 2007; Appleton et al., 2001). However, microbiome studies in diabetic cats have produced more subtle results. While large-scale taxonomic differences between diabetic and healthy cats have not been consistently observed, this does not negate a microbial role. Instead, the link may lie in functional rather than compositional changes, such as differences in microbial metabolite production or shifts occurring in the small intestine—a site less commonly sampled (Bell et al., 2014). Supporting this, diabetic cats exhibit significant intestinal remodeling, including inflammation, mucosal thickening, and muscular hypertrophy, suggesting that dysbiosis may contribute to disease indirectly through chronic intestinal inflammation and metabolic stress (Esteves-Monteiro et al., 2025).

8.5.5 Atopic Dermatitis

While metabolic disorders primarily involve the gut microbiome, skin microbiome dysbiosis is a defining feature of Canine Atopic Dermatitis (cAD), a common and chronic inflammatory skin disease affecting roughly 10% of dogs (Hillier & Griffin, 2001). cAD is characterized by intense pruritus (itching), recurrent infections, and immune hypersensitivity to environmental allergens. The condition shares many immunological and microbiological similarities with human atopic dermatitis (AD), making it a valuable natural model for studying host–microbe–immune interactions in chronic inflammation (Bradley et al., 2016).

During disease flares, atopic dogs exhibit a profound reduction in microbial diversity and a dramatic overgrowth of opportunistic bacteria, most notably Staphylococcus pseudintermedius, which can dominate the skin microbiota and displace commensal species (Bradley et al., 2016; Pierezan et al., 2016). Concurrently, Corynebacterium species often increase in abundance, compounding the inflammatory state. This dysbiosis is closely linked to epidermal barrier dysfunction—measured by elevated transepidermal water loss (TEWL)—and forms a vicious feedback loop: a weakened barrier allows allergens and microbes to penetrate deeper, triggering inflammation, which in turn worsens microbial imbalance. Moreover, staphylococcal toxins further damage the barrier and amplify immune activation (Bradley et al., 2016; Williams & Gallo, 2015).

Therapeutic interventions for cAD, such as antimicrobial or anti-inflammatory treatments, often restore microbial diversity and reduce staphylococcal overgrowth in parallel with clinical improvement and normalization of barrier function (Bradley et al., 2016). These findings reinforce a broader principle emerging from microbiome research across body sites and species: disease often arises not from the invasion of a single pathogen, but from the collapse of a balanced microbial ecosystem.

Whether in the gut during IBD and metabolic disease or on the skin during AD, a recurring theme emerges: loss of microbial diversity and ecological stability undermines host resilience. This collapse eliminates the "colonization resistance" normally provided by a healthy, diverse microbial community, creating ecological space for opportunistic, proinflammatory organisms to flourish. Thus, therapeutic strategies must evolve beyond pathogen eradication toward restoring ecological balance through dietary modulation, probiotics, prebiotics, synbiotics, or microbiome transplantation. This holistic approach acknowledges that health is not merely the absence of pathogens but the presence of a stable, functionally diverse microbial ecosystem that supports host homeostasis and immune equilibrium.

8.5.6 Impact of Antibiotics on the Microbiome

Antibiotics are essential tools in veterinary medicine, crucial for the treatment and prevention of bacterial infections in dogs, cats, and other companion animals. However, their therapeutic benefits come with significant unintended consequences. Most antibiotics, particularly those administered orally, are broad-spectrum, meaning they do not discriminate between pathogenic bacteria and beneficial commensal microbes. This indiscriminate activity often leads to dysbiosis, a disruption of the host microbial ecosystem that can impair gut homeostasis, compromise immune function, and increase susceptibility to secondary infections or chronic diseases (Connelly et al., 2020). The magnitude and duration of these effects can vary depending on the class of antibiotic, dosage, treatment length, and the host's baseline microbiome composition.

8.5.7 Evidence from Canine Studies

Research in dogs has highlighted the profound and lasting impact of antibiotics on the gut microbiome. For example, a 14-day course of metronidazole, one of the most commonly prescribed antimicrobials in veterinary practice, caused a significant reduction in microbial richness and the depletion of key bacterial taxa, including members of the phylum Fusobacteria (Pilla et al., 2020). These compositional shifts were accompanied by functional metabolic changes, such as an increase in fecal lactate levels and a marked reduction in beneficial secondary bile acids, which are important for lipid digestion, cholesterol metabolism, and the maintenance of gut epithelial integrity (Pilla et al., 2020). Notably, the microbiome and metabolome did not fully recover even four weeks after the cessation of antibiotics, indicating that antibiotic-induced dysbiosis can have long-term consequences for host physiology and gastrointestinal health (Pilla et al., 2020).

8.5.8 Evidence in Feline Studies

Studies in free-roaming urban cats have revealed even more complex and paradoxical effects of antibiotic therapy. While antibiotic administration can reduce the overall microbial diversity and decrease the relative abundance of certain potential zoonotic pathogens, it can also destabilize the microbial community in ways that may be detrimental to the host (Xie et al., 2025). In untreated cats, higher microbial diversity is typically associated with lower pathogen abundance, suggesting that a diverse microbiome confers colonization resistance against opportunistic microbes. However, following antibiotic-induced disruption, this relationship may invert, with higher diversity correlating with a greater presence of pathogens during the recovery phase (Xie et al., 2025). This paradox indicates that after broad-spectrum antibiotic use, the microbial ecosystem becomes ecologically unstable, allowing opportunistic recolonization by both beneficial and harmful bacteria.

8.5.9 Strategies to Mitigate Antibiotic-Induced Dysbiosis

Given the significant and potentially long-lasting effects of antibiotics on the microbiome, there is growing interest in strategies to minimize collateral damage while maintaining therapeutic efficacy. One common approach is the coadministration of probiotics, which can help restore microbial diversity, support the recovery of beneficial taxa, and normalize metabolic function during and after antibiotic treatment (Kim et al., 2025).

Emerging and more targeted strategies include the use of orally delivered beta-lactamase enzymes, designed to specifically inactivate beta-lactam antibiotics within the lower gastrointestinal tract. This approach protects the colonic microbiota from antibiotic exposure while preserving the systemic efficacy of the drug against pathogenic bacteria (Connelly et al., 2020). Other strategies under investigation include prebiotic supplementation to promote the growth of commensal microbes and precision microbiome therapeutics, which aim to selectively replenish key bacterial taxa lost during antibiotic therapy.

Overall, these findings underscore the importance of judicious antibiotic use in veterinary practice. Recognizing antibiotics as double-edged swords, veterinarians are increasingly focused on balancing the necessity of antimicrobial therapy with strategies to preserve microbial diversity and ecosystem function, thereby safeguarding both short-term treatment success and long-term host health.

8.6 SHARED MICROBIOME BETWEEN HUMANS AND COMPANION ANIMALS

8.6.1 Bidirectional Microbial Exchange

The intimate cohabitation of humans and their companion animals within domestic environments creates a dynamic interface for bidirectional microbial exchange (Song et al., 2013; Hernandez et al., 2022). This sharing is not random. Evidence shows that human gut microbiomes are more similar to those of their own dogs than to dogs from other households, indicating the existence of a household-specific microbial signature (Coelho et al., 2018). While microbial transfer occurs across multiple body sites, the skin microbiome is the most significant and well-documented site of exchange (Song et al., 2013).

Dog ownership has been shown to substantially increase the sharing of skin microbiota among cohabiting humans and pets. Bacteria from the canine oral cavity (e.g., Methylophilaceae, a Betaproteobacteria) and soil-associated microbes from paws (e.g., Actinobacteria) are readily detectable on the skin of owners (Song et al., 2013). This exchange is primarily contact-driven, occurring via petting, licking, and shedding of fur. Interestingly, the effect of dogs on human skin microbiomes is much stronger than on oral or gut microbiomes, suggesting that direct physical contact is a key determinant of microbial sharing (Song et al., 2013).

Pets, particularly dogs, function as "microbial ferries," transporting environmental microbes from outdoor spaces into the domestic environment. Through interactions with soil, water, vegetation, and other animals, dogs acquire a diverse microbial repertoire, which is then introduced into the home and to their human companions. This continuous reintroduction of environmental microbial diversity may counteract some of the negative immunological consequences of modern, highly sanitized living conditions (Song et al., 2013; Azad et al., 2013; Tun et al., 2017). The concept of the pet-owning household as a single integrated microbial ecosystem has emerged from these findings, highlighting the interconnectedness of humans, pets, and their shared environment.

8.6.2 The Hygiene Hypothesis and Immune Development

The hygiene hypothesis, first proposed by Strachan in 1989, suggests that the rise in allergic and autoimmune diseases in developed nations is linked to reduced microbial exposure during early life (Strachan, 1989; Bloomfield et al., 2016). Humans coevolved with a diverse microbial environment, and early-life exposure to these microbes is essential for "training" the immune system to tolerate harmless antigens while responding appropriately to pathogens (Okada et al., 2010). Modern sanitation, urbanization, and widespread antibiotic use have limited these microbial exposures, potentially predisposing individuals to immune dysregulation (Okada et al., 2010; Bloomfield et al., 2016).

Epidemiological evidence supports the role of pet ownership as a protective factor in early life. Infants raised in households with pets exhibit greater gut microbial richness and diversity compared to infants without pets (Tun et al., 2017; Azad et al., 2013). Specific bacterial families, such as Peptostreptococcaceae, are overrepresented in pet-exposed infants, indicating that pets actively contribute novel microbial species that shape the developing microbiome (Azad et al., 2013). This exposure reduces the risk of allergic diseases, including asthma and AD, providing empirical support for the hygiene hypothesis (Okada et al., 2010).

The concept has been refined into the "old friends" or "microflora" hypothesis, emphasizing the evolutionary importance of exposure to commensal microorganisms that coexisted with humans over millennia. By acting as conduits for environmental and animal-associated microbes, pets provide critical inputs for immune education and maturation, promoting immunotolerance and reducing the likelihood of allergic sensitization (Azad et al., 2013).

8.7 ZOONOTIC PATHOGENS

Despite the benefits of microbial sharing, close human–animal contact also poses risks for the transmission of zoonotic pathogens (Stull et al., 2015). Pets can harbor bacterial, viral, parasitic, and fungal agents, which can be transmitted to humans via bites, scratches, saliva, urine, feces, aerosols, or arthropod vectors such as fleas and ticks (Stull et al., 2015).

Common bacterial zoonoses include *Bartonella henselae* (cat scratch disease), *Campylobacter jejuni*, *Pasteurella multocida*, and *Capnocytophaga canimorsus*. These pathogens can cause gastrointestinal illness, wound infections, or, in severe cases, sepsis (Stull et al., 2015). There is also growing concern regarding multidrug-resistant (MDR) bacteria in pets, including MRSA, ESBL-producing *E. coli*, and *Clostridioides difficile*, which can be transferred bidirectionally between humans and pets in domestic and clinical environments (Stull et al., 2015). The risk of zoonotic infection can be mitigated through proper hygiene, parasite control, and routine veterinary care, emphasizing the importance of responsible pet ownership.

8.7.1 Therapeutic and Health Implications

The human–animal bond extends beyond emotional companionship, offering measurable mental and physical health benefits. Animal-Assisted Therapy (AAT), which incorporates structured interactions with animals—most commonly dogs—has been shown to reduce stress, anxiety, and depressive symptoms, improve social interaction, and strengthen the therapeutic alliance in diverse populations, including hospital patients and individuals with PTSD (Szewczyk et al., 2023).

The benefits of pets may operate through a synergistic psycho-microbiological framework. On the "top-down" pathway, interaction with pets reduces stress and creates a physiological environment conducive to gut and immune health. On the "bottom-up" pathway, pets enrich the human microbiome through the transfer of commensal and environmental microbes, promoting immune system maturation and tolerance (Song et al., 2013; Azad et al., 2013). These pathways reinforce each other, establishing a positive feedback loop: lower stress enhances microbial stability, while a diverse microbiome supports immune function and may even influence neurological and mood regulation. This integrated perspective suggests that the therapeutic value of companion animals lies in their ability to simultaneously soothe the mind and diversify the microbial ecosystem, underscoring the profound interconnectedness of humans, pets, and their shared microbial environment.

ACKNOWLEDGMENTS

The author (ST) acknowledges Sharpe-Strumia Research Foundation, W.W. Smith Charitable Trust and Abraham Thomas Foundation for their generous support in providing the resources that made this work possible.

REFERENCES

Abdolghanizadeh S, Salmeh E, Mirzakhani F, Soroush E, Siadat SD, Tarashi S. Microbiota insights into pet ownership and human health. *Res Vet Sci*. 2024 May;171:105220.

Aizpurua O, Botnen AB, Eisenhofer R, Odriozola I, Santos-Bay L, Bjørnsen MB, Gilbert MTP, Alberdi A. Functional insights into the effect of feralisation on the gut microbiota of cats worldwide. *Mol Ecol*. 2025;34(6):e17695.

Alessandri G, Argentini C, Milani C, Turroni F, Cristina Ossiprandi M, van Sinderen D, Ventura M. Catching a glimpse of the bacterial gut community of companion animals: A canine and feline perspective. *Microb Biotechnol*. 2020 Nov;13(6):1708–1732.

AlShawaqfeh MK, Wajid B, Minamoto Y, Markel M, Lidbury JA, Steiner JM, Serpedin E, Suchodolski JS. A dysbiosis index to assess microbial changes in fecal samples of dogs with chronic inflammatory enteropathy. *FEMS Microbiol Ecol*. 2017;93(11):fix136.

An JU, Mun SH, Kim WH, Seong JK, Park K, Cho S. Dynamics of the canine gut microbiota of a military dog birth cohort. *Front Microbiol*. 2025 Mar 24;16:1481567.

Appleton DJ, Rand JS, Sunvold GD. Insulin sensitivity decreases with obesity, and lean cats with low insulin sensitivity are at greatest risk of glucose intolerance with weight gain. *J Feline Med Surg*. 2001;3(4):211–228.

Azad MB, Konya T, Maughan H, Guttman DS, Field CJ, Sears MR, Becker AB, Scott JA, Kozyrskyj AL. Infant gut microbiota and the hygiene hypothesis of allergic disease: impact of household pets and siblings on microbiota composition and diversity. *Allergy Asthma Clin Immunol*. 2013 Apr 22;9(1):15.

Bai H, Liu T, Wang S, et al. Variations in gut microbiome and metabolites of dogs with acute diarrhea in poodles and Labrador retrievers. *Arch Microbiol*. 2023;205(3):97.

Bell ET, Suchodolski JS, Isaiah A, et al. Faecal microbiota of cats with insulin-treated diabetes mellitus. *PLoS One*. 2014;9(10):e108729.

Bermingham EN, Young W, Butowski CF, Moon CD, Maclean PH, Rosendale D, Cave NJ, Thomas DG. The fecal microbiota in the domestic cat (Felis catus) is influenced by interactions between age and diet; a five year longitudinal study. *Front Microbiol*. 2018 Jun 19;9:1231.

Black K. The relationship between companion animals and loneliness among rural adolescents. *J Pediatr Nurs*. 2012;27:103–112.

Bloomfield SF, Rook GA, Scott EA, Shanahan F, Stanwell-Smith R, Turner P. Time to abandon the hygiene hypothesis: new perspectives on allergic disease, the human microbiome, infectious disease prevention and the role of targeted hygiene. *Perspect Public Health*. 2016 Jul;136(4):213-24.

Bradley CW, Morris DO, Rankin SC, et al. Longitudinal evaluation of the skin microbiome and association with microenvironment and treatment in canine atopic dermatitis. *J Invest Dermatol*. 2016;136(6):1182–1190.

Brown A. About half of U.S. pet owners say their pets are as much a part of their family as a human member. *Pew Research Center*. 2023. https://www.pewresearch.org/short-reads/2023/07/07/about-half-us-of-pet-owners-say-their-pets-are-as-much-a-part-of-their-family-as-a-human-member/

Brunkwall L, Orho-Melander M. The gut microbiome as a target for prevention and treatment of hyperglycaemia in type 2 diabetes: From current human evidence to future possibilities. *Diabetologia*. 2017;60(6):943–951.

Cai Y, Tang H, Xiang G, et al. Deciphering of differences in gut microbiota and plasma metabolites profile between non-obese and obese Golden Retrievers dogs. *Front Microbiol*. 2024;15:1514633.

Caugant DA, Levin BR, Selander RK. Distribution of multilocus genotypes of Escherichia coli within and between host families. *J Hyg*. 1984;92:377–384.

Coelho LP, Kultima JR, Costea PI, Fournier C, Pan Y, Czarnecki-Maulden G, Hayward MR, Forslund SK, Schmidt TSB, Descombes P, Jackson JR, Li Q, Bork P. Similarity of the dog and human gut microbiomes in gene content and response to diet. *Microbiome*. 2018 Apr 19;6(1):72.

Connelly S, Fanelli B, Hasan NA, et al. SYN-007, an orally administered beta-lactamase enzyme, protects the gut microbiome from oral amoxicillin/clavulanate without adversely affecting antibiotic systemic absorption in dogs. *Microorganisms*. 2020;8(2):152.

Cornell University. Inflammatory Bowel Disease. 2018. https://www.vet.cornell.edu/departments-centers-and-institutes/cornell-feline-health-center/health-information/feline-health-topics/inflammatory-bowel-disease

Cuscó A, Sánchez A, Altet L, Ferrer L, Francino O. Individual signatures define canine skin microbiota composition and variability. *Front Vet Sci*. 2017 Feb 6;4:6.

Dorn ES, Tress B, Suchodolski JS, et al. Bacterial microbiome in the nose of healthy cats and in cats with nasal disease. *Plos One*. 2017;12(6):e0180299.

Dotson MJ, Hyatt EM. Understanding dog–human companionship. *J Business Res*. 2008;61:457–466.

Driscoll CA, Macdonald DW, O'Brien SJ. From wild animals to domestic pets, an evolutionary view of domestication. *Proc Natl Acad Sci U S A*. 2009 Jun 16;106 Suppl 1(Suppl 1):9971–9978.

Esteves-Monteiro M, Baptista CS, Cardoso-Coutinho D, et al. Exploring gastrointestinal health in diabetic cats: Insights from owner surveys, ultrasound, and histopathological analysis. *Vet Sci*. 2025;12(6):529.

Flowers L, Grice EA. The skin microbiota: Balancing risk and reward. *Cell Host Microbe*. 2020;28(2):190–200.

Freeman A. *Feeding the Sled Dogs – [Internet]*. Wildernessclassroom.org. 2015 [cited 2025 Nov 12].

Galibert F, Quignon P, Hitte C, André C. Toward understanding dog evolutionary and domestication history. *C R Biol*. 2011 Mar;334(3):190–196.

Gallo RL. Human skin is the largest epithelial surface for interaction with microbes. *J Invest Dermatol*. 2017;137(6):1213–1214.

Ganz HH, Jospin G, Rojas CA, et al. The kitty microbiome project: Defining the healthy fecal "core microbiome" in pet domestic cats. *Vet Sci*. 2022;9(11):635.

Garrigues Q, Apper E, Chastant S, et al. Gut microbiota development in the growing dog: A dynamic process influenced by maternal, environmental and host factors. *Front Vet Sci [Internet]*. 2022;9:964649.

Hernandez J, Rhimi S, Kriaa A, et al. Domestic environment and gut microbiota: Lessons from pet dogs. *Microorganisms*. 2022;10(5):949.

Hesselmar B, Aberg N, Aberg B, Eriksson B, Bjorksten B. Does early exposure to cat or dog protect against later allergy development? *Clin Exp Allergy*. 1999;29:611–617.

Hillier A, Griffin CE. The ACVD task force on canine atopic dermatitis (I): Incidence and prevalence. *Vet Immunol Immunopathol*. 2001;81(3–4):147–151.

Honneffer JB, Minamoto Y, Suchodolski JS. Microbiota alterations in acute and chronic gastrointestinal inflammation of cats and dogs. *World J Gastroenterol*. 2014;20(44):16489–16497.

Huang Z, Pan Z, Yang R, et al. The canine gastrointestinal microbiota: Early studies and research frontiers. *Gut Microbes*. 2020;11(4):635–654.

Jarett JK, Kingsbury DD, Dahlhausen KE, Ganz HH. Best practices for microbiome study design in companion animal research. *Front Vet Sci*. 2021 Apr 9;8:644836.

Jergens AE. Feline idiopathic inflammatory bowel disease: What we know and what remains to be unraveled. *J Feline Med Surg*. 2012;14(7):445–458.

Kieler IN, Mølbak L, Hansen LL, Hermann-Bank ML, Bjornvad CR. Overweight and the feline gut microbiome - a pilot study. *J Anim Physiol Anim Nutr (Berl)*. 2016 Jun;100(3):478-84.

Kim J, An J-U, Kim W, et al. Differences in the gut microbiota of dogs (Canis lupus familiaris) fed a natural diet or a commercial feed revealed by the Illumina MiSeq platform. *Gut Pathog*. 2017;9:68.

Kim J-H, An J-H, Lee J-H, et al. Changes in lactate-related fecal microbiome in hyperlactatemia diabetic dogs. *In Vivo*. 2023;37(2):696–701.

Kim S-J, Chung H-C, Park S-Y, et al. Beneficial effects of probiotics on dysbiosis of gut microbiota induced by antibiotic treatment in healthy dogs. *Res Vet Sci*. 2025;191:105674.

Koukourikos K, Georgopoulou A, Kourkouta L, et al. Benefits of animal assisted therapy in mental health. *Int J Caring Sci*. 2019;12(3):1898–1905.

Laia NL, Barko PC, Sullivan DR, et al. Longitudinal analysis of the rectal microbiome in dogs with diabetes mellitus after initiation of insulin therapy. *Plos One*. 2022;17(9):e0273792.

Leverett K, Manjarín R, Laird E, Valtierra D, Santiago-Rodriguez TM, Donadelli R, Perez-Camargo G. Fresh food consumption increases microbiome diversity and promotes changes in bacteria composition on the skin of pet dogs compared to dry foods. *Animals (Basel)*. 2022;12(15):1881.

Levine GN, Allen K, Braun LT, Christian HE, Friedmann E, Taubert KA, Thomas SA, Wells DL, Lange RA; American heart association council on clinical cardiology; Council on cardiovascular and stroke nursing. Pet ownership and cardiovascular risk: A scientific statement from the American heart association. *Circulation*. 2013 Jun 11;127(23):2353–2363.

Li Q, Lauber CL, Czarnecki-Maulden G, et al. Effects of the dietary protein and carbohydrate ratio on gut microbiomes in dogs of different body conditions. *mBio*. 2017;8(1):e01703–e01716.

Litonjua AA, Milton DK, Celedon JC, Ryan L, Weiss ST, Gold DR. A longitudinal analysis of wheezing in young children: The independent effects of early life exposure to house dust endotoxin, allergens, and pets. *J Allergy Clin Immunol*. 2002;110:736–742.

Lyu Y, Su C, Verbrugghe A, et al. Past, present, and future of gastrointestinal microbiota research in cats. *Front Microbiol*. 2020;11:1661.

Ma X, Brinker E, Graff EC, et al. Whole-genome shotgun metagenomic sequencing reveals distinct gut microbiome signatures of obese cats. *Microbiol Spectr*. 2022;10(3):e0083722.

Marsilio S. Feline chronic enteropathy. *J Small Anim Pract*. 2021;62(6):409–419.

Masuoka H, Shimada K, Kiyosue-Yasuda T, et al. Transition of the intestinal microbiota of dogs with age. *Biosci Microbiota Food Health*. 2017;36(1):27–31.

Meason-Smith C, Diesel A, Patterson AP, Older CE, Johnson TJ, Mansell JM, Suchodolski JS, Rodrigues Hoffmann A. Characterization of the cutaneous mycobiota in healthy and allergic cats using next generation sequencing. *Vet Dermatol*. 2017 Feb;28(1):71-e17.

Nugent WR, Daugherty L. A measurement equivalence study of the family bondedness scale: Measurement equivalence between cat and dog owners. *Front Vet Sci*. 2022;8:812922.

Okada H, Kuhn C, Feillet H, Bach JF. The 'hygiene hypothesis' for autoimmune and allergic diseases: an update. *Clin Exp Immunol*. 2010 Apr;160(1):1-9.

Older CE, Diesel AB, Lawhon SD, Queiroz CRR, Henker LC, Rodrigues Hoffmann A. The feline cutaneous and oral microbiota are influenced by breed and environment. *PLoS One*. 2019 Jul 30;14(7):e0220463.

Older CE, Hoffmann AR, Diesel AB. The feline skin microbiome: Interrelationship between health and disease. *J Feline Med Surgery*. 2023;25(7):1098612X231180231.

Páez-Triana L, Luna N, Ramirez AL, et al. Exploring bacterial and eukaryotic communities in the gut microbiota of urban and rural cats (Felis catus) in Colombia. *Vet Res Commun*. 2025;49(6):312.

Pavlíčková Z, Pafčo B, Ilík V, O'Brien Andersen L, Jirků M, Brožová K, Modrý D, Kadlecová O, Stensvold CR, Jirků K. Shaping the human gut microbiota: the role of canine companionship, lifestyle choices, and Blastocystis sp. *One Health*. 2025 Jan 25;20:100979.

Peimbert M, Alcaraz LD. Where environmental microbiome meets its host: Subway and passenger microbiome relationships. *Mol Ecol*. 2023;32(10):2602–2618.

Pereira AM, Clemente A. Dogs' microbiome from tip to toe. *Top Companion Anim Med*. 2021 Nov;45:100584.

Pierezan F, Olivry T, Paps JS, et al. The skin microbiome in allergen-induced canine atopic dermatitis. *Vet Dermatol*. 2016;27(5):332–e82.

Pilla R, Gaschen FP, Barr JW, et al. Effects of metronidazole on the fecal microbiome and metabolome in healthy dogs. *J Vet Intern Med*. 2020;34(5):1853–1866.

Pilla R, Suchodolski JS. The gut microbiome of dogs and cats, and the influence of diet. *Vet Clin North Am Small Anim Pract*. 2021 May;51(3):605–621.

Reddy KE, Kim HR, Jeong JY, So KM, Lee S, Ji SY, Kim M, Lee HJ, Lee S, Kim KH, Kim M. Impact of breed on the fecal microbiome of dogs under the same dietary condition. *J Microbiol Biotechnol*. 2019 Dec 28;29(12):1947–1956.

Reese AT, Chadaideh KS, Diggins CE, et al. Effects of domestication on the gut microbiota parallel those of human industrialization. *Elife*. 2021;10:e60197.

Rodrigues Hoffmann A, Patterson AP, Diesel A, Lawhon SD, Ly HJ, Stephenson CE, et al. The skin microbiome in healthy and allergic dogs. *PLoS ONE*. 2014;9(1):e83197.

Scoresby KJ, Strand EB, Ng Z, Brown KC, Stilz CR, Strobel K, Barroso CS, Souza M. Pet ownership and quality of life: A systematic review of the literature. *Vet Sci*. 2021 Dec 16;8(12):332.

Serpell JA. Anthropomorphism and anthropomorphic selection-beyond the 'cute response.' *Society & Animals: Journal of Human-Animal Studies*. 2002;10(4):437–454.

Sigudu T, Qekwana D, Oguttu J. A retrospective descriptive study of staphylococcus species isolated from canine specimens submitted to a diagnostic laboratory in South Africa, 2012-2017. *Animals (Basel)*. 2024 Apr 26;14(9):1304.

Slingerland LI, Robben JH, van Haeften TW, et al. Insulin sensitivity and beta-cell function in healthy cats: Assessment with the use of the hyperglycemic glucose clamp. *Horm Metab Res*. 2007;39(5):341–346.

Smith TA, Srikanth K, Huson HJ. Comparative population genomics of arctic sled dogs reveals a deep and complex history. *Genome Biol Evolution*. 2024 Aug 28;16(9):evae190

Song SJ, Lauber C, Costello EK, et al. Cohabiting family members share microbiota with one another and with their dogs. *Elife*. 2013;2:e00458.

Štempelová L, Micenková L, Andrla P, et al. The skin microbiome on healthy and inflammatory altered canine skin determined by next generation sequencing. *Front Microbiol*. 2025;16:1528747.

Strachan DP. Hay fever, hygiene, and household size. *BMJ*. 1989 Nov 18;299(6710):1259-60.

Stull JW, Brophy J, Weese JS. Reducing the risk of pet-associated zoonotic infections. *CMAJ*. 2015 Jul 14;187(10):736-743.

Suchodolski JS, Foster ML, Sohail MU, Leutenegger C, Queen EV, Steiner JM, et al. The fecal microbiome in cats with diarrhea. *PLoS ONE*. 2015;10(5):e0127378.

Sun N, Xie L, Chao J, et al. Study on the correlation between aggressive behavior and gut microbiota and serum serotonin (5-HT) in working dogs. *Vet Sci*. 2025;12(6):526.

Szewczyk D, Fiega J, Michalska M, Milena, Zurek U, Lubaszka Z, Zuzanna, Sikorska E. Therapeutic role of animals: a comprehensive literature review on the prevalent forms and species in animal-assisted interventions. *Journal of Education, Health and Sport*. 2023;45:215-235.

Tancredi D, Cardinali I. Being a dog: A review of the domestication process. *Genes (Basel)*. 2023 Apr 27;14(5):992.

Tress B, Dorn ES, Suchodolski JS, Nisar T, Ravindran P, Weber K, Hartmann K, Schulz BS. Bacterial microbiome of the nose of healthy dogs and dogs with nasal disease. *PLoS One*. 2017 May 1;12:e0176736.

Trinh P, Zaneveld JR, Safranek S, Rabinowitz PM. One health relationships between human, animal, and environmental microbiomes: A mini-review. *Front Public Health*. 2018 Aug 30;6:235.

Tun HM, Konya T, Takaro TK, Brook JR, Chari R, Field CJ, Guttman DS, Becker AB, Mandhane PJ, Turvey SE, Subbarao P, Sears MR, Scott JA, Kozyrskyj AL; CHILD Study Investigators. Exposure to household furry pets influences the gut microbiota of infant at 3-4 months following various birth scenarios. *Microbiome*. 2017 Apr 6;5(1):40.

Turnbaugh PJ, Bäckhed F, Fulton L, et al. Diet-induced obesity is linked to marked but reversible alterations in the mouse distal gut microbiome. *Cell Host Microbe*. 2008;3(4):213–223.

Turnbaugh PJ, Ley RE, Hamady M, et al. The human microbiome project. *Nature*. 2007;449(7164):804–810.

Wang D, Russel WA, Macdonald KM, De Leon VM, Ay A, Belanger KD. Analysis of the gut microbiome in sled dogs reveals glucosamine- and activity-related effects on gut microbial composition. *Front Vet Sci*. 2024 Feb 7;11:1272711.

Williams MR, Gallo RL. The role of the skin microbiome in atopic dermatitis. *Curr Allergy Asthma Rep*. 2015;15(11):65.

Wisdom JP, Saedi GA, Green CA. Another breed of "service" animals: STARS study findings about pet ownership and recovery from serious mental illness. *Am J Orthopsychiatry*. 2009 Jul;79(3):430–436.

Wu W-C, Pan Y-F, Zhou W-D, Liao Y-Q, Peng M-W, Luo G-Y, Xin G-Y, Peng Y-N, An T, Li B, Luo H, Barrs VR, Beatty JA, Holmes EC, Zhao W, Shi M, Shu Y. Meta-transcriptomic analysis of companion animal infectomes reveals their diversity and potential roles in animal and human disease. *mSphere*. 2024 Aug 28;9(8):e0043924.

Xie A, Zhang Y, Tang Z, et al. Gut microbiome disruption in free-roaming cats: Do antibiotics reduce or restructure zoonotic risk? *One Health*. 2025;21:101186.

Yeon K-D, Kim S-M, Kim J-H. Association between gut microbiota and metabolic health and obesity status in cats. *Animals (Basel)*. 2024;14(17):2524.

You I, Kim MJ. Comparison of gut microbiota of 96 healthy dogs by individual traits: Breed, age, and body condition score. *Animals (Basel)*. 2021 Aug 18;11(8):2432.

CHAPTER NINE

The Iguanid Microbiome and Its Physiological Significance

Kwanho C. Ki and Erin L. Lewis

9.1 INTRODUCTION

Reptiles are an incredibly diverse group of ectotherms, including snakes and lizards (order Squamata), turtles (order Testudines), crocodilians (order Crocodillia), and tuataras (order Rynchocephalia). Organisms in class Reptilia are evolutionarily, ecologically, and physiologically diverse. Ranging from the extant *Sphenodon punctalus*, Tuataras, of New Zealand, to *Rhineura floridana*, the Florida worm lizard; these animals can be found on every continent except for Antarctica. Reptiles inhabit deserts, forests, grasslands, wetlands, freshwater, marine, and even urban areas. These different environmental pressures select for a number of different physiological and dietary adaptations; thereby giving rise to equally diverse microorganisms that colonize the intestinal tracts of different species of reptiles.

Generally, the core bacteria that makes up the microbiome of reptiles are from the phyla Firmicutes, Proteobacteria, and Bacteroidota (Hoffbeck et al., 2023a). The relative ratios of these bacterial groups vastly differ across reptilian species. The phyla of core bacteria in the microbiome of reptiles are on par with the general microbiome of endothermal vertebrates (Wexler & Goodman, 2017). There is much evidence supporting the idea that these commensal microorganisms play an important role in digestion, immunity, endocrine signaling, and more (Hoffbeck et al., 2023b; Konstantinidis et al., 2020; O'Sullivan et al., 2013; Shi et al., 2017; Siddiqui et al., 2022). While there is a wealth of knowledge on this blossoming field investigating the microbiome, most research uses humans and mice as their model species. Reptiles are often overlooked as nonmodel organisms, thus literature on reptilian microbiomes is sparse in comparison to their mammalian counterparts. This chapter will describe the microbiome of reptiles while applying research gathered from other species as well.

At the genus level, Bacteroides are the most common bacterial groups found inhabiting the intestinal tract of reptiles (Hoffbeck et al., 2023a). Bacteroides are notably a commensal found in the gut of most vertebrates. Members of this genus aid in metabolism of diet-derived glycans and oligosaccharides which allows for the utilization of these sugars for both the host and other microbes (Patrick, 2022; Zafar & Saier, 2021). Additionally, experiments using germ-free mice has demonstrated the importance of Bacteroides in proper development of immune tissues like Peyer's patches and mesenteric lymph nodes, reduction of infection by other pathogens, and promoting immune tolerance (Wexler & Goodman, 2017). However, this bacterial group is not as altruistic as they may seem. Bacteroides are also ferocious competitors in the microbial landscape, with some species releasing antimicrobial toxins that target other gut commensals, thereby allowing this group of bacteria to modulate the microbiome (Zafar & Saier, 2021). Several strains have also been identified as opportunistically pathogenic such as *B. fragilis* and *B. thetaiotaomicron*, which can cause infections when transported to other sites in

DOI: 10.1201/9781003638568-9

the case a ruptured appendix or injuries (Patrick, 2022; H. M. Wexler, 2007).

9.1.1 Iguanids

Within class Reptilia, squamates are the largest order and are a highly diverse paraphyletic clade. This order includes snakes, legless lizards, geckos, chameleons, and iguanas. Due to the diversity of squamates, this chapter will focus on describing the known microbiome of the family Iguanidae. Iguanas are distinct in that most iguanas are herbivorous while other squamates are generally carnivores or omnivores. The difference in diet between herbivores and omnivores/carnivores greatly affects the composition of iguanids' microbial community due to the symbiotic relationship between herbivorous hosts and their gut microbes that aid in digesting complex plant polysaccharides (Burr et al., 2020a; Cresci & Bawden, 2015).

There are ~44 iguanid species with some debate about whether certain species are a subclass of others. Iguanids are categorized by nine major living genera: rock iguanas (Cyclura), green iguanas (Iguana), Fiji iguanas (Brachylophus), Galapagos iguanas (Amblyrhynchus, Colonophus), spiny-tailed iguanas (Ctenosaura, Cachrynx), desert iguanas (Dipsosaurus), and chuckwallas (Sauromalus) (Böhm et al., 2013; Etheridge & de Queiroz, 1988; Köhler et al., 2000; Malone et al., 2017). Most of these iguanas are herbivorous and have specialized intestinal tracts for digesting fibrous leaves, stems, and fruits (Iverson, 1982). Rock iguanas, such as *Cyclura pinguis*, *Cyclura cornuta*, *Cyclura ricordii*, and *Cyclura cychlura* prefer floral plants and fruits, (Iverson et al., 2016; Knapp et al., 2013; Michell, 1999; Pasachnik & Martin-Velez, 2017). Green iguana (*Iguana iguana*), Fiji iguanas (*Brachlyophus vitiensis*), and chuckwallas (*Sauromalus obesus*, *Sauromalus hispidus*, *Sauromalus varius*) are strictly herbivores, eating a range of leafy vegetation, stems, and seeds (Morrison et al., 2007; Nagy, 1973; Rand et al., 1990; Sylber, 1988). Marine iguanas, *Amblyrhnchys cristatus*, notably graze on red and green algae on the ocean floor, while land iguanas, *Colonophus pallidus*, are folivores that opportunistically eat carrion like dead rats and birds (Christian et al., 1984; Vasco et al., 2022). Desert iguanas, *Dipsosaurus dorsalis*, are omnivores and different populations have been found to have different percentages of plant and animal matter in their intestines suggesting that they are opportunists (Dibble et al., 2008). Most spiny-tailed iguanas such as *Ctenosaura hemilopha*, *Ctenosaura oedirhina*, *Ctenosaura palearis*, *Ctenosaura similis* are opportunistically omnivorous and have been observed to eat leaves, flowers, fruit, insects, crabs, and even bats (Blázquez & Rodríguez-Estrella, 2007; Coti & Ariano-Sánchez, 2008; Gandola & Hendry, 2011; Krysko et al., 2009; Queiroz, 1987). Finally, there is almost no literature describing *Cachyrnx defensor* and *Cachrynx alfredschmidti*, and very little is known about their ecology.

To date, there are only a handful of studies exploring the composition of the microbiome in iguanas. Among them, Galapagos iguanas (*A. cristatus* and *C. pallidus*) are well studied. The intestinal tract of *A. cristatus*, from most abundant to least, is dominated by the phyla Firmicutes, Proteobacteria, Bacteroidetes, and Actinobacteria (Hong et al., 2011; Vasco et al., 2022). While the microbiome of land iguanas (*Conolophus subcristatus*, *C. pallidus*, *Conolophus marthae*), are also dominated by bacteria from phyla Firmicutes, Bacteroidetes, and Proteobacteria, there are substantially higher abundance of Proteobacteria and lower Bacteroidetes in land iguanas than marine iguanas (Hong et al., 2011, 2015). This is likely due to the dietary differences between these two species where marine iguanas exclusively feed on algae in the ocean and land iguanas on terrestrial plants. There is likely an evolutionary advantage for the natural selection of *Bacteroides* spp. in the marine iguana's gut microbiome as Bacteroides can catabolize complex polysaccharides only found in algae, the main food source of their host. Bacteria from the phyla Firmicutes were found in both land and marine iguanas and bacteria in this phylum is known for their ability to break down plant polysaccharides like hemicelluloses and pectin for host metabolism (Hong et al., 2011, 2015). A recent study characterized the gut microbiome of *A. cristatus* and corroborates the core phyla that makes up their gut microbiome found in the previous 2011 and 2015 studies (Vasco et al., 2022). However, in this new study, there are different levels of abundance across the core phyla than previously determined. The authors attribute this difference in abundance to a difference in sample sites, seasonality, and sequencing technology (Vasco et al., 2022). Furthermore, the functional microbiome reveals that marine iguanas rely on these microbes for metabolism of polysaccharides, proteins, and vitamins. As expected, the microbes that inhabit the intestinal tracts of both land and marine iguanas have a mutualistic relationship based on host dietary needs.

The microbiome of Fijian crested iguanas, *Brachylophus vitiensis*, is dominated by the phyla Firmicutes, Proteobacteria, Actinobacteria, and Bacteroidetes, from most abundant to least abundant (Eliades et al., 2021). All iguanas also had one *Oscillospira* spp. and *Phascolarctobacterium* spp. (Eliades et al., 2021). Most Oscillospira can metabolize plant fibers into short-chain fatty acids like butyrate and are common in the digestive tract of herbivores and omnivores. Oscillospira are slow-growing bacteria and are generally found in the colon of animals that have slower transit time (i.e. herbivores). The slow digestion of folivores of *B. vitiensis* presents a suitable environment for Oscillospira to be cultured and in return, Oscillospira can breakdown polysaccharides for host usage (Gophna et al., 2017; Yang et al., 2021). On the other hand, *Phascolarctobacterium* was first isolated in the Koala, an herbivorous mammal, but is interestingly linked to high fat diets and the digestion of animal products (Lau et al., 2018; Wu et al., 2017). It is unclear what role Phascolarctobacterium plays in the microbiome of *B. vitiensis*, or whether they are simply a gut commensal.

The gut microbiome of the green iguana, *I. iguana*, is dominated by the same four phyla as other iguanas, Firmicutes, Actinobacteria, Bacteroidetes, and Proteobacteria (Ki et al., 2024). All other journal articles reviewed in this chapter thus far involve samples from free-living iguanas in their native habitat, however, the juvenile green iguanas used in this 2024 study are laboratory animals and originally captured as invasives from Florida. Thus, it is interesting that despite the controlled and anthropomorphized environments/diet, the dominating phyla are still the same. This is likely an evolutionary product of selecting gut symbionts needed to help ferment and digest cellulose and hemicellulose (Troyer, 1984). However, it is still important to note that the microbiome of free-living *I. iguana* may very well be different from captive animals. For example, in crocodile lizards, *Shinisaurus crocodilurus*, captive populations of lizards from different reserves fed different diets had significantly different microbiome composition. In one population, they had significantly higher levels of Fusobacteria while the other had higher levels of Elizabethkingia, Halomonas, Morganella, and Salmonella (Jiang et al., 2017). As the diet of captive iguanas is substantially different from the diet of free-living iguanas, future research will need to confirm the microbiome of free-living green iguanas in their native habitats.

The Bahamian rock iguana gut microbiome varies across sex and populations and is influenced by tourism-related supplemental feeding (Lewis, 2026). Across both studied subspecies, several taxa were differentially abundant between males and females, with families such as Ruminococcaceae, Lachnospiraceae, Clostridiales vadinBB60 group, Corynebacteriaceae, and Bacteroidaceae commonly represented. Populations on islands with higher tourism exposure exhibited increased alpha and beta diversity, suggesting a more varied diet and more distinct microbial communities among individuals, likely driven by provisioned foods. These patterns suggest that dietary shifts related to tourism alter gut microbial communities, potentially affecting individual physiology and population health.

There are currently no publications describing the microbiome of desert iguanas, chuckwallas, or ctenosaurs. As chuckwallas are strict herbivores like green iguanas and Fiji iguanas, it may be inferred that they have similar microbial symbionts that support digestion and detoxification of plant matter. Desert iguanas and ctenosaurs are similar in that they are both generalists, eating both plants and animals. These two genera of iguanas would be an interesting route of exploration as they are most likely to have a different microbiome profile compared to other iguanids due to their omnivorous diets. In a naturally omnivorous lizard, *Liolaemus ruibali*, it was found that experimentally feeding them plant-rich diets resulted in gut communities with higher diversity and higher abundance of bacteria from the genera Lactobacillus, Oscillospira, and Desulfovibrio (Kohl et al., 2016). Perhaps in the omnivorous desert iguanas and ctenosaurs, it can be expected that their gut microbiome is more homogenous (lower diversity) and there is less of a reliance on microbes associated with fermentation like Oscillospira and Desulfovibrio.

9.2 ESTABLISHMENT OF MICROBIOME

The establishment of the gut microbiome in many species, including iguanas, is a dynamic process influenced by a combination of maternal, environmental, dietary, and social factors. Together these shape the composition, diversity, and functionality of microbial communities throughout the animal's life, which in turn affect health and survival.

Gut microbiome establishment begins prior to hatching, with microbial exposure coming from multiple environmental and maternal sources. The initial colonization of the gut microbiome in iguanas may be influenced by maternal transfer of microbes, which can occur through either vertical transmission (directly from the mother) or horizontal transmission (from the nest environment or maternal feces). Eggs and nesting sites can harbor microbes from the mother and surrounding soil, providing early colonists to the gastrointestinal tract (Kohl, 2012; Warne et al., 2019). While maternal transfer has not been extensively studied in oviparous reptiles, evidence from related species, such as Sceloporus lizards, suggests that mothers can transmit microbial communities to their offspring (Bunker et al., 2021; Bunker & Weiss, 2024). Additionally, while limited, there is some evidence for vertical transmission of gut microbiomes in birds (Shterzer et al., 2023). While this entire field of research is limited thus far, there is potential to provide insight into phenotypic evolution (Murphy et al., 2023). Such transmission could play a critical role in establishing the hatchling gut with beneficial microbes, establishing the foundation for microbial diversity and functionality that supports digestion, immune development, and overall health throughout early life.

The establishment of the gut microbiome in iguanas is profoundly shaped by environmental factors, particularly during early life stages. The gut microbiome is initially colonized through a variety of factors, including exposure to its specific surroundings, including habitat, diet, and social interactions (Dominguez-Bello et al., 2010; Maritan et al., 2024). As hatchlings, iguanas are exposed to microbiota from their mother's skin, nesting environment, and other external sources, such as water, soil, and plant material (Liu et al., 2021). These environmental exposures are crucial for the early development of a diverse gut microbiome, which is essential for digestion, immune function, and overall health (Kau et al., 2011). Specific dietary preference, such as plants, leaves, and fruits, also play a significant role in shaping the microbial community, as different dietary components can promote or inhibit the growth of different bacterial taxa (Ley et al., 2008; David et al., 2014; Ki et al., 2024). Furthermore, environmental conditions like temperature have been found to influence the microbial composition across a number of species, by affecting things like microbial survival/growth and enzyme activity (Sepulveda & Moeller, 2020). As individuals mature, they are increasingly exposed to the broader ecological community, including other animals, which can introduce new microbial species into their gut (Sarkar et al., 2020). A study in barn swallows demonstrated that the stress response, gut microbial diversity, and sexual signals are all correlated with social interactions, providing evidence for the role of social dynamics in gut microbiome establishment, where body contact was associated with stress reactivity and gut microbial diversity (Levin et al., 2016). The balance between microbial colonization and immune responses during early development is crucial in establishing a functional microbiome that supports nutrition and disease resistance, similar to how the gut microbiome aids immune function and resilience in other reptiles (Siddiqui et al., 2022). Differences between wild and captive animals have provided further evidence for how much environmental conditions can change the gut microbiome (Eliades et al., 2021; Dallas & Warne, 2022). However, a study in marine iguanas found highly conserved gut microbial composition where differences between sites (some of which are on different islands) were linked to specific functions, not to taxa (Vasco et al., 2022). These observations place further emphasis on the need for further research into the iguana gut microbiome and complexities involved in establishment.

There has been documented evidence for coprophagy in the establishment of the gut microbiome in herbivorous, hatchling iguanas (*Iguana iguana*). This involves the acquisition of key fermentative microbes through the ingestion of fecal material, which facilitates the development of a functional hindgut fermentation system necessary for digesting plant material (Troyer, 1984). Hatchlings reared in isolation fail to develop the full complement of hindgut microbes and exhibit slower growth and reduced digestive efficiency, whereas those provided with fresh adult feces quickly establish complex microbial communities and grow at rates comparable to wild hatchlings (Troyer, 1982). This behavioral acquisition of microbes ensures that offspring inherit a microbiome adapted to their herbivorous diet and local environment, supporting digestion, nutrient absorption, and the maturation of immune function. There is some evidence for similar behavior in other species, including birds (Dunbar et al., 2024).

9.3 MICROBIOME AND PHYSIOLOGY

In the last two decades, there has been mounting evidence in support of the microbiome's ability to modulate and even direct important physiological variables like glucocorticoids, oxidative stress markers, and immunity (Alexander & Turnbaugh, 2020; Burr et al., 2020b; Hori et al., 2020; Iizumi et al., 2017; Kim, 2018; Konstantinidis et al., 2020; O'Sullivan et al., 2013; Tao et al., 2020). Some notable bacterial strains and their roles in physiology include species from the Bacteroides family, which is considered a core member of the reptilian and iguanian microbiome (Hoffbeck et al., 2023b). *B. fragilis* can inhibit IL-17, activate CD4 + T-cells, and increase the activity of regulatory T-cells (H. M. Wexler, 2007; Zimmerman & Tracy, 1989). Aside from immune regulation, the genome of *B. fragilis* can produce proteins that assist host cells in fighting against oxidative stress agents (Zafar & Saier, 2021). *B. vulgatus* decreased expression of immune factors like IL-6, IL-1B, and TNF-a (Liu et al., 2022). While the complex relationships between physiology and the microbiome are slowly becoming elucidated, very few studies are in reptiles, signifying a still unknown realm that needs further consideration.

Research has consistently highlighted the complex relationships between the gut microbiome and overall health, particularly emphasizing how diet-related disruptions can affect physiological systems. For instance, studies have shown that dysbiosis (an imbalance in gut microbial composition) is linked to altered glucocorticoid levels, often favoring stress-tolerant or pathogenic microbes over beneficial commensals (Levin et al., 2016; Amini-Khoei et al., 2019; Xie et al., 2019). Growing evidence suggests that the relationship between glucocorticoids and the gut microbiome is multidirectional, each capable of influencing the other through multiple physiological pathways. Glucocorticoids, such as cortisol or corticosterone (CORT), are central mediators of the stress response (Romero et al., 2009) and play key roles in energy metabolism, immune regulation, and inflammation (Sapolsky et al., 2000, Cain & Cidlowski, 2017); impacting health in complex ways, including playing a role in the synchronization of the internal circadian system (Oster et al., 2017). Changes in glucocorticoid levels can directly alter gut physiology by affecting intestinal permeability, mucus production, and gut motility (Meddings & Swain, 2000; Bai et al., 2007; Adcock & Mumby, 2016: Filaretova & Bagaeva, 2016), all of which have potential to influence microbial colonization and community structure, though more research is needed to definitively link these (Collins et al., 2012, Cain & Cidlowski, 2017, Chakaroun et al., 2020). For example, glucocorticoid signaling has been shown to influence intestinal barrier function by modulating tight junctions, affecting intestinal permeability and microbiota composition, and in conditions like sepsis or ARDS, this pathway helps restore vascular integrity and intestinal barrier function (Meduri & Psarra, 2025).

Experimental studies further support this relationship by demonstrating that manipulation of baseline glucocorticoid levels can lead to measurable changes in gut microbial communities. For example, experimentally increased CORT levels in birds resulted in reduced abundance of certain bacterial taxa, suggesting that chronic stress may suppress microbial groups important for digestion or immune modulation (Noguera et al., 2018). Similar patterns have been observed in mammalian systems, where stress-induced elevations in glucocorticoids are linked to dysbiosis and increased susceptibility to inflammation and metabolic dysfunction (Xie et al., 2019). Conversely, the gut microbiome can also influence glucocorticoid regulation through interactions with the hypothalamic–pituitary–adrenal (HPA) axis. Microbial metabolites, such as short-chain fatty acids, and microbial modulation of immune signaling pathways have been shown to affect HPA axis activity, altering glucocorticoid release (Sharon et al., 2014, Van de Wouw et al., 2018). Disruption of the gut microbiome, particularly early in life, has been associated with exaggerated glucocorticoid responses to stress, indicating that a stable and diverse microbial community may be essential for normal stress regulation (Tetel et al., 2018). Collectively, these findings indicate that glucocorticoids and the gut microbiome form a connected bidirectional relationship, where stress physiology can reshape microbial communities, and microbial composition can, in turn, influence endocrine responses. This feedback relationship highlights the potential for environmental stressors (e.g. dietary changes, habitat disturbance, or human interaction) to indirectly affect animal health by altering those glucocorticoid/microbiome dynamics.

Other physiological metrics such as oxidative stress have been linked to gut microbiota (Semenova et al., 2024). For example, in one study looking at human infants, researchers found correlations between oxidative stress markers (specifically levels of F2-isoprostane), and abundance of specific microbial taxa (Cai et al., 2019). Additionally, high starch diets have also been found to increase oxidative stress through microbes (through short-chain fatty acids) in rats (Vaziri et al., 2014). Additionally oxidative stress via arterial dysfunction has been shown to be mediated by the gut microbiome (Brunt et al., 2019), providing evidence for the importance of microbiome–physiology relationships in disrupted systems.

In terms of links between the gut microbiome and immunity, there are several studies in reptiles to reference. Green iguanas, *I. iguana*, evolved to be folivores that consume a relatively low sugar diet. Experimentally manipulating their diet to have substantially higher levels of sugar have revealed that this diet not only altered their gut microbial communities but also reduced immune agglutination and lysis when faced with an immune challenge (Ki et al., 2024). Aside from dietary induced changes to the gut microbiome, the reptile microbiome is also sensitive to heat (Moeller et al., 2020). For example, a desert lizard, *Eremias multiocellata*, exposed to long term warming had increased microbial diversity and an increase in serum antibacterial activity. The authors found that this enhanced antibacterial activity is linked to the higher abundance of *B. fragilis* (Yang et al., 2024). In mammals, this interplay between the gut microbiome and the immune system is, in part, explained by microbially produced signaling molecules like short-chain fatty acids (Kasubuchi et al., 2015; Smith et al., 2013). However, these molecules have yet to be confirmed to play a similar role in reptiles, presenting yet another direction of exploration in this field (Siddiqui et al., 2022). From the mere studies that exist, it is known that the microbiome and immunity are linked in reptiles but much more research needs to be done before we can determine what those relationships are and the mechanism behind their interactions.

REFERENCES

Adcock, I. M., & S. Mumby. (2017). Glucocorticoids. In C. P. Page & P. J. Barnes (Eds.), *Pharmacology and Therapeutics of Asthma and COPD* (pp. 171–196). Cham: Springer International.

Alexander, M., & Turnbaugh, P. J. (2020). Deconstructing mechanisms of diet-microbiome-immune interactions. *Immunity*, 53(2), 264–276. https://doi.org/10.1016/j.immuni.2020.07.015

Amini-Khoei, H., E. Haghani-Samani, M. Beigi, A. Soltani, G. R. Mobini, S. Balali-Dehkordi, A. Haj-Mirzaian, M. Rafieian-Kopaei, A. Alizadeh, & M. R. Hojjati. (2019). On the role of corticosterone in behavioral disorders, microbiota composition alteration and neuroimmune response in adult male mice subjected to maternal separation stress. *International Immunopharmacology*, 66, 242–250.

Bai, C. H., S. Y. Song, & Y. D. Kim. (2007). Effect of glucocorticoid on the MUC4 gene in nasal polyps. *The Laryngoscope*, 11, 2169–2173.

Blázquez, M. C., & Rodríguez-Estrella, R. (2007). Microhabitat selection in diet and trophic ecology of a spiny-tailed iguana *Ctenosaura hemilopha*. *Biotropica*, 39(4), 496–501. https://doi.org/10.1111/j.1744-7429.2007.00294.x

Böhm, M., Collen, B., Baillie, J. E. M., Bowles, P., Chanson, J., Cox, N., Hammerson, G., Hoffmann, M., Livingstone, S. R., Ram, M., Rhodin, A. G. J., Stuart, S. N., van Dijk, P. P., Young, B. E., Afuang, L. E., Aghasyan, A., García, A., Aguilar, C., Ajtic, R., ... Zug, G. (2013). The conservation status of the world's reptiles. *Biological Conservation*, 157, 372–385. https://doi.org/10.1016/j.biocon.2012.07.015

Brunt, V. E., R. A. Gioscia-Ryan, J. J. Richey, M. C. Zigler, L. M. Cuevas, A. Gonzalez, Y. Vázquez-Baeza, M. L. Battson, A. T. Smithson, & A. D. Gilley. (2019). Suppression of the gut microbiome ameliorates age-related arterial dysfunction and oxidative stress in mice. *The Journal of Physiology*, 597, 2361–2378.

Bunker, M., G. Elliott, H. Heyer-Gray, M. Martin, A. Arnold, & S. Weiss. (2021). Vertically transmitted microbiome protects eggs from fungal infection and egg failure. *Animal Microbiome*, 3, 43.

Bunker, M. E., & S. L. Weiss. (2024). The reproductive microbiome and maternal transmission of microbiota via eggs in Sceloporus virgatus. *FEMS Microbiology Ecologyi*, 100, fiae011.

Burr, A. H. P., Bhattacharjee, A., & Hand, T. W. (2020a). Nutritional modulation of the microbiome and immune response. *The Journal of Immunology*, 205(6), 1479–1487. https://doi.org/10.4049/jimmunol.2000419

Burr, A. H. P., Bhattacharjee, A., & Hand, T. W. (2020b). Nutritional modulation of the microbiome and immune response. *The Journal of Immunology*, 205(6), 1479–1487. https://doi.org/10.4049/jimmunol.2000419

Cai, C., Z. Zhang, M. Morales, Y. Wang, E. Khafipour, & J. Friel. (2019). Feeding practice influences gut microbiome composition in very low birth weight preterm infants and the association with oxidative stress: A prospective cohort study. *Free Radical Biology and Medicine*, 142, 146–154.

Cain, D. W., & J. A. Cidlowski. (2017). Immune regulation by glucocorticoids. *Nature Reviews Immunology*, 17, 233–247.

Chakaroun, R. M., L. Massier, & P. Kovacs. (2020). Gut microbiome, intestinal permeability, and tissue bacteria in metabolic disease: perpetrators or bystanders? *Nutrients*, 12, 1082.

Christian, K. A., Tracy, C. R., & Porter, W. P. (1984). Diet, digestion, and food preferences of Galapagos land iguanas. Herpetologica, 40(2), 205–212. https://about.jstor.org/terms

Collins, S. M., M. Surette, & P. Bercik. (2012). The interplay between the intestinal microbiota and the brain. *Nature Reviews Microbiology*, 10, 735–742.

Coti, P., & Ariano-Sánchez, D. (2008). Ecology and traditional use of Guatemalan black iguana (*Ctenosaura palearis*) in the dry forests of the Motagua Valleu, Guatemala. *Iguana*, 15, 143–149.

Cresci, G. A., & Bawden, E. (2015). Gut microbiome: What we do and don't know. *Nutrition in Clinical Practice*, 30(6), 734–746. https://doi.org/10.1177/0884533615609899

Dallas, J. W., & R. W. Warne. (2022). Captivity and animal microbiomes: Potential roles of microbiota for influencing animal conservation. *Microbial Ecology*, 1–19.

David, L. A., C. F. Maurice, R. N. Carmody, D. B. Gootenberg, J. E. Button, B. E. Wolfe, A. V. Ling, A. S. Devlin, Y. Varma, & M. A. Fischbach. (2014). Diet rapidly and reproducibly alters the human gut microbiome. *Nature*, 505, 559–563.

Dibble, C. J., Smith, G. R., & Lemos-Espinal, J. A. (2008). Diet and sexual dimorphism of the desert iguana, dipsosaurus dorsalis, from Sonora, Mexico. *Western North American Naturalist*, 68(4), 521–523.

Dominguez-Bello, M. G., E. K. Costello, M. Contreras, M. Magris, G. Hidalgo, N. Fierer, & R. Knight. (2010). Delivery mode shapes the acquisition and structure of the initial microbiota across multiple body habitats in newborns. *Proceedings of the National Academy of Sciences*, 07, 11971–11975.

Dunbar, A., B. Drigo, S. P. Djordjevic, E. Donner, & B. J. Hoye. (2024). Impacts of coprophagic foraging behaviour on the avian gut microbiome. *Biological Reviews*, 99, 582–597.

Eliades, S. J., Brown, J. C., Colston, T. J., Fisher, R. N., Niukula, J. B., Gray, K., Vadada, J., Rasalato, S., & Siler, C. D. (2021). Gut microbial ecology of the critically endangered Fijian crested iguana (*Brachylophus vitiensis*): Effects of captivity status and host reintroduction on endogenous microbiomes. *Ecology and Evolution*, 11(9), 4731–4743. https://doi.org/10.1002/ece3.7373

Etheridge, R., & de Queiroz, K. (1988). A phylogeny of iguanidae. In R. Estes & G. Pregill (Eds.), *Phylogenetic Relationships of the Lizard Families: Essays Commemorating Charles L.Camp* (pp. 283–367). Stanford University Press.

Filaretova, L., & T. Bagaeva. (2016). The realization of the brain-gut interactions with corticotropin-releasing factor and glucocorticoids. *Current Neuropharmacology*, 14, 876–881.

Gandola, R., & Hendry, C. (2011). Ctenosaura oedirhina Roatán spiny-tailed iguana diet. *Herpetological Review*, 42(3), 428–429. https://doi.org/10.2305/IUCN.UK.2019-2.RLTS.T44191A122558520.en

Gophna, U., Konikoff, T., & Nielsen, H. B. (2017). Oscillospira and related bacteria – From metagenomic species to metabolic features. *Environmental Microbiology*, 19(3), 835–841. https://doi.org/10.1111/1462-2920.13658

Hoffbeck, C., Middleton, D. M. R. L., Nelson, N. J., & Taylor, M. W. (2023a). 16S rRNA gene-based meta-analysis of the reptile gut microbiota reveals environmental effects, host influences and a limited core microbiota. *Molecular Ecology*, 32(22), 6044–6058. https://doi.org/10.1111/mec.17153

Hoffbeck, C., Middleton, D. M. R. L., Nelson, N. J., & Taylor, M. W. (2023b). 16S rRNA gene-based meta-analysis of the reptile gut microbiota reveals environmental effects, host influences and a limited core microbiota. *Molecular Ecology*, 32(22), 6044–6058. https://doi.org/10.1111/mec.17153

Hong, P. Y., Mao, Y., Ortiz-Kofoed, S., Shah, R., Cann, I., & Mackie, R. I. (2015). Metagenomic-based study of the phylogenetic and functional gene diversity in Galápagos land and marine iguanas. *Microbial Ecology*, 69(2), 444–456. https://doi.org/10.1007/s00248-014-0547-6

Hong, P. Y., Wheeler, E., Cann, I. K. O., & Mackie, R. I. (2011). Phylogenetic analysis of the fecal microbial community in herbivorous land and marine iguanas of the Galápagos Islands using 16S rRNA-based pyrosequencing. *ISME Journal*, 5(9), 1461–1470. https://doi.org/10.1038/ismej.2011.33

Hori, T., Matsuda, K., & Oishi, K. (2020). Probiotics: A dietary factor to modulate the gut microbiome, host immune system, and gut–brain interaction. *Microorganisms*, 8(9), 1–24. https://doi.org/10.3390/microorganisms8091401

Iizumi, T., Battaglia, T., Ruiz, V., & Perez Perez, G. I. (2017). Gut microbiome and antibiotics. *Archives of Medical Research*, 48(8), 727–734. https://doi.org/10.1016/j.arcmed.2017.11.004

Iverson, J. B. (1982). Adaptations to herbivory in iguanine lizards. In G. M. Burghardt & A. S. Rand (Eds.). *Iguanas of the World: Their Behavior, Ecology and Conservation* (pp. 60–76). Noyes Publications. Park Ridge, New Jersey.

Iverson, J. B., Grant, T. D., Knapp, C. R., & Pasachnik, Stesha A. (2016). Iguanas: Biology, systematics, and conservation. *Herpetological Conservation and Biology*, 11(June), 1–264. www.herpconbio.org/contents_vol11_Monograph6.html

Jiang, H.Y., Ma, J. E., Li, J., Zhang, X. J., Li, L. M., He, N., Liu, H.Y., Luo, S.Y., Wu, Z. J., Han, R. C., & Chen, J. P. (2017). Diets alter the gut microbiome of crocodile lizards. *Frontiers in Microbiology*, 8(Oct), 1–11. https://doi.org/10.3389/fmicb.2017.02073

Kau, A. L., P. P. Ahern, N. W. Griffin, A. L. Goodman, & J. I. Gordon. (2011). Human nutrition, the gut microbiome and the immune system. *Nature*, 474, 327–336.

Kasubuchi, M., Hasegawa, S., Hiramatsu, T., Ichimura, A., & Kimura, I. (2015). Dietary gut microbial metabolites, short-chain fatty acids, and host metabolic regulation. *Nutrients*, 7(4), 2839–2849. https://doi.org/10.3390/nu7042839

Ki, K. C., Lewis, E. L., Wu, E., Oliaro, F. J., Aubry, L. M., Knapp, C. R., Kapheim, K. M., Denardo, D., & French, S. S. (2024). High sugar diet alters immune function and the gut microbiome in juvenile green iguanas (*Iguana iguana*). *Journal of Experimental Biology*, 227(13). https://doi.org/10.1242/jeb.246981

Kim, C. H. (2018). Immune regulation by microbiome metabolites. *Immunology*, 154(2), 220–229. https://doi.org/10.1111/imm.12930

Knapp, C. R., Hines, K. N., Zachariah, T. T., Perez-Heydrich, C., Iverson, J. B., Buckner, S. D., Halach, S. C., Lattin, C. R., & Romero, L. M. (2013). Physiological effects of tourism and associated food provisioning in an endangered iguana. *Conservation Physiology*, 1, 1–12. https://doi.org/10.1093/conphys/cot032

Kohl, K. D. (2012). Diversity and function of the avian gut microbiota. *Journal of Comparative Physiology B*, 182, 591–602.

Kohl, K. D., Brun, A., Magallanes, M., Brinkerhoff, J., Laspiur, A., Acosta, J. C., Bordenstein, S. R., & Caviedes-Vidal, E. (2016). Physiological and microbial adjustments to diet quality permit facultative herbivory in an omnivorous lizard. *Journal of Experimental Biology*, 219(12), 1903–1912. https://doi.org/10.1242/jeb.138370

Köhler, G., Schroth, W., & Streit, B. (2000). Systematics of the Ctenosaura group of lizards (Reptilia: Sauria: Iguanidae). *Amphibia Reptilia*, 21(2), 177–191. https://doi.org/10.1163/156853800507363

Konstantinidis, T., Tsigalou, C., Karvelas, A., Stavropoulou, E., Voidarou, C., & Bezirtzoglou, E. (2020). Effects of antibiotics upon the gut microbiome: A review of the literature. *Biomedicines*, 8(11), 1–15. https://doi.org/10.3390/biomedicines8110502

Krysko, K. L., Larson, K.W., Diep, D., & Abellana, E. (2009). Diet of the nonindigenous black spiny-tailed iguana, *Ctenosaura similis* (Gray 1831) (Sauria: Iguanidae), in southern Florida. *Biological Sciences*, 72(1), 48–58. www.jstor.org/stable/24321725

Lau, S. K. P., Teng, J. L. L., Chiu, T. H., Chan, E., Tsang, A. K. L., Panagiotou, G., Zhai, S. L., & Woo, P. C.Y. (2018). Differential microbial communities of omnivorous and herbivorous cattle in southern China. *Computational and Structural Biotechnology Journal*, 16, 54–60. https://doi.org/10.1016/j.csbj.2018.02.004

Levin, I. I., D. M. Zonana, B. K. Fosdick, S. J. Song, R. Knight, & R. J. Safran. (2016). Stress response, gut microbial diversity and sexual signals correlate with social interactions. *Biology Letters*, 12, 20160352.

Lewis, E. L. 2026. *Effects of Tourist Feeding on the Gut Microbiome and Ecophysiology of Endangered Northern Bahamian Rock Iguana (Cyclura cychlura)*. All Graduate Theses and Dissertations, Fall 2023 to Present. Utah State University.

Ley, R. E., M. Hamady, C. Lozupone, P. J. Turnbaugh, R. R. Ramey, J. S. Bircher, M. L. Schlegel, T. A. Tucker, M. D. Schrenzel, & R. Knight. (2008). Evolution of mammals and their gut microbes. *Science*, 320, 1647–1651.

Liu, L., Xu, M., Lan, R., Hu, D., Li, X., Qiao, L., Zhang, S., Lin, X., Yang, J., Ren, Z., & Xu, J. (2022). Bacteroides vulgatus attenuates experimental mice colitis through modulating gut microbiota and immune responses. *Frontiers in Immunology*, 13, 1036196. https://doi.org/10.3389/fimmu.2022.1036196

Liu, W., Z. Sun, C. Ma, J. Zhang, C. Ma, Y. Zhao, H. Wei, S. Huang, & H. Zhang. (2021). Exposure to soil environments during earlier life stages is distinguishable in the gut microbiome of adult mice. *Gut Microbes*, 13, 1830699.

Malone, C. L., Reynoso, V. H., & Buckley, L. (2017). Never judge an iguana by its spines: Systematics of the Yucatan spiny tailed iguana, *Ctenosaura defensor* (Cope, 1866). *Molecular Phylogenetics and Evolution*, 115, 27–39. https://doi.org/10.1016/j.ympev.2017.07.010

Maritan, E., A. Quagliariello, E. Frago, T. Patarnello, & M. E. Martino. (2024). The role of animal hosts in shaping gut microbiome variation. *Philosophical Transactions of the Royal Society B*, 379, 20230071.

Michell, N. C. (1999). Effect of introduced ungulates on density, dietary preferences, home range, and physical condition of the iguana (*Cyclura pinguis*) on Anegada. *Herpetologica*, 22(1), 7–17. www.jstor.org/stable/3893064?seq=1&cid=pdf-

Moeller, A. H., Ivey, K., Cornwall, M. B., Herr, K., Rede, J., Taylor, E. N., & Gunderson, A. R. (2020). The lizard gut microbiome changes with temperature and is associated with heat tolerance. *Applied and Environmental Microbiology*, 86(17). https://doi.org/10.1128/AEM.01181-20

Morrison, C., Osborne, T., Harlow, P. S., Thomas, N., Biciloa, P., & Niukula, J. (2007). Diet and habitat preferences of the Fijian crested iguana (*Brachylophus vitiensis*) on Yadua Taba, Fiji: Implications for conservation. *Australian Journal of Zoology*, 55(6), 341–350. https://doi.org/10.1071/ZO07062

Meddings, J. B., & M. Swain. (2000). Environmental stress–induced gastrointestinal permeability is mediated by endogenous glucocorticoids in the rat. *Gastroenterology*, 119, 1019–1028.

Meduri, G. U., & A.-M. G. Psarra. (2025). The glucocorticoid system: A multifaceted regulator of mitochondrial function, endothelial homeostasis, and intestinal barrier integrity. *Seminars in Respiratory and Critical Care Medicine*, 47(1), 32-46.

Murphy, K. M., S. M. Le, A. E. Wilson, & D. A. Warner. (2023). The microbiome as a maternal effect: a systematic review on vertical transmission of microbiota. *Integrative and Comparative Biology*, 63, 597–609.

Nagy, K. A. (1973). Behavior, diet and reproduction in a desert lizard, Sauromalus obesus. 1973(1), 93–102. www.jstor.org/stable/1442363?seq=1&cid=pdf-

Noguera, J. C., M. Aira, M. Pérez-Losada, J. Domínguez, & A. Velando. (2018). Glucocorticoids modulate gastrointestinal microbiome in a wild bird. *Royal Society Open Science*, 5, 171743.

Oster, H., E. Challet, V. Ott, E. Arvat, E. R. de Kloet, D.-J. Dijk, S. Lightman, A. Vgontzas, & E. Van Cauter. (2017). The functional and clinical significance of the 24-hour rhythm of circulating glucocorticoids. *Endocrine Reviews*, 38, 3–45.

O'Sullivan, A., He, X., McNiven, E. M. S., Haggarty, N. W., Lönnerdal, B., & Slupsky, C. M. (2013). Early diet impacts infant rhesus gut microbiome, immunity, and metabolism. *Journal of Proteome Research*, 12(6), 2833–2845. https://doi.org/10.1021/pr4001702

Pasachnik, S. A., & Martin-Velez, V. (2017). An evaluation of the diet of *Cyclura iguanas* in the Dominican Republic. *The Herpetological Bulletin*, 140, 6–12.

Patrick, S. (2022). A tale of two habitats: Bacteroides fragilis, a lethal pathogen and resident in the human gastrointestinal microbiome. *Microbiology (United Kingdom)*, 168(4). https://doi.org/10.1099/mic.0.001156

Queiroz, K. de. (1987). A new spiny-tailed iguana from Honduras, with comments on relationships within *Ctenosaura* (Squamata: Iguania). *American Society of Ichthyologists and Herpetologists (ASIH)*, 1987(4), 892–902.

Rand, A. S., Dugan, B. A., Monteza, H., & Vianda, D. (1990). The diet of a generalized folivore: *Iguana iguana* in Panama. *Journal of Herpetology* 24(2), 211–214. www.jstor.org/stable/1564235

Romero, L. M., M. J. Dickens, & N. E. Cyr. (2009). The reactive scope model—A new model integrating homeostasis, allostasis, and stress. *Hormones and Behavior*, 55, 375–389.

Sapolsky, R. M., L. M. Romero, & A. U. Munck. (2000). How do glucocorticoids influence stress responses? Integrating permissive, suppressive, stimulatory, and preparative actions. *Endocrine Reviews*, 21, 55–89.

Sarkar, A., S. Harty, K. V.-A. Johnson, A. H. Moeller, E. A. Archie, L. D. Schell, R. N. Carmody, T. H. Clutton-Brock, R. I. Dunbar, & P. W. Burnet. (2020). Microbial transmission in animal social networks and the social microbiome. *Nature Ecology & Evolution*, 4, 1020–1035.

Semenova, N., N. Garashchenko, S. Kolesnikov, M. Darenskaya, & L. Kolesnikova. (2024). Gut microbiome interactions with oxidative stress: Mechanisms and consequences for health. *Pathophysiology*, 31, 309–330.

Sepulveda, J., & A. H. Moeller. (2020). The effects of temperature on animal gut microbiomes. *Frontiers in Microbiology*, 11, 384.

Sharon, G., N. Garg, J. Debelius, R. Knight, P. C. Dorrestein, & S. K. Mazmanian. (2014). Specialized metabolites from the microbiome in health and disease. *Cell Metabolism*, 20, 719–730.

Shi, N., Li, N., Duan, X., & Niu, H. (2017). Interaction between the gut microbiome and mucosal immune system. *Military Medical Research*, 4(1), 1–7. https://doi.org/10.1186/s40779-017-0122-9

Shterzer, N., N. Rothschild, Y. Sbehat, J. Dayan, D. Eytan, Z. Uni, & E. Mills. (2023). Vertical transmission of gut bacteria in commercial chickens is limited. *Animal Microbiome*, 5, 50.

Siddiqui, R., Maciver, S. K., & Khan, N. A. (2022). Gut microbiome–immune system interaction in reptiles. *Journal of Applied Microbiology*, 132(4), 2558–2571. https://doi.org/10.1111/jam.15438

Smith, P. M., Howitt, M. R., Panikov, N., Michaud, M., Gallini, C. A., Bohlooly-y, M., Glickman, J. N., & Garrett, W. S. (2013). The microbial metabolites short-chain fatty acids, regulate colonic Treg cell homeostasis. *Science*, 341(August), 569–574.

Sylber, C. K. (1988). Feeding habits of the lizards *Sauromalus varius* and *S. hispidus* in the Gulf of California. *Journal of Herpetology*, 22(4), 413–424

Tao, C., Zhang, Q., Zeng, W., Liu, G., & Shao, H. (2020). The effect of antibiotic cocktails on host immune status is dynamic and does not always correspond to changes in gut microbiota. *Applied Microbiology and Biotechnology*, 104(11), 4995–5009. https://doi.org/10.1007/s00253-020-10611-1

Tetel, M. J., G. J. De Vries, R. C. Melcangi, G. Panzica, & S. M. O'Mahony. (2018). Steroids, stress and the gut microbiome-brain axis. *Journal of Neuroendocrinology*, 30, e12548.

Troyer, K. (1982). Transfer of fermentative microbes between generations in a herbivorous lizard. *Science*, 216, 540–542.

Troyer, K. (1984). Structure and function of the digestive tract of a herbivorous lizard *Iguana iguana*. *Physiological Zoology*, 57(11), 1–8.

Van de Wouw, M., M. Boehme, J. M. Lyte, N. Wiley, C. Strain, O. O'Sullivan, G. Clarke, C. Stanton, T. G. Dinan, & J. F. Cryan. (2018). Short-chain fatty acids: Microbial metabolites that alleviate stress-induced brain–gut axis alterations. *The Journal of Physiology*, 596, 4923–4944.

Vasco, K., Guevara, N., Mosquera, J., Zapata, S., & Zhang, L. (2022). Characterization of the gut microbiome and resistome of Galapagos marine iguanas (*Amblyrhynchus cristatus*) from uninhabited islands. *Animal Microbiome*, 4(1), 65. https://doi.org/10.1186/s42523-022-00218-4

Vaziri, N. D., S.-M. Liu, W. L. Lau, M. Khazaeli, S. Nazertehrani, S. H. Farzaneh, D. A. Kieffer, S. H. Adams, & R. J. Martin. (2014). High amylose resistant starch diet ameliorates oxidative stress, inflammation, and progression of chronic kidney disease. *PloS one*, 9, e114881.

Warne, R. W., L. Kirschman, & L. Zeglin. (2019). Manipulation of gut microbiota during critical developmental windows affects host physiological performance and disease susceptibility across ontogeny. *Journal of Animal Ecology* 88, 845–856.

Wexler, A. G., & Goodman, A. L. (2017). An insider's perspective: Bacteroides as a window into the microbiome. *Nature Microbiology*, 2. Nature Publishing Group. https://doi.org/10.1038/nmicrobiol.2017.26

Wexler, H. M. (2007). Bacteroides: The good, the bad, and the nitty-gritty. *Clinical Microbiology Reviews*, 20(4) 593–621. https://doi.org/10.1128/CMR.00008-07

Wu, F., Guo, X., Zhang, J., Zhang, M., Ou, Z., & Peng, Y. (2017). Phascolarctobacterium faecium abundant colonization in human gastrointestinal tract. *Experimental and Therapeutic Medicine*, 14(4), 3122–3126. https://doi.org/10.3892/etm.2017.4878

Xie, X., Q. Xiao, Z. Xiong, C. Yu, J. Zhou, & Z. Fu. (2019). Crocin-I ameliorates the disruption of lipid metabolism and dysbiosis of the gut microbiota induced by chronic corticosterone in mice. *Food & Function*, 10, 6779–6791.

Yang, J., Li, Y., Wen, Z., Liu, W., Meng, L., & Huang, H. (2021). Oscillospira – A candidate for the next-generation probiotics. *Gut Microbes*, 13(1), 1987783. https://doi.org/10.1080/19490976.2021.1987783

Yang, J., Liu, W., Han, X., Hao, X., Yao, Q., & Du, W. (2024). Gut microbiota modulation enhances the immune capacity of lizards under climate warming. *Microbiome*, 12(1), 1–15. https://doi.org/10.1186/s40168-023-01736-2

Zafar, H., & Saier, M. H. (2021). Gut bacteroides species in health and disease. *Gut Microbes*, 13(1), 1–20. https://doi.org/10.1080/19490976.2020.1848158

Zimmerman, L. C., & Tracy, C. R. (1989). Interactions between the environment and ectothermy and herbivory in reptiles. *Physiological Zoology*, 62(2), 374–409. https://doi.org/10.1086/physzool.62.2.30156176

CHAPTER TEN

Assessing the Influence of Microorganisms on Rodent Well-Being and Health

Fu Yin, Jinfeng Zhao, Huiyan Xu, and Longxian Zhang

10.1 INTRODUCTION

The mammalian microbiota, comprising a highly complex and diverse community of bacteria, archaea, and eukaryotes residing on body surfaces and within cavities, plays an indispensable role in maintaining host health (Canton et al., 2024). Advances in next-generation sequencing (NGS) technologies have revolutionized our understanding of microbial composition and functional gene content across humans and model animals. These studies have uncovered significant associations between gut microbiota alterations and various diseases, behavioral patterns, and developmental processes (Ezenwa et al., 2012; Pennisi, 2013; Quince et al., 2017).

Rodents have high research value for microbial research. According to statistics, rodents represent nearly half of the current mammalian diversity with more than 2,261 species organized into 474 genera (Figure 10.1).These small to medium-sized placentals have spread over all continents (except Antarctica) and most islands, where they occupy virtually all terrestrial ecosystems from tropical rainforests and deserts to the arctic tundra (Fabre et al., 2012). Inevitably, most human settlements are accompanied by rodent activities. As vectors for pathogens, numerous outbreaks of human epidemics throughout history have been linked to pathogenic microorganisms carried by rodents. Evidence has shown that rodents can harbor zoonotic pathogens such as Hantavirus (causing hemorrhagic fever with renal syndrome), *Yersinia pestis* (the causative agent of plague), and Leptospira (Jahan et al., 2021; Keesing & Ostfeld, 2024), underscoring the urgency of understanding their pathogen reservoirs for zoonosis prevention.

Concurrently, murine models have become indispensable for mechanistic studies of gut microbiota. Due to limitations of human research, the details behind many of these processes are still unknown. Therefore, murine models have become crucial in gut microbiota research for gaining mechanistic insights into gut flora establishment and upkeep (Hildebrand et al., 2013). Mouse models have been used to study the interaction of intestinal microbes, dietary effects, disease development, and the impact of microbial therapies when large scale studies became feasible due the development of molecular and high-throughput approaches (Hugenholtz & de Vos, 2018). Moreover, rodents and specifically mice have become the most studied disease models for pharmaceutical research (Vandamme, 2014). While rodents are widely used in pharmaceutical research, this section specifically focuses on microbiota–host interactions, excluding direct drug-related mechanisms to maintain thematic clarity.

10.2 COMPOSITION AND FUNCTION OF GUT MICROBIOTA IN RODENTS

The rodents intestinal ecosystem hosts a diverse microbial community predominantly composed of the phyla Firmicutes, Bacteroidetes, Proteobacteria, Actinobacteria, and Verrucomicrobia which

DOI: 10.1201/9781003638568-10

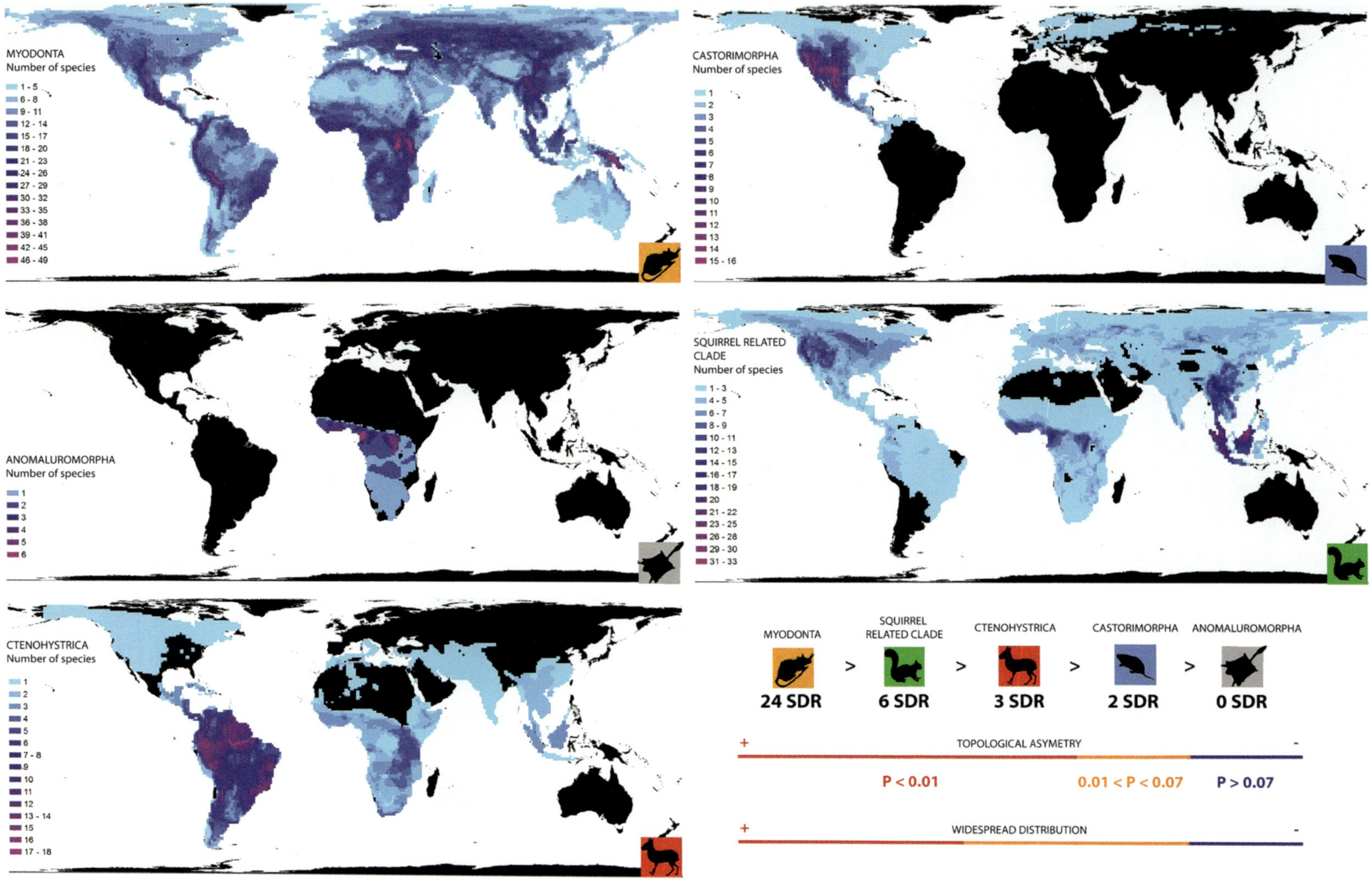

Figure 10.1 Current distribution maps of the major rodent clades. Color gradient represents species richness – a warmer color indicates higher richness. Black corresponds to areas where the group is not present. The maximum number of species in a cell (warmest color) for each clade is 49 (Myodonta), 16 (Castorimorpha), 6 (Anomaluromorpha), 33 (squirrel-related clade), and 18 (Ctenohystrica) respectively. Distribution width and topological asymmetry are indicated on the lower right part of the figure, together with the number of significant SDR (shifts in diversification rate). (Fabre et al., 2012.)

collectively constitute almost 90% of the gut microbiota (Pan et al., 2018; Yang & Chun, 2021; Zhu et al., 2021). Distinct microbial composition patterns are observed across intestinal segments: the upper gastrointestinal tract (e.g., duodenum and jejunum) is primarily colonized by Firmicutes-associated genera such as Lactobacillus and Streptococcus, whereas the distal gut (e.g., cecum and colon) exhibits greater biodiversity with elevated abundances of Bacteroides, Ruminococcus, and other taxa affiliated with Bacteroidetes and Proteobacteria (Suzuki and Nachman, 2016; Xu et al., 2021) (Figure 10.2).

The mouse gut harbors an exceptionally diverse microbiota. Through high-throughput sequencing and metagenomic binning, researchers have reconstructed 42,245 nonredundant prokaryotic genomes belonging to 1,524 species, which encode over 11.7 million protein sequences (Kim et al., 2024). In contrast, only 541 prokaryotic species were identified in the rat gut, encompassing approximately 2.6 million genes (Xiao et al., 2015), with subsequent studies expanding this number to 5.1 million (Pan et al., 2018). The abundant gut microbes and numerous functional genes also play a significant role in rodents. Gut microbes are enriched in metabolic pathways related to carbohydrate and amino acid metabolism, polysaccharide fermentation, and biosynthesis of short-chain fatty acids (SCFAs) such as acetate and butyrate. These SCFAs serve as critical energy substrates for intestinal epithelial cells and contribute to mucosal barrier integrity (Darnaud et al., 2021). Rodent microbial diversity extends even further in wild populations. For instance, in wild *Rattus norvegicus*, *Rattus tanezumi*, and *Rattus andamanensis* from Hong Kong, 1,191 species-level prokaryotic genomes were assembled, of which 733 showed no similarity to existing genomic databases, suggesting novel bacterial species (Wen et al., 2025). Compared to the rich diversity of wild rodent species, there is a significant gap in research on the gut microbiota of wild rodents. It is likely that many more microbial species and functional genes in the rodent gut remain undiscovered.

Humans and rodents can harbor similar microorganisms, with approximately 10% of mouse gut bacterial species also found in the human intestine (Kim et al., 2024). However, certain pathogenic microbes can be transmitted to humans through rodents as vectors. As significant carrier animals, rodents can spread various zoonotic pathogens through direct contact, contamination by excreta, or vector-borne

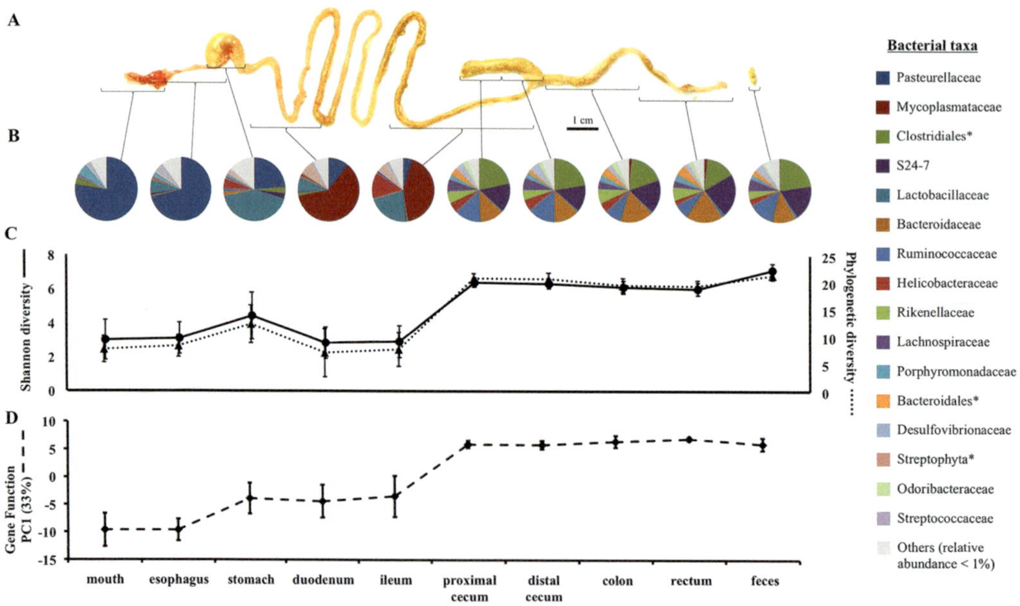

Figure 10.2 Spatial heterogeneity of microbial composition along the gastrointestinal (GI) tract. Ten samples were collected along the GI tract per individual (A). The averages of relative abundance of bacterial families (B), the microbial diversity measures (C), and the first principle component value from predicted gene function categories (D) across the GI tract are shown. The error bars are standard deviations. Asterisks denote unclassified families within the listed order. (Suzuki and Nachman, 2016.)

transmission. History has witnessed multiple major pandemics associated with rodents. For instance, the bacterium *Yersinia pestis*, carried by rodents, caused severe plague outbreaks in Asia and Europe, resulting in tens of millions of deaths. Viruses such as Hantavirus, Lassa virus, and Junin virus carried by rodents have also led to serious epidemics and human fatalities. Additionally, rodents host multiple zoonotic parasites, including Cryptosporidium, *Giardia duodenalis*, and *Toxoplasma gondii*, posing ongoing and substantial threats to public health (Table 10.1).

10.3 THE FACTORS INFLUENCING GUT MICROBIOTA IN RODENTS

Mammalian microbiotas are acquired both vertically from mother to offspring and horizontally among nonkin through social interactions and shared environments. The transmission of gut microbiota in mice primarily relies on vertical transmission, while horizontal transmission also occurs for certain microbes (Moeller et al., 2018; Wanelik et al., 2023). This divergence in transmission modes is closely linked to bacterial taxonomy: for instance, within the Firmicutes phylum, the genus Bacillus

TABLE 10.1
Zoonotic pathogenic microorganisms carried by rodents

Kingdom	Genus	Species	Reference
Bacteria	*Leptonema*	*L. interrogans*	Boey et al., 2019
		L. borgpetersenii	Perez et al., 2011
	Yersinia	*Y. pestis*	Barbieri et al., 2020
	Bartonella	*Bar. elizabethae*	Krugel et al., 2022
		Bar. grahamii	
	Borreliella	*Bor. burgdorferi*	Martello et al., 2019
	Rickettsia	*R. typhi*	Rymaszewska and Piotrowski, 2024 Perec-Matysiak et al., 2025
		R. felis	
		R. heilongjiangensis	
		R. conorii	
		R. rickettsii	
		R. slovaca	
	Orientia	*Ori. tsutsugamushi*	Ponnusamy et al., 2022
	Anaplasma	*A. capra*	Xue et al., 2024
		A. phagocytophilum	
	Ehrlichia	*Ehr. chaffeensis*	
Virus	*Orthohantavirus*	*Ort. seoulense*	Taylor et al., 2025
		Ort. hantanense	

(Continued)

Table 10.1 (Continued)

Kingdom	Genus	Species	Reference
		Ort. puumalaense	
		Ort. dobravaense	
		Ort. sinnombreense	
		Ort. andesense	
		Ort. tulaense	
		Ort. negraense	
		Ort. chocloense	
	Mammarenavirus	*M. lassaense*	Duvignaud et al., 2024
		M. choriomeningitidis	Camp et al., 2025
		M. juninense	Lendino et al., 2024
		M. machupoense	
		M. guanaritoense	
		M. brazilense	
		M. wenzhouense	
Protozoa	*Giardia*	*G. duodenalis*	Hunter-Barnett and Viney, 2024
	Entamoeba	*Ent. histolytica*	
	Toxoplasma	*T. gondii*	
	Cryptosporidium	*C. hominis*	Zhang et al., 2022
		C. parvum	
		C. meleagridis	
		C. viatorum	
		C. ubiquitum	

predominantly spreads vertically, whereas the genus Bacteroides tends to propagate horizontally. Transmission pathways are further associated with metabolic traits—obligate anaerobes depend on the host's hypoxic environment, favoring vertical transmission, while aerotolerant or aerobic bacteria can spread horizontally via environmental vectors. Notably, spore-forming bacteria like Bacillus enhance environmental survival through spores, yet their transmission strategies do not exhibit a clear preference for vertical or horizontal modes (Moeller et al., 2018).

Host genetics (species identity) of rodents dominates over a shared environment in predicting gut microbiota composition. Although most rodent species share similar bacterial phyla in their gut microbiota, the relative abundances of these phyla vary significantly across different species. In laboratory rats, Firmicutes and Bacteroidetes account for approximately 75.9% and 10.8%, respectively, while in laboratory mice, these proportions are about 77.0% and 14.0%. In contrast, wild rodents exhibit greater fluctuations in gut microbial composition, with some even hosting unique bacterial taxa. For instance, although the subterranean rodents *Eospalax cansus* and *Eospalax rothschildi* have similar gut microbial profiles, their bacterial proportions differ markedly—*E. cansus* shows significantly higher abundances of Firmicutes and Bacteroidetes than *E. rothschildi*. The capybara (*Hydrochoerus hydrochaeris*) has comparable proportions of Firmicutes and Bacteroidetes in its cecum (34.5% and 32.2%), which is similar to the patterns observed in *Apodemus peninsulae* and *Myodes rufocanus*. However, the latter two species exhibit higher abundances of Proteobacteria and Actinobacteria, with *M. rufocanus* even having a higher proportion of Proteobacteria than Firmicutes. Additionally, the gut microbiota of the *Spalax leucodon* is dominated by Bacteroidetes (over 50%), while Firmicutes constitute only about 31%, which similar to *Myodes glareolus* (Table 10.2). Specifically, individual's microbiota was on average more similar to conspecifics living elsewhere than to members of a closely related species living in the same location. Moreover, while environment (capture site) did shape the microbiota, this effect was largely within species; heterospecific rodents converged somewhat in gut microbiota composition when living in sympatry, but this was insufficient to override the strong influence of species identity (Knowles et al., 2019; Neha et al., 2025; Wang et al., 2022).

The intestinal microbiota of mice undergoes dynamic adjustments following colonization, with changes closely tied to host age. Postweaning, the gut microbial community exhibits two distinct states. During the first 10 days after weaning, the microbial composition remains relatively unstable and continuously evolves as the mice mature. By day 15 postweaning, the community stabilizes and maintains this equilibrium until adulthood. Intriguingly, although the bacterial taxa present in these two states are identical, their relative abundances differ significantly. This shift in microbial structure may correlate with the gradual maturation of the immune system in postweaning mice (Figure 10.3) (Schloss et al., 2012). However, the diversity of gut microbiota is positively correlated with increasing age, primarily due to the accumulation of bacteria belonging to the phylum Bacteroidetes, and these accumulated bacteria do not exhibit significant decline under natural conditions (Alam et al., 2021; Fenn et al., 2023).

The composition of gut microbiota is strongly influenced by habitat, leading to significant differences in microbial profiles among wild rodents of the same species across laboratory, zoo, and natural environments. A longitudinal study on *Microtus agrestis* revealed a marked decline in gut microbiota alpha-diversity within just 1–2 days after transitioning from the wild to the lab, characterized by a significant increase in the relative abundance of Bacteroidetes and a decrease in Firmicutes. In contrast, wild *Microtus agrestis* exhibited a significant increase in alpha-diversity over 129 days, demonstrating a pattern of continuous microbial species accumulation throughout the animal's lifespan. Specifically, Firmicutes taxa showed significantly higher variation compared to Bacteroidetes (Wanelik et al., 2023). Similarly, the gut microbial composition of *Peromyscus maniculatus* differed significantly across environments, with free-living individuals carrying lower bacterial loads than captive ones (Diaz et al., 2023). In wild house mice, gut microbiota varied with altitude: obligate anaerobes generally correlated positively with elevation, while most other microbial groups correlated negatively. The most pronounced altitudinal covariation was observed in bacteria of the genus Prevotella. Environmental pollution also impacts the gut microbiome of wild rodents. In *Myodes glareolus*, exposure to higher metal loads was associated with increased gut microbiota alpha-diversity and altered community composition (beta-diversity). Multimetal exposure and elevated levels of specific metals, such as Cd, Hg, Pb, and Se, were linked to changes in the abundance of microbial taxa, particularly those within the families Clostridiales vadinBB60 group, Desulfovibrionaceae, Lachnospiraceae, Muribaculaceae, and Ruminococcaceae.

Diet is crucial for shaping and maintaining the personalized gut microbiome of rodents. By controlling dietary components such as moisture, crude protein, crude fat, crude fiber, crude ash, calcium, and phosphorus in mice, significant differences

TABLE 10.2
Differences in the proportion of gut microbiota in rodents

Rodent	Proportion of phylum				Ref
	Firmicutes	Bacteroidota	Proteobacteria	Actinobacteria	
Apodemus peninsulae	31.0%	33.1%	19.7%	13.7%	Cao et al., 2025
Apodemus flavicollis	67.0%	27.0%	4.0%	<1%	Kreisinger et al., 2015
Eospalax cansus	70.0%	22.8%	-	-	Zou et al., 2021
Eospalax rothschildi	51.3%	35.9%	-	-	Zou et al., 2021
Sprague-Dawley rats	75.9%	10.83%	6.77%	-	Pan et al., 2018
Hydrochoerus hydrochaeris	34.5%	32.2%	8.2%	-	Garcia-Amado et al., 2012
C57BL/6 Mice	75.0%	18.1%	<1%	<1%	Rosshart et al., 2017
Mus musculus domesticus	39.1%	25.0%	31.4%	<1%	
Myodes glareolus	24.0%	55.0%	3.0%	-	Lavrinienko et al., 2020
	22.0%	73.0%	2.0%	-	Brila et al., 2021
Myodes rufocanus	26.4%	23.8%	29.4%	12.4%	Cao et al., 2025
Peromyscus maniculatus	57.1%	34.2%	1.44%	-	Diaz et al., 2023
Spalax leucodon	31–32%	50–65%	3.0%	-	Sibai et al., 2020
Petaurista philippensis grandis	62.36–100%	0–7.2%	-	2.16-4.46%	Liu et al., 2020
Petaurista alborufus lena					
Trogopterus xanthipes					
Pteromys volans orii					

– represents the unknown.

were observed in the levels of *Limosilactobacillus reuteri*, *Parabacteroides distasonis*, and *Akkermansia muciniphila* among different groups (Yang et al., 2025). Mice on a high-fat diet exhibited an increase in Firmicutes, Proteobacteria, and Verrucomicrobia, alongside a decrease in Bacteroidetes. Additionally, a high-fat diet caused extensive spatial reorganization of the microbiota, leading to dense colonization within the intervillous zone. Meanwhile, a high-calorie diet showed a significant negative correlation with *Lactobacillus intestinalis* (Fraumene et al., 2018; Lecomte et al., 2015), while an increase in Prevotella abundance was associated with higher dietary fiber intake (Gálvez et al., 2020; Wu & Kasper, 2020; Tomas et al., 2016). Overall, the gut microbiota composition in rodents is influenced by a combination of factors including species, living environment, and diet.

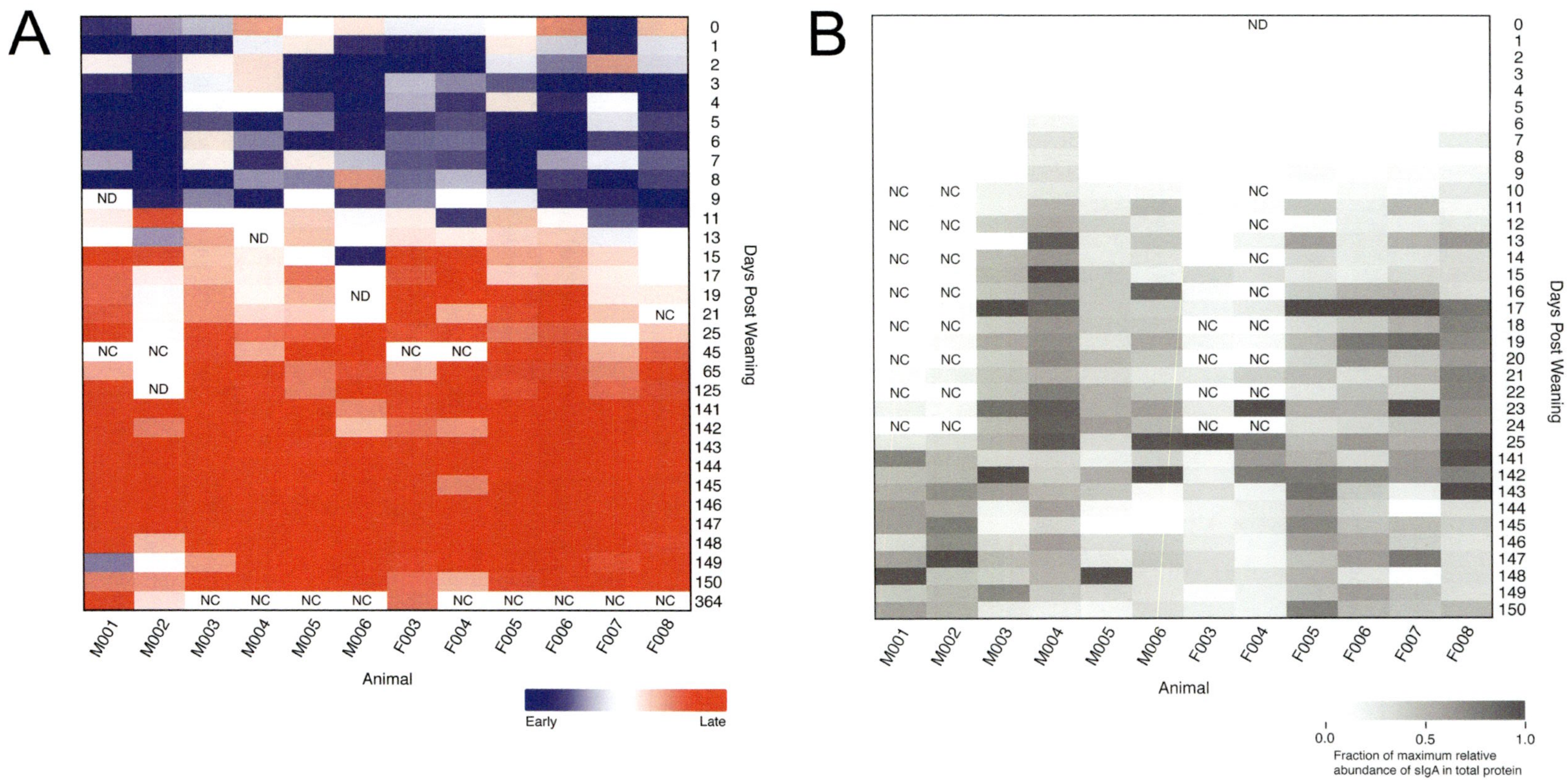

Figure 10.3 (a) Relative difference in root mean squared θ_{YC} distances to the early (0–9 dpw) and late (141–150 dpw) samples for each animal. (b) Relative percentage of secretory IgA as total protein in the feces of each animal over the first 150 dpw. Rectangles with "NC" indicate that the samples were not characterized and the sample with "ND" did not yield useable data (Schloss et al., 2012). Although the initial acquisition of gut microbiota in rodents primarily relies on vertical transmission, external factors also have a significant impact on shaping the gut microbiota. In wild mice, the maternal influence on gut microbiota composition diminishes as offspring mature, while the relative impact of social interactions increases. However, the specific factors driving these age-related changes in the intestinal microbial communities of wild rodents remain unclear.

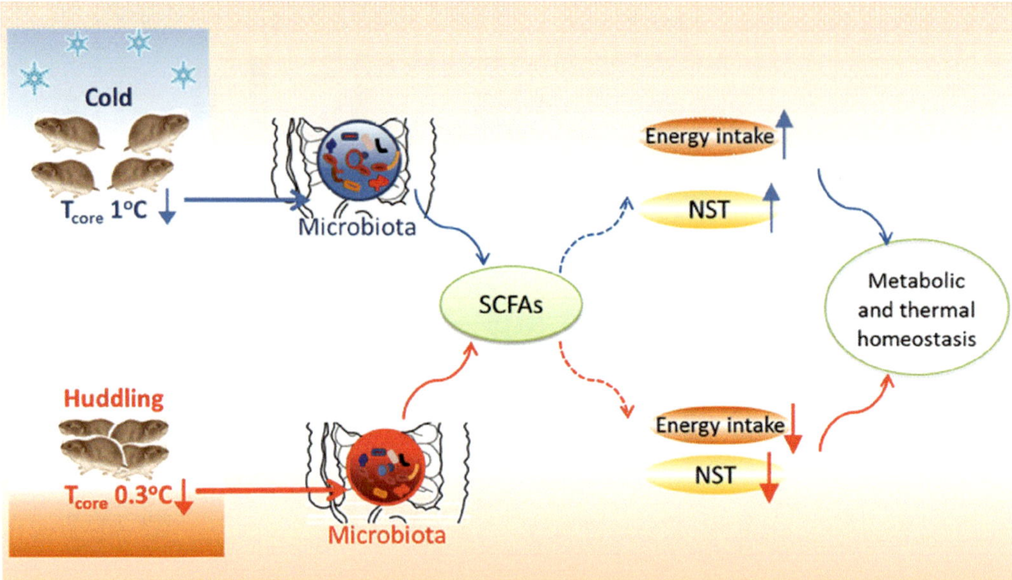

Figure 10.4 Schematic model that cold- or huddling-induced remodeling of gut microbiota orchestrates host metabolic and thermal homeostasis. SCFAs, short-chain fatty acids; NST, nonshivering thermogenesis. (Zhang et al., 2018.)

10.4 THE INTERACTION BETWEEN GUT MICROBIOTA AND RODENT BEHAVIOR

The shaping of gut microbiota in animals is influenced by host factors, while conversely, gut microbiota actively modulates host behavioral patterns. The gastrointestinal tract serves as a critical interface where dense networks of nerve endings and immune cells interact dynamically. Host behavioral states and emotional responses can induce compositional shifts in gut microbial communities, while conversely, microbial stimulation within the gut can reciprocally modify host behavioral patterns. This bidirectional interaction culminates in the well-established concept of the microbiota–gut–brain axis, which has been extensively documented to influence host neurobehavioral outcomes (Rieder et al., 2017).

Brandt's voles (*Lasiopodomys brandtii*) are widely distributed across the grasslands of Inner Mongolia (China), eastern Mongolia, and the southeastern Transbaikal region of Russia. Their habitat is characterized by extreme winter cold, aridity, and deep-frozen soils, with vegetation dominated by grasses such as *Leymus chinensis* and *Stipa krylovii*. Studies reveal that huddling behavior in Brandt's voles enhances gut microbiome diversity, whereas cold-exposed individuals exhibit a higher abundance of bacteria producing SCFAs, likely as an adaptive response to energy demands in harsh conditions (Figure 10.4) (Zhang et al., 2018). Intriguingly, gut microbiota also reciprocally regulates dietary preferences in rodents. For instance, germ-free mice colonized with gut microbiomes from wild rodents with distinct feeding strategies showed marked differences in voluntary food selection. These dietary preferences correlated with variations in circulating essential amino acids, bacterial tryptophan metabolism, and intestinal morphology (Trevelline & Kohl, 2022).

The intestinal microbiota plays a critical role in modulating cognitive functions in rodents. Studies have shown that decreased microbial diversity and abundance from weaning can impair adult behaviors and disrupt key neuromodulators of the microbiota–gut–brain axis, suggesting that postweaning dysregulation of this axis may contribute to disorders associated with cognitive deficits and anxiety. (Desbonnet et al., 2015). Notably, *Methanobrevibacter smithii* interacts with the gut bacterial ecosystem and positively influences cognition by modulating butyrate, histidine, phenylalanine, and linoleic acid metabolism. These findings highlight the importance of early-life microbial balance in maintaining neurodevelopmental trajectories and cognitive health (Fumagalli et al., 2025).

The gut microbiota exerts regulatory effects on emotional responses in rodents. During fear and anxiety states, rodents exhibit suppressed activity of the central amygdala, which subsequently inhibits Brunner's gland function. This impairment disrupts glandular secretion and reduces populations of beneficial Lactobacilli, ultimately compromising host immune defenses (Chang et al., 2024). Conversely, microbial metabolites directly influence mood regulation. For instance, excessive indole—a signaling molecule derived from gut microbiota—exacerbates anxiety-like behaviors in rats, including increased hesitation in novel environments and prolonged immobility in stress tests. This effect may involve vagus nerve activation and hepatic conversion of indole into neuroactive derivatives that penetrate the brain (Jaglin et al., 2018). In contrast, SCFAs, produced through bacterial fermentation of dietary fiber, demonstrate protective effects by alleviating behavioral and physiological consequences of social stress (van de Wouw et al., 2018). These bidirectional interactions highlight the role of the rodent microbiota as a key regulator of emotional state in neuroendocrine, immune and metabolic pathways (Figure 10.5).

10.5 THE MICROBIOTA PROTECTS RODENTS AGAINST PATHOGEN INVASION

The mammalian microbiome has coevolved with its host over millions of years through natural selection, establishing a symbiotic relationship indispensable for host physiology. Normal gut microbiota serves dual protective roles: it prevents colonization by exogenous pathogens and suppresses the overgrowth of resident pathogenic strains. Simultaneously, these microbial communities critically contribute to the development and functional maturation of the host's immune system, regulating both innate and adaptive immunity. Through sustained microbial–epithelial crosstalk, gut microbes enhance intestinal barrier integrity, modulate immune cell differentiation (including B and T lymphocytes), and fine-tune inflammatory responses.

The gut microbiota of rodents is highly diverse and relatively stable, playing a crucial role in maintaining intestinal health by preventing pathogen colonization. Conversely, the absence of gut microbiota significantly compromises host resistance to infections. For instance, germ-free mice exhibit heightened susceptibility to intestinal pathogens compared to their conventionally

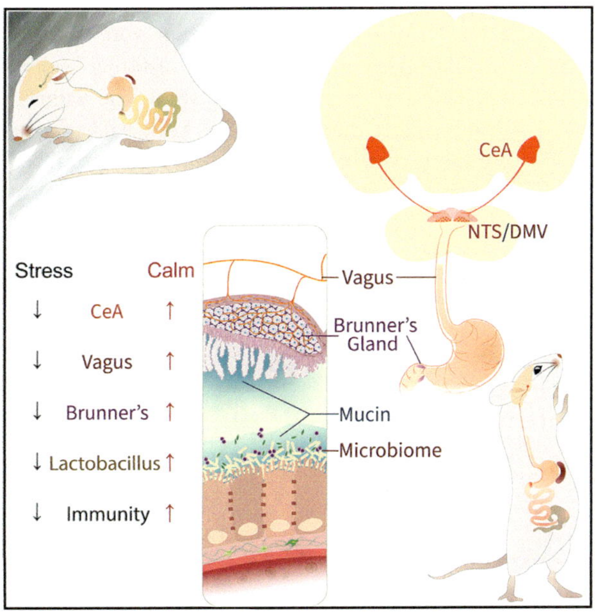

Figure 10.5 Secretions from the duodenal glands of Brunner enhance host defense by promoting gastrointestinal Lactobacilli proliferation. Chronic stress inhibited a brain-vagus nerve circuit that stimulates glandular secretion, thereby suppressing Lactobacillus populations and systemic immunity. (Chang et al., 2024.)

colonized counterparts (Kamada et al., 2012). A key protective mechanism lies in nutrient competition between commensal microbes and pathogens. Specifically, overlapping nutrient utilization profiles between the microbiota and invading pathogens determine colonization resistance, as demonstrated by microbiota-mediated depletion of dietary amino acids that restricts *Citrobacter rodentium* proliferation (Caballero-Flores et al., 2020; Spragge et al., 2023).

Wild rodents exhibit superior pathogen resistance compared to laboratory mice, a distinction strongly linked to their distinct gut microbiota composition. The more diverse microbial communities in wild rodents confer enhanced protection, as evidenced by laboratory mice transplanted with wild-derived microbiota showing reduced inflammation and improved survival during influenza virus challenges (Figure 10.6) (Rosshart et al., 2017). Notably, gut microbiota dysbiosis frequently correlates with disease progression. During colitis development, for example, abnormal expansion of Bacteroidaceae and Enterobacteriaceae coincides with depletion of beneficial commensal bacteria (Osaka et al., 2017).

The gut microbiota of rodents plays a pivotal role in promoting immune system maturation and homeostasis by stimulating both mucosal and systemic immunity. In weaned mice, the gut-associated lymphoid tissue (GALT) is critically activated through microbial diversity, which enhances systemic immunity and sustains immune activation (Bemark et al., 2024). Notably, increased gut microbial diversity correlates with superior antibody production capacity in weaned mice, while alterations in microbiota composition directly modulate systemic immune responses (Lamousé-Smith et al., 2011). Intestinal commensals further extend their immunomodulatory effects beyond the gut, facilitating lung mucosal immunity development and improving pulmonary infection resistance (Gray et al., 2017), correspondingly, mice with pulmonary fungal infections demonstrate concurrent gut microbiota dysbiosis and intestinal inflammation, highlighting the systemic interplay between microbial communities and immune function (Figure 10.7) (Popovic et al., 2023).

10.6 GUT MICROBIOTA HELPS RODENTS ADAPT TO THE ENVIRONMENT

Wild rodents exhibit broad distribution and remarkable adaptability, enabling survival in diverse environments. Their gut microbiota plays a vital role by enhancing host health, disease resistance, and environmental adaptation. These microbial communities aid in detoxifying harmful compounds and may influence lifespan, directly impacting rodents' survival and population expansion in the wild. Host phylogeny and diet primarily shape microbial diversity, with herbivorous species harboring richer communities than omnivores. Habitat characteristics, such as fragmented forests or anthropogenic disturbances, further modulate microbiota composition, often outweighing host species effects. Collectively, these adaptive mechanisms underscore the microbiota's critical role in rodent ecological success.

Unlike laboratory mice, wild rodents endure substantial survival pressures, with food acquisition posing a major challenge. Most wild rodents are herbivores or omnivores, and their specialized gut microbiota facilitates the digestion of plant fibers, insect chitins, and toxins, enabling efficient resource utilization and niche occupation in

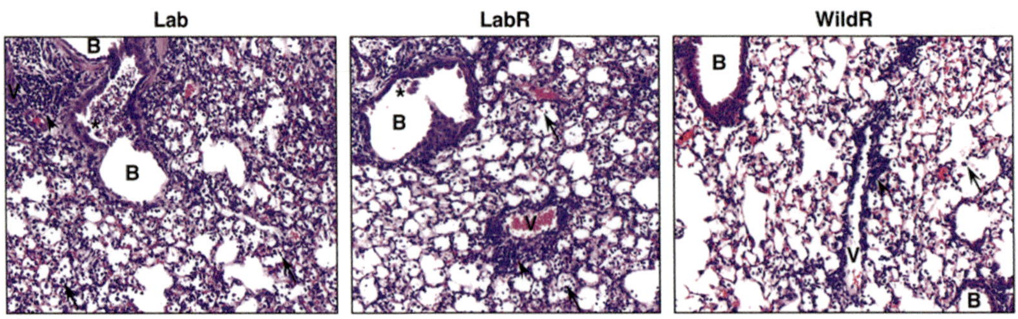

Figure 10.6 Representative lung histology 7 days post IAV infection (original magnification 40x). Lab, n=18; LabR, n=18; WildR, n=18. (Rosshart et al., 2017.)

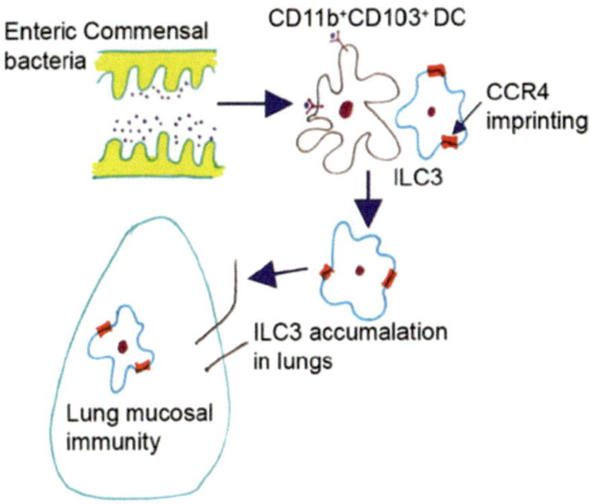

Figure 10.7 The resistance of newborn mice to pneumonia relied on commensal bacteria-directed ILC3-influx into the lungs, which mediated IL-22-dependent host resistance to pneumonia during this developmental window. (Gray et al., 2017.)

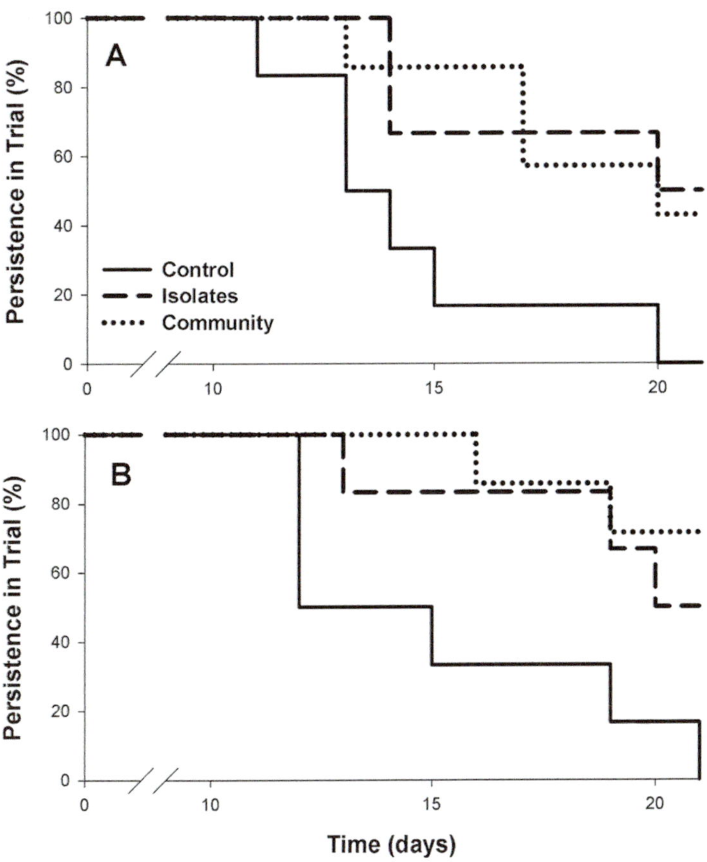

Figure 10.8 Persistence of lab rats on tannin-rich diets. (A) Persistence following bacterial isolates from woodrat feces transplants. (B) Persistence on tannin-rich diets after 6 weeks on a tannin-free diet. The rats were fed powdered rat chow with increasing amounts of tannic acid (3%, 6% and 9% for 3 days each from days 0–9) and then held on a diet of 12% tannic acid for 12 days (days 10–21). (Kohl et al., 2016.)

ecosystems (Dearing & Weinstein, 2022; Kohl et al., 2022). Herbivorous rodents exhibit greater species diversity and harbor more fiber-degrading genera (e.g., *Ruminococcus*) in their gut microbiota compared to insectivores, along with a higher abundance of genes linked to monosaccharide and complex fiber metabolism. In contrast, insectivorous rodents show elevated gene abundance related to chitin degradation (Figure 10.8) (Kohl et al., 2022).

Herbivorous rodents have relatively easy access to food, however, plants defend themselves against herbivory through the production of myriad plant secondary compounds (PSCs), which can act as toxins, digestive inhibitors, and diuretics. Several species of herbivorous woodrats (genus Neotoma) that naturally feed on a variety of dietary toxins, such as *Juniperus monosperma*, *Lupinus acticus*, *Calocedrus decurrens* and *Opuntia* spp., which produce primary class of toxin like terpenes, alkaloids and oxalate. Herbivorous woodrats are able to feed on these plants primarily because their gut microbiota can detoxify through fermentation. Bacillus, Enterococcus, and Clostridium have been shown to degrade toxic phenolics (tannins) (Kohl et al., 2016), and Clostridium, Enterococcus, Lactobacillus are capable of consuming and degrading large quantities of dietary oxalate (Miller et al., 2014).

Although herbivorous rodents have relatively accessible food sources, plants produce PSCs, such as terpenes, alkaloids, and oxalates, as chemical defenses against herbivory. For example, several Neotoma woodrat species consume toxin-rich plants like *Juniperus monosperma*, *Lupinus acticus*, *Calocedrus decurrens*, and *Opuntia* spp. Their ability to thrive on these plants stems from gut microbiota-mediated detoxification through fermentation (Bai et al., 2022). Wild house mice exhibited the strongest patterns of covariation with altitude for bacteria in the genus Prevotella, and the relative abundance of Prevotella increased dramatically above 2,000 m in elevation (Figure 10.9) (Suzuki et al., 2019). Wild house mice exhibited the strongest patterns of covariation with altitude for bacteria in the genus Prevotella, and the relative abundance of Prevotella increased dramatically above 2,000 m in elevation (Scholier et al., 2024). Specific bacterial taxa, including Bacillus, Enterococcus, and Clostridium, degrade phenolic toxins like tannins and oxalates (Scholier et al., 2024). These microbial adaptations not only enhance nutrient extraction but also neutralize plant-derived toxins, underscoring their indispensable role in rodent survival and ecological dominance.

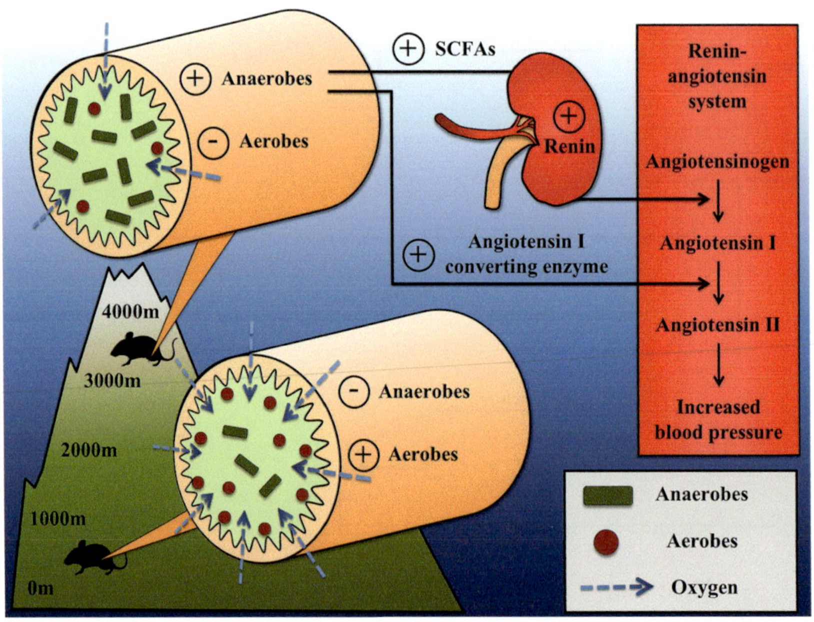

Figure 10.9 A proposed mechanism of the microbiota-mediated regulation of blood pressure in response to atmospheric oxygen at high altitudes. "Anaerobes" refer to obligate anaerobes and "Aerobes" refer to all other oxygen requirements (i.e. facultative anaerobes, aerotolerant anaerobes, microaerophiles, and obligate aerobes). (Suzuki et al., 2019.)

10.7 SUMMARY

The study of gut microbiota in rodents is an important research direction in both biomedical and ecological fields. Rodents are chosen as model organisms due to their high similarity to humans in physiological structure, as well as their ease of genetic manipulation and experimental control. As one of the most numerous and widely distributed groups of mammals, the composition and functional changes of rodent gut microbiota reflect the complex interactions between the host and the environment. With the development of multiomics technologies such as metagenomics, transcriptomics, and metabolomics, future research will be able to reveal the potential functions and regulatory mechanisms of microbial communities more systematically under different host and environmental conditions. By studying the composition and function of gut microbiota, we can better understand the interactions between the host and microorganisms, and provide new insights and strategies for human health.

REFERENCES

Alam MS, Gangiredla J, Hasan NA, Barnaba T, Tartera C, (2021). Aging-induced dysbiosis of gut microbiota as a risk factor for increased infection. *Front Immunol* 12:672353.

Bai X, Liu G, Yang J, Zhu J, Wang Q, Zhou Y, Gu W, La L, Li X, (2022). Changes in the gut microbiota of rats in high-altitude hypoxic environments. *Microbiol Spectr* 10:e0162622.

Barbieri R, Signoli M, Cheve D, Costedoat C, Tzortzis S, Aboudharam G, Raoult D, Drancourt M, (2020). Yersinia pestis: the natural history of plague. *Clin Microbiol Rev* 34:e00044–19.

Bemark M, Pitcher MJ, Dionisi C, Spencer J, (2024). Gut-associated lymphoid tissue: a microbiota-driven hub of B cell immunity. *Trends Immunol* 45:211–223.

Boey K, Shiokawa K, Rajeev S, (2019). Leptospira infection in rats: a literature review of global prevalence and distribution. *PLoS Negl Trop Dis* 13:e0007499.

Brila I, Lavrinienko A, Tukalenko E, Ecke F, Rodushkin I, Kallio ER, Mappes T, Watts PC, (2021). Low-level environmental metal pollution is associated with altered gut microbiota of a wild rodent, the bank vole (*Myodes glareolus*). *Sci Total Environ* 790:148224.

Caballero-Flores G, Pickard JM, Fukuda S, Inohara N, Nunez G, (2020). An enteric pathogen subverts colonization resistance by evading competition for amino acids in the gut. *Cell Host Microbe* 28:526–533.

Camp JV, Nowotny N, Aberle SW, Redlberger-Fritz M, (2025). Retrospective screening for zoonotic viruses in encephalitis cases in austria, 2019-2023, reveals infection with lymphocytic choriomeningitis virus but Not with rustrela virus or tahyna virus. *Viruses* 17:300.

Canton R, De Lucas Ramos P, Garcia-Botella A, Garcia-Lledo A, Hernandez-Sampelayo T, Gomez-Pavon J, Gonzalez Del Castillo J, Martin-Delgado MC, Martin Sanchez FJ, Martinez-Selles M, Molero Garcia JM, Moreno Guillen S, Rodriguez-Artalejo FJ, Reigadas E, Del Campo R, Serrano S, Ruiz-Galiana J, Bouza E, (2024). Human intestinal microbiome: role in health and disease. *Rev Esp Quimioter* 37:438–453.

Cao J, Wang S, Ding R, Liu Y, Yuan B, (2025). Comparative analyses of the gut microbiome of two sympatric rodent species, Myodes rufocanus and Apodemus peninsulae, in northeast China based on metagenome sequencing. *PeerJ* 13:e19260.

Chang H, Perkins MH, Novaes LS, Qian F, Zhang T, Neckel PH, Scherer S, Ley RE, Han W, de Araujo IE, (2024). Stress-sensitive neural circuits change the gut microbiome via duodenal glands. *Cell* 187:5393–5412.

Darnaud M, De Vadder F, Bogeat P, Boucinha L, Bulteau AL, Bunescu A, Couturier C, Delgado A, Dugua H, Elie C, Mathieu A, Novotna T, Ouattara DA, Planel S, Saliou A, Srutkova D, Yansouni J, Stecher B, Schwarzer M, Leulier F, Tamellini A, (2021). A standardized gnotobiotic mouse model harboring a minimal 15-member mouse gut microbiota recapitulates SOPF/SPF phenotypes. *Nat Commun* 12:6686.

Dearing MD, Weinstein SB, (2022). Metabolic enabling and detoxification by mammalian gut microbes. *Annu Rev Microbiol* 76:579–596.

Desbonnet L, Clarke G, Traplin A, O'Sullivan O, Crispie F, Moloney RD, Cotter PD, Dinan TG, Cryan JF, (2015). Gut microbiota depletion from early adolescence in mice: implications for brain and behaviour. *Brain Behav Immun* 48:165–173.

Diaz J, Redford KH, Reese AT, (2023). Captive and urban environments are associated with distinct gut microbiota in deer mice (*Peromyscus maniculatus*). *Biol Lett* 19:20220547.

Duvignaud A, Etafo IC, Jaspard M, Salau Q, Serra B, Kareem AJ, Juchet S, Jegede TO, Gabillard D, Abidoye AT, Le Gal C, Abejegah C, Owhin S, Okwaraeke K, Doutchi M, Vihundira JK, Besong-Lache RM, Seri B, Bererd-Camara M, Salam APA, Olayinka A, Horby P, Ogbaini-Emovon E, Duraffour S, Ahmed LA, Günther S, Adedosu AN, Anglaret X, Malvy D, Lang HJ, Ayodeji OO, (2024). Presentation and outcomes of lassa fever in children in nigeria: a prospective cohort study (lascope). *J Pediat Inf Dis Soc* 13:513–522.

Ezenwa VO, Gerardo NM, Inouye DW, Medina M, Xavier JB, (2012). Microbiology. Animal behavior and the microbiome. *Science* 338:198–199.
Fabre PH, Hautier L, Dimitrov D, Douzery EJ, (2012). A glimpse on the pattern of rodent diversification: a phylogenetic approach. *BMC Evol Biol* 12:88.
Fenn J, Taylor C, Goertz S, Wanelik KM, Paterson S, Begon M, Jackson J, Bradley J, (2023). Discrete patterns of microbiome variability across timescales in a wild rodent population. *BMC Microbiol* 23:87.
Fraumene C, Manghina V, Cadoni E, Marongiu F, Abbondio M, Serra M, Palomba A, Tanca A, Laconi E, Uzzau S, (2018). Caloric restriction promotes rapid expansion and long-lasting increase of *Lactobacillus* in the rat fecal microbiota. *Gut Microbes* 9:104–114.
Fumagalli A, Castells-Nobau A, Trivedi D, Garre-Olmo J, Puig J, Ramos R, Ramió-Torrentà L, Pérez-Brocal V, Moya A, Swann J, Martin-Garcia E, Maldonado R, Fernández-Real JM, Mayneris-Perxachs J, (2025). Archaea methanogens are associated with cognitive performance through the shaping of gut microbiota, butyrate and histidine metabolism. *Gut Microbes* 17:2455506.
Gálvez EJC, Iljazovic A, Amend L, Lesker TR, Renault T, Thiemann S, Hao LX, Roy U, Gronow A, Charpentier E, Strowig T, (2020). Distinct polysaccharide utilization determines interspecies competition between intestinal *Prevotella* spp. *Cell Host Microbe* 28:838–852.
Garcia-Amado MA, Godoy-Vitorino F, Piceno YM, Tom LM, Andersen GL, Herrera EA, Dominguez-Bello MG, (2012). Bacterial diversity in the cecum of the world's largest living rodent (*Hydrochoerus hydrochaeris*). *Microb Ecol* 63:719–725.
Gray J, Oehrle K, Worthen G, Alenghat T, Whitsett J, Deshmukh H, (2017). Intestinal commensal bacteria mediate lung mucosal immunity and promote resistance of newborn mice to infection. *Sci Transl Med* 9:eaaf9412.
Hildebrand F, Nguyen TL, Brinkman B, Yunta RG, Cauwe B, Vandenabeele P, Liston A, Raes J, (2013). Inflammation-associated enterotypes, host genotype, cage and inter-individual effects drive gut microbiota variation in common laboratory mice. *Genome Biol* 14:R4.
Hugenholtz F, de Vos WM, (2018). Mouse models for human intestinal microbiota research: a critical evaluation. *Cell Mol Life Sci* 75:149–160.
Hunter-Barnett S, Viney M, (2024). Gut protozoa of wild rodents – a meta-analysis. *Parasitology* 151:594–605.
Jaglin M, Rhimi M, Philippe C, Pons N, Bruneau A, Goustard B, Dauge V, Maguin E, Naudon L, Rabot S, (2018). Indole, a signaling molecule produced by the gut microbiota, negatively impacts emotional behaviors in rats. *Front Neurosci* 12:216.
Jahan NA, Lindsey LL, Larsen PA, (2021). The role of peridomestic rodents as reservoirs for zoonotic foodborne pathogens. *Vector-Borne Zoonot* 21:133–148.
Kamada N, Kim YG, Sham HP, Vallance BA, Puente JL, Martens EC, Nunez G, (2012). Regulated virulence controls the ability of a pathogen to compete with the gut microbiota. *Science* 336:1325–1329.
Keesing F, Ostfeld RS, (2024). Emerging patterns in rodent-borne zoonotic diseases. *Science* 385:1305–1310.
Kim N, Kim CY, Ma J, Yang S, Park DJ, Ha SJ, Belenky P, Lee I, (2024). MRGM: an enhanced catalog of mouse gut microbial genomes substantially broadening taxonomic and functional landscapes. *Gut Microbes* 16:2393791.
Knowles SCL, Eccles RM, Baltrunaite L, (2019). Species identity dominates over environment in shaping the microbiota of small mammals. *Ecol Lett* 22:826–837.
Kohl KD, Dieppa-Colon E, Goyco-Blas J, Peralta-Martinez K, Scafidi L, Shah S, Zawacki E, Barts N, Ahn Y, Hedayati S, Secor SM, Rowe MP, (2022). Gut microbial ecology of five species of sympatric desert rodents in relation to herbivorous and insectivorous feeding strategies. *Integr Comp Biol* 62:237–251.
Kohl KD, Stengel A, Dearing MD, (2016). Inoculation of tannin-degrading bacteria into novel hosts increases performance on tannin-rich diets. *Environ Microbiol* 18:1720–1729.
Kreisinger J, Bastien G, Hauffe HC, Marchesi J, Perkins SE, (2015). Interactions between multiple helminths and the gut microbiota in wild rodents. *Philos T R Soc B* 370:20140295.
Krugel M, Krol N, Kempf VAJ, Pfeffer M, Obiegala A, (2022). Emerging rodent-associated Bartonella: a threat for human health? *Parasit Vectors* 15:113.
Lamousé-Smith ES, Tzeng A, Starnbach MN, (2011). The intestinal flora is required to support antibody responses to systemic immunization in infant and germ free mice. *Plos One* 6:e27662.
Lavrinienko A, Tukalenko E, Mousseau TA, Thompson LR, Knight R, Mappes T, Watts PC, (2020). Two hundred and fifty-four metagenome-assembled bacterial genomes from the bank vole gut microbiota. *Sci Data* 7:312.
Lecomte V, Kaakoush NO, Maloney CA, Raipuria M, Huinao KD, Mitchell HM, Morris MJ, (2015). Changes in gut microbiota in rats fed a high fat diet correlate with obesity-associated metabolic parameters. *PLoS One* 10:e0126931.
Lendino A, Castellanos AA, Pigott DM, Han BA, (2024). A review of emerging health threats from zoonotic New World mammarenaviruses. *BMC Microbiol* 24:115.

Liu PY, Cheng AC, Huang SW, Lu HP, Oshida T, Liu WH, Yu HT, (2020). Body-size scaling is related to gut microbial diversity, metabolism and dietary niche of arboreal folivorous flying squirrels. *Sci Rep* 10:7809.

Martello E, Mannelli A, Grego E, Ceballos LA, Ragagli C, Stella MC, Tomassone L, (2019). Borrelia burgdorferi sensu lato and spotted fever group rickettsiae in small rodents and attached ticks in the Northern Apennines, Italy. *Ticks Tick Borne Dis* 10:862–867.

Miller AW, Kohl KD, Dearing MD, (2014). The gastrointestinal tract of the white-throated Woodrat (*Neotoma albigula*) harbors distinct consortia of oxalate-degrading bacteria. *Appl Environ Microbiol* 80:1595–1601.

Moeller AH, Suzuki TA, Phifer-Rixey M, Nachman MW, (2018). Transmission modes of the mammalian gut microbiota. *Science* 362:453–457.

Neha SA, Hanson JD, Wilkinson JE, Bradley RD, Phillips CD, (2025). Impacts of host phylogeny, diet, and geography on the gut microbiome of rodents. *PLoS One* 20:e0316101.

Osaka T, Moriyama E, Arai S, Date Y, Yagi J, Kikuchi J, Tsuneda S, (2017). Meta-analysis of fecal microbiota and metabolites in experimental colitic mice during the inflammatory and healing phases. *Nutrients* 9:1329.

Pan H, Guo R, Zhu J, Wang Q, Ju Y, Xie Y, Zheng Y, Wang Z, Li T, Liu Z, Lu L, Li F, Tong B, Xiao L, Xu X, Li R, Yuan Z, Yang H, Wang J, Kristiansen K, Jia H, Liu L, (2018). A gene catalogue of the Sprague-Dawley rat gut metagenome. *Gigascience* 7:055.

Pennisi E, (2013). How do microbes shape animal development? *Science* 340:1159–1160.

Perec-Matysiak A, Bunkowska-Gawlik K, Tomassone L, Hildebrand J, (2025). Wild mammals as hosts of a molecular evidence-based review. *Int J Parasitol-Par* 28:101128.

Perez J, Brescia F, Becam J, Mauron C, Goarant C, (2011). Rodent abundance dynamics and leptospirosis carriage in an area of hyper-endemicity in New Caledonia. *PLoS Negl Trop Dis* 5:e1361.

Ponnusamy L, Garshong R, McLean BS, Wasserberg G, Durden LA, Crossley D, Apperson CS, Roe RM, (2022). Rickettsia felis and other rickettsia species in chigger mites collected from wild rodents in north carolina, USA. *Microorganisms* 10:1342.

Popovic D, Kulas J, Tucovic D, Aleksandrov AP, Malesevic A, Glamoclija J, Brdaric E, Bajic SS, Golic N, Mirkov I, Tolinacki M, (2023). Gut microbial dysbiosis occurring during pulmonary fungal infection in rats is linked to inflammation and depends on healthy microbiota composition. *Microbiol Spectr* 11:e0199023.

Quince C, Walker AW, Simpson JT, Loman NJ, Segata N, (2017). Shotgun metagenomics, from sampling to analysis. *Nat Biotechnol* 35:833–844.

Rieder R, Wisniewski PJ, Alderman BL, Campbell SC, (2017). Microbes and mental health: a review. *Brain Behav Immun* 66:9–17.

Rosshart SP, Vassallo BG, Angeletti D, Hutchinson DS, Morgan AP, Takeda K, Hickman HD, McCulloch JA, Badger JH, Ajami NJ, Trinchieri G, Pardo-Manuel de Villena F, Yewdell JW, Rehermann B, (2017). Wild mouse gut microbiota promotes host fitness and improves disease resistance. *Cell* 171:1015–1028.

Rymaszewska A, Piotrowski M, (2024). Rickettsia species: genetic variability, vectors, and rickettsiosis-a review. *pathogens* 13:661.

Schloss PD, Schubert AM, Zackular JP, Iverson KD, Young VB, Petrosino JF, (2012). Stabilization of the murine gut microbiome following weaning. *Gut Microbes* 3:383–393.

Scholier T, Lavrinienko A, Kallio ER, Watts PC, Mappes T, (2024). Effects of past and present habitat on the gut microbiota of a wild rodent. *Proc Biol Sci* 291:20232531.

Sibai M, Altuntas E, Yildirim B, Ozturk G, Yildirim S, DemircanT, (2020). Microbiome and longevity: high abundance of longevity-linked muribaculaceae in the gut of the long-living rodent spalax leucodon. *OMICS* 24:592–601.

Spragge F, Bakkeren E, Jahn MT, BN Araujo E, Pearson CF, Wang X, Pankhurst L, Cunrath O, Foster KR, (2023). Microbiome diversity protects against pathogens by nutrient blocking. *Science* 382:eadj3502.

Suzuki TA, Martins FM, Nachman MW, (2019). Altitudinal variation of the gut microbiota in wild house mice. *Mol Ecol* 28:2378–2390.

Suzuki TA, Nachman MW, (2016). Spatial heterogeneity of gut microbial composition along the gastrointestinal tract in natural populations of house mice. *PLoS One* 11:e0163720.

Taylor SL, Schmaljohn CS, Williams EP, Jonsson CB, (2025). Pathogenicity and virulence of rodent-borne orthohantaviruses. *Virulence* 16:2553784.

Tomas J, Mulet C, Saffarian A, Cavin JB, Ducroc R, Regnault B, Kun Tan C, Duszka K, Burcelin R, Wahli W, Sansonetti PJ, Pédron T, (2016). High-fat diet modifies the PPAR-γ pathway leading to disruption of microbial and physiological ecosystem in murine small intestine. *Proc Natl Acad Sci U S A* 113:E5934–E5943.

Trevelline BK, Kohl KD, (2022). The gut microbiome influences host diet selection behavior. *Proc Natl Acad Sci U S A* 119:e2117537119.

van de Wouw M, Boehme M, Lyte JM, Wiley N, Strain C, O'Sullivan O, Clarke G, Stanton C, Dinan TG, Cryan JF, (2018). Short-chain fatty acids: microbial metabolites that alleviate stress-induced brain-gut axis alterations. *J Physiol-London* 596:4923–4944.

Vandamme TF, (2014). Use of rodents as models of human diseases. *J Pharm Bioallied Sci* 6:2–9.

Wanelik KM, Raulo A, Troitsky T, Husby A, Knowles SCL, (2023). Maternal transmission gives way to social transmission during gut microbiota assembly in wild mice. *Anim Microbiome* 5:29.

Wang Z, Zhang C, Li G, Yi X, (2022). The influence of species identity and geographic locations on gut microbiota of small rodents. *Front Microbiol* 13:983660.

Wen C, Guan J, Uea-Anuwong T, Shang J, Peng C, Tang X, Magouras I, Sun Y, Li F, (2025). Dissecting the gut microbial communities and resistomes of wild rats from different ecological areas in Hong Kong. *Environ Res* 283:122108.

Wu M, Kasper DL, (2020). Fiber sets up the battleground for intestinal prevotella. *Cell Host Microbe* 28:776–777.

Xiao L, Feng Q, Liang S, Sonne SB, Xia Z, Qiu X, Li X, Long H, Zhang J, Zhang D, Liu C, Fang Z, Chou J, Glanville J, Hao Q, Kotowska D, Colding C, Licht TR, Wu D, Yu J, Sung JJ, Liang Q, Li J, Jia H, Lan Z, Tremaroli V, Dworzynski P, Nielsen HB, Backhed F, Dore J, Le Chatelier E, Ehrlich SD, Lin JC, Arumugam M, Wang J, Madsen L, Kristiansen K, (2015). A catalog of the mouse gut metagenome. *Nat Biotechnol* 33:1103–1108.

Xu C, Liu J, Gao J, Wu X, Cui C, Wei H, Peng J, Zheng R, (2021). The effect of functional fiber on microbiota composition in different intestinal segments of obese mice. *Int J Mol Sci* 22:6525.

Xue J, Chen SS, Jian R, Chen GQ, Qin X, Lu M, Wang W, Xie GC, Du L, Li K, Guo WP, (2024). Great genetic diversity of vector-borne bacteria and protozoan in wild rodents from Guangxi, China. *PLoS Negl Trop Dis* 18:e0012159.

Yang J, Chun J, (2021). Taxonomic composition and variation in the gut microbiota of laboratory mice. *Mamm Genome* 32:297–310.

Yang Z, Zhang Z, Jiang S, Li A, Song H, Zhang J, (2025). Diet shapes and maintains the personalized native gut microbiomes in mice. *J Sci Food Agric* 105:2987–3000.

Zhang K, Fu Y, Li J, Zhang L, (2022). Public health and ecological significance of rodents in Cryptosporidium infections. *One Health* 14:100364.

Zhang XY, Sukhchuluun G, Bo TB, Chi QS, Yang JJ, Chen B, Zhang L, Wang DH, (2018). Huddling remodels gut microbiota to reduce energy requirements in a small mammal species during cold exposure. *Microbiome* 6:103.

Zhu J, Ren H, Zhong H, Li X, Zou Y, Han M, Li M, Madsen L, Kristiansen K, Xiao L, (2021). An expanded gene catalog of mouse gut metagenomes. *mSphere* 6:e01119–01120.

Zou Y, Liang N, Zhang X, Han C, Nan X, (2021). Functional differentiation related to decomposing complex carbohydrates of intestinal microbes between two wild zokor species based on 16SrRNA sequences. *BMC Vet Res* 17:216.

CHAPTER ELEVEN

The Marine Microbiome

Olivia Lutz and Sunil Thomas

11.1 INTRODUCTION

11.1.1 Scope and Significance of Marine Microbiome

The ocean microbiome constitutes the most abundant, diverse, and functionally indispensable biological system on Earth, comprising vast assemblages of bacteria, archaea, viruses, and protists that collectively underpin the stability of the planet's biosphere. These microorganisms are responsible for nearly 50% of global primary production and play a central role in the biogeochemical cycles that regulate Earth's climate, atmospheric composition, and nutrient availability (Falkowski et al., 2008; Moran, 2015). Far from being marginal components of marine ecosystems, marine microorganisms account for approximately 90% of total oceanic biomass. Among them, marine viruses dominate numerically, with an estimated global abundance of $\sim 10^{30}$ particles, making them the most numerous biological entities on the planet (Suttle, 2005, 2007). Together, these microbial communities function as the invisible foundation of ocean life and planetary habitability.

Marine microorganisms adopt three major lifestyles in the ocean: a planktonic lifestyle in which cells are free-living in the water column, a sessile lifestyle characterized by attachment to surfaces and growth within biofilms, and symbiotic lifestyles, such as intracellular associations between bacteria and marine hosts. Among these, the biofilm-associated lifestyle is particularly widespread and ecologically significant. Marine biofilms form when microorganisms attach to a wide variety of natural and artificial substrates immersed in seawater. These substrates include rock surfaces in subtidal and intertidal zones, hydrothermal vent chimneys in the deep ocean, the surfaces and tissues of marine animals, organic particles within marine sediments, ship hulls, and increasingly abundant microplastics, which represent a growing environmental concern. As a result of this broad substrate availability, biofilms are distributed nearly ubiquitously throughout the global ocean (Lu et al., 2023).

Biofilms are estimated to account for approximately 40% of the total microbial biomass in the ocean, highlighting their substantial contribution to marine microbial life. A typical marine biofilm community is highly complex, often consisting of hundreds to thousands of microbial species, including bacteria, archaea, fungi, and microeukaryotes. This diversity generally exceeds that of the surrounding planktonic microbial communities in adjacent seawater, reflecting the heterogeneous and resource-rich nature of the biofilm habitat (Lu et al., 2023).

From a physiological and structural perspective, marine biofilms exhibit pronounced spatial heterogeneity. Gradients of nutrients, oxygen, pH, and signaling molecules develop within the biofilm matrix, creating stratified microenvironments that support diverse metabolic strategies and microbial interactions. These micro-scale gradients promote cooperation, competition, and communication among community members, contributing to the stability and resilience of biofilm ecosystems.

Importantly, marine biofilms play critical roles in oceanic biogeochemical cycles. They are actively

DOI: 10.1201/9781003638568-11

involved in the transformation of dissolved organic matter, the remineralization of particulate organic material, and the mediation of nutrient fluxes from surface waters to the deep ocean. Through these processes, biofilms influence carbon and nutrient cycling, energy flow, and the overall functioning of marine ecosystems, underscoring their fundamental importance in the global ocean (Lu et al., 2023).

Marine microorganisms actively drive the transformation and flux of key elements—including carbon, nitrogen, sulfur, phosphorus, and trace metals—through highly specialized metabolic pathways. These processes regulate nutrient regeneration, control the availability of energy sources for higher trophic levels, and modulate feedbacks between the ocean and atmosphere. By mediating processes such as photosynthesis, respiration, nitrogen fixation, nitrification, denitrification, sulfur oxidation, and metal redox cycling, the ocean microbiome exerts a first-order control on Earth's climate system and long-term geochemical balance (Falkowski et al., 2008). In this sense, marine microbes are not passive inhabitants but active engineers of the Earth system, shaping environmental conditions that allow complex life to persist.

The influence of the marine microbiome spans the full depth and breadth of the global ocean. In the sunlit surface waters, picocyanobacteria such as Prochlorococcus and Synechococcus dominate primary production, collectively fixing an estimated 4–8 Gt of carbon per year and forming the energetic base of pelagic food webs (Giovannoni, 2017). At the opposite extreme, in the aphotic deep ocean and hadal trenches, chemolithoautotrophic microbial communities sustain ecosystems independent of solar energy by oxidizing inorganic compounds such as hydrogen sulfide, methane, and ammonia (Dick, 2019). This remarkable metabolic versatility allows life to thrive across extreme gradients of pressure, temperature, and nutrient availability, highlighting the adaptive capacity and ecological reach of marine microorganisms.

A central mechanism through which the ocean microbiome influences global climate is the sequestration of carbon via the Biological Carbon Pump (BCP). Through the production, aggregation, and sinking of organic matter from the surface to the deep ocean, the BCP generates approximately two-thirds of the vertical gradient in dissolved inorganic carbon and sequesters an estimated 1,300 (±230) Pg of carbon in the deep ocean reservoir (Worden et al., 2015). This process effectively removes carbon dioxide from the atmosphere on timescales ranging from decades to millennia, acting as a major buffer against rapid climate change.

Complementing the BCP, the Microbial Carbon Pump (MCP) represents a distinct and equally critical microbial pathway for long-term carbon storage. Through complex microbial transformations, labile organic carbon is converted into refractory dissolved organic carbon that resists microbial degradation for thousands of years. This process contributes approximately 0.2 Pg C yr^{-1} to oceanic carbon storage and constitutes a persistent, microbially mediated sink for carbon in the marine environment (Jiao et al., 2010). Together, the BCP and MCP illustrate how microbial processes govern both the magnitude and longevity of carbon sequestration in the ocean.

Collectively, these findings underscore the ocean microbiome as a central regulator of Earth's biogeochemical cycles and climate system. By linking microbial metabolism to global-scale processes, marine microorganisms act as the primary drivers of ocean productivity, elemental cycling, and long-term carbon storage. Understanding the structure, function, and dynamics of the ocean microbiome is therefore essential for predicting how marine ecosystems and the Earth system will respond to ongoing environmental change.

11.1.2 Historical Context and the Microbial Ocean Paradigm

Over the past four decades, our understanding of marine microbial ecology has undergone a profound and transformative shift. A pivotal moment occurred with the recognition of the microbial loop by Azam et al. (1983), which fundamentally altered traditional oceanographic paradigms. This concept demonstrated that dissolved organic carbon, previously thought to be largely lost from classical food webs, is efficiently reincorporated into higher trophic levels through bacterial uptake and subsequent grazing by microzooplankton. In oligotrophic marine ecosystems, this microbial pathway can account for 50% or more of total carbon flux, underscoring the central role of microbes in marine carbon cycling (Worden et al., 2015).

The field advanced further with the rapid development and widespread adoption of molecular and genomic techniques. Methods such as 16S rRNA gene sequencing and metagenomics revealed an unprecedented level of microbial diversity in the oceans, overturning earlier assumptions based on culture-dependent approaches. These tools led to the discovery of dominant yet previously unrecognized microbial lineages, most notably the SAR11 clade. Members of this group, exemplified by *Pelagibacter ubique*, are now known to constitute approximately 25–50% of bacterioplankton in surface waters worldwide, making SAR11 one of the most abundant organisms on the planet (Giovannoni, 2017).

As a result of these discoveries, the perception of marine microbes has evolved from that of simple heterotrophic decomposers to highly adapted organisms exhibiting diverse and sophisticated ecological strategies. This complexity is reflected in the remarkable range of microbial genome architectures, spanning from the highly streamlined genomes of oligotrophic bacteria such as Pelagibacter ubique to the extraordinarily large and metabolically versatile genomes of giant marine viruses, which can exceed 2 Mb in size and encode extensive metabolic networks (Endo et al., 2020). These findings highlight the intricate evolutionary pressures shaping microbial life in the ocean and the tight coupling between genomic content and ecological function.

Large-scale, coordinated oceanographic research programs have been instrumental in consolidating these insights. Initiatives such as the Tara Oceans expedition—which generated 7.2 terabases of sequence data from 243 samples collected across 68 global locations; the Hawaii Ocean Time-series; and the Bermuda Atlantic Time-series Study have produced foundational, long-term datasets linking microbial diversity to environmental gradients. Analyses of these datasets have revealed that temperature is the primary global driver of epipelagic microbial community composition, exerting a stronger influence than nutrient availability or geographic distance (Sunagawa et al., 2020). Together, these advances have reshaped marine microbial ecology into a data-rich, systems-level discipline, positioning microbes as central architects of ocean biogeochemistry and ecosystem function.

11.2 MICROBIAL DIVERSITY AND DISTRIBUTION

11.2.1 Prokaryotic Communities (Bacteria and Archaea)

Marine bacterioplankton communities are characterized by a strongly uneven abundance distribution, in which a relatively small number of highly successful taxa dominate numerically, coexisting with a vast "rare biosphere" composed of thousands of low-abundance species. This structure is a defining feature of oceanic microbial ecosystems and reflects both strong environmental selection and the capacity of rare taxa to act as a reservoir of functional and genetic diversity that may become ecologically important under changing conditions.

Among dominant lineages, the SAR11 clade (family Pelagibacteraceae) represents the most abundant and ecologically successful group of bacteria in the marine environment. Members of SAR11 account for approximately 25–50% of surface bacterioplankton and around 20% of microbial communities in the mesopelagic zone. Conservative global estimates suggest the existence of $\sim 2.4 \times 10^{28}$ SAR11 cells, making this lineage one of the most numerous groups of organisms on Earth. The extraordinary success of SAR11 is largely attributed to extreme genomic streamlining, with genome sizes of approximately 1.3 Mb, which minimizes metabolic costs while maximizing efficiency. This streamlined architecture supports high-affinity nutrient uptake systems, reduced regulatory complexity, and high surface area-to-volume ratios, enabling SAR11 cells to thrive in oligotrophic waters where nutrients are scarce and competition is intense (Giovannoni, 2017).

Photosynthetic cyanobacteria underpin much of marine primary production and represent another cornerstone of oceanic microbial communities. Prochlorococcus, the smallest and most abundant photosynthetic organism known, reaches peak densities exceeding 200,000 cells mL^{-1} in equatorial regions and is estimated to contribute between 13% and 48% of global photosynthetic oxygen production. This genus is largely confined to the latitudinal band between 40°N and 40°S and occupies the upper 150–200 m of the water column, where it is finely adapted to gradients in light and nutrient availability. In contrast, Synechococcus exhibits a broader ecological distribution, extending into coastal, temperate, and polar waters, and contributes approximately 35–50% of total chlorophyll-a in oligotrophic surface oceans. Collectively, these picocyanobacteria fix an estimated 4–8 Gt of carbon per year, corresponding to roughly 8.5–16.7% of total oceanic net primary productivity, underscoring their central role in the global carbon cycle (Giovannoni, 2017).

With increasing depth, archaeal lineages become progressively more prominent in marine microbial assemblages. Thaumarchaeota, also known as Marine Group I archaea, dominate communities in the mesopelagic and peak in the bathypelagic zone (approximately 1,000–4,000 m). Many members of this group are ammonia-oxidizing archaea, playing a pivotal role in nitrification and the broader marine nitrogen cycle. Distinct ecotypes adapted to High-Ammonia Concentration (HAC) and Low-Ammonia Concentration (LAC) conditions exhibit consistent depth-related and basin-scale distribution patterns, reflecting fine-scale niche differentiation in the deep ocean (Salazar & Sunagawa, 2017).

In addition to these dominant groups, heterotrophic bacterial taxa display strong associations with specific ecological niches. The family Rhodobacteraceae ranks among the most abundant lineages in free-living bacterioplankton, particularly in surface and coastal waters, where its members exhibit versatile metabolic capabilities. In contrast, members of the class Flavobacteriia dominate particle-associated microbial fractions, where they specialize in the degradation of complex organic matter. Their success is linked to extensive repertoires of carbohydrate-active enzymes that enable efficient breakdown of polysaccharides derived from phytoplankton, detritus, and other organic particles (Moran, 2015).

At the global scale, marine bacterial diversity displays pronounced biogeographic structuring shaped by environmental gradients and ocean circulation. Distinct microbial provinces align with major oceanographic features such as temperature fronts, upwelling systems, and large-scale current patterns. These observations indicate that marine microbial biogeography arises from the combined influence of contemporary environmental selection and historical dispersal processes, jointly determining community composition and diversity across ocean basins (Salazar & Sunagawa, 2017).

11.2.2 Eukaryotic Microbes (Diatoms, Dinoflagellates, Coccolithophores)

Diatoms (phylum Bacillariophyta) represent the single most important group of phytoplankton in terms of oceanic carbon cycling. Collectively, they contribute approximately 40% of total marine primary productivity and are responsible for a similar proportion of particulate organic carbon export via the biological pump. Global diatom productivity is estimated to reach up to ~26 Pg C yr^{-1}, underscoring their outsized influence on the sequestration of atmospheric CO_2. A defining feature of diatoms is their siliceous frustules, which provide mineral ballast that enhances the density and sinking velocity of marine snow aggregates and zooplankton fecal pellets. This ballasting effect promotes efficient transport of organic carbon from the surface ocean to depth, thereby strengthening long-term carbon storage in the deep sea (Tréguer et al., 2018). Dominant diatom genera such as Skeletonema, Chaetoceros, Thalassiosira, and Fragilariopsis structure phytoplankton communities across diverse marine environments, from coastal upwelling systems to polar oceans. Their biogeographic distributions and species composition are strongly regulated by temperature, macronutrient supply, and trace metal availability, with iron limitation exerting a particularly strong control in high-nutrient, low-chlorophyll (HNLC) regions (Moore et al., 2013).

Dinoflagellates represent one of the most ecologically versatile groups of marine microbial eukaryotes, exhibiting an extraordinary range of nutritional strategies. Species span the full trophic spectrum from strict autotrophy to obligate heterotrophy, with mixotrophy—combining photosynthesis and prey ingestion—being especially widespread. This trophic plasticity allows dinoflagellates to persist across highly variable environmental conditions and confers competitive advantages in stratified or nutrient-poor waters (Stoecker et al., 2017). Empirical observations from the northern Gulf of Alaska indicate that mixotrophic taxa contribute between 38% and 61% of total ciliate plus dinoflagellate biomass, depending on seasonal conditions. However, dinoflagellates also include numerous harmful algal bloom (HAB)-forming species that pose severe ecological, economic, and public health challenges. Notable examples include *Alexandrium* spp., which produce saxitoxins responsible for paralytic shellfish poisoning; *Karenia brevis*, the causative agent of brevetoxin-associated neurotoxic shellfish poisoning; and toxigenic Pseudo-nitzschia species that synthesize domoic acid, leading to amnesic shellfish poisoning (Guillou et al., 2008).

Coccolithophores, particularly the cosmopolitan species *Emiliania huxleyi*, are the most prolific calcifying organisms in the modern ocean. These phytoplankton form extensive blooms visible from space and contribute substantially to the production of biogenic calcite. Combined global calcification by coccolithophores and foraminifera is estimated at 1.4–1.5 Pg C yr^{-1}, linking coccolithophores directly to both the biological and carbonate pumps. However, ongoing ocean acidification represents a significant threat to coccolithophore calcification. Elevated atmospheric CO_2 leads to reduced seawater pH and carbonate ion availability, which in turn decreases intracellular Ca^{2+} concentrations, alters coccolith surface chemistry, and compromises the mechanical integrity of calcite plates. Paleoceanographic records provide compelling evidence for this sensitivity: fossil coccolithophores from the Paleocene–Eocene Thermal Maximum exhibit markedly reduced calcification, suggesting that coccolithophore responses to future ocean chemistry changes may be profound (Ridgwell & Schmidt, 2010).

Beyond these dominant bloom-forming taxa, recent advances in high-throughput genomics and transcriptomics have begun to illuminate the immense functional diversity of marine microbial eukaryotes. Large-scale initiatives such as the Marine Microbial Eukaryote Transcriptome Sequencing Project (MMETSP) have generated comprehensive gene expression datasets across a broad range of protistan lineages, providing unprecedented insights into their metabolic capabilities, stress responses, and ecological strategies (Keeling et al., 2014). Subsequent comparative analyses revealed striking convergence in functional repertoires among distantly related eukaryotic plankton groups that dominate the sunlit ocean. This convergence suggests that similar environmental constraints, such as light availability, nutrient limitation, and grazing pressure, drive the evolution of analogous metabolic solutions across phylogenetically distinct taxa (Delmont et al., 2022). Collectively, these findings emphasize that functional traits, rather than phylogenetic affiliation alone, are critical determinants of the ecological roles and biogeochemical impacts of marine microbial eukaryotes.

11.2.3 Marine Viruses: Abundance and Diversity

Marine viruses constitute the most abundant biological entities in the ocean and represent a dominant force shaping marine microbial ecosystems. In surface waters, viral abundances typically reach $1.64 \pm 0.71 \times 10^7$ virus-like particles (VLPs) per milliliter, decreasing with depth to approximately $6.50 \pm 2.26 \times 10^5$ VLPs mL^{-1} in waters below 2,000 m. Correspondingly, the virus-to-bacteria ratio (VBR) varies markedly across depth gradients, averaging around 10 in highly productive surface waters and exceeding 100 in the deep ocean, where bacterial abundances are comparatively low (Suttle, 2005, 2007). These patterns highlight the pervasive influence of viruses throughout the entire water column, from sunlit surface layers to the bathypelagic realm.

Taxonomically, marine viruses exhibit extraordinary diversity that far surpasses that of their cellular hosts. Advances in high-throughput sequencing and large-scale viromic surveys have dramatically expanded known viral diversity. The Global Ocean Viromes 2.0 dataset alone identified 195,728 distinct viral populations, revealing extensive biogeographic structure and functional heterogeneity across ocean basins. Complementary efforts from the Tara Oceans expedition uncovered more than 140,000 previously unknown RNA virus species, with communities dominated by members of the order Picornavirales, underscoring the vast and largely unexplored diversity of marine RNA viruses (Gregory et al., 2019). Together, these datasets have transformed marine virology from a descriptive discipline into a data-rich field capable of resolving global patterns in viral ecology and evolution.

Bacteriophages are numerically dominant within marine viral assemblages, typically outnumbering their bacterial hosts by roughly an order of magnitude. Through lytic infection, phages are estimated to kill approximately 20% of marine microbial biomass each day, exerting strong top-down control on microbial populations (Weinbauer, 2004; Suttle, 2007). This process not only regulates microbial abundance and diversity through selective predation—often described as "killing the winner"—but also drives the release of cellular contents back into the dissolved organic matter pool. This viral shunt redirects carbon and nutrients away from higher trophic levels and retains them within the microbial loop, with major consequences for marine food-web structure and biogeochemical cycling.

An especially dynamic and rapidly expanding area of marine virology concerns giant viruses belonging to the phylum Nucleocytoviricota. These large double-stranded DNA viruses possess genome sizes ranging from approximately 100 kb to an extraordinary 2.77 Mb and can encode up to 900 proteins, including genes involved in central metabolic pathways previously thought to be exclusive to cellular organisms. Within this group, members of the family Phycodnaviridae infect ecologically and biogeochemically important phytoplankton taxa such as *Emiliania huxleyi*, Ostreococcus, Micromonas, and Bathycoccus. Viral infection of these hosts can abruptly terminate phytoplankton blooms, alter carbon export efficiency, and influence nutrient regeneration, thereby exerting a powerful influence on ocean productivity and elemental cycling (Endo et al., 2020).

Beyond their immediate ecological effects, marine viruses are major drivers of microbial evolution. Viral infection imposes intense selective pressure on microbial populations, promoting the evolution of resistance mechanisms such as receptor modification, restriction–modification systems, and CRISPR–Cas immunity. In response, viruses rapidly evolve counterstrategies, fueling a continuous evolutionary arms race between phages and their hosts. This coevolutionary dynamic accelerates genetic diversification in both viral and microbial communities and facilitates horizontal gene transfer, reshaping metabolic capabilities and ecological traits (Middelboe & Brussaard, 2017). Collectively, these processes have profound implications for the long-term maintenance of marine microbial diversity and for the capacity of ocean ecosystems to adapt to environmental change and anthropogenic perturbations.

11.3 BIOGEOCHEMICAL CYCLING

11.3.1 Carbon Cycling and the Biological Pump

The relative contributions of viral-mediated carbon pathways in the ocean are dynamic and likely vary over diel cycles, as viral infection efficiency and burst size are strongly influenced by host metabolic state, growth phase, and light availability. These temporal fluctuations introduce an additional layer of complexity into marine carbon cycling, particularly in surface waters where photosynthesis drives strong day–night contrasts in microbial activity.

11.3.1.1 The Viral Shunt and Carbon Redirection

The viral shunt concept, first articulated by Wilhelm and Suttle (1999) and Fuhrman (1999), describes a key mechanism by which viral lysis redirects organic carbon from particulate pathways into the dissolved organic matter (DOM) pool. It is estimated that viral lysis diverts approximately 25% of oceanic surface primary productivity away from particulate organic matter export and into DOM, thereby substantially altering the efficiency of carbon sequestration in the ocean. Quantitatively, this process releases approximately 0.2–3.2 μg C L^{-1} in offshore waters and 0.5–3.4 μg C L^{-1} in coastal systems, reflecting spatial variability in productivity and microbial abundance. On a global scale, viral lysis is estimated to release on the order of 145 Gt C annually in tropical and subtropical oceans alone, highlighting its immense biogeochemical significance (Danovaro et al., 2011).

The mechanistic basis of the viral shunt can be traced through the infection cycle of bacteriophages. Following successful attachment and genome injection, lytic viruses hijack host biosynthetic machinery to produce viral progeny, culminating in host cell rupture. This lysis releases cellular constituents, including proteins, lipids, nucleic acids, and polysaccharides, directly into the surrounding seawater. Because these materials are predominantly small, soluble, and readily bioavailable, they are incorporated into the DOM pool rather than forming sinking particles. This diversion effectively short-circuits the classical biological pump by preventing organic carbon from being exported to depth via sinking biomass (Wilhelm & Suttle, 1999; Fuhrman, 1999).

The immediate consequence of this process is a localized enrichment of labile DOM, which reduces resource limitation for surviving heterotrophic bacteria and stimulates rapid microbial regrowth. A broad spectrum of substrates becomes simultaneously available, enhancing bacterial production and accelerating carbon recycling within the surface ocean (Jiao et al., 2010). As a result, a greater fraction of fixed carbon is respired back to CO_2 in the upper ocean rather than being sequestered at depth, leading to a net reduction in carbon export efficiency.

Counterbalancing this process is the complementary viral shuttle hypothesis, which proposes that viral lysis can also enhance carbon export under certain conditions. Viral infection stimulates the release of sticky organic polymers, including transparent exopolymer particles (TEPs), which act as biological adhesives that promote aggregation of cells, detritus, and colloidal material into larger particles. These aggregates, commonly referred to as marine snow, sink more rapidly and efficiently transport organic carbon to deeper waters (Danovaro et al., 2011). Whether viral activity ultimately suppresses or enhances carbon export therefore depends on the balance between DOM production and TEP-mediated aggregation. This balance is regulated by host community composition, viral life strategies, infection dynamics, and ambient nutrient conditions.

Despite its recognized importance, accurately quantifying the relative contributions of viral shunt and shuttle pathways remains a major methodological challenge. Most estimates of viral lysis rates are derived from dilution experiments, frequency-of-infected-cells approaches, or virus-to-bacteria ratios measured in surface waters. Extrapolating these measurements to the mesopelagic and deep ocean—where viral–host interactions remain poorly characterized—introduces considerable uncertainty. Moreover, laboratory-derived lysis stoichiometries may not accurately reflect in situ conditions, where host physiology, nutrient limitation, and viral diversity differ substantially from controlled experiments (Weinbauer, 2004; Suttle, 2007). Consequently, current estimates of viral carbon redirection should be regarded as order-of-magnitude approximations, and comprehensive field-based validation across depth gradients remains a critical research priority.

11.3.1.2 Primary Production and Export

The BCP is a central component of the ocean's role in regulating atmospheric CO_2, creating approximately two-thirds of the surface-to-deep gradient in dissolved inorganic carbon and sequestering an estimated 1,300 (±230) Pg C in the deep ocean (Worden et al., 2015). This pump is driven by the fixation of CO_2 by phytoplankton in surface waters, followed by the transfer of organic matter to depth through sinking particles, zooplankton fecal pellets, and vertically migrating organisms.

Heterotrophic bacteria play a pivotal role within this system, processing approximately half of net primary production. Through respiration, viral lysis, and protozoan grazing, organic carbon and nutrients are continuously recycled, sustaining microbial growth and secondary production (Fuhrman, 1999; Sherr & Sherr, 2002). The microbial loop functions as a critical conduit that returns dissolved organic carbon to higher trophic levels through bacterial incorporation, contributing 50% or more of total carbon flux in oligotrophic ecosystems where classical grazing food webs are less efficient (Worden et al., 2015).

In parallel with the BCP, the MCP represents a distinct and complementary pathway for long-term carbon storage. The MCP transforms labile, bioavailable organic carbon into refractory dissolved organic carbon (RDOC) that resists microbial degradation on timescales of centuries to millennia. This process contributes approximately 0.2 Pg C yr^{-1} to long-term oceanic carbon sequestration. Biomarker analyses based on D-amino acids suggest that bacterially derived RDOC accounts for roughly 25% of the total marine RDOC pool (Jiao et al., 2010). Under future climate warming scenarios, enhanced stratification and increased partitioning of primary production into DOM may further amplify the relative importance of the MCP compared to particle-based export pathways (Worden et al., 2015).

Together, the interactions among the viral shunt, viral shuttle, microbial loop, and MCP underscore the central role of microbes and viruses in regulating the efficiency, directionality, and longevity of carbon storage in the ocean.

11.3.2 Nitrogen Cycling

Marine nitrogen cycling is governed by a tightly coupled cascade of microbially mediated redox transformations, each constrained by distinct thermodynamic, physiological, and ecological factors. At the entry point of biologically available nitrogen into the ocean lies nitrogen fixation, the reduction of atmospheric dinitrogen (N_2) to ammonium (NH_4^+). This process is restricted to diazotrophic microorganisms because it relies on the nitrogenase enzyme, which is irreversibly inhibited by molecular oxygen. As a consequence, active nitrogen fixation is confined either to organisms capable of protecting nitrogenase from oxygen exposure—such as Trichodesmium through spatial and temporal segregation of photosynthesis—or to low-oxygen microenvironments, including diatom–diazotroph associations (Dyhrman et al., 2006). Despite its central importance, global marine nitrogen fixation remains poorly constrained, with current estimates converging around ~140 Tg N yr^{-1}. Considerable uncertainty persists due to methodological challenges in measuring fixation rates in situ and the need to extrapolate sparse regional observations across heterogeneous ocean basins (Moore et al., 2013).

Once incorporated into biomass, fixed nitrogen is rapidly redistributed through the water column via nitrification, a two-step aerobic process in which ammonium is oxidized first to nitrite and then to nitrate. These reactions are carried out by distinct functional guilds—ammonia-oxidizing archaea and bacteria followed by nitrite-oxidizing bacteria—because no single organism efficiently catalyzes both steps. Nitrification thus represents a key transformation linking regenerated nitrogen to the nitrate pool that fuels new primary production, particularly below the euphotic zone (Salazar & Sunagawa, 2017).

In regions where oxygen concentrations fall below approximately 10 µmol L^{-1}, such as oxygen-deficient zones (ODZs), nitrogen cycling shifts toward anaerobic pathways. Under these conditions, nitrate replaces oxygen as the terminal electron acceptor, enabling denitrifying bacteria to reduce nitrate sequentially to N_2, thereby permanently removing bioavailable nitrogen from the ocean. Anammox (anaerobic ammonium oxidation) provides an energetically efficient alternative pathway by directly coupling ammonium oxidation to nitrite reduction without requiring molecular oxygen. This metabolic shortcut explains why anammox dominates nitrogen loss in the most oxygen-depleted environments. Current estimates suggest that anammox alone accounts for approximately 29% of fixed nitrogen loss in open-ocean ODZs under conditions of complete organic matter utilization (Breitburg et al., 2018). As ocean deoxygenation progresses under climate change, the spatial extent of ODZs is expected to increase, potentially accelerating nitrogen loss via anammox and altering global nitrogen inventories.

Marine nitrogen cycling is also tightly linked to climate regulation through the production of nitrous oxide (N_2O), a potent greenhouse gas and ozone-depleting substance. The oceans contribute approximately 30% of global atmospheric N_2O emissions, primarily through microbial nitrification and denitrification pathways operating at low oxygen concentrations (Breitburg et al., 2018). Adding further complexity, anaerobic methane-oxidizing archaea (ANME) have been shown to couple methane oxidation to nitrate reduction, revealing unexpected metabolic flexibility beyond the classical sulfate-dependent pathway and highlighting the interconnectedness of carbon and nitrogen cycles (Haroon et al., 2013). In the deep ocean, Thaumarchaeota dominate ammonia oxidation, exerting long-range control over nitrogen speciation and availability throughout the water column (Salazar & Sunagawa, 2017).

11.3.3 Sulfur Cycling: DMSP, DMS, and Climate Feedback

Sulfur cycling in the marine environment is strongly influenced by the production and transformation of dimethylsulfoniopropionate (DMSP), a sulfur-containing compound synthesized predominantly by dinoflagellates, prymnesiophytes (notably *Emiliania huxleyi* and Phaeocystis), and chrysophytes. These lineages evolved DMSP as a multifunctional osmolyte adapted to the high-salinity, high-light, and oxidative stress conditions characteristic of surface-ocean environments. DMSP stabilizes cellular membranes, scavenges reactive oxygen species, and protects enzymatic function—roles that diatoms fulfill through alternative biochemical strategies, explaining their comparatively low DMSP production (Curson et al., 2017). In blooms dominated by strong DMSP producers, DMSP can account for up to 10% of total fixed carbon, representing a substantial biochemical investment (Hopkins et al., 2023).

Following phytoplankton cell lysis or active exudation, DMSP enters the dissolved pool where it is rapidly processed by heterotrophic bacteria, principally members of *Ruegeria pomeroyi* and the SAR11 clade. Bacterial degradation proceeds via two competing enzymatic pathways. The cleavage pathway, mediated by DMSP lyases, converts DMSP into dimethyl sulfide (DMS) and acrylate, while the demethylation pathway sequentially removes methyl groups to yield methanethiol, retaining sulfur within the microbial food web. The dominance of one pathway over the other depends on bacterial community composition and local sulfur demand, with sulfur-limited conditions favoring demethylation (Hopkins et al., 2023).

Globally, oceanic DMS emissions are estimated at 18–31 Tg S yr^{-1}, accounting for approximately half of all biogenic sulfate entering the atmosphere. Yet less than 10% of the DMS produced in seawater ultimately reaches the atmosphere. This discrepancy arises from a kinetic competition between microbial and photochemical consumption of DMS in the mixed layer—which occurs on timescales of hours to days—and air–sea gas exchange, which is constrained by wind speed and gas transfer velocity (Hopkins et al., 2023). As a result, only DMS produced very near the surface or during high-wind events contributes significantly to atmospheric sulfate loading.

This bottleneck lies at the heart of ongoing debate surrounding the CLAW hypothesis, which proposes that DMS-derived sulfate aerosols enhance cloud condensation nuclei formation, increasing planetary albedo and exerting a negative climate feedback. The sensitivity of DMS flux to microbial processing, wind regimes, and surface-ocean chemistry renders this feedback highly uncertain under future climate scenarios. Ocean acidification may further suppress DMSP production—by up to 50% at pH 7.49—while warming-induced stratification may favor diatom-dominated communities that produce comparatively little DMSP, thereby weakening the sulfur–climate linkage at its source (Hopkins et al., 2023).

11.3.4 Phosphorus and Trace Metal Dynamics

Phosphorus cycling in the ocean is shaped by microbial adaptation to chronic phosphate scarcity, particularly in oligotrophic gyres. In these environments, phytoplankton and heterotrophic bacteria rely heavily on dissolved organic phosphorus (DOP) as an alternative nutrient source. Key taxa such as Synechococcus, Prochlorococcus, and diverse heterotrophic bacteria express alkaline phosphatases (PhoA, PhoD, and PhoX) that hydrolyze phosphoesters, liberating inorganic phosphate for assimilation. Additionally, the C–P lyase pathway enables the utilization of phosphonates, which are enriched in persistently phosphate-depleted regions (Dyhrman et al., 2006). In oligotrophic waters, DOP concentrations exceed inorganic phosphate by factors of 5–10, while polyphosphate pools reach 3–7 nmol equivalents per liter in tropical oceans, underscoring the importance of noncanonical phosphorus reservoirs (Moore et al., 2013).

Trace metals further constrain marine productivity by defining narrow windows of enzymatic functionality. Iron limitation affects approximately 40% of the surface ocean, including the Southern Ocean and HNLC regions of the tropical and subarctic Pacific. To acquire iron, marine bacteria synthesize high-affinity siderophores such as amphibactins (~9 pM) and ferrioxamines (1–2 pM), with over 99% of dissolved iron complexed to organic ligands (Vraspir & Butler, 2009). Beyond iron, trace metals such as zinc, cobalt, and nickel are essential cofactors in microbial metabolism: zinc-dependent enzymes constitute nearly half (47.2%) of bacterial metallomes, cobalt is required for vitamin B_{12}–dependent reactions, and nickel-dependent superoxide dismutase enables oxidative stress tolerance in Prochlorococcus (Moore et al., 2013). These tightly regulated metal requirements create complex biogeochemical feedbacks that link trace element availability to microbial community structure, nutrient cycling, and ocean productivity.

11.3.5 The Silica Pump

The biological silica pump is a central yet often underappreciated component of marine biogeochemical cycling, operating primarily through the activities of diatoms. These organisms actively take up dissolved silicic acid [$Si(OH)_4$] from seawater via specialized silicic acid transporters (SITs), which exhibit high specificity and regulation in response to ambient silica availability. Once internalized, silicic acid is polymerized within intracellular silica deposition vesicles, where it is assembled into the intricate siliceous frustules that characterize diatom cell walls. Upon cell death, viral lysis, or zooplankton grazing, these frustules are released into the water column and exported to depth, linking surface productivity to benthic silica burial.

On a global scale, gross biogenic silica production by diatoms is estimated at approximately 2.40×10^{14} mol Si yr^{-1}. A substantial fraction of this biogenic silica ultimately accumulates in marine sediments, with the Southern Ocean representing the dominant global sink, accounting for roughly 4.5×10^{14} g SiO_2 yr^{-1} (Nelson et al., 1995; Tréguer & De La Rocha, 2013). This disproportionate role reflects the high diatom productivity of Southern Ocean ecosystems, sustained by upwelling of silica-rich deep waters and relatively weak grazing pressure compared to lower latitudes.

Beyond their role in silica cycling, diatom frustules exert a powerful influence on the oceanic carbon cycle. The dense, mineralized nature of silica provides effective ballast to organic aggregates, increasing the sinking velocity of marine snow and zooplankton fecal pellets. This ballasting effect enhances the efficiency of the BCP by promoting rapid transfer of particulate organic carbon from the surface ocean to depth, thereby reducing remineralization in the upper water column (Tréguer et al., 2018). Consequently, the silica pump and carbon pump are tightly coupled, with shifts in diatom abundance and silicification directly influencing long-term carbon sequestration.

Future climate projections suggest that the strength and spatial distribution of the silica pump may change substantially over the coming century. Warming-induced stratification is expected to reduce nutrient resupply to surface waters in many ocean regions, favoring smaller phytoplankton such as picocyanobacteria over heavily silicified diatoms. As a result, diatom contributions to global primary production are projected to decline across much of the ocean, with the notable exception of the Southern Ocean, where persistent upwelling may partially offset stratification effects (Tréguer et al., 2018). Such shifts could weaken both silica export and the associated carbon ballast effect, with cascading consequences for deep-ocean carbon storage.

At the terminus of the silica pump, sedimentary processes further modulate the fate of exported material. Sediment diagenesis proceeds through a well-defined sequence of terminal electron acceptors—oxygen, nitrate, manganese(IV), iron(III), sulfate, and ultimately carbon dioxide—reflecting progressively lower energy yields for organic matter oxidation. Rates of organic carbon remineralization span more than three orders of magnitude across this redox cascade. In Arctic shelf sediments, for example, sulfate reduction alone accounts for approximately 35% of total organic matter remineralization, underscoring the importance of anaerobic microbial metabolism in regulating benthic carbon and nutrient fluxes (Jørgensen & Kasten, 2006).

Crucially, the biogeochemical cycles described throughout this chapter—carbon, nitrogen, sulfur, phosphorus, and silica—are not governed by abiotic chemistry alone. They are actively structured and reshaped by ecological interactions within microbial communities, particularly virus–host dynamics. Viral infection determines which microbial populations persist, which are lysed, and how rapidly nutrients are regenerated and redistributed. Through selective mortality, altered aggregation processes, and accelerated recycling of limiting elements, viruses exert a subtle yet pervasive control over the efficiency, coupling, and long-term balance of the ocean's major elemental cycles.

11.4 VIRUS–HOST DYNAMICS

11.4.1 Lytic versus Lysogenic Infection Strategies

The balance between lytic and lysogenic viral infection strategies is a critical regulator of microbial mortality, nutrient recycling, and the magnitude of viral carbon shunting in marine ecosystems. Environmental conditions—particularly host abundance, growth rate, and nutrient availability—strongly influence which strategy predominates. Lysogeny is favored in oligotrophic environments characterized by low microbial densities and limited resources, where viral replication through immediate lysis would be inefficient. Under these conditions, viral genomes persist as prophages integrated into host chromosomes, allowing long-term maintenance of viral genetic material without killing the host. In contrast, lytic infection dominates nutrient-replete environments where host cells are abundant and rapidly dividing, enabling high viral replication rates and extensive host cell lysis. This ecological switching underpins the classic "kill-the-winner" dynamic, in which viruses selectively target the most competitive and abundant microbial taxa, preventing competitive exclusion and stabilizing community structure (Thingstad & Lignell, 1997; Knowles et al., 2016).

Through density-dependent predation, viruses impose frequency-dependent selection on microbial populations, maintaining high levels of taxonomic and functional diversity. When a particular bacterial lineage becomes dominant, it experiences increased viral pressure, leading to population collapse and release of resources that can be exploited by less abundant taxa. This dynamic not only regulates microbial diversity but also modulates the partitioning of organic carbon between higher trophic levels and the DOM pool, thereby influencing the efficiency of the BCP.

Lysogenic interactions are far more prevalent in the ocean than previously assumed. More than 40% of cultured marine bacterial isolates harbor inducible prophages, with lysogenic frequencies generally higher in offshore oligotrophic waters compared to productive coastal regions (Paul, 2008; Knowles et al., 2016). Prophage induction can be triggered by environmental stressors such as ultraviolet radiation, oxidative stress, or nutrient limitation. In cyanobacteria, prophage induction exhibits pronounced diel periodicity, with UV-induced lysis peaking during periods of maximum solar irradiance. This timing can generate afternoon pulses of DOM, fueling heterotrophic bacterial activity and reinforcing tight coupling between viral activity and daily biogeochemical rhythms.

Beyond their role as latent parasites, prophages can confer significant ecological advantages to their hosts. These benefits include immunity to superinfection by closely related phages, enhancement of host fitness through auxiliary metabolic genes, and the production of toxins or effector proteins that suppress competing microbial lineages (Paul, 2008). As a result, lysogeny represents not merely a survival strategy for viruses under unfavorable conditions, but a mutualistic or conditionally beneficial interaction that reshapes microbial ecological niches and evolutionary trajectories.

11.4.2 Auxiliary Metabolic Genes and Host Manipulation

One of the most striking discoveries in marine virology is the widespread presence of auxiliary metabolic genes (AMGs) within viral genomes, particularly among cyanophages infecting the globally dominant picocyanobacteria Prochlorococcus and Synechococcus. These AMGs do not support viral structure or replication directly; instead, they reprogram host metabolism during infection to optimize the cellular environment for viral genome synthesis and virion production.

Among the most extensively studied AMGs are photosynthesis-related genes. The photosystem II core gene psbA is present in approximately 88% of surveyed cyanophage genomes, while psbD occurs in roughly 50%, with notable enrichment in Synechococcus-infecting myoviruses (Lindell et al., 2007). By supplementing or replacing damaged host photosystem components, these phage-encoded genes maintain photosynthetic electron transport during infection, preventing photoinhibition and sustaining ATP and NADPH production. This metabolic support prolongs host viability, extending the duration of viral replication and maximizing viral burst size.

Cyanophage AMGs extend well beyond photosynthesis to encompass central carbon and nutrient metabolism. Genes such as CP12, which inhibits the Calvin cycle and redirects carbon flux toward the pentose phosphate pathway, enhance the supply of reducing power and nucleotide precursors required for viral DNA synthesis. Additional genes involved in carbon metabolism, including talC (transaldolase), further optimize carbon flow toward biosynthetic pathways. To overcome nutrient limitation in oligotrophic environments, cyanophages also encode phosphate acquisition genes such as pstS, phoH, and phoA, enabling infected hosts to scavenge phosphorus more effectively and sustain viral replication under severe nutrient stress (Lindell et al., 2007).

Global ecogenomic analyses have revealed that AMGs are not restricted to cyanophages but are widespread across diverse marine viral communities. The large-scale study by Roux et al. (2016) demonstrated that globally abundant ocean viruses encode a broad array of genes influencing carbon metabolism, nitrogen transformations, and sulfur cycling.These findings suggest that viruses function as mobile genetic reservoirs, redistributing metabolic capabilities across microbial lineages and actively reshaping host physiology at ecosystem scales. Through AMG-mediated host manipulation, marine viruses exert a direct and dynamic influence on elemental cycling, blurring the distinction between ecological interactions and biochemical regulation in the ocean.

11.4.3 Giant Viruses (Nucleocytoviricota)

Giant viruses belonging to the phylum Nucleocytoviricota constitute one of the most rapidly expanding and conceptually transformative frontiers in marine virology. These large double-stranded DNA viruses challenge traditional definitions of viruses by virtue of their extraordinary genome sizes, which range from approximately 100 kb to 2.77 Mb and encode up to ~900 proteins. Collectively spanning at least six orders and more than 30 families, Nucleocytoviricota display levels of genetic complexity that rival those of parasitic eukaryotes and blur the boundary between viral and cellular life.

Among the most ecologically consequential members of this group are the Phycodnaviridae, which infect globally important phytoplankton taxa including the coccolithophore Emiliania huxleyi and prasinophyte algae such as Ostreococcus, Micromonas, and Bathycoccus. These hosts dominate primary production across vast regions of the ocean, and viral infection events therefore exert a disproportionate influence on phytoplankton bloom dynamics, community succession, and biogeochemical fluxes. Viral lysis of phytoplankton cells can abruptly terminate blooms, redirecting fixed carbon from particulate biomass into dissolved pools and altering the timing and magnitude of carbon export to the deep ocean (Endo et al., 2020).

A particularly striking example of viral metabolic innovation is provided by Emiliania huxleyi virus EhV-86, which has acquired and retained an entire sphingolipid biosynthesis pathway from its host. This represents the first documented case of complete metabolic pathway transfer between a eukaryote and a virus (Monier et al., 2009). Viral sphingolipid production is not merely a relic of horizontal gene transfer but plays an active role during infection, modulating host membrane composition, regulating programmed cell death, and synchronizing host population collapse during bloom termination. This viral control over lipid signaling pathways directly links infection dynamics to large-scale carbon export events associated with coccolithophore blooms.

Metatranscriptomic studies have further revealed that marine Nucleocytoviricota actively express genes involved in central carbon metabolism, light-harvesting processes, oxidative stress mitigation, and nutrient acquisition during infection. Unexpectedly, these viruses also encode and express genes for cytoskeletal proteins, including myosin and kinesin, suggesting sophisticated manipulation of host intracellular transport systems to facilitate viral assembly and egress (Endo et al., 2020). Such findings underscore that giant viruses are not passive parasites but highly integrated metabolic engineers capable of restructuring host physiology in ways that propagate upward to ecosystem-scale processes.

Recent expansions in genomic sampling and environmental sequencing indicate that giant viruses are far more abundant, diverse, and metabolically versatile than previously recognized. Their cumulative impact on phytoplankton mortality, organic matter transformation, and elemental cycling suggests that Nucleocytoviricota should be regarded as integral components of marine food webs and biogeochemical models, rather than as biological curiosities.

11.4.4 Pelagiphages and SAR11 Interactions

Pelagiphages—viruses that infect members of the SAR11 clade—occupy a uniquely influential position in marine ecosystems due to the overwhelming numerical dominance of their hosts. SAR11 bacteria account for approximately 20–50% of planktonic microbial cells in surface oceans worldwide, making them the most abundant free-living organisms on the planet. Consequently, viral predation targeting this single lineage has outsized implications for global carbon flow, nutrient regeneration, and microbial community structure (Giovannoni, 2017).

The hypothesis that pelagiphages regulate SAR11 populations through a classic kill-the-winner mechanism was first supported by large-scale metagenomic analyses. The podovirus HTVC010P emerged as the most abundant known pelagiphage globally, reaching concentrations of $\sim 10^5$ genome copies per milliliter in surface waters. Subsequent metagenomic surveys uncovered a remarkably diverse assemblage of SAR11-infecting viruses, spanning multiple viral families and far exceeding the diversity captured through culture-based isolation. These findings indicate that pelagiphage predation pressure on SAR11 is both intense and phylogenetically broad, ensuring tight top-down control over this dominant bacterial clade (Zhao et al., 2013).

Despite strong evidence for pervasive viral regulation, key uncertainties remain regarding the physiological consequences of pelagiphage infection. One particularly intriguing discovery is the presence of so-called "zombie cells"—phage-infected SAR11 cells that lack detectable ribosomes. This observation suggests that ribosomal RNA may be actively degraded and repurposed as a nucleotide source for viral genome synthesis, representing a novel and extreme form of metabolic hijacking (Zhao et al., 2013). The downstream effects of this process on DOM composition, microbial remineralization rates, and nutrient availability have yet to be quantified, leaving an important gap in our understanding of virus-mediated carbon cycling.

Ongoing metagenomic and single-cell studies continue to reveal previously uncharacterized pelagiphage lineages, indicating that current estimates of viral diversity and infection rates likely underestimate the true scale of SAR11–virus interactions. As these datasets expand, it is becoming increasingly clear that virus–host dynamics at the level of dominant microbial clades do not merely influence microbial community composition, but actively shape the structure and efficiency of the broader pelagic food web. Through their regulation of SAR11, pelagiphages exert a subtle yet profound control over the fate of organic carbon in the surface ocean and its eventual transfer to deeper waters.

11.5 TROPHIC INTERACTIONS AND MICROBIAL FOOD WEBS

11.5.1 Heterotrophic Protist Grazing

Heterotrophic and mixotrophic protists exert a dominant top-down control on marine bacterial populations, representing one of the principal mortality agents shaping microbial biomass and turnover in the ocean. In productive environments, protistan grazers consume an estimated 28–62% of daily prokaryotic biomass, a rate sufficient to balance bacterial production and stabilize standing stocks despite rapid microbial growth (Sherr & Sherr, 2002). Heterotrophic nanoflagellates (HNFs) are particularly effective bacterivores, exhibiting strong prey selectivity: they disproportionately graze heterotrophic bacteria (accounting for ~51–75% of grazing encounters) and the abundant cyanobacterium Prochlorococcus (78–93% exposure), whereas ciliates preferentially target picoeukaryotic phytoplankton and larger prey items.

Field measurements from frontal and upwelling zones reveal grazing rates approaching 1.05 d^{-1}, closely matched to bacterial growth rates of approximately 1.33 d^{-1}, underscoring a tight coupling between microbial production and mortality (Sherr & Sherr, 2002; Steinberg & Landry, 2017). This near-equilibrium dynamic maintains relatively constant bacterial biomass while sustaining rapid carbon and nutrient flux through the microbial loop. Rather than allowing bacterial accumulation, protist grazing ensures continual recycling of organic matter into higher trophic levels and regenerated nutrient pools.

The central role of protistan predation in aquatic microbial food webs was firmly established by foundational studies demonstrating that flagellate grazing constitutes the dominant loss term for marine bacteria, exceeding viral lysis in many regions and seasons (Sherr & Sherr, 2002). Importantly, warming and enhanced stratification appear to amplify trophic cascades, shifting ecosystem regulation from bottom-up nutrient limitation toward stronger top-down control by grazers. Depth-resolved observations further demonstrate that protist grazing is not confined to surface waters: active heterotrophic protist communities persist throughout the mesopelagic and bathypelagic zones. Recent work has highlighted significant grazing impacts on deep-sea bacterial assemblages, including those associated with hydrothermal vents, where protists modulate chemosynthetically driven microbial production (Medina et al., 2017). These findings extend the influence of protist grazing across the full depth of the ocean, reinforcing its role in global biogeochemical cycling.

11.5.2 Mixotrophy: Blurring Trophic Boundaries

The widespread prevalence of mixotrophy—the capacity of single protist cells to combine photosynthesis with phagotrophy—fundamentally challenges traditional representations of marine food webs and carries far-reaching biogeochemical consequences. Mixotrophic protists simultaneously fix inorganic carbon and consume prey, allowing them to flexibly balance energy acquisition, nutrient uptake, and survival across fluctuating environmental conditions. This trophic plasticity enables mixotrophs to enhance total carbon fixation, suppress prokaryotic abundance through grazing, and facilitate more efficient transfer of organic matter to higher trophic levels (Stoecker et al., 2017; Flynn et al., 2019).

Ecosystem and trait-based modeling studies demonstrate that incorporating mixotrophy into plankton food webs increases trophic transfer efficiency, shifts biomass toward larger mean organism size, and enhances vertical carbon export relative to systems constrained to strict autotroph–heterotroph separation (Ward & Follows, 2016). By directly linking primary production to grazing within a single organism, mixotrophs reduce energetic losses associated with multistep food chains, thereby strengthening the BCP.

Empirical observations corroborate these theoretical predictions. Mixotrophic haptophytes have been identified as dominant bacterial grazers in oligotrophic coastal waters, where nutrient scarcity favors organisms capable of supplementing photosynthesis with prey-derived nitrogen and phosphorus (Unrein et al., 2014). In the oligotrophic Atlantic Ocean, direct measurements of bacterivory combined with photosynthetic activity reveal that mixotrophy underpins a substantial fraction of ecosystem metabolism (Hartmann et al., 2012). As a result, functional classifications of marine protists increasingly emphasize mechanisms of energy and nutrient acquisition rather than rigid trophic labels.

This shift has catalyzed a new ecological paradigm in which the traditional dichotomy between autotrophs and heterotrophs is recognized as an oversimplification. Instead, trophic flexibility is now understood to be a defining feature of marine protist ecology, with mixotrophy representing a dominant and adaptive strategy in nutrient-poor oceans (Mitra et al., 2016; Flynn et al., 2019).

11.5.3 Parasitic Protists (Syndiniales) and Top-Down Control

Parasitic protists constitute a powerful but historically underappreciated form of top-down regulation in marine ecosystems. Among them, the Syndiniales (marine alveolates) form an exclusively parasitic dinoflagellate lineage characterized by extraordinary genetic diversity, global distribution, and ecological impact (Guillou et al., 2008). These parasites infect a wide range of planktonic hosts, including dinoflagellates, radiolarians, and other protists, thereby influencing community composition and bloom dynamics.

Species within the genus Amoebophrya are particularly notable for their capacity to terminate blooms of harmful and nonharmful dinoflagellates. Their infection cycle is rapid and efficient, involving a 2–3-day intracellular trophont phase during which the parasite consumes host cellular contents, followed by host lysis and release of hundreds of free-living dinospores. These infective stages persist for 1–2 days in the water column, enabling swift transmission during periods of high host density. Infection prevalence in dense blooms can reach 75–79%, imposing strong mortality pressure independent of grazing or nutrient depletion (Guillou et al., 2008).

By rapidly converting host biomass into parasitic propagules and DOM, Syndiniales infections alter the partitioning of organic carbon and nutrients within the microbial food web. Data from the Tara Oceans expedition identified Syndiniales lineages among those most strongly associated with carbon export in oligotrophic regions, suggesting that parasitism may enhance aggregation processes or promote sinking flux through altered particle composition. These findings highlight parasitic protists as critical, yet often overlooked, agents of biogeochemical transformation, operating alongside grazers and viruses to regulate plankton populations and reshape the efficiency of the biological pump.

11.5.4 HABs: Ecology and Bacterial Associations

HABs represent one of the most disruptive manifestations of plankton ecology, exerting severe ecological, public health, and economic consequences through the production of potent secondary metabolites. These blooms can lead to massive fish kills, contamination of seafood, hypoxia, and long-term ecosystem degradation. Among the most well-characterized toxins are saxitoxins, produced by dinoflagellate genera Alexandrium, Gymnodinium, and Pyrodinium. Saxitoxins bind to and block voltage-gated sodium channels in nerve and muscle cells, preventing action potential propagation and causing paralytic shellfish poisoning (PSP). Even trace concentrations can be lethal, making PSP one of the most dangerous marine intoxications.

Domoic acid, responsible for amnesic shellfish poisoning (ASP), is synthesized by at least 26 confirmed toxigenic species within the diatom genus Pseudo-nitzschia. Acting as a high-affinity agonist of kainate and AMPA-type glutamate receptors, domoic acid induces neuronal excitotoxicity, leading to seizures, permanent memory loss, and death in severe cases. Brevetoxins, produced by the dinoflagellate Karenia brevis, are lipid-soluble, polycyclic polyether compounds assembled by modular polyketide synthase pathways. These toxins bind voltage-gated sodium channels but, unlike saxitoxins, force them to remain persistently open, causing uncontrolled neuronal firing and neurotoxic shellfish poisoning. The ecological and economic consequences of HABs are exemplified by the 2015 Northeast Pacific "warm blob" bloom, which resulted in widespread fisheries closures and losses approaching $100 million, underscoring the vulnerability of coastal economies to bloom events (Holser et al., 2022).

The initiation, maintenance, and termination of HABs are strongly modulated by bacterial–algal interactions, which function as hidden drivers of bloom ecology. Many HAB-forming phytoplankton engage in mutualistic relationships with specific bacterial taxa. Members of the Roseobacter clade, for example, are frequently associated with Karenia brevis and other dinoflagellates, supplying essential metabolites such as vitamins B_1 (thiamine) and B_{12} (cobalamin), phytohormones, and bioavailable trace metals that enhance algal growth and toxin production. In return, algae release dissolved organic carbon that fuels bacterial metabolism, reinforcing bloom persistence.

Conversely, blooms can be actively terminated by algicidal bacteria, including species of Marinobacter, Pseudomonas, Pseudoalteromonas, and Alteromonas. These bacteria produce lytic enzymes, reactive oxygen species, or secondary metabolites that disrupt algal membranes or interfere with cellular metabolism, leading to bloom collapse. The balance between mutualistic and antagonistic bacterial associations therefore plays a decisive role in determining HAB longevity, toxicity, and spatial extent. Climate change is expected to exacerbate HAB frequency and severity by expanding the thermal niche of toxigenic species, intensifying water column stratification, and altering bacterial community composition in ways that may preferentially support bloom-forming algae (Guillou et al., 2008).

11.5.5 Diel Rhythmicity in Marine Microbial Communities

Marine microbial ecosystems operate under strong diel (24-hour) rhythms, analogous in many respects to the circadian dynamics observed in host-associated microbiomes. These rhythms are driven primarily by the solar cycle and propagate across multiple trophic levels, synchronizing metabolism, growth, and ecological interactions throughout the water column. The cyanobacterium Prochlorococcus provides one of the most striking examples: populations exhibit globally synchronized cell cycles, with DNA replication initiating in the late afternoon and cytokinesis occurring near dusk. This remarkable coordination is governed by endogenous circadian oscillators homologous to the KaiABC clock system first described in Synechococcus (Johnson et al., 2017).

At the molecular level, diel regulation is pervasive. Approximately 80% of Prochlorococcus transcripts display significant diel periodicity, with genes involved in photosynthesis and light harvesting peaking near dawn, and those associated with DNA replication, cell division, and repair reaching maximal expression in the late afternoon (Waldbauer et al., 2012). These oscillations ensure that energetically demanding processes are optimally timed to light availability, maximizing growth efficiency in oligotrophic environments.

Diel rhythms are further amplified by diel vertical migration (DVM) of zooplankton—the largest synchronized animal movement on Earth. Each night, billions of organisms ascend to surface waters to feed, returning to deeper layers at dawn. This daily migration generates temporally pulsed inputs of grazing pressure, fecal pellet production, and DOM release that reverberate through microbial communities. In the mesopelagic zone, bacterial abundance can fluctuate two- to threefold over a 24-hour cycle in response to organic matter subsidies delivered by migrating zooplankton (Steinberg et al., 2008).

Viruses also participate in diel structuring of microbial dynamics. Lytic infection rates of cyanobacteria peak during daylight hours, when host metabolic activity and nucleotide pools are highest, thereby maximizing viral replication efficiency (Aylward et al., 2017). These synchronized infection cycles contribute to predictable daily pulses of DOM through viral lysis, reinforcing tight coupling between phototrophic production and heterotrophic consumption.

The biogeochemical consequences of diel rhythmicity are substantial. Emissions of DMS exhibit pronounced diel variation linked to light-dependent DMSP cleavage and bacterial sulfur metabolism (Galí et al., 2013). Nitrogen fixation by Trichodesmium peaks during daylight, while heterotrophic respiration and nutrient remineralization follow distinct, phase-shifted temporal patterns. As a result, biogeochemical flux estimates derived from single time-point sampling can be strongly biased. Accurately resolving marine carbon, nitrogen, and sulfur budgets therefore requires explicit consideration of diel variability, reinforcing the need for high-frequency sampling and time-resolved modeling in future ocean biogeochemistry studies.

11.6 SYMBIOSES IN MARINE SYSTEMS

11.6.1 Coral Reef Holobionts

Coral reefs are sustained by one of the most intricate symbiotic systems in the marine biosphere: the coral holobiont, a functional metaorganism composed of the cnidarian animal host, photosynthetic dinoflagellates of the family Symbiodiniaceae, diverse bacterial and archaeal communities, fungi, and an abundant viral assemblage. Rather than acting as passive passengers, these microbial partners are integral to host nutrition, immunity, stress tolerance, and biogeochemical exchange. Collectively, microorganisms can constitute up to 35% of total holobiont biomass, occupying spatially structured microhabitats including coral tissues, the gastrovascular cavity, the calcium carbonate skeleton, and the surface mucus layer, each of which selects for distinct microbial consortia.

The hologenome concept reframes corals not as solitary organisms but as evolutionary units whose phenotypes emerge from interactions among host genes, Symbiodiniaceae genomes, and associated microbiomes (Bourne et al., 2016).This integrated perspective helps explain the extraordinary plasticity of coral responses to environmental change, as shifts in symbiont identity or microbial community structure can rapidly alter metabolic capacity, stress resistance, and disease susceptibility without requiring host genetic change.

A defining feature of coral holobiont functioning is the presence of strong diel metabolic rhythms. Photosynthetically derived carbon is translocated from Symbiodiniaceae to the coral host primarily during daylight hours, fueling respiration, mucus production, and growth. Paradoxically, calcification rates often peak at night, when reduced photosynthetic activity and changes in internal carbonate chemistry favor calcium carbonate deposition. Thermal stress disrupts these finely tuned temporal dynamics, and growing evidence suggests that circadian desynchronization between host and symbionts may precede visible bleaching, analogous to microbiome-driven dysbiosis observed in stressed mammalian systems. Such temporal misalignment may therefore represent an early warning indicator of holobiont instability.

11.6.1.1 *Symbiodiniaceae Diversity and Thermal Tolerance*

The ecological performance of coral holobionts is strongly influenced by the taxonomic identity of their dinoflagellate symbionts. Following comprehensive taxonomic revision, the family Symbiodiniaceae now comprises 11 recognized genera—Symbiodinium, Breviolum, Cladocopium, Durusdinium, Effrenium, Fugacium, Gerakladium, Halluxium, Freudenthalidium, Philozoon, and Miliolidium—each exhibiting distinct physiological traits, ecological niches, and thermal tolerances (LaJeunesse et al., 2018).

Among these, *Durusdinium trenchii* is widely recognized for conferring enhanced thermal tolerance to coral hosts, increasing bleaching thresholds by approximately 1–1.5°C. During thermal stress events exceeding 34°C, Durusdinium densities often increase as thermally sensitive genera such as Cladocopium decline. However, this resilience comes at a cost: corals dominated by Durusdinium typically exhibit reduced growth and calcification rates under nonstress conditions. This tradeoff highlights a central dilemma of coral adaptation to climate change—short-term survival through symbiont switching may compromise long-term reef accretion and structural integrity (LaJeunesse et al., 2018).

11.6.1.2 *Bacterial Communities and Endozoicomonas*

Bacterial symbionts perform essential metabolic and regulatory functions within the coral holobiont. Among them, Endozoicomonas (Gammaproteobacteria, Oceanospirillales) stands out as a ubiquitous and often dominant coral-associated taxon, forming dense intracellular and extracellular aggregates within host tissues. Its consistent presence across coral species and geographic regions has led to its designation as a core symbiont (Bourne et al., 2016).

Genomic analyses reveal that Endozoicomonas species possess versatile metabolic capabilities consistent with facultative symbiosis. Endozoicomonas acroporae, for example, encodes the dddD gene, enabling the conversion of dimethylsulfoniopropionate (DMSP) into DMS, thereby linking coral-associated sulfur cycling to carbon metabolism and potentially to climate-relevant aerosol formation. These bacteria also encode nitrate reductases involved in nitrogen cycling and pathways for synthesizing amino acids and B-complex vitamins that may supplement host and symbiont nutrition. Genome sizes ranging from 4.09 to 6.69 Mb further support metabolic flexibility rather than obligate dependence on the host (Bourne et al., 2016).

11.6.1.3 *Coral Diseases and Dysbiosis*

Coral diseases have increased dramatically in prevalence and severity over recent decades, reflecting the destabilization of holobiont–microbiome interactions under environmental stress. Stony coral tissue loss disease (SCTLD), first documented near Miami in 2014, has emerged as one of the most destructive coral diseases on record, causing mortality rates approaching 99% in highly susceptible species such as *Colpophyllia natans*, *Montastraea cavernosa*, and *Orbicella faveolata*. Microbiome analyses consistently reveal enrichment of Rhodobacterales and Peptostreptococcales–Tissierellales in diseased tissues, including Clostridiales taxa encoding alpha-toxins implicated in tissue necrosis (Meyer et al., 2019).

Black band disease represents another classic example of microbiome-driven pathology. This disease is characterized by a migrating mat dominated by filamentous cyanobacteria that advances across coral surfaces at rates of 3 mm to 1 cm per day. Beneath the mat, intense microbial respiration creates hypoxic and sulfide-rich microenvironments, fueled by sulfate-reducing bacteria such as Desulfovibrio. The resulting chemical conditions are lethal to coral tissues, illustrating how microbial metabolism can directly mediate disease progression (Sweet & Bulling, 2017).

11.6.1.4 Bleaching Mechanisms at the Microbiome Level

Coral bleaching emerges from the convergence of physiological stress, oxidative damage, and microbiome disruption. Elevated temperature and irradiance impair Symbiodiniaceae photosystems and disrupt host mitochondrial electron transport, leading to excessive production of reactive oxygen species (ROS). Measured superoxide concentrations in corals can reach ~120 nM, among the highest recorded in marine environments (Lesser, 2011). When antioxidant defenses are overwhelmed, oxidative damage triggers symbiont expulsion or loss of pigmentation.

Microbiome-level responses follow a characteristic progression. Initial environmental stress causes subtle shifts in microbial community composition, with increased abundance of opportunistic taxa while overall holobiont function remains intact. Continued stress permits pathogenic or opportunistic microbes to outcompete native symbionts, leading to functional dysbiosis. Ultimately, sustained disruption culminates in bleaching, tissue loss, and potential mortality (Bourne et al., 2016). This framework underscores bleaching not as a single event but as a multistage ecological collapse of the holobiont system.

11.6.2 Sponge Microbiomes and the Sponge Loop

Marine sponges represent another cornerstone of benthic holobiont ecology, harboring some of the most taxonomically and functionally diverse microbiomes known. Sponge-associated communities span up to 52 bacterial phyla and four archaeal phyla, with microbial cells comprising as much as 35% of sponge biomass. Sponges are broadly categorized into high microbial abundance (HMA) and low microbial abundance (LMA) types, with HMA sponges hosting microbial densities 2–4 orders of magnitude higher than LMA species.

HMA sponges are dominated by lineages such as Chloroflexi, Actinobacteria, Acidobacteria, and Candidatus Poribacteria, a sponge-specific candidate phylum that can constitute up to 75% of individual microbiomes. These microbes drive the sponge loop, a process in which sponges efficiently assimilate DOM and convert it into particulate detritus that reenters reef food webs. Rates of DOM uptake by cryptic sponges may rival gross reef primary productivity and exceed traditional microbial loop fluxes.

Sponge microbiomes also play critical roles in nitrogen cycling, particularly through nitrification that removes host-excreted ammonia, thereby stabilizing nutrient dynamics in oligotrophic reef environments. The global convergence of sponge microbiome structure suggests ancient coevolutionary relationships and highlights sponges as reservoirs of novel metabolic pathways with significant potential for natural product discovery (Thomas et al., 2016).

11.6.3 Hydrothermal Vent Symbioses: Riftia and Chemosynthetic Partners

Hydrothermal vent ecosystems exemplify the extreme endpoint of host–microbe metabolic integration. The giant tubeworm *Riftia pachyptila* lacks a digestive system and depends entirely on its intracellular symbiont, *Candidatus Endoriftia persephone*, a sulfur-oxidizing gammaproteobacterium that can reach densities exceeding 10^9 cells per gram of trophosome tissue. The host supplies all substrates required for chemosynthesis—oxygen, carbon dioxide, and hydrogen sulfide—via a highly specialized hemoglobin capable of binding oxygen and sulfide simultaneously without inhibition.

Within the trophosome, symbiont populations exhibit pronounced cell size differentiation, with smaller cells prioritizing rapid division and larger cells specializing in carbon fixation through endoreduplication. This functional partitioning enhances overall symbiotic efficiency and maximizes carbon transfer to the host (Dubilier et al., 2008).

Across marine ecosystems, similar symbioses have evolved independently in multiple animal phyla, each harnessing chemosynthetic bacteria oxidizing sulfur, methane, or hydrogen. These partnerships demonstrate that microbial metabolism can substitute entirely for photosynthesis as the energetic foundation of complex ecosystems, underscoring the central role of microbes in structuring life across the ocean—from sunlit reefs to the darkest depths (Dubilier et al., 2008).

11.7 EXTREME AND DEEP-SEA MICROBIOMES

11.7.1 Chemolithoautotrophy and Carbon Fixation Pathways

Hydrothermal vent ecosystems are sustained by chemolithoautotrophic primary production rather than photosynthesis, supporting dense biological communities in perpetual darkness. These systems are dominated by hyperthermophilic and thermophilic Archaea—particularly members of the Archaeoglobales, Thermococcales, Methanococcales, and Desulfurococcales—and by Bacteria including Aquificae and Campylobacterota. Many of these lineages occupy deeply branching positions on the tree of life, reinforcing the view that chemolithoautotrophy represents an ancient metabolic strategy likely present on early Earth (Dick, 2019).

Among bacterial groups, the Aquificales are emblematic vent inhabitants, functioning as obligate or facultative lithoautotrophs that exploit reduced inorganic compounds such as molecular hydrogen, elemental sulfur, sulfide, and thiosulfate as electron donors. These organisms thrive in steep thermal and chemical gradients characteristic of vent mixing zones, where reduced hydrothermal fluids encounter oxygenated seawater.

Carbon fixation in vent environments proceeds through four major biochemical pathways, each favored under specific redox and energetic constraints. The Calvin–Benson–Bassham (CBB) cycle is the most widespread, operating in taxa such as Thiomicrospira, Beggiatoa, and the intracellular symbionts of vestimentiferan tubeworms. The reverse tricarboxylic acid (rTCA) cycle, which is more energy-efficient under low-oxygen conditions, predominates among Aquificales and Thermoproteales in anoxic or microoxic niches. The 3-hydroxypropionate bicycle supports mixotrophic lifestyles, while the Wood–Ljungdahl pathway, the most ancient and energetically conservative carbon fixation route, is characteristic of methanogens and acetogens inhabiting hydrogen-rich environments (Dick, 2019). Collectively, these pathways illustrate the remarkable metabolic innovation fostered by extreme geochemical settings.

11.7.2 Sulfur-Oxidizing Bacteria

Sulfur oxidation constitutes a central metabolic process at hydrothermal vents, with sulfur-oxidizing bacteria (SOB) partitioning ecological niches along redox gradients and fluid flow regimes. Sulfurovum species dominate surfaces exposed to diffuse venting, where sulfide concentrations are moderate and oxygen is intermittently available. In contrast, the SUP05 clade prevails within hydrothermal plumes, forming free-living bacterioplankton populations capable of rapid dispersal (Dick, 2019).

SUP05 bacteria exhibit exceptional metabolic flexibility, adjusting their energy metabolism in response to fluctuating availability of sulfur compounds, hydrogen, and oxygen. Metatranscriptomic analyses reveal that genes involved in sulfur and hydrogen oxidation are most highly expressed proximal to vent sources, underscoring tight coupling between geochemistry and microbial activity. Viral interactions further modulate these processes. Notably, the majority of viral genomes recovered from vent plumes encode sulfur-metabolizing genes homologous to those of SUP05, indicating that viruses actively reshape sulfur cycling through horizontal gene transfer and metabolic reprogramming of their hosts (Anantharaman et al., 2014). These findings highlight viruses as integral players in vent biogeochemistry rather than passive predators.

11.7.3 Anaerobic Methane Oxidation (ANME Archaea)

Anaerobic methane-oxidizing archaea (ANME) serve as critical biogeochemical sentinels, preventing large-scale methane release from marine sediments into the ocean and atmosphere. Three major ANME lineages have been described, each occupying distinct ecological niches. ANME-1 predominates in deep sulfate–methane transition zones and exhibits thermotolerance, as observed in the hydrothermally influenced sediments of the Guaymas Basin. ANME-2, the most globally distributed group, dominates cold seep environments and typically forms tight consortia with sulfate-reducing bacteria such as Desulfosarcina and Desulfococcus. ANME-3 is most abundant in mud volcanoes, where it is associated with Desulfobulbus partners.

Together, these communities consume approximately 80% of the methane generated in deep-sea sediments, with sulfate-driven anaerobic oxidation filtering up to 90% of methane diffusing through sulfate–methane transition zones (Jørgensen & Kasten, 2006). Beyond canonical sulfate coupling, ANME archaea display unexpected metabolic versatility, including methane oxidation linked to nitrate reduction (Haroon et al., 2013) and iron-dependent pathways. This metabolic breadth enhances the resilience of methane filtering under variable environmental conditions. However, warming-induced destabilization of methane hydrates or sedimentary redox regimes could disrupt these microbial barriers, amplifying greenhouse gas emissions with global climate implications.

11.7.4 Extremophile Adaptations: Piezophiles and Thermophiles

Life at hydrothermal vents and in the deep ocean requires extensive physiological and molecular adaptations to extreme pressure and temperature. Obligate piezophiles such as *Colwellia marinimaniae* MTCD1 exhibit optimal growth at pressures between 70 and 120 MPa, with documented growth at 140 MPa—the highest pressure tolerance recorded for any organism—following isolation from amphipods in the Mariana Trench. Similarly, *Pyrococcus yayanosii* CH1 represents the first known obligate piezophilic hyperthermophilic archaeon, growing optimally at 52 MPa and 98°C (Dick, 2019).

Adaptations to these extremes occur at multiple biological levels. Membrane composition is modified through increased unsaturated fatty acids in bacteria and tetraether lipid monolayers in archaea, enhancing membrane stability under pressure and heat. Proteins exhibit reduced internal voids and increased conformational flexibility, maintaining catalytic function despite compression. Cells also accumulate small organic molecules known as piezolytes, such as glutamate, which stabilize macromolecular structures. These adaptations define the biochemical boundaries of life and inform astrobiological models of habitability.

11.7.5 Organic Falls: Whale Falls and Bone Microbiomes

Large organic inputs to the deep sea, such as whale carcasses, generate ephemeral yet ecologically transformative habitats known as whale falls. These systems progress through well-defined successional stages: an initial mobile scavenger phase, followed by an enrichment opportunist stage dominated by heterotrophic invertebrates and microbes, and ultimately a sulfophilic stage characterized by chemosynthesis-based communities resembling those at vents and seeps (Smith et al., 2015).

Bone-eating annelids of the genus Osedax exemplify extreme symbiotic specialization within these habitats. Lacking a mouth or digestive tract, Osedax worms rely entirely on endosymbiotic Oceanospirillales bacteria—primarily Profundimonas—housed in specialized root tissues. The worms secrete protons and carbonic anhydrase to dissolve bone mineral, while symbionts metabolize liberated lipids via the glyoxylate cycle (Goffredi et al., 2005; Smith et al., 2015).

Whale fall bone microbiomes are taxonomically diverse, dominated by Gammaproteobacteria and Campylobacteria with extensive gene repertoires encoding collagenases, peptidases, and glycosidases necessary for bone matrix degradation. These habitats not only recycle organic matter but may function as dispersal stepping stones for chemosynthetic fauna across the deep ocean.

11.7.6 Hadal Zone Communities

The hadal zone, extending below 6,000 m, represents one of the least explored and most extreme marine environments. Trench ecosystems harbor distinct microbial assemblages enriched in dozens of bacterial and archaeal operational taxonomic units absent from abyssal plains (Peoples et al., 2019). At depths exceeding 10,400 m, hydrocarbon-degrading bacteria such as Oleibacter, Thalassolituus, and Alcanivorax become increasingly abundant, suggesting selective pressure for lipid and hydrocarbon utilization in these isolated environments.

Beyond the water column, the marine subsurface constitutes the largest microbial habitat on Earth, hosting approximately 80% of global archaeal and bacterial biomass. Dormant bacterial endospores alone are estimated to number 2.5×10^{28} to 1.9×10^{29} in the upper kilometer of sediment, with total endospore biomass exceeding that of all vegetative microbial cells (D'Hondt et al., 2019). These deep biosphere communities persist at the edge of thermodynamic feasibility, sustained by vanishingly small energy fluxes derived from ancient organic matter and water–rock interactions (D'Hondt et al., 2019; Jørgensen & Boetius, 2007). Collectively, they underscore the extraordinary persistence of life and redefine the spatial and energetic limits of the biosphere.

11.8 VERTEBRATE-ASSOCIATED MICROBIOMES

11.8.1 Cetacean Gut and Respiratory Microbiomes

Cetaceans host highly specialized microbiomes that reflect both their evolutionary history and extreme dietary strategies in marine environments. Gut microbiome composition in cetaceans is strongly structured by feeding mode and host phylogeny. Baleen whales, which consume vast quantities of zooplankton such as krill and copepods, harbor gut microbial communities dominated by Firmicutes (53.9–91.2%) and Bacteroidetes (0.9–32.4%), together accounting for 80–97% of recovered bacterial sequences. Despite their carnivorous diets, these microbiomes exhibit striking functional convergence with those of terrestrial herbivores, particularly in the enrichment of genes encoding glycoside hydrolases and fermentative pathways capable of degrading chitin-rich polysaccharides derived from crustacean exoskeletons (Sanders et al., 2015). This functional analogy underscores how microbial metabolism enables baleen whales to efficiently exploit nutritionally challenging prey at immense scales.

In contrast, odontocetes such as bottlenose dolphins possess gut microbiomes adapted to a piscivorous lifestyle. These communities are enriched in Gammaproteobacteria and exhibit extremely low relative abundances of Bacteroidetes, often below 1% (Sanders et al., 2015). Such compositional differences reflect divergent digestive strategies and gut physiologies, highlighting how microbial assemblages mirror ecological niches even among closely related marine mammals.

Cetacean respiratory microbiomes offer a rare, noninvasive window into marine mammal health. Unlike most mammals, cetacean airways are anatomically separated from the oral cavity, allowing relatively uncontaminated sampling of the respiratory tract via exhaled breath condensate, or "blow." Studies of humpback whales reveal a conserved core respiratory microbiome comprising approximately 25 taxa, including members related to Porphyromonas, Psychrobacter, Tenacibaculum, and Staphylococcus, shared across individuals from geographically distinct ocean basins. Beyond this shared core, individuals maintain stable, host-specific "intra-core" microbial signatures persisting for at least eight months (Apprill et al. 2017). This temporal stability supports the use of blow microbiome profiling as a promising biomarker framework for monitoring respiratory health, disease, and stress in free-ranging cetaceans.

11.8.2 Pinniped Microbiomes

Pinnipeds—including seals, sea lions, and walruses—harbor gut microbiomes that reflect both marine carnivory and terrestrial ancestry. Across species, pinniped gut communities are dominated by Firmicutes, Proteobacteria, Bacteroidetes, Actinobacteria, and Fusobacteria. Comparative studies of Australian sea lions demonstrate that while Firmicutes dominate in both wild (76.4%) and captive (61.4%) individuals, captivity is associated with a marked increase in Proteobacteria, which contribute nearly threefold more to captive microbiomes (29.3%) than to those of wild animals (10.6%) (Nelson et al., 2013).

Fine-scale taxonomic differences further distinguish between these populations. Operational taxonomic units within Clostridiaceae and Ruminococcaceae account for a substantial proportion of community dissimilarity between wild and captive animals, indicating that diet, environmental exposure, and human management profoundly reshape microbial assemblages. Such shifts may have functional consequences, as these taxa are often involved in fermentation, short-chain fatty acid production, and gut barrier maintenance. These findings raise important considerations for animal welfare and conservation, suggesting that microbiome perturbations in captivity could influence immune function, metabolism, and disease susceptibility in pinniped populations.

11.8.3 Seabird Guano and Nutrient Cycling

Seabirds act as powerful biological vectors connecting marine and terrestrial biogeochemical cycles through the deposition of guano. With an estimated global population of approximately 840 million individuals spanning 320 species, seabirds collectively excrete about 591 gigagrams of nitrogen and 99 gigagrams of phosphorus annually—nutrient fluxes comparable in magnitude to the combined contributions of all rivers worldwide. Of this nitrogen pool, approximately 72.5 gigagrams per year is readily solubilizable and thus immediately available for microbial and plant uptake (Otero et al., 2018).

Upon deposition, guano-derived organic compounds, particularly uric acid, are rapidly metabolized by soil and sediment microbial communities. Microbial mineralization converts uric acid into ammonium and nitrate, fueling primary productivity in otherwise nutrient-limited coastal and island ecosystems. In Hawaiian montane forests, for example, isotopic analyses indicate that up to 32% of soil nitrogen originates from marine sources delivered via seabird colonies (Otero et al., 2018).

This microbially mediated transformation of seabird-derived nutrients exemplifies the far-reaching ecological influence of marine organisms beyond ocean boundaries. By subsidizing terrestrial food webs and enhancing soil fertility, seabird–microbe interactions illustrate how animal microbiomes and associated microbial processes integrate marine and terrestrial ecosystems into a coupled biogeochemical continuum.

11.9 ANTHROPOGENIC IMPACTS OF MARINE MICROBIAL SYSTEMS

11.9.1 Climate Change and Ocean Acidification

Ocean warming is driving coherent, predictable restructuring of marine microbial communities at basin to global scales. Long-term metagenomic observations provide compelling evidence that rising temperatures act as a primary ecological filter, favoring streamlined, small-genome microorganisms adapted to oligotrophic conditions. An 11-year time-series from the southern California Current revealed clear climate-driven succession: cooler, nutrient-replete periods were dominated by large-genome taxa with greater metabolic versatility, whereas warmer stratified conditions selected for taxa such as Prochlorococcus and Pelagibacter, characterized by reduced genomes and highly efficient nutrient uptake systems. Machine learning projections based on these trends suggest that under no-policy climate scenarios, approximately 32.44% of the surface ocean may undergo ecological regime shifts by 2100, including poleward redistribution of dominant taxa, altered photosynthetic carbon fixation rates, and reduced community-level nutrient metabolism capacity (Danovaro et al., 2011).

Ocean acidification compounds warming-driven change by directly altering microbial physiology and ecosystem function. Experimental and field studies indicate that elevated pCO_2 reduces bacterial production by 17–37% while paradoxically increasing taxonomic diversity. High-nucleic-acid-content bacteria—often associated with rapid growth—are selectively suppressed under low pH, resulting in reduced bacterial growth efficiency. Consequently, a larger fraction of dissolved organic carbon (DOC) is respired as CO_2 rather than converted into microbial biomass, weakening carbon sequestration efficiency. Biofilm communities exposed to acidified conditions also shift from autotrophic toward heterotrophic dominance, further reinforcing respiratory carbon loss. Observations from natural CO_2 seep systems demonstrate partial adaptive potential over evolutionary timescales; however, coral-associated microbial communities decline sharply at these sites, indicating that holobiont systems may be less resilient than free-living microbial assemblages (Danovaro et al., 2011).

11.9.2 The Plastisphere

The proliferation of plastic debris has created a novel marine habitat: the plastisphere. Microbial communities colonizing plastic surfaces differ fundamentally from surrounding seawater, forming dense, structured biofilms within hours of immersion. Early colonizers include diatoms, filamentous cyanobacteria (Phormidium, Rivularia), and Gammaproteobacteria—particularly Oceanospirillaceae and Alteromonadaceae—which together can comprise up to 59.1% of total plastisphere biomass within a single day (Zettler et al., 2013). These assemblages rapidly develop metabolic complexity, incorporating heterotrophs, phototrophs, and hydrocarbon-degrading specialists.

Of particular concern is the consistent enrichment of Vibrio species on microplastic surfaces, raising the possibility that plastics function as vectors for pathogenic bacteria and associated virulence genes, thereby linking ecosystem disturbance to human health risk (Oberbeckmann & Labrenz, 2020). Conversely, plastisphere communities also include taxa with potential remediation capacity. Alcanivorax species show strong enrichment on polyethylene and polypropylene, and recent studies report plastisphere biofilms accelerating polyethylene degradation by up to 40% while simultaneously degrading petroleum hydrocarbons.

Despite these promising observations, critical uncertainties remain. Most plastisphere studies rely on laboratory mesocosms or short-term deployments that cannot fully replicate the complexity of open-ocean conditions. Long-term exposure to UV radiation, fluctuating pressure and temperature regimes, and the chemical heterogeneity of weathered plastics likely alter both community composition and degradation rates. Bridging the gap between controlled experimental results and real-world dynamics remains essential before plastisphere processes can be confidently incorporated into marine carbon cycling models or public health risk assessments (Oberbeckmann & Labrenz, 2020).

11.9.3 Oil Spills and Hydrocarbon Degradation

Marine microbial communities demonstrate extraordinary responsiveness to acute hydrocarbon contamination. The 2010 Deepwater Horizon oil spill released an estimated 4.1–4.4 million barrels of crude oil, triggering rapid and dramatic microbial blooms. In contaminated beach sands, microbial abundance increased by an order of magnitude, with Alcanivorax species alone comprising up to 10% of the total community by July 2010 (Mason et al., 2012).

Hydrocarbon degradation is partitioned among specialized taxa according to substrate chemistry. Alcanivorax and Marinobacter preferentially metabolize aliphatic hydrocarbons, while polycyclic aromatic hydrocarbons (PAHs) are degraded primarily by Alteromonas, Cycloclasticus, and Colwellia (King et al., 2015). This functional partitioning highlights the metabolic plasticity and resilience of marine microbiomes, which can rapidly reorganize to mitigate pollution impacts—although often at the cost of transient ecosystem disruption.

11.9.4 Eutrophication and Hypoxic Zones

Nutrient enrichment from agricultural runoff and wastewater discharge has driven a dramatic global expansion of eutrophication-induced hypoxic "dead zones." From fewer than 10 documented cases in 1960, more than 415 hypoxic regions are now recognized worldwide, with the Baltic Sea alone containing seven of the ten largest. In the Gulf of Mexico, seasonal hypoxic zones routinely span 5,000–22,000 km^2 (Diaz & Rosenberg, 2008).

Hypoxia profoundly restructures sediment microbial communities. Sulfate-reducing Desulfobacteraceae dominate oxygen-depleted sediments, producing sulfide while simultaneously releasing phosphate from sediments, thereby reinforcing eutrophication feedback loops. On a global scale, expanding oxygen minimum zones now account for 35–50% of marine nitrogen loss through denitrification and anammox, with cascading effects on productivity, greenhouse gas emissions, and food-web stability (Breitburg et al., 2018).

11.9.5 Pharmaceutical Pollution

Pharmaceutical contaminants represent a chronic and growing selective pressure on marine microbial ecosystems. Sulfonamides, fluoroquinolones, macrolides, and tetracyclines dominate marine contamination profiles, with sulfamethoxazole detected in over 71% of seawater samples worldwide. Concentrations reach extreme values—up to 332,440 ng L^{-1}—particularly in heavily urbanized coastal regions such as China (Wilkinson et al., 2022).

Even at low concentrations, antibiotics alter microbial metabolism, suppressing respiration and primary production within biofilms while favoring resistant taxa. These selective pressures are intensified by cooccurring stressors, including heavy metals and microplastics, which can synergistically enhance resistance selection and gene exchange (Martinez, 2009).

11.9.6 Antibiotic Resistance Gene Proliferation

Marine environments function as vast reservoirs and evolutionary incubators for antibiotic resistance genes (ARGs). Approximately 65% of marine ARGs are actively expressed, with over 90% of high-risk ARGs exhibiting particularly high expression probabilities. In coastal aquaculture systems, sulfonamide resistance genes (sul1, sul2) reach extraordinary abundances of 3.5×10^7 to 6.5×10^{10} copies per milliliter (Cuadrat et al., 2020).

ARG abundance correlates strongly with mobile genetic elements, including integrons, plasmids, and transposons, while hypoxic conditions enhance horizontal gene transfer. Machine learning models predict that over 97% of marine ARGs will alter their expression patterns by 2100, with high-risk ARGs increasingly concentrated in low-latitude regions. These dynamics underscore the ocean's role as a global conduit linking environmental change to antimicrobial resistance risks with direct implications for human health (Hatosy & Martiny, 2015).

11.9.7 Ballast Water and Aquaculture

Global shipping and aquaculture further accelerate microbial redistribution and ecological change. Ballast water alone transports an estimated 10,000 species daily, including pathogenic bacteria. *Vibrio cholerae* O1 and O139 strains attach to copepods during transport, facilitating long-distance dispersal. The 1991 cholera outbreak in Peru—resulting in more than 10,000 deaths over three years—has been attributed to ballast water-mediated introduction from South Asia (Ruiz et al., 2000; Takahashi et al., 2008).

Aquaculture operations contribute additional pressures by releasing large quantities of organic waste. Scottish salmon farms discharged approximately 13,000 tons of nitrogen and 1,800 tons of phosphorus in 2019 alone. Beneath fish cages, sediments develop highly specialized microbial communities dominated by Epsilonproteobacteria, Bacteroidetes, and Clostridiales. Cable bacteria establish electrical networks that couple sulfide oxidation to oxygen or nitrate reduction, providing microbially mediated detoxification that partially mitigates environmental damage (Keeley et al., 2019).

Together, ballast water transport and aquaculture intensification exemplify how human activity reshapes marine microbial biogeography, function, and evolutionary trajectories on a planetary scale.

11.10 METHODOLOGICAL FRONTIERS IN MARINE MICROBIAL ECOLOGY

11.10.1 Metagenomics and Metatranscriptomics

Shotgun metagenomics has become the cornerstone of modern marine microbial ecology, enabling comprehensive, cultivation-independent characterization of community composition and functional potential without the primer biases inherent to amplicon sequencing. Large-scale initiatives illustrate the transformative scope of this approach. The KAUST Metagenome Analysis Platform analyzed 2,102 ocean metagenomes comprising 308.6 million gene clusters, revealing an immense reservoir of previously unknown genetic diversity. Ultra-deep sequencing efforts—reaching 180 Gb per sample—enabled recovery of 173 metagenome-assembled genomes (MAGs) from the elusive SAR202 clade, including 154 novel species and 104 new genera. Globally, reconstruction of 2,631 draft MAGs from open-ocean datasets underscores how metagenomics has reshaped understanding of marine microbial diversity, metabolic potential, and biogeography (Tully et al., 2018).

While metagenomics captures community potential, metatranscriptomics reveals which genes are actively expressed in situ, offering direct insights into microbial physiology and ecosystem function. This approach faces significant technical challenges, particularly for prokaryotes whose mRNAs lack poly(A) tails and exhibit extremely short half-lives—median transcript persistence in Prochlorococcus is approximately 2.4 minutes. Despite these obstacles, advances in sampling and preservation have enabled robust community-scale transcriptomic analyses. The Environmental Sample Processor (ESP) allows autonomous, in situ filtration and RNALater preservation, producing high-quality RNA even after 30 days at room temperature. Metatranscriptomic reconstruction of marine eukaryotic plankton communities has illuminated the functional diversity of protists, revealing complex diel, seasonal, and nutrient-responsive expression patterns that were previously invisible through DNA-based surveys alone (Keeling et al., 2014; Vorobev et al., 2020).

11.10.2 Single-Cell Genomics and Environmental DNA

Single-cell genomics bypasses cultivation by isolating and sequencing individual cells, generating single amplified genomes (SAGs) that capture lineages often absent from MAG-based reconstructions. The GORG-Tropics database exemplifies the power of this approach, containing 12,715 SAGs derived from tropical and subtropical euphotic-zone samples. Technical workflows typically employ fluorescence-activated cell sorting using SYTO-9 staining, followed by multiple displacement amplification with phi29 DNA polymerase. While individual SAGs often recover only ~20% of a genome due to amplification bias, coassembly of SAGs from closely related cells increases genome completeness to 68–74%, providing high-resolution insights into rare and uncultivated taxa (Tully et al., 2018).

Environmental DNA (eDNA) represents a complementary, noninvasive approach for biodiversity assessment across trophic levels. Direct shotgun sequencing of seawater yields only ~0.34% eukaryotic reads, limiting taxonomic resolution. In contrast, tree-of-life metabarcoding using multiple genetic markers (12S, 16S, 18S, COI) dramatically increases detection efficiency. Using 20-fold less sequencing effort than shotgun approaches, eDNA metabarcoding detected 287 families spanning major eukaryotic divisions. In kelp forest ecosystems, eDNA recovered 76% of fish species observed by visual census, and mitochondrial DNA persistence exceeding 20 days in cold Antarctic waters highlights the method's sensitivity for low-abundance taxa (Deiner et al., 2017). eDNA has thus emerged as a transformative tool for large-scale, cost-effective monitoring of marine biodiversity.

11.10.3 Long-Read Sequencing Technologies

Long-read sequencing technologies have addressed long-standing challenges in metagenome assembly caused by repetitive elements, structural variation, and genomic islands. PacBio HiFi sequencing produces reads of 10–20 kb with >99.9% accuracy, while Oxford Nanopore R10.4 chemistry achieves up to 99.996% consensus accuracy at 50× coverage. These advances substantially improve MAG contiguity and completeness, particularly for the flexible genome fraction that often harbors ecologically and biotechnologically relevant genes (Tully et al., 2018).

Beyond genome assembly, nanopore-based protein sequencing represents an emerging frontier. Direct electrical detection of amino acid signatures offers the potential to bypass translation inference entirely, opening new avenues for characterizing active proteomes in environmental samples where transcript–protein coupling may be weak (Afshar Bakshloo et al., 2022).

11.10.4 Marine Microbial Observatories

Long-term microbial observatories provide the temporal and spatial context essential for interpreting genomic data. The Hawaii Ocean Time-series (HOT) and Bermuda Atlantic Time-series Study (BATS) together generated over 5 terabases of sequence data from 610 samples spanning major ocean basins. The Tara Oceans expedition (2009–2013) expanded this scope dramatically, sampling 210 stations across 20 biogeographic provinces and producing over 70 terabases of data from more than 500 samples.

Key outputs include the Ocean Microbial Reference Gene Catalog v2, comprising 46.8 million nonredundant genes, over 35,000 microbial species, and approximately 150,000 eukaryotic operational taxonomic units. Among the most influential findings from these systematic efforts was the demonstration that temperature is the primary driver of epipelagic microbial community composition at global scales, providing a unifying framework linking microbial ecology to climate change (Sunagawa et al., 2020).

11.10.5 Cultivation Breakthroughs

Despite the power of omics approaches, cultivation remains indispensable for mechanistic experimentation. Innovative cultivation strategies have dramatically expanded the range of isolates available for study. Dilution-to-extinction methods enabled isolation of Candidatus Pelagibacter ubique (SAR11), achieving 48% novel species recovery compared to only 9% using traditional plating techniques.

The isolation chip (iChip) further revolutionized cultivation by enclosing individual cells in diffusion chambers that allow exchange of growth factors with natural environments while preventing overgrowth by fast-growing competitors. Perhaps the most striking example of cultivation perseverance is the isolation of *Candidatus Prometheoarchaeum syntrophicum*, an Asgard archaeon central to hypotheses on eukaryotic origins. This achievement required over 12 years of enrichment culture, underscoring both the difficulty and the scientific payoff of cultivating slow-growing, syntrophy-dependent microbes (Nichols et al., 2010; Imachi et al., 2020).

11.10.6 Bioinformatic Challenges and Standards

Bioinformatic analysis remains a major bottleneck in marine microbial ecology. Assembly tools exhibit tradeoffs: MEGAHIT achieves median chromosomal coverage of 81.9%, while metaSPAdes attains 76.8% coverage but excels at plasmid reconstruction. Binning algorithms—including MetaBAT2, CONCOCT, MaxBin2, and newer machine learning–based tools such as SemiBin—use distinct clustering strategies, often yielding complementary results.

Taxonomic assignment has been standardized through genome-based frameworks such as GTDB-Tk, which classifies organisms based on phylogenomic context rather than marker genes alone. The MiMAG standards define high-quality MAGs as those exceeding 90% completeness, containing <5% contamination, and possessing rRNA genes and at least 18 tRNAs. Broader reporting frameworks, including MIMARKS and MIxS, have substantially improved data comparability, reproducibility, and long-term utility across studies (Yilmaz et al., 2011).

11.10.7 Temporal Sampling Considerations

Recognition of diel rhythmicity in marine microbial communities has profound methodological implications. Temporal variability in gene expression, growth, and mortality can exceed spatial variation, yet most oceanographic surveys rely on single time-point sampling. Notably, the Tara Oceans expedition collected samples predominantly during daylight hours, potentially biasing community composition toward phototroph-active states and underrepresenting nocturnal processes.

Future observatories must incorporate autonomous, high-frequency sampling capable of resolving subdiel dynamics. Minimum information standards should explicitly require reporting of local sampling time relative to the solar cycle, paralleling emerging best practices in circadian microbiome research that mandate standardized collection times (Thaiss et al., 2016). Integrating temporal resolution into marine microbial studies is essential for accurately quantifying biogeochemical fluxes and predicting ecosystem responses to environmental change.

11.11 APPLICATIONS AND INTERVENTIONS

11.11.1 Microbiome-Based Coral Restoration (BMCs)

Microbiome-based coral restoration has emerged as a promising, mechanism-driven strategy to enhance the resilience of coral reefs facing accelerating climate stress. The Beneficial Microorganisms for Corals (BMCs) framework adapts principles from plant growth–promoting rhizobacteria to coral holobionts, emphasizing the targeted manipulation of coral-associated microbial communities to improve host health and stress tolerance. Candidate probiotic consortia, including *Pseudoalteromonas galatheae*, *Cobetia amphilecti*, *Halomonas* spp., and *Sutclifiella* spp., are selected based on functional traits such as antagonism against coral pathogens (notably *Vibrio coralliilyticus*), ROS scavenging via catalase and peroxidase activity, and siderophore production that enhances iron availability under oligotrophic conditions (Peixoto et al., 2017).

Field trials in the Red Sea demonstrated that repeated probiotic inoculation over a three-month period successfully enriched target microbial taxa within the microbiomes of *Pocillopora verrucosa* without detectable spillover effects on surrounding seawater microbial communities. These findings indicate that microbiome manipulation can be spatially constrained and ecologically stable at reef scales. Complementary mesocosm experiments further showed that BMC-treated *Mussismilia hispida* corals exhibited approximately 40% higher survival under thermal stress, maintained photosynthetic efficiency, and displayed enhanced degradation of dimethylsulfoniopropionate (DMSP), a compound implicated in oxidative stress mitigation and sulfur cycling (Rosado et al., 2019).

Beyond probiotic addition, coral microbiome transplantation from thermally tolerant donor colonies represents a parallel strategy for enhancing resilience. Recipient corals exposed to heat stress (up to 34°C) exhibited reduced bleaching severity and successfully incorporated donor-specific bacterial taxa, suggesting partial transfer of stress-tolerant microbiome functions. Collectively, these studies support the concept that coral thermal tolerance is, in part, a microbiome-mediated trait. However, long-term persistence of introduced microbes, ecosystem-level impacts, and evolutionary consequences of repeated microbiome manipulation remain open questions that must be addressed before large-scale deployment.

11.11.2 Bioremediation Potential

Marine microorganisms possess extraordinary metabolic versatility, positioning them as key agents for the natural attenuation of marine pollution. The response of indigenous microbial communities to the Deepwater Horizon oil spill provided a striking demonstration of this capacity, with rapid enrichment of hydrocarbon-degrading taxa capable of metabolizing petroleum compounds across oxic and anoxic zones. Among these, *Alcanivorax borkumensis* emerged as a dominant obligate hydrocarbonoclastic bacterium, specializing in the degradation of branched and linear alkanes through coordinated monooxygenase, dioxygenase, dehydrogenase, and hydratase pathways (King et al., 2015).

More recently, attention has turned to the "plastisphere"—microbial communities colonizing plastic debris—which appear capable of accelerating polymer weathering and facilitating partial biodegradation. These findings suggest a potential microbial contribution to mitigating the growing burden of marine plastic pollution, although complete mineralization of synthetic polymers remains rare and slow (Oberbeckmann & Labrenz, 2020).

Despite these encouraging observations, a substantial gap persists between naturally occurring bioremediation following acute pollution events and the feasibility of engineered, scalable bioremediation strategies. The Deepwater Horizon response was driven by a massive, sudden influx of hydrocarbons that selected for preexisting degraders; it was not the result of deliberate microbial augmentation. Laboratory-derived degradation rates for Alcanivorax and related taxa are typically measured under nutrient-replete, temperature-controlled conditions that poorly reflect the spatial heterogeneity, nutrient limitation, and physical dispersal characteristic of open-ocean and deep-sea environments (King et al., 2015).

Consequently, the effectiveness of bioaugmentation—intentional introduction of pollutant-degrading microbes—remains largely unproven under realistic field conditions. Key uncertainties include whether introduced strains can compete with native communities, maintain functional activity over time, and achieve degradation rates comparable to natural bloom responses. Until controlled, ecosystem-scale field trials validate both efficacy and ecological safety, claims regarding engineered marine bioremediation should be interpreted cautiously and framed as long-term research objectives rather than deployable solutions.

11.11.3 Marine Mammals as Sentinel Species

Marine mammals function as powerful sentinel species for assessing ocean health due to their long lifespans, high trophic positions, and substantial lipid reserves that bioaccumulate anthropogenic contaminants. These characteristics integrate environmental exposure over broad spatial and temporal scales, making marine mammals particularly sensitive indicators of ecosystem-level stressors (Apprill, 2017). Microbiome studies have revealed widespread occurrence of antibiotic-resistant bacteria in marine mammals, including methicillin-resistant *Staphylococcus aureus* (MRSA). In bottlenose dolphins, approximately 25% of fecal *Escherichia coli* isolates exhibit resistance to one or more antibiotics, reflecting terrestrial runoff, wastewater inputs, and global dissemination of antimicrobial resistance genes (Apprill, 2017).

Comparative analyses further demonstrate that marine habitats exert strong selective pressures on gut microbiome composition, often overriding host phylogeny. Marine carnivores display convergent microbiome structures distinct from their terrestrial relatives, shaped by high-protein diets, saline environments, and exposure to marine pathogens. These consistent microbiome signatures provide a noninvasive means of monitoring population health, nutritional status, and exposure to pollutants. As microbiome baselines for healthy populations become better defined, deviations from these norms may serve as early warning indicators of ecosystem degradation, reinforcing the value of marine mammals as integrative sentinels of ocean health in a rapidly changing world.

11.12 FUTURE DIRECTIONS

11.12.1 Key Emergent Insights

Over the past decade, marine microbiome research has been fundamentally reshaped by the convergence of methodological innovations. Single-cell genomics has opened access to vast reservoirs of previously uncultured microbial diversity, autonomous and time-resolved observatories have generated petascale datasets spanning entire ocean basins, and long-read sequencing technologies have resolved mobile genetic elements and complex genomic architectures that underpin horizontal gene transfer and rapid adaptation (Sunagawa et al., 2020; Tully et al., 2018). Together, these advances have shifted the field from descriptive inventories toward a systems-level understanding of the ocean as a dynamic microbial network, in which viruses, protists, and prokaryotes interact continuously under the constraints of nutrient availability, physical mixing, and increasing anthropogenic pressure.

Several emergent insights now compel a reassessment of how ocean ecosystems function. First, the magnitude of virus-mediated carbon fluxes—the viral shunt and shuttle pathways—has been shown to be quantitatively central to marine carbon cycling, with an estimated ~145 Gt of carbon released annually in tropical oceans alone. This scale of flux requires explicit representation in Earth system models seeking to project carbon sequestration and climate feedbacks (Wilhelm and Suttle, 1999; Danovaro et al., 2011). Second, the discovery that marine viruses encode AMGs capable of sustaining host photosynthesis and nutrient acquisition during infection has revealed viruses as active regulators of primary productivity rather than passive mortality agents (Lindell et al., 2007; Roux et al., 2016). Third, recognition that mixotrophy dominates protistan trophic strategies has undermined the long-standing dichotomy between autotrophy and heterotrophy, forcing biogeochemical models to account for flexible metabolic modes that blur traditional food-web boundaries (Flynn et al., 2019; Ward and Follows, 2016). Finally, the identification of anaerobic methane-oxidizing archaeal (ANME) consortia as biological filters removing up to 90% of sedimentary methane highlights microbial communities as critical gatekeepers of greenhouse gas fluxes—gatekeepers that may be destabilized by ocean warming and altered redox regimes (Haroon et al., 2013; Jørgensen & Kasten, 2006).

11.12.2 Persistent Knowledge Gaps

Despite these advances, major knowledge gaps continue to constrain predictive understanding. The overwhelming majority of fish-associated microorganisms—approximately 97%—remain uncultured, limiting insight into host–microbiome interactions and trophic transmission pathways across marine food webs. Similarly, long-term metagenomic time-series exceeding a decade are exceedingly rare, hindering the ability to distinguish transient variability from true regime shifts in microbial community structure and function (Sunagawa et al., 2020). In coral reef systems, the causative agent of SCTLD remains unresolved despite extensive microbiome surveys, underscoring the limitations of correlation-based approaches in the absence of experimental validation (Meyer et al., 2019).

Among the most consequential gaps is the near-complete lack of understanding of deep-sea viral ecology. While surface-ocean viral communities have been extensively catalogued, the mesopelagic and bathypelagic realms—where most exported organic carbon is remineralized—remain largely unexplored virologically. Fundamental parameters such as virus-to-host ratios, burst sizes, infection strategies, and the balance between lytic and lysogenic cycles are poorly constrained below the photic zone (Suttle, 2007; Weinbauer, 2004). Because viral lysis at depth directly determines how much carbon is recycled in the water column versus sequestered in sediments, this gap has immediate implications for global carbon cycle models and climate projections.

Coral disease causation represents another urgent frontier. For SCTLD in particular, consistent microbial signatures—such as enrichment of Rhodobacterales and Peptostreptococcales–Tissierellales in lesion tissues—have been documented, yet Koch's postulates have not been satisfied, leaving the etiological agent formally unidentified (Meyer et al., 2019; Sweet & Bulling, 2017). Whether SCTLD arises from a single pathogen, a dysbiotic microbial consortium, or an interaction between host immune dysfunction and opportunistic colonizers remains unknown. Each scenario implies fundamentally different intervention strategies, from targeted antimicrobial treatments to microbiome restoration. Resolving this question will require controlled infection experiments using defined microbial consortia, coupled with high-frequency, longitudinal sampling that captures the earliest stages of disease onset.

A third persistent gap lies in the translation of microbial bioremediation from laboratory demonstrations to reliable field-scale applications. While hydrocarbon degradation by taxa such as Alcanivorax and Cycloclasticus, and plastic-associated microbial communities, has been convincingly demonstrated under controlled conditions, real-world environments introduce nutrient limitation, spatial heterogeneity, hydrodynamic mixing, and competing microbial processes that are rarely captured in experimental systems (King et al., 2015; Oberbeckmann & Labrenz, 2020). Bridging this gap will require rigorously designed, replicated field trials that simultaneously quantify degradation rates, microbial community dynamics, and broader ecosystem impacts.

Finally, predicting microbial responses to multiple, interacting stressors remains a fundamental challenge. Most experimental studies examine single stressors—warming, acidification, hypoxia, or pollution—in isolation, yet future oceans will experience these pressures concurrently. Emerging evidence suggests that combined stressors can interact synergistically, producing outcomes that exceed the sum of individual effects. For example, hypoxia can enhance horizontal gene transfer, potentially accelerating the spread of antibiotic resistance already driven by pharmaceutical contamination (Hatosy & Martiny, 2015; Cuadrat et al., 2020). Capturing such dynamics will require factorial experimental designs and trait-based ecological models capable of representing microbial adaptation under compounding stress regimes.

11.12.3 Research Priorities

Looking forward, several research priorities emerge as critical for advancing both fundamental understanding and applied outcomes. First, the development of standardized microbial indicators for ecosystem health is essential, enabling microbiomes to be incorporated into routine monitoring and early warning systems. Second, predictive frameworks that link microbiome composition and functional traits to environmental drivers must be strengthened, allowing robust forecasts of microbial community responses to climate change. Third, microbiome-based interventions—including probiotics and microbiome transplantation—must be rigorously evaluated under realistic field conditions to assess their efficacy, stability, and unintended ecological consequences (Peixoto et al., 2017; Rosado et al., 2019).

Ocean biogeochemical modeling is increasingly moving toward trait-based approaches that explicitly connect microbial genetic diversity to ecosystem function. By integrating genomic information with physiological traits and rate measurements, these frameworks offer a pathway to translate the immense metabolic diversity revealed by metagenomics into quantitative predictions at ecosystem and global scales (Coles et al., 2017). Advancing and validating such models represents a central priority for embedding microbiome science within Earth system modeling efforts.

Finally, international initiatives such as the Ocean Biomolecular Observing Network (OBON), endorsed by the UN Ocean Decade, provide a platform for integrating microbiome monitoring into global ocean observation systems. Seawater microbiomes already show promise as sensitive indicators of environmental change, with specific taxa responding predictably to temperature, chlorophyll-a, and particulate organic carbon gradients. As recognition grows that many marine microbiomes have diverged substantially from historical baselines, conservation strategies must evolve to incorporate active microbiome management alongside traditional protection measures. Embracing marine microbiomes as integral components of ocean ecosystems will be essential for sustaining ocean health in an era of rapid environmental change (Bourne et al., 2016).

11.13 CONCLUSIONS

The marine microbiome constitutes both the largest continuous ecosystem on Earth and one of its most powerful biogeochemical engines. This review has synthesized current knowledge spanning molecular-scale virus–host interactions, community-level ecological organization, and global patterns of microbial biogeography and biogeochemical cycling. Collectively, marine microorganisms comprise roughly 90% of ocean biomass and mediate the chemical transformations that sustain planetary habitability, including the processing of approximately 50 petagrams of carbon each year through the biological pump. Without these microbial processes, ocean productivity, atmospheric composition, and climate regulation as we know them would not be possible.

Across the global ocean, environmental gradients in temperature, depth, light, and nutrient availability impose strong structuring forces on microbial communities, generating predictable biogeographic patterns from polar to tropical regions and from the surface ocean to the hadal zone. Superimposed on these gradients are host-associated microbiomes inhabiting organisms ranging from corals and sponges to fish and marine mammals. These host–microbe partnerships underpin essential functions including nutrition, immune defense, development, and resilience to environmental stress. The growing acceptance of the holobiont concept—viewing hosts and their associated microbiomes as integrated ecological units—has fundamentally reshaped understanding of marine ecosystem dynamics, particularly under rapid climate-driven change.

A critical emerging dimension of marine microbiome ecology is the recognition that microbial communities are not only spatially organized but also profoundly temporally structured. Diel rhythmicity is evident across multiple scales, from synchronized cell division in Prochlorococcus to daily oscillations in viral infection, metabolite production, nutrient regeneration, and organic matter flux driven by vertical migration of zooplankton and fish. Despite mounting evidence, this temporal organization remains largely absent from biogeochemical and ecosystem models. Analogous to the bidirectional circadian–microbiome interactions now recognized as fundamental to mammalian physiology, marine host–microbiome systems likely depend on finely tuned temporal coordination. Disruption of these rhythms by warming, acidification, or pollution may represent an early and underappreciated signal of ecosystem dysfunction. Developing time-resolved sampling strategies and chronobiological frameworks tailored to marine systems is therefore a pressing priority.

Climate vulnerability permeates marine microbiome ecology at every level of organization. Coral holobionts face compounding pressures from ocean warming, bleaching-associated microbiome dysbiosis, and the emergence of novel diseases. Ocean acidification threatens calcifying plankton such as coccolithophores and foraminifera, with cascading effects on carbon export and sulfur cycling, including potential reductions in DMS emissions that influence cloud formation and climate feedbacks. Expanding oxygen minimum zones are reshaping nitrogen cycling on a global scale, while the rapid proliferation and redistribution of antibiotic resistance genes—projected to intensify and shift toward low latitudes by 2100—create intersecting challenges at the nexus of climate change, ecosystem health, and human well-being.

The methodological revolution in marine microbiology has been transformative. Metagenomics, single-cell genomics, long-read sequencing, and autonomous observatories have revealed extraordinary microbial diversity, metabolic versatility, and ecological complexity previously inaccessible to observation. Yet these advances also highlight persistent gaps. Most marine microorganisms remain uncultured, mechanisms governing microbiome assembly, resilience, and stability are incompletely understood, temporal dynamics are still undersampled, and predictive models struggle to capture microbial responses to multiple, interacting stressors. Addressing these gaps will require coordinated, long-term, and interdisciplinary efforts that integrate molecular biology, ecology, oceanography, and modeling.

Ultimately, understanding how marine microbiomes respond to anthropogenic pressures—from plastic pollution and pharmaceutical contamination to climate-driven restructuring of communities—will be central to improving climate projections and guiding effective ocean conservation in the Anthropocene. Emerging applications of microbiome science, including probiotic approaches for coral restoration and microbial strategies for bioremediation, offer promising tools but demand cautious, evidence-based deployment. Integrating chronobiological principles into marine microbiome research—recognizing these systems as temporally organized, adaptive networks rather than static assemblages—will be essential for building predictive frameworks that reflect the true complexity of ocean ecosystems. As environmental change accelerates, the marine microbiome must be recognized not merely as an object of scientific inquiry, but as a foundational component of Earth's life support system, deserving both intensive study and proactive protection.

ACKNOWLEDGMENTS

The author (ST) acknowledges Sharpe-Strumia Research Foundation, W.W. Smith Charitable Trust, and the Abraham Thomas Foundation for their generous support in providing the resources that made this work possible.

REFERENCES

Afshar Bakshloo, M., Kasianowicz, J.J., Pastoriza-Gallego, M., Mathé, J., Daniel, R., Piguet, F., Oukhaled, A. (2022). Nanopore-Based Protein Identification. *J Am Chem Soc*. Feb 16;144(6):2716–2725.

Anantharaman, K., Duhaime, M.B., Breier, J.A., Wendt, K.A., Toner, B.M., & Dick, G.J. (2014). Sulfur oxidation genes in diverse deep-sea viruses. *Science*, 344(6185), 757–760.

Apprill, A. (2017). Marine animal microbiomes: Toward understanding host–microbiome interactions in a changing ocean. *Frontiers in Marine Science*, 4, 222.

Apprill, A., Miller, C.A., Moore, M.J., Durban, J.W., Fearnbach, H., & Barrett-Lennard, L.G. (2017). Extensive Core Microbiome in Drone-Captured Whale Blow Supports a Framework for Health Monitoring. *Systems*. Oct 10;2(5):e00119-17.

Aylward, F.O., Boeuf, D., Mende, D.R., Wood-Charlson, E.M., Vislova, A., Eppley, J.M., Romano, A.E. & DeLong, E.F. (2017). Diel cycling and long-term persistence of viruses in the ocean's euphotic zone. *Proceedings of the National Academy of Sciences*, 114, 11446–11451.

Azam, F., Fenchel, T., Field, J., Gray, J., Meyer-Reil, L. & Thingstad, F. (1983). The ecological role of water-column microbes in the sea. *Marine Ecology Progress Series*, 10, 257–263.

Bourne, D.G., Morrow, K.M. & Webster, N.S. (2016). Insights into the coral microbiome: Underpinning the health and resilience of reef ecosystems. *Annual Review of Microbiology*, 70, 317–340.

Breitburg, D. et al. (2018). Declining oxygen in the global ocean and coastal waters. *Science*, 359, eaam7240.

Coles, V.J. et al. (2017). Ocean biogeochemistry modeled with emergent trait-based genomics. *Science*, 358, 1149–1154.

Cuadrat, R.R.C. et al. (2020). Global ocean resistome revealed: Exploring antibiotic resistance gene abundance and distribution in TARA Oceans samples. *GigaScience*, 9, giaa046.

Curson, A. et al. (2017). Dimethylsulfoniopropionate biosynthesis in marine bacteria and identification of the key gene in this process. *Nature Microbiology*, 2, 17009.

D'Hondt, E.D. et al. (2019). Subsurface life and its detection in a subseafloor drilling site. *Proceedings of the National Academy of Sciences*, 116, 7361–7366.

Danovaro, R. et al. (2011). Marine viruses and global climate change. *FEMS Microbiology Reviews*, 35, 993–1034.

Deiner, K. et al. (2017). Environmental DNA (eDNA)—A revolution in biodiversity assessments. *Trends in Ecology & Evolution*, 32, 840–851.

Delmont, T.O. et al. (2022). Functional repertoire convergence of distantly related eukaryotic plankton lineages abundant in the sunlit ocean. *Cell Genomics*, 2, 100123.

Diaz, R.J. & Rosenberg, R. (2008). Spreading dead zones and consequences for marine ecosystems. *Science*, 321, 926–929.

Dick, G.J. (2019). The microbiomes of deep-sea hydrothermal vents: Distributed globally, shaped locally. *Nature Reviews Microbiology*, 17, 271–283.

Dubilier, N., Bergin, C. & Lott, C. (2008). Symbiotic diversity in marine animals: The art of harnessing chemosynthesis. *Nature Reviews Microbiology*, 6, 725–740.

Dyhrman, S. et al. (2006). Phosphonate utilization by the globally important marine diazotroph Trichodesmium. *Nature*, 439, 68–71.

Endo, H. et al. (2020). Biogeography of marine giant viruses reveals their interplay with eukaryotes and ecological functions. *Nature Ecology & Evolution*, 4, 1639–1649.

Falkowski, P.G., Fenchel, T. & Delong, E.F. (2008). The microbial engines that drive Earth's biogeochemical cycles. *Science*, 320, 1034–1039.

Flynn, K.J. et al. (2019). Mixotrophic protists and a new paradigm for marine ecology: Where does plankton research go now? *Journal of Plankton Research*, 41, 375–391.

Fuhrman, J.A. (1999). Marine viruses and their biogeochemical and ecological effects. *Nature*, 399, 541–548.

Galí, M., Simó, R., Vila-Costa, M., Ruiz-González, C., Gasol, J.M. & Matrai, P. (2013). Diel patterns of oceanic dimethylsulfide (DMS) cycling: Microbial and physical drivers. *Global Biogeochemical Cycles*, 27, 620–636.

Giovannoni, S.J. (2017). SAR11 bacteria: The most abundant plankton in the oceans. *Annual Review of Marine Science*, 9, 231–255.

Goffredi, S.K. et al. (2005). Evolutionary innovation: Bone-eating marine worms with bacterial endosymbionts. *Nature*, 434, 1–4.

Gregory, A.C. et al. (2019). Marine DNA viral macro- and microdiversity from pole to pole. *Cell*, 177, 1109–1123.

Guillou, L. et al. (2008). Widespread occurrence and genetic diversity of marine parasitoids belonging to Syndiniales. *Environmental Microbiology*, 10, 3349–3365.

Haroon, M.F. et al. (2013). Anaerobic oxidation of methane coupled to nitrate reduction in a novel archaeal lineage. *Nature*, 500, 567–570.

Hartmann, M. et al. (2012). Mixotrophic basis of Atlantic oligotrophic ecosystems. *Proceedings of the National Academy of Sciences*, 109, 5756–5760.

Hatosy, S.W. & Martiny, A.C. (2015). The ocean as a reservoir of antibiotic resistance genes. *mBio*, 6, e00790–15.

Holser RR, Keates TR, Costa DP, Edwards CA. (2022). Extent and magnitude of subsurface anomalies during the Northeast Pacific blob as measured by animal-borne sensors. *Journal of Geophysical Research: Oceans*, 127(7), e2021JC018356.

Hopkins, F.E. et al. (2023). The biogeochemistry of marine dimethylsulfide. *Nature Reviews Earth & Environment*, 4, 361–376.

Imachi, H. et al. (2020). Isolation of an archaeon at the prokaryote–eukaryote interface. *Nature*, 577, 519–525.

Jiao, N. et al. (2010). Microbial production of recalcitrant dissolved organic matter: Long-term carbon storage in the global ocean. *Nature Reviews Microbiology*, 8, 593–599.

Johnson, C.H., Zhao, C., Xu, Y. & Mori, T. (2017). Timing the day: What makes bacterial clocks tick? *Nature Reviews Microbiology*, 15, 232–242.

Jørgensen, B.B. & Boetius, A. (2007). Metabolism of subsurface-biosphere microorganisms. *Nature*, 451, 684–687.

Jørgensen, B.B. & Kasten, S. (2006). Sulfur Cycling and Methane Oxidation. In H. D. Schulz & M. Zabel (Eds.), *Marine Geochemistry* (pp. 271–309). Springer.

Keeley, N. et al. (2019). Resilience of dynamic coastal benthic ecosystems in response to large-scale finfish farming. *Aquaculture & Environment Interactions*, 11, 161–179.

Keeling, P.J. et al. (2014). The marine microbial eukaryote transcriptome sequencing project (MMETSP). *PLoS Biology*, 12, e1001889.

King, G.M. et al. (2015). Microbial responses to the Deepwater Horizon oil spill. *Annual Review of Marine Science*, 7, 377–401.

Knowles, B. et al. (2016). Lytic to temperate switching of viral communities. *Nature*, 531, 466–470.

LaJeunesse, T.C. et al. (2018). Systematic revision of symbiodiniaceae highlights the antiquity and diversity of coral endosymbionts. *Current Biology*, 28, 2570–2580.

Lesser, M. P. (2011). Coral Bleaching: Causes and Mechanisms. In Z. Dubinsky & N. Stambler (Eds.), *Coral Reefs: An Ecosystem in Transition*. Dordrecht: Springer.

Lindell, D. et al. (2007). Genome-wide expression dynamics of a marine virus and host reveal features of co-evolution. *Nature*, 449, 83–86.

Lu, J., Shu, Y., Zhang, H., Zhang, S., Zhu, C, Ding, W. & Zhang, W. (2023). The landscape of global ocean microbiome: From bacterioplankton to biofilms. *International Journal of Molecular Science*, 24(7), 6491.

Martinez, J.L. (2009). Environmental pollution by antibiotics and by antibiotic resistance determinants. *Environmental Pollution*, 157, 2893–2902.

Mason, O.U. et al. (2012). Metagenome, metatranscriptome and single-cell sequencing reveal microbial response to Deepwater Horizon oil spill. *The ISME Journal*, 6, 1715–1727.

Medina, L.E. et al. (2017). A review of protist grazing below the photic zone emphasizing studies of oxygen-depleted water columns. *Frontiers in Marine Science*, 4, 105.

Meyer, J.L. et al. (2019). Microbial community shifts associated with stony coral tissue loss disease. *Frontiers in Microbiology*. Sep 24;10:2244.

Middelboe, M. & Brussaard, C.P.D. (Eds.). (2017). *Marine Viruses: Key Players in Ecosystem Processes*. Springer.

Mitra, A. et al. (2016). Defining planktonic protist functional groups on mechanisms for energy and nutrient acquisition. *Protist*, 167, 106–120.

Monier, A. et al. (2009). Horizontal gene transfer of an entire metabolic pathway between a eukaryotic alga and its DNA virus. *Genome Research*, 19, 1441–1449.

Moore, C.M. et al. (2013). Processes and patterns of oceanic nutrient limitation. *Nature Geoscience*, 6, 701–710.

Moran, M.A. (2015). The global ocean microbiome. *Science*, 350, aac8455.

Nelson, D.M. et al. (1995). Production and dissolution of biogenic silica in the ocean: Revised global estimates. *Global Biogeochemical Cycles*, 9, 359–372.

Nelson, T.M. et al. (2013). Diet and phylogeny shape the gut microbiota of Antarctic seals. *Environmental Microbiology*, 15, 1132–1145.

Nichols, D. et al. (2010). Use of iChip for high-throughput in situ cultivation of "uncultivable" microbial species. *Applied and Environmental Microbiology*, 76, 2445–2450.

Oberbeckmann, S. & Labrenz, M. (2020). Marine microbial assemblages on microplastics: Diversity, adaptation, and role in degradation. *Annual Review of Marine Science*, 12, 209–232.

Otero, X.L. et al. (2018). Seabird colonies as important global drivers in the nitrogen and phosphorus cycles. *Nature Communications*, 9, 246.

Paul, J. (2008). Prophages in marine bacteria: Dangerous molecular time bombs or the key to survival in the seas? *The ISME Journal*, 2, 579–589.

Peixoto, R.S. et al. (2017). Beneficial microorganisms for corals (BMC): Proposed mechanisms for coral health and resilience. *Frontiers in Microbiology*, 8, 341.

Peoples, L.M. et al. (2019). Microbial community diversity within sediments from two hadal trenches. *Frontiers in Microbiology*, 10, 347.

Ridgwell, A. & Schmidt, D. (2010). Past constraints on the vulnerability of marine calcifiers to massive carbon dioxide release. *Nature Geoscience*, 3, 196–200.

Rosado, P.M. et al. (2019). Marine probiotics: Increasing coral resistance to bleaching through microbiome manipulation. *The ISME Journal*, 13, 921–936.

Roux, S. et al. (2016). Ecogenomics and potential biogeochemical impacts of globally abundant ocean viruses. *Nature*, 537, 689–693.

Ruiz, G.M. et al. (2000). Global spread of microorganisms by ships. *Nature*, 408, 49–50.

Salazar, G. & Sunagawa, S. (2017). The global structure of marine plankton communities. *Nature Communications*, 8, 15619.

Sanders, J.G. et al. (2015). Baleen whales host a unique gut microbiome with similarities to both carnivores and herbivores. *Nature Communications*, 6, 8285.

Sherr, E.B. & Sherr, B.F. (2002). Significance of predation by protists in aquatic microbial food webs. *Antonie Van Leeuwenhoek*, 81, 293–308.

Smith, C.R. et al. (2015). Whale-fall ecosystems: Recent insights into ecology, paleoecology, and evolution. *Annual Review of Marine Science*, 7, 571–596.

Steinberg, D.K. & Landry, M.R. (2017). Zooplankton and the ocean carbon cycle. *Annual Review of Marine Science*, 9, 413–444.

Steinberg, D.K., Van Mooy, B.A.S., Buesseler, K.O., Boyd, P.W., Kobari, T. & Karl, D.M. (2008). Bacterial vs. zooplankton control of sinking particle flux in the ocean's twilight zone. *Limnology and Oceanography*, 53, 1327–1338.

Stoecker, D.K. et al. (2017). Mixotrophy in the marine plankton. *Annual Review of Marine Science*, 9, 311–335.

Sunagawa, S. et al. (2020). Tara Oceans: Towards global ocean ecosystems biology. *Nature Reviews Microbiology*, 18, 428–445.

Suttle, C.A. (2005). Viruses in the sea. *Nature*, 437, 356–361.

Suttle, C.A. (2007). Marine viruses – major players in the global ecosystem. *Nature Reviews Microbiology*, 5, 801–812.

Sweet, M.J. & Bulling, M.T. (2017). On the importance of the microbiome and pathobiome in coral health and disease. *Frontiers in Marine Science*, 4, 9

Takahashi, C.K. et al. (2008). Ballast water: A review of the impact on world public health. *Journal of Venomous Animals and Tropical Diseases*, 14, 393–408.

Thaiss, C.A., Levy, M., Korem, T., Dohnalová, L., Shapiro, H., Jaitin, D.A., David, E., Winter, D.R., Gury-BenAri, M., Tatirovsky, E., Tuganbaev, T., Federici, S., Zmora, N., Zeevi, D., Dori-Bachash, M., Pevsner-Fischer, M., Kartvelishvily, E., Brandis, A., Harmelin, A., Shibolet, O., Halpern, Z., Honda, K., Amit, I., Segal, E., & Elinav, E. (2016). Microbiota Diurnal Rhythmicity Programs Host Transcriptome Oscillations. *Cell*. Dec 1;167(6):1495-1510.e12.

Thingstad, T.F. and Lignell, R. (1997). Theoretical models for the control of bacterial growth rate, abundance, diversity and carbon demand. *Aquatic Microbial Ecology*, 13, 19–27.

Thomas, T. et al. (2016). Diversity, structure and convergent evolution of the global sponge microbiome. *Nature Communications*, 7, 11870.

Tréguer, P. et al. (2018). Influence of diatom diversity on the ocean biological carbon pump. *Nature Geoscience*, 11, 27–37.

Tréguer, P. & De La Rocha, C.L. (2013). The world ocean silica cycle. *Annual Review of Marine Science*, 5, 477–501.

Tully, B.J., Graham, E.D. & Heidelberg, J.F. (2018). The reconstruction of 2,631 draft metagenome-assembled genomes from the global oceans. *Scientific Data*, 5, 170203.

Unrein, F. et al. (2014). Mixotrophic haptophytes are key bacterial grazers in oligotrophic coastal waters. *The ISME Journal*, 8, 164–176.

Vorobev, A. et al. (2020). Transcriptome reconstruction and functional analysis of eukaryotic marine plankton communities via high-throughput metagenomics and metatranscriptomics. *Genome Research*, 30, 647–659.

Vraspir, J.M. & Butler, A. (2009). Chemistry of marine ligands and siderophores. *Annual Review of Marine Science*, 1, 43–63.

Waldbauer, J.R., Rodrigue, S., Coleman, M.L. & Chisholm, S.W. (2012). Transcriptome and proteome dynamics of a light-dark synchronized bacterial cell cycle. *PLoS ONE*, 7, e43432.

Ward, B.A. & Follows, M.J. (2016). Marine mixotrophy increases trophic transfer efficiency, mean organism size, and vertical carbon flux. *Proceedings of the National Academy of Sciences*, 113, 2958–2963.

Weinbauer, M.G. (2004). Ecology of prokaryotic viruses. *FEMS Microbiology Reviews*, 28, 127–181.

Wilhelm, S.W. & Suttle, C.A. (1999). Viruses and nutrient cycles in the sea. *BioScience*, 49, 781–788.

Wilkinson, J.L. et al. (2022). Pharmaceutical pollution of the world's rivers. *Proceedings of the National Academy of Sciences*, 119, e2113947119.

Worden, A.Z. et al. (2015). Rethinking the marine carbon cycle: Factoring in the multifarious lifestyles of microbes. *Science*, 347, 1257594.

Yilmaz, P. et al. (2011). Minimum information about a marker gene sequence (MIMARKS) and minimum information about any (x) sequence (MIxS) specifications. *Nature Biotechnology*, 29, 415–420.

Zettler, E.R., Mincer, T.J., Amaral-Zettler, L.A. (2013). Life in the "plastisphere": Microbial communities on plastic marine debris. *Environ Science Technol*. 47(13), 7137–46.

Zhao, Y. et al. (2013). Abundant SAR11 viruses in the ocean. *Nature*, 494, 357–360.

INDEX

A
Alpaca, 107, 108
Amplicon sequencing, 2, 5, 64, 109, 132, 231
Anaerobic methane-oxidizing archaea, 216, 226, 235
Antibiotic resistance, 8, 40, 58, 128, 136, 139, 230, 236, 237
Aspergillosis, 134, 135
Atopic dermatitis, 162, 175

B
Bacteriocins, 1, 13, 15, 55, 60, 82
Bacteriophage, 9, 35, 39, 40, 42, 62, 106, 214
Bacteroidetes, 6–10, 55, 77, 80, 96, 100–103, 108, 109, 111, 113, 115, 116, 120, 128–131, 133, 159–170, 174, 180, 183, 184, 192, 194, 197, 198, 231
Ballast water, 231
Biogeochemical cycle, 92, 209, 210, 218, 229
Biological carbon pump, 210
Bovine milk, 53, 55–58, 67
Buffalo, 11, 92, 93, 99–102, 112

C
Camel, 93, 104–108
Candidiasis, 135
Canine microbiome, 158, 159, 161, 162, 166, 167
Carbon cycling, 210, 212, 214, 220, 230, 235
Colostrum, 15, 32, 78, 81
Companion animal, 27, 75, 77, 82, 157, 158, 160, 172–177
Coprophagy, 36, 185
Coral, 223–225, 229, 233, 235, 237
Cow, 3, 6, 11, 12, 36, 38, 41, 42, 54, 57, 61, 62, 66, 67, 96, 100
Cryptococcus, 135, 136

D
Diatom, 212, 216–218, 222, 229
Diel cycle, 214
Diel rhythms, 223

E
Enterococcus, 2, 15, 39, 55–57, 59, 60, 65, 82, 107, 119, 120, 136, 161, 168, 174, 204
Environmental DNA (eDNA), 231, 232

F
Fecal Microbiota Transplantation (FMT), 1, 12, 13, 39
Feline microbiome, 168, 171, 172
Firmicutes, 6–10, 55, 77, 80, 96, 97, 99–103, 105, 108, 109, 111–116, 120, 128–131, 133, 159–170, 173, 174, 182–184, 192, 194, 195, 197, 198, 228

G
Giant virus, 214, 219, 220
Goat, 9, 11, 27, 30, 37, 77, 92, 93, 105, 112–115

H
Hadal zone, 227, 237
Holobiont, 26, 43, 91, 223–225, 229, 233, 237

Horizontal transmission, 185, 195
Horse, 9, 27, 28, 30, 92, 109–111
Hypoxic zone, 230

I
Iguana, 183–185, 187

K
Ketosis, 36, 41

L
Lactobacillus, 2, 3, 8, 10, 11, 28, 30, 34, 36, 37, 39, 40, 55–58, 65, 66, 77, 80, 82, 101, 102, 107, 109–111, 116, 117, 120, 121, 128, 129, 160, 162, 167, 169, 170, 184, 194, 198, 201, 204
Lactococcus, 55–57, 60, 105, 106, 109, 115
Leuconostoc, 57, 59, 66
Long-read sequencing, 2–4, 18, 232, 235, 237
Lysogeny, 218, 219

M
Marine virus, 209, 211, 213, 214, 219, 235
Metagenomics, 1–3, 5, 11, 12, 16–18, 36, 38, 64, 76, 96, 118, 150, 154, 205, 211, 231, 236
Metaproteomics, 4, 5, 76, 85
Microbiome engineering, 1, 12–14, 40, 42
Milk microbiome, 11, 53, 55, 64, 66, 67, 106, 107
Mixotrophy, 212, 221, 222, 235

N
Nutritional modulation, 13, 14, 38, 40

O
Ocean microbiome, 209, 210
One Health, 16, 17, 26, 41, 42, 43, 92, 94, 173

P
Parasitic protists, 222
Pediococcus, 57, 60
Pelagiphages, 220, 221
Phage therapy, 13, 15
Pharmaceutical pollution, 230
Pig, 3, 5, 9, 10, 12, 13, 26, 28, 30, 32–41, 93, 116–119
Pinnipeds, 228
Plastisphere, 229, 230, 234
Postbiotics, 1, 14, 76, 83
Poultry litter, 149–155
Probiotics, 1, 10–14, 29, 25, 38–43, 56, 57, 59, 60, 67
Propionibacterium, 6, 12, 57–59, 128
Proteobacteria, 6–9, 55, 77, 80, 96, 97, 99, 102, 103, 105, 109, 111–113, 115, 116, 120, 128–133, 159–170, 173, 174, 182–184, 192, 194, 197, 198, 228
Psychrotrophic bacteria, 60, 61, 65

R
Reproductive tract, 8, 32, 121
Respiratory tract, 1, 8, 18, 135, 136, 163, 172, 228
Rodent gut microbiota, 205
Ruminants, 5, 7, 9–12, 14, 26–30, 32, 34, 37, 38, 40, 42, 75, 77, 78, 81–83, 94, 95, 98, 100, 105, 108, 109, 112, 115, 121

S
Seabird guano, 229
Sheep, 9, 26, 27, 30, 32, 37, 40, 77, 92, 93, 105, 108, 115, 116
Shotgun metagenomics, 2, 3, 5, 64, 231
Shrimp, 27, 29, 30, 35, 37
Silica pump, 217, 218
Skin, 1, 4, 6, 35, 36, 40, 58, 133, 159, 162–166, 169–172, 175, 176, 185
Spoilage, 53, 55, 57–62, 65–67
Sponge, 225, 237
Staphylococcus aureus, 6, 53, 57, 60, 61, 67, 107, 234
Streptococcus, 7, 8, 10, 36, 55–58, 102, 104–107, 109–111, 120, 128, 151, 160–163, 165, 194
Sulfur cycling, 216, 219, 224, 226, 233, 237

T
3-nitrooxypropanol, 98

V
Vertical transmission, 185, 195, 196

W
Whale falls, 227
Working dogs, 161–163

Y
Yak, 93, 102–104
Yeast, 53, 54, 62, 66, 67, 82, 133, 135, 136

Z
Zoonotic pathogens, 16, 27, 61, 94, 128, 133, 141, 158, 176, 177, 192, 194